Handbook of Northwestern Plants

Handbook of Northwestern Plants

revised edition

Helen M. Gilkey

La Rea J. Dennis

Oregon State University Press

Corvallis

♾ This paper meets the requirements of ANSI/NISO Z39.48-1992 (Permanence of Paper).

Library of Congress Cataloging-in-Publication Data
Gilkey, Helen Margaret, 1886-
Handbook of northwestern plants / Helen M. Gilkey, La Rea J. Dennis.—Rev. ed.
p. cm.
Includes bibliographical references (p.).
ISBN 0-87071-490-2
1. Botany—Oregon. 2. Botany—Washington (State) 3. Plants—Identification. I. Johnston, L. D. (La Rea Dennis) II. Title.
QK144 .G5 2001
581.9795—dc21

2001000147

 First OSU Press edition 2001. Fourth printing 2017.
Printed in the United States of America

Oregon State University Press
121 The Valley Library
Corvallis OR 97331
541-737-3166 •fax 541-737-3170
www.osupress.oregonstate.edu

Contents

Introduction

This book is intended for botany students and individuals willing to learn to use keys and basic botanical scientific terminology. It is the outgrowth of *A Spring Flora of Northwestern Oregon*; published in 1929 by Dr. Helen M. Gilkey for use in Field Botany courses. In 1947 she expanded the work so that it would be of use year round to a wider audience. It was published as *Handbook of Northwest Flowering Plants*; with a second edition appearing in 1951. When nomenclatural changes again made an extensive revision necessary I had the privilege of co-authoring the 1967 edition which came out under the title *Handbook of Northwestern Plants*, a name change made necessary when the ferns and fern allies were added. Sadly, Dr. Gilkey died in 1972. I revised the nomenclature in the Handbook in 1973 and 1980. In 1999, when the 1980 edition went out of print, and many nomenclatural changes were necessary, Warren Slesinger of the OSU Press persuaded me to do this present revision.

I have updated the nomenclature to conform to current taxonomic literature to the best of my ability, revised some of the descriptions and keys, numbered all but the shortest keys, added a few species, and arranged the genera and species in alphabetical order within the families. Distributional information of species and synonymy have been kept at a minimum in order to keep the book a reasonable size. Line drawings were chosen over photographs because they allow for the illustration of key characters.

Area Covered

The area covered in this book is roughly that of the rather natural floristic unit from the summit of the Cascade Range to the coastline of Washington and of Oregon as far south as the Umpqua Divide, about the southern limit of Lane County, Oregon. Northern Washington, in its more unique botanical aspects, has been excluded.

Plant Classification and Nomenclature

The development of botany as a science was coincident, in its beginning, with that of medicine. Plants were early employed as remedies for various ills. As a knowledge of disease and its cure grew more specific, it was natural that some attempt at classification of the plants employed be made. From the beginning of human existence, plants and animals have been given common names which are useful locally. The difficulty lies in the fact that these names are purely local. A widespread plant such as *Hypericum perforatum* L. will have several common names; in Oregon alone it is known as Klamath Weed, Goat Weed, and Common St. John's wort; to the Chinese it is *Chin Ssu T'ao Shu*; to the Germans, *Johanniskraut* and to the French, *Millepertuis perfore*. Yet botanists throughout the world would immediately recognize the name *Hypericum perforatum* L. Also the same common name is often applied to different plants in different parts of the country, and there are many plants without common names. Obviously a system needed to be devised. Consequently, International

Botanical Congresses have met at regular intervals since 1867 for the purpose, in part, of establishing and stabilizing botanical *nomenclature* (a system of naming plants) on an international basis. In order to maintain the integrity of the system, the rules are published following the convening of each congress, under the title *International Code of Botanical Nomenclature.* As a beginning, for the purpose of uniformity, Latin was adopted as the official language, and each plant was to be assigned a name consisting of a *binomial,* i.e., two associated words. The first word is the *genus* (or generic name); the second, the *species* (or specific epithet). Technically these are followed by the name of the person who bestowed the specific epithet. Thus, our common Red-flowering currant is known as *Ribes sanguineum* Pursh (genus, *Ribes*; species, *sanguineum,* and Pursh for Frederick Pursh, who first described the species).

In earlier editions of this Handbook, *Ribes* was in the family Saxifragaceae, in this edition it is in a segregate family, the Grossulariaceae. It must be noted that while definite international rules govern the naming of plants and must be followed, in certain cases an author is free to choose between two names. This choice involves the question of splitting a previously named family into several separate families and authors are free to use their own judgment. While I have always been considered a "lumper" and not a "splitter," I felt in this case it was reasonable to remove the woody genera from the Saxifragaceae. Similar situations may occur in the treatment of genera, which explains why Oregon grape may be listed correctly under either *Berberis* or *Mahonia.*

Taxonomists are often criticized for continually changing the names of plants. It must be remembered that, in spite of the antiquity of the science, new information is continually being accumulated and earlier classifications must be revised. Also the plants involved are living, dynamic, and often fluctuating populations and, as such, are not always interpreted in exactly the same manner by all taxonomists. It is necessary to keep the names as stable as possible while allowing for a difference of opinion among specialists.

Keys

Identifications of unknown plants are usually made by means of keys. Keys present the user with a progressive sequence of choices, and, by always making the correct choice, one will arrive at the name of the unknown plant. Normally the first step is to determine the family to which the plant belongs by using the family key. If the family is already known, or when one has arrived at a family in the key, the description of the family should be read to determine if the plant could belong there. If a mistake has been made, it will usually show up at this point. If the family appears to be correct, then the key provided should be used to determine the genus and, following the genus description, the key to species. Upon arriving at a species name, the species description should be read carefully and compared with the plant in question.

The following points should be kept in mind when using a key:

1. Read both choices given, for although the first choice may sound good, the second choice is sometimes better.

2. Be certain that the terminology is understood; look up any term about which you are uncertain.

3. When measurements are required, make them accurately.

Selected References

Of special value in our area: *Vascular Plants of the Pacific Northwest*, Hitchcock, C. L., A. Cronquist, M. Ownbey, and J. W. Thompson. 5 vols.1955, 1959, 1961, 1964, 1969. University of Washington Press, Seattle, WA. *Flora of the Pacific Northwest*, Hitchcock, C. L. and A. Cronquist. 1973. University of Washington Press, Seattle, WA. *The Jepson Manual: Higher Plants of California*, Hickman, J. C., editor. 1993. University of California Press, Berkeley, CA. *Flora of North America North of Mexico.* Flora North America Editorial Committee, eds. 1993+. Oxford University Press, New York, NY. When completed the flora will consist of 30 vols. Also still useful, but out of print, is: *A Manual of the Higher Plants of Oregon*, Peck, M. E. 1961. Binfords and Mort, publishers, Portland, OR. There are other more local floras which are very good for the areas they cover, for example: *Wildflowers of the Western Cascades*, Ross, Robert and Henrietta Chambers. 1988. Timber Press, Portland, OR. *Flora of Mount Rainier National Park*, Biek, D. 1999. Oregon State University Press, Corvallis, OR. *Wildflowers of the Columbia Gorge*, Jolley, Russ. 1988. Oregon Historical Society Press, Portland, OR.

Acknowledgements

I cannot adequately express my gratitude to Richard R. Halse and Robert F. Obermire for reviewing the entire manuscript and helping with the proof reading. Their suggestions have undoubtedly made this book more useful. I am also indebted to The Oregon Flora Project and to George Argus, Henrietta L. Chambers, Kenton L. Chambers, Rhoda Love, and Scott Sundberg, who gave freely of their time and expertise. I would like to express my appreciation to Elizabeth Waldron, Lynda M. Ciuffetti, Dianne L. Simpson, and Typha nmi Johnston, who have been a constant source of encouragement to me. I would be remiss if I did not thank my many other friends, too numerous to name (for fear I would inadvertently omit one), who encouraged me and sometimes distracted me from my work on this revision. Thanks are also due to the staff of the OSU Press, especially: Jeffrey Grass, Warren Slesinger, Jo Alexander, Mary Braun, Tom Booth, and Pennie Coe. I would like to take this opportunity to thank my many students who over the years have made suggestions for improvements in keys and descriptions. Yes, I was listening to you and I did keep notes and I have tried to incorporate many of those suggestions into this edition. Thanks are still due all those persons who assisted in preparation of the earlier editions: Maxwell Doty, James R. Estes, Beulah G. Gilkey, Weldon K. Johnston, Harry K. Phinney, Garland Powell, Frank H. Smith, Ronald Tyrl and to the artists: Cathrine Davis Young Feikert, Daisy R. Overlander, Alleda Burlage, Fern Duncan, and Patricia Packard, who in addition to Helen Gilkey produced the line drawings.

La Rea J. Dennis
May 2000

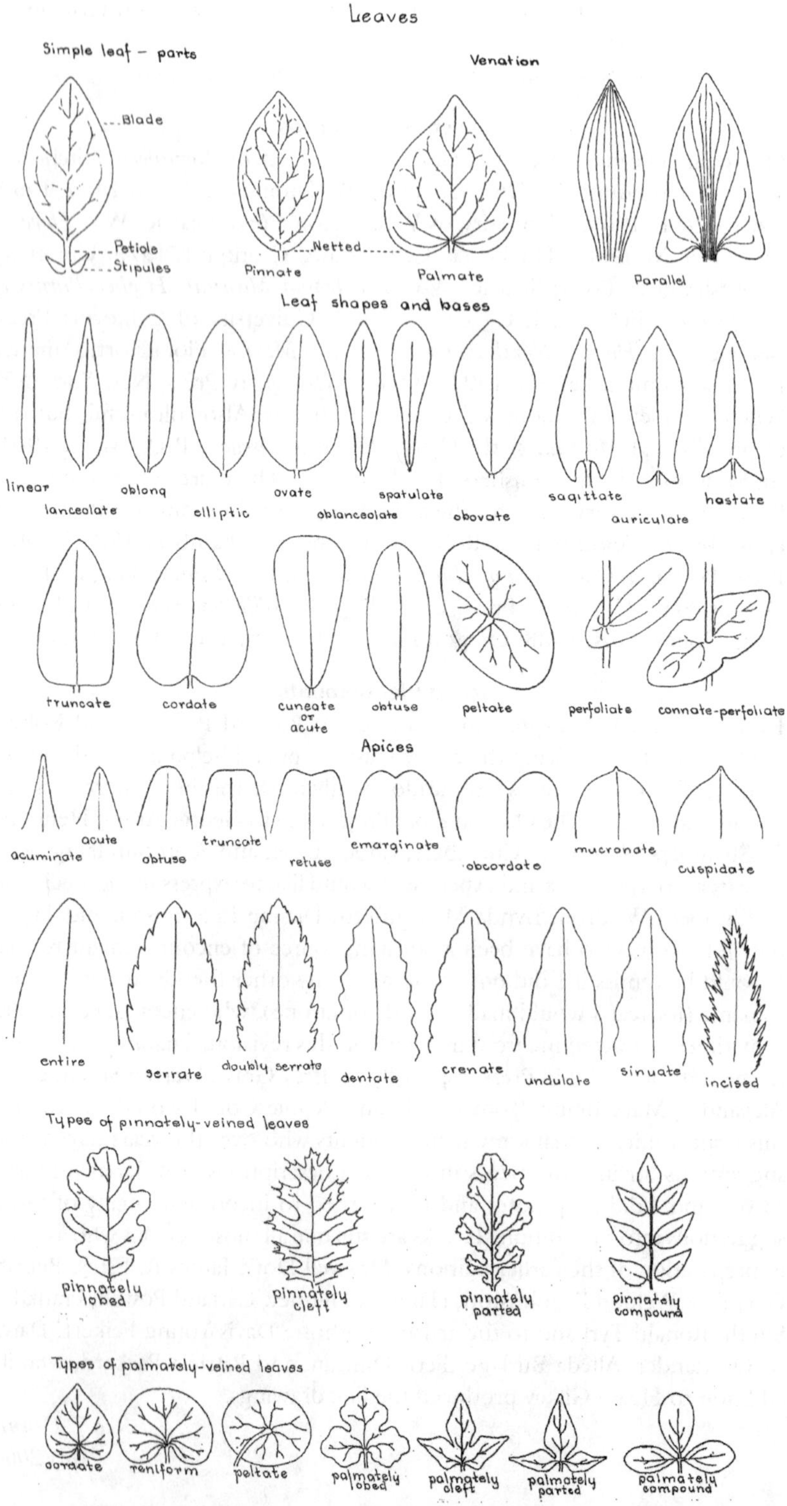

Leaves

Simple leaf – parts

Blade
Petiole
Stipules

Venation

Netted
Pinnate
Palmate
Parallel

Leaf shapes and bases

linear
lanceolate
oblong
elliptic
ovate
oblanceolate
spatulate
obovate
sagittate
auriculate
hastate

truncate
cordate
cuneate or acute
obtuse
peltate
perfoliate
connate-perfoliate

Apices

acuminate
acute
obtuse
truncate
retuse
emarginate
obcordate
mucronate
cuspidate

entire
serrate
doubly serrate
dentate
crenate
undulate
sinuate
incised

Types of pinnately-veined leaves

pinnately lobed
pinnately cleft
pinnately parted
pinnately compound

Types of palmately-veined leaves

cordate
reniform
peltate
palmately lobed
palmately cleft
palmately parted
palmately compound

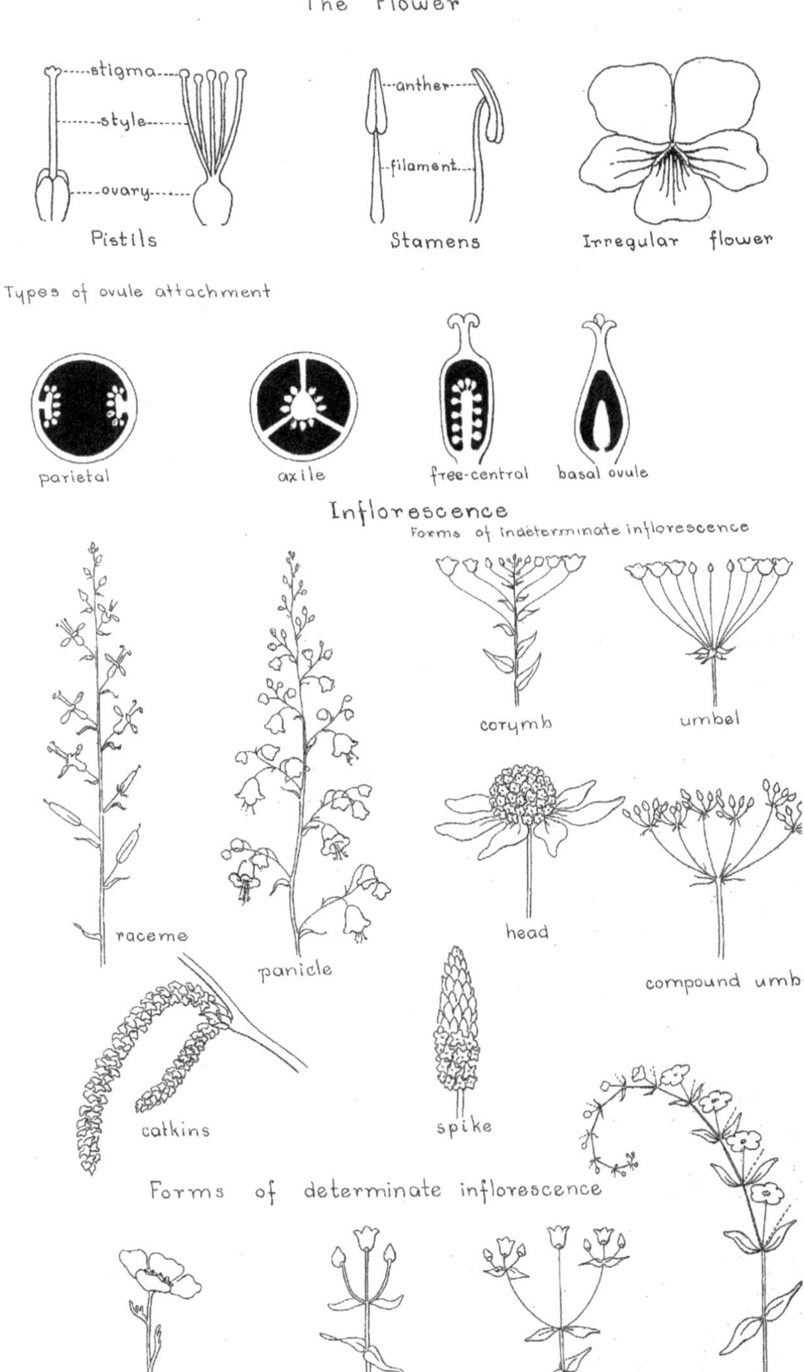
The Flower
stigma
style
ovary
Pistils
anther
filament
Stamens
Irregular flower
Types of ovule attachment
parietal
axile
free-central
basal ovule
Inflorescence
Forms of indeterminate inflorescence
corymb
umbel
raceme
panicle
head
compound umb
catkins
spike
Forms of determinate inflorescence
solitary
cyme
compound cyme
scorpoid cym

Families

Analytical Key

1a Plants not producing seeds or flowers; reproducing by spores; gametophyte and sporophyte independent at maturity.. **1. FERNS AND FERN ALLIES**

1b Plants typically reproducing by seeds, bearing flowers or seed-bearing cones; gametophyte reduced and dependent upon the sporophyte

2a Ovules not enclosed in ovaries; seeds borne in cone-like structures; not producing flowers; leaves needle-like, subulate or scale-like; trees or shrubs, evergreen (ours)... **II. GYMNOSPERMS**

2b Ovules enclosed in an ovary; seeds borne in fruits derived from the ovary; producing flowers.. **ANGIOSPERMS**

3a Vascular bundles typically scattered throughout the stem; flowers parts usually in 3's, rarely in 2's or 4's, never in 5's; leaves typically parallel-veined... **III. MONOCOTS**

3b Vascular elements arranged within a cylinder around the pith or in bundles arranged in a circle; flower parts typically in 4's or 5's, rarely in 3's; leaves typically with pinnate or palmate venation .... **IV. DICOTS**

I. Ferns and Fern Allies

1a Stems conspicuously jointed, hollow except at the nodes, the reduced leaves joined into a toothed sheath at the nodes; sporangia borne on the under side of the scales in a terminal cone-like spike .... **EQUISETACEAE p. 30**

1b Stems not jointed

2a Aquatic or subaquatic plants; spores of two sizes

3a Plants free-floating on still water or occasionally stranded in mud; leaves minute, 2-lobed, imbricate.............................. **AZOLLACEAE p. 28**

3b Plants rooted, sometimes submerged, but never free-floating

4a Leaves 4-foliolate, clover-like (ours); sporangia in bony sporocarps borne on the rhizome separate from the leaves.. **MARSILEACEAE p. 29**

4b Leaves narrowly linear or subulate; sporangia enclosed in the enlarged bases of the leaves.................................. **ISOETACEAE p. 32**

2b Land plants, or at least essentially terrestrial; sporangia borne on the under surface or the margins of the leaves, on specialized leaf segments, in the leaf axils or in terminal strobili

5a Stems elongated, freely branching, closely covered with scale-like entire leaves; sporangia axillary or in terminal strobili

6a Spores of two sizes, some sporangia producing 4 megaspores, others numerous microspores .... **SELAGINELLACEAE p. 35**

6b Spores of one size, minute and numerous .... **LYCOPODIACEAE p. 33**

5b Stems usually very short in comparison with the leaves, usually unbranched; leaves usually large and variously divided

7a Sterile and fertile leaf segments distinctly different, but attached to a common stalk; new leaves not coiled in the bud .... **OPHIOGLOSSACEAE p. 19**

7b Leaves either fertile or sterile; sporangia borne on the under surface or the margins of the leaves; new leaves usually coiled in the bud
8a Leaves markedly dimorphic, the fertile leaves erect and the sterile ones spreading or ascending
9a The sterile and fertile leaves pinnatifid to once-pinnate............ **BLECHNACEAE p. 23**
9b The sterile and fertile leaves 2-4-pinnate............ (*Cryptogramma*) **PTERIDACEAE p. 19**
8b Leaves not markedly dimorphic
10a Rhizomes and stipes with felt-like hairs, lacking scales; under surface of leaf velvety-tomentose; large, coarse ferns............ **DENNSTAEDTIACEAE p. 22**
10b Rhizomes and stipes, if bearing hairs, then also bearing scales
11a Stipes, at least the lower portions, dark-colored and shiny, often also scaly
12a Stipes with a single vascular bundle; sporangia often hidden by reflexed margins of the leaf-segments (false indusia) true indusia lacking............ **PTERIDACEAE p. 19**
12b Stipes with 2 or more vascular bundles, at least near the base
13a Leaves once-pinnate .... **ASPLENIACEAE p. 23**
13b Leaves more than once-pinnate............ **DRYOPTERIDACEAE p. 24**
11b Stipes light colored, although often covered with brownish scales
14a Pinnae auriculate on the upper side at the base; indusia peltate............ (*Polystichum*) **DRYOPTERIDACEAE p. 24**
14b Pinnae not auriculate; indusia, if present, not peltate
15a Leaf blades with needle-like hairs, at least along the midrib............ **THELYPTERIDACEAE p. 22**
15b Leaf blades without needle-like hairs
16a Leaves once-pinnate or merely pinnatifid; indusia absent **POLYPODIACEAE p. 27**
16b Leaves more than once-pinnate; indusia present or absent .... **DRYOPTERIDACEAE p. 24**

II. Gymnosperms

Trees or shrubs, evergreen (ours); leaves needle-like, subulate, or scale-like; ovules borne naked on scales arranged spirally to form a woody or fleshy cone, or in pulpy berry-like or drupe-like structures.

1a Seeds borne in woody cones
2a Foliage leaves generally linear, borne singly or in bundles of 2-5; cone scales alternate, overlapping, each subtended by a bract...... **PINACEAE p. 37**
2b Foliage leaves and cone scales generally scale-like, opposite or whorled, overlapping; cone scales without bracts .... **CUPRESSACEAE p. 44**
1b Seeds borne in fleshy cones or surrounded by a fleshy aril
3a Seed 1, surrounded by a red fleshy aril; leaves appearing 2-ranked............ **TAXACEAE p. 37**
3b Seeds 1-3, borne in bluish fleshy berry-like cones; leaves mostly in 3's (ours) .... (*Juniperus*) **CUPRESSACEAE p. 44**

Angiosperms

Ovules enclosed in an ovary; plants bearing true flowers typically consisting of stamens and pistils generally surrounded by sepals and petals.

III. Monocots

Leaves commonly parallel-veined; flower parts generally in 3's (rarely 2's or 4's); vascular bundles typically scattered through the stem; embryo usually with one seed leaf.

1a Plants with vegetative parts submerged or floating
  2a Diminutive free-floating, thallus-like plants; with or without rootlets .... **LEMNACEAE p. 61**
  2b Plants rooted in the subsoil, mostly submerged, but sometimes with floating leaves; plants differentiated into roots, stems and leaves
    3a Ovary inferior; leaves usually whorled or all basal .... **HYDROCHARITACEAE p. 58**
    3b Ovary superior
      4a Leaves all basal .... **ALISMATACEAE p. 56**
      4b Leaves, at least in part, cauline
        5a Flowers perfect
          6a Perianth segments 4 .... **POTAMOGETONACEAE p. 49**
          6b Perianth segments absent; fruit borne on a long slender stipe .... **RUPPIACEAE p. 49**
        5b Flowers unisexual
          7a Flowers in dense globose unisexual heads, the staminate usually borne above the pistillate .... **SPARGANIACEAE p. 47**
          7b Flowers variously arranged but never in dense globose unisexual heads
            8a Leaves alternate; flowers in a spadix subtended by a spathe; marine species .... **ZOSTERACEAE p. 54**
            8b Leaves opposite, subopposite or appearing whorled; flowers axillary
              9a Pistils 4-5 (ours), stigma 1 per pistil .... **ZANNICHELLIACEAE p. 48**
              9b Pistil 1, stigmas 2-4 .... **NAJADACEAE p. 48**

1b Plants not floating nor submerged, although sometimes growing in bogs or shallow water
  10a Inflorescence a spadix surrounded by a spathe .... **ARACEAE p. 61**
  10b Inflorescence not a spadix surrounded by a spathe
    11a Perianth absent (pistils and stamens sometimes subtended by bracts or bristles); flower parts mostly unequal in number
      12a Flowers unisexual, borne in globose heads or dense elongated spikes
        13a Flowers in dense globose unisexual heads .... **SPARGANIACEAE p. 47**
        13b Inflorescence a dense long cylindrical spike with staminate flowers borne in the upper portion of the spike .... **TYPHACEAE p. 46**

12b Flowers often perfect, borne in the axils of dry chaffy bracts, arranged in spikes or spikelets
14a Leaves 2-ranked; stems (culms) mostly terete and hollow except at the nodes; fruit a caryopsis .... **POACEAE p. 60**
14b Leaves 3-ranked or reduced to basal sheaths; stems (culms) usually solid, often triangular in cross-section; fruit an achene .... **CYPERACEAE p. 60**
11b Perianth present, its segments in 2, rarely 1, series; parts of the flower often equal in number, typically in 3's
15a Ovary inferior
16a Stamens 3; ovary 3-loculed; perianth usually regular.......... **IRIDACEAE p. 80**
16b Stamens 1 or 2, joined with the style; ovary 1-loculed; perianth irregular .... **ORCHIDACEAE p. 83**
15b Ovary superior
17a Perianth segments distinct, usually scarious, brown or greenish
18a Pistil 1; fruit a 3-many-seeded capsule .... **JUNCACEAE p. 65**
18b Pistils 3 or 6; fruits 1-seeded.......... **JUNCAGINACEAE p. 55**
17b Perianth segments, if distinct, at least the inner segments white or colored
19a Pistil one, often 3-lobed.......... **LILIACEAE p. 65**
19b Pistils more than one .... **ALISMATACEAE p. 56**

IV. Dicots

Leaves commonly netted-veined; flower parts generally in 4's or 5's (rarely 3's or 6's); vascular bundles typically arranged in concentric layers in the stem; embryo usually with 2 seed leaves.

1a Petals none
2a Trees or shrubs; at least some of the flowers in catkins or catkin-like spikes
3a Fruit a several-many-seeded capsule; seeds with a tuft of hairs.......... **SALICACEAE p. 91**
3b Fruit a nut, nutlet or drupaceous, 1-seeded; seeds without a tuft of hairs
4a Plants aromatic; fruits (in ours) waxy; staminate catkins short; leaves resinous-dotted.......... **MYRICACEAE p. 96**
4b Plants not strongly aromatic; fruits not waxy; staminate catkins elongated
5a Both staminate and pistillate flowers in catkins; fruit a winged nutlet .... **BETULACEAE p. 97**
5b Only staminate flowers in catkins; pistillate flowers solitary or clustered
6a Nut at maturity enclosed by leaf-like involucre.......... (*Corylus*) **BETULACEAE p. 97**
6b Nut borne in scaly cup-like or a spiny bur-like involvucre.......... **FAGACEAE p. 100**

2b Herbs, shrubs or trees; flowers not in catkins
  7a Ovary inferior
    8a Leaves alternate
      9a Calyx 3-lobed; ovary 6-loculed.......... **ARISTOLOCHIACEAE p. 104**
      9b Calyx lobes 5, rarely 4; ovary 1-4-loculed
        10a Parasitic on roots; calyx lobes petaloid; leaves entire; fruit a 1-seeded drupe .... **SANTALACEAE p. 102**
        10b Not parasitic; fruit a capsule or a follicle .... **SAXIFRAGACEAE p. 189**
    8b Leaves opposite or whorled at least below
      11a Parasitic on tree branches .... **VISCACEAE p. 103**
      11b Not parasitic and not growing on tree branches
        12a At least the lower leaves whorled; partially submerged aquatics
          13a Leaves all entire .... **HIPPURIDACEAE p. 281**
          13b Leaves, at least those submerged, filiformly dissected .... **HALORAGIDACEAE p. 279**
        12b Leaves opposite
          14a Calyx tube not fused with the ovary for the full length of the ovary.. **SAXIFRAGACEAE p. 189**
          14b Calyx tube fused with the ovary for the full length of the ovary
            15a Stamens 12; ovary 6-loculed .... **ARISTOLOCHIACEAE p. 104**
            15b Stamens 4; ovary 1-5-loculed ..... **ONAGRACEAE p. 272**
  7b Ovary superior
    16a Trees, shrubs or vines
      17a Leaves alternate, linear
        18a Leaves with scarious sheathing stipules; fruit an achene .... **POLYGONACEAE p. 105**
        18b Leaves without sheathing stipules; fruit drupaceous................ .... **EMPETRACEAE p. 255**
      17b Leaves opposite, not linear
        19a Leaves with brown to silvery scales on the lower surface .... **ELAEAGNACEAE p. 271**
        19b Leaves without brown or silver scales
          20a Pistils more than 1.................... **RANUNCULACEAE p. 144**
          20b Pistil 1; fruit a samara
            21a Stamens 2; style 1; fruit a single samara......................... .... **OLEACEAE p. 318**
            21b Stamens 3-12; styles 2, distinct or united at the base; fruit a double samara .... **ACERACEAE p. 257**
    16b Herbs
      22a Leaves all reduced to non-green alternate scales borne along the stem .... **ERICACEAE p. 299**
      22b Leaves not all reduced to non-green alternate scales
        23a Leaves with scarious sheathing stipules, except *Eriogonum*, in which the flowers are surrounded by an involucre .... **POLYGONACEAE p. 105**
        23b Stipules, if present, not sheathing (see also *Eriogonum* if flowers in involucrate heads)

24a Calyx absent (in Euphorbiaceae there may be a calyx-like involucre with gland-bearing lobes, enclosing simple staminate flowers and a stalked, 3-lobed pistil)
- 25a Terrestrial plants
  - 26a Flowers unisexual.................................... **EUPHORBIACEAE p. 251**
  - 26b Flowers perfect............................ (*Achlys*) **BERBERIDACEAE p. 161**
- 25b Aquatic plants; leaves submerged and/or floating
  - 27a Leaves entire.. **CALLITRICHACEAE p. 253**
  - 27b Leaves finely dissected ....**CERATOPHYLLACEAE p. 144**

24b Calyx present
- 28a Pistils more than 1
  - 29a Flowers with a floral cup ....**ROSACEAE p. 205**
  - 29b Flowers without a floral cup................... **RANUNCULACEAE p. 144**
- 28b Pistil 1
  - 30a Flowers unisexual; calyx present only in the staminate flowers or calyx-like structure atypical
    - 31a Leaves simple...................................... **EUPHORBIACEAE p. 251**
    - 31b Leaves finely dissected ....**CERATOPHYLLACEAE p. 144**
  - 30b Flowers perfect, or if unisexual, both staminate and pistillate flowers with a typical calyx
    - 32a Herbage with stinging hairs; flowers minute ....**URTICACEAE p. 101**
    - 32b Herbage without stinging hairs
      - 33a Leaves compound
        - 34a Floral cup present.................................**ROSACEAE p. 205**
        - 34b Floral cup absent...................... **RANUNCULACEAE p. 144**
      - 33b Leaves simple
        - 35a Plants commonly scurfy or leaves reduced to opposite scales borne on fleshy stems..................... **CHENOPODIACEAE p. 116**
        - 35b Plants not scurfy, nor leaves reduced to scales borne on fleshy stems
          - 36a Sepals free
            - 37a Ovary 3-5-loculed...................... **AIZOACEAE p. 120**
            - 37b Ovary 1-loculed, at least above
              - 38a Leaves alternate......... **AMARANTHACEAE p. 119**
              - 38b Leaves opposite or appearing whorled .... **CARYOPHYLLACEAE p. 128**
          - 36b Sepals fused or calyx lobes borne on the floral tube
            - 39a Calyx petaloid
              - 40a Flowers in terminal heads or umbels, subtended by an involucre .... **NYCTAGINACEAE p. 121**
              - 40b Flowers solitary in the leaf axils ....(*Glaux*) **PRIMULACEAE p. 314**
            - 39b Calyx not petaloid
              - 41a Stamens one or two................**ROSACEAE p. 205**
              - 41b Stamens three or more; leaves opposite .... **CARYOPHYLLACEAE p. 128**

1b Petals present; calyx usually present, but sometimes variously modified or reduced
42a Petals free from each other, or only slightly joined at the base
43a Ovary inferior
44a Trees or shrubs (occasionally low trailing)
45a Stamens more numerous than the petals
46a Leaves opposite; fruit a capsule .... **HYDRANGEACEAE p. 204**
46b Leaves alternate; fruit a pome or pome-like .... **ROSACEAE p. 205**
45b Stamens of the same number as the petals or fewer
47a Stamens opposite the petals; petals hooded .... **RHAMNACEAE p. 260**
47b Stamens alternate with the petals; petals not hooded
48a Leaves opposite or whorled; petals and stamens 4.. **CORNACEAE p. 298**
48b Leaves alternate; petals and stamens usually 5 (rarely 4)
49a Flowers borne in racemose or paniculate umbels .... **ARALIACEAE p. 282**
49b Flowers not borne in racemose or paniculate umbels
50a Leaves palmately veined.. **GROSSULARIACEAE p. 201**
50b Leaves pinnately veined..........**ROSACEAE p. 205**
44b Herbs
51a Petals 12-30; stamens 50 or more.. **NYMPHAEACEAE p. 141**
51b Petals less than 10
52a Styles 2-7
53a Flowers not in umbels
54a Sepals 2 ....**PORTULACACEAE p. 122**
54b Sepals or calyx lobes 4-6, usually 5... **SAXIFRAGACEAE p. 189**
53b Flowers in simple, compound or capitate umbels
55a Fruit dry (a schizocarp); umbels compound, simple or capitate; petals inflexed at the tips....**APIACEAE p. 283**
55b Fruit fleshy (a berry); umbels simple, these single or borne in panicles .... **ARALIACEAE p. 282**
52b Style, if present, only one
56a Style none; stigma often plumose.. **HALORAGIDACEAE p. 279**
56b Style present
57a Sepals 2
58a Petals more than 2; plants more or less fleshy... **PORTULACACEAE p. 122**
58b Petals 2; plants not fleshy... **ONAGRACEAE p. 272**
57b Sepals 4-7
59a Flower cluster surrounded by showy white or pinkish bracts; fruit a drupe.........**CORNACEAE p. 298**
59b Inflorescence not surrounded by showy bracts; fruit not a drupe

60a Flowers in umbels; fruit a berry .... **ARALIACEAE p. 282**
60b Flowers not in umbels; fruit a capsule................. **ONAGRACEAE p. 272**
43b Ovary superior
61a Flowers with a floral tube; stamens borne on the floral tube or on an evident disk
62a Herbs
63a Sepals 2, more or less fused and shed as a unit; leaves ternately-compound.............................. (*Eschscholzia*) **PAPAVERACEAE p. 163**
63b Sepals 4 or more
64a Stipules absent; stamens less than 15
65a Style 1 .... **LYTHRACEAE p. 271**
65b Styles more than 1 .... **SAXIFRAGACEAE p. 189**
64b Stipules usually present, if absent, stamens 15 or more
66a Carpels 2 or 3; stamens 5-10 ...... **SAXIFRAGACEAE p. 189**
66b Carpels 1 or more than 3, or if rarely 3, the stamens more than 10 .... **ROSACEAE p. 205**
62b Trees, shrubs or woody vines
67a Ovary 2-winged; fruit a double samara .... **ACERACEAE p. 257**
67b Ovary not 2-winged; fruit not a double samara
68a Stamens usually many, borne on a calyx-like floral tube, not on a disk .... **ROSACEAE p. 205**
68b Stamens borne on a prominent disk or on a disk lining the base of the calyx
69a Stamens as many as the petals and opposite them; petals hooded .... **RHAMNACEAE p. 260**
69b Stamens if equal in number to the petals then alternate with them
70a Leaves simple........................... **CELASTRACEAE p. 257**
70b Leaves compound .... **ANACARDIACEAE p. 256**
61b Flowers without a floral tube or an evident disk (perianth and stamens arising directly from the receptacle or sometimes the petals connected at the base of the united filaments)
71a Stamens more than 10
72a Pistils 2 to many, simple, distinct
73a Leaves, or at least some of them peltate, floating on the surface of the water .... (*Brasenia*) **CABOMBACEAE p. 143**
73b Leaves not peltate .... **RANUNCULACEAE p. 144**
72b Pistil 1
74a Leaves opposite or some whorled
75a Sepals 5; stamens in 3-5 clusters.. .... **HYPERICACEAE p. 265**
75b Sepals 2 or 3; stamens not in clusters................................ .... (*Meconella*) **PAPAVERACEAE p. 163**
74b Leaves alternate or all basal
76a Leaves hooded, all basal; carnivorous bog plants .... **SARRACENIACEAE p. 185**
76b Leaves not hooded; plants not carnivorous
77a Ovary deeply 5-many-lobed, the lobes separating at maturity into 1-seeded fruits................. **MALVACEAE p. 262**
77b Ovary not deeply 5-many-lobed

78a Pistil simple .... **RANUNCULACEAE p. 144**
78b Pistil compound
79a Ovary more than 1-loculed; sepals 4 or more
80a Aquatic herbs with floating leaves ....
.... (*Nuphar*) **NYMPHAEACEAE p. 141**
80b Shrubs ....**ROSACEAE p. 205**
79b Ovary 1-loculed; sepals 2 or 3
81a Placentation free-central or basal ....**PORTULACACEAE p. 122**
81b Placentation parietal ....**PAPAVERACEAE p. 163**
71b Stamens 10 or fewer
82a Pistils more than 1, distinct or nearly so
83a Herbage strongly succulent; pistils subtended by scale-like glands .....
.... **CRASSULACEAE p. 187**
83b Herbage not succulent; glands, if present, not subtending the pistils
84a Aquatics; sepals and petals 3; stamens 6 ....
.... (*Cabomba*) **CABOMBACEAE p. 143**
84b Without this combination of characters
85a Pistils uni-ovulate, usually exceeding the petals and sepals in number; fruit achenes .... **RANUNCULACEAE p. 144**
85b Pistils multi-ovulate; fruit never an achene..............................
.... **SAXIFRAGACEAE p. 189**
82b Pistil 1
86a Leaves with scarious sheathing stipules, except in *Erigonum*, in which the flowers are surrounded by an involucre .... **POLYGONACEAE p. 105**
86b Stipules, if present, not sheathing (see also *Eriogonum*, if flowers surrounded by an involucre)
87a Flowers strongly irregular
88a Flowers papilionaceous (or petal only 1 in *Amorpha*); stamens 10, usually united in 1 or 2 sets; leaves generally compound ....
.... **FABACEAE p. 223**
88b Flowers not papilionaceous, but often spurred
89a Sepals actually 3, but often apparently 4; petals 5 with the upper petal notched, the others united in 2 pairs, 1 of each pair much smaller than the other............ **BALSAMINACEAE p. 259**
89b Sepals 2 or 5
90a Sepals 5; petals 5, the lowest spurred............................
....**VIOLACEAE p. 267**
90b Sepals 2; petals 4, in 2 dissimilar pairs ....
.... **FUMARIACEAE p. 164**
87b Flowers regular or nearly so
91a Flowers with 4 sepals, 4 petals and usually 6 stamens (4 long and 2 short), rarely only 4; ovary 2-loculed, or if 1-loculed, then not stipitate; fruit usually a silique or silicle..................................
.... **BRASSICACEAE p. 167**
91b Without the above combination of characters
92a Leaves compound
93a Pistil simple; sepals 6-9 and petals 6 ....
.... **BERBERIDACEAp. 161**
93b Pistil compound
94a Fruit a capsule; leaflets 3 .... **OXALIDACEAE p. 246**

94b Ovary separating at maturity into 1-seeded fruits; leaves pinnately compound.. **LIMNANTHACEAE p. 255**
92b Leaves simple
95a Ovary more than 1-loculed
96a Stipules present
97a Styles united around an elongated axis, free only at the tip....... ....**GERANIACEAE p. 247**
97b Styles distinct.................................. **SAXIFRAGACEAE p. 189**
96b Stipules absent
98a Flowers imperfect, or the species polygamous; sepals and petals 3 each... **EMPETRACEAE p. 255**
98b Flowers perfect
99a Styles 2 or 5 .... **SAXIFRAGACEAE p. 189**
99b Style 1
100a Leaves dissected.................. **LIMNANTHACEAE p. 255**
100b Leaves entire or merely toothed or reduced to scales.... .... **ERICACEAE p. 299**
95b Ovary 1-loculed, at least above
101a Calyx of 2-3 distinct sepals
102a Placentation free-central or basal ....**PORTULACACEAE p. 122**
102b Placentation parietal...... (*Meconella*) **PAPAVERACEAE p. 163**
101b Calyx not of 2-3 distinct sepals
103a Leaves opposite or appearing whorled ....**CARYOPHYLLACEAE p. 128**
103b Leaves basal or alternate, or sometimes reduced to scales
104a Leaves reduced to non-green scales borne along the stem; sepals free .... **ERICACEAE p. 299**
104b Leaves not reduced to scales
105a Leaves alternate, stipules present; perianth parts undifferentiated............................. **POLYGONACEAE p. 105**
105b Leaves basal
106a Calyx funnel-form, 5-toothed ....**PLUMBAGINACEAE p. 313**
106b Calyx saucer-shaped, 5-parted; leaves sticky-glandular; carnivorous plants of bogs or swamps (ours).... ....**DROSERACEAE p. 186**
42b Petals united at least at the base
107a Stamens more than 5
108a Corolla urn-shaped, tubular, cup-shaped or bell-shaped; style 1; anthers often opening by pores................................ **ERICACEAE p. 299**
108b Corolla not urn-shaped nor tubular, commonly with the petals joined only at the base
109a Pistils 4-5; stamens 10.............(*Sedum*) **CRASSULACEAE p. 187**
109b Pistil 1
110a Corolla irregular
111a Flowers not papilionaceous; petals 4, in 2 dissimilar pairs; leaves dissected .... **FUMARIACEAE p. 164**
111b Flowers papilionaceous; petals 5 .... **FABACEAE p. 223**

110b Corolla regular
112a Stamens 6-9, the same number as the petals and opposite them .... **PRIMULACEAE p. 314**
112b Stamens 10 or more, more numerous than the petals
113a Stamens 10, united at the base; leaves 3-foliolate .... **OXALIDACEAE p. 246**
113b Stamens more than 10, united into a tube................................. .... **MALVACEAE p. 262**
107b Stamens 5 or less
114a Ovary inferior
115a Stamens united to each other, either by the filaments or the anthers
116a Flowers in a head surrounded by an involucre; calyx modified into a pappus or lacking; fruit 1-seeded.............. **ASTERACEAE p. 399**
116b Flowers not in involucrate heads
117a Stems tendril-bearing; flowers unisexual; leaves palmate......... .... **CUCURBITACEAE p. 393**
117b Stems not tendril bearing; flowers perfect............................ .... **CAMPANULACEAE p. 394**
115b Stamens distinct. (See also Asteraceae, if stamens 5 and flowers in involucrate heads and fruit 1-seeded.)
118a Leaves alternate
119a Leaves compound; corolla either bearded or bearing "scales" .... **MENYANTHACEAE p. 321**
119b Leaves simple; corolla neither bearded nor bearing "scales"... .... **CAMPANULACEAE p. 394**
118b Leaves opposite or whorled
120a Stamens 1-3; flowers usually slightly irregular; erect herbs...... .... **VALERIANACEAE p. 390**
120b Stamens usually 4 or 5
121a Flowers in dense involucrate spikes; leaves prickly .... **DIPSACACEAE p. 393**
121b Flowers usually not in spikes, but, if so, the involucre absent or the leaves not prickly
122a Trees, shrubs or woody twining or creeping vines with opposite leaves .... **CAPRIFOLIACEAE p. 386**
122b Herbs with whorled leaves............ **RUBIACEAE p. 383**
114b Ovary superior
123a Corolla more or less irregular
124a Leaves alternate or basal, or only the lower leaves opposite
125a Plants without green leaves; leaves reduced to alternate scales; root parasites .... **OROBANCHACEAE p. 379**
125b Plants with green leaves
126a Ovary 1-loculed; carnivorous plants................................ .... **LENTIBULARIACEAE p. 377**
126b Ovary 2-loculed................ **SCROPHULARIACEAE p. 356**
124b Leaves all opposite or whorled
127a Ovary deeply 4-lobed, separating into 4 nutlets; style usually arising from the base of the lobes; stem square in cross-section....... .... **LAMIACEAE p. 345**

127b Ovary not deeply 4-lobed (if separating into nutlets, doing so only at maturity); style arising from the top of the ovary
128a Mature ovary separating into 2-4 one-seeded nutlets.. **VERBENACEAE p. 344**
128b Mature ovary not separating into nutlets; fruit a capsule .... **SCROPHULARIACEAE p. 356**
123b Corolla regular
129a Pistils 3-5; petals only slightly fused at the base .... **CRASSULACEAE p. 187**
129b Pistils 1 or 2
130a Ovaries two, free below but united into a single style; plants with milky juice
131a Stamens and stigmas united, the column bearing hood-like appendages; sepals usually reflexed............. **ASCLEPIADACEAE p. 323**
131b Stamens scarcely united to the stigma; sepals usually not reflexed .... **APOCYNACEAE p. 321**
130b Ovary solitary, although often deeply lobed and compound
132a Ovary deeply 4-lobed, separating at maturity into nutlets .... **BORAGINACEAE p. 338**
132b Ovary not separating into nutlets, typically not 4-lobed
133a Stamens opposite the corolla lobes
134a Sepals 2 .... **PORTULACACEAE p. 122**
134b Sepals or calyx lobes 4-5
135a Style 1; seeds more than 1 per capsule.. **PRIMULACEAE p. 314**
135b Styles 5; seed one .... **PLUMBAGINACEAE p. 313**
133b Stamens alternate with the corolla lobes
136a Twining or trailing herbs, either with normal green leaves or without green color and with leaves reduced to scales
137a Leafless, non-green parasites .... **CUSCUTACEAE p. 326**
137b Plants with normal green foliage... **CONVOLVULACEAE p. 324**
136b Not twining or trailing herbs
138a Corolla scarious, 4-lobed; flowers minute in bracteate spikes; leaves usually all basal, venation often appearing parallel .... **PLANTAGINACEAE p. 381**
138b Without the above combination of characters
139a Ovary 1-loculed, rarely appearing 2-loculed by the intrusion of the placentae; placentation parietal
140a Leaves mainly alternate, if opposite, then not entire; sepals usually united only at the base; plants usually pubescent........... **HYDROPHYLLACEAE p. 334**
140b Leaves opposite or whorled, mostly entire; calyx usually 4-5-toothed .... **GENTIANACEAE p. 319**
139b Ovary more than 1-loculed; placentation axile
141a Ovary 4-5-loculed............... **ERICACEAE p. 299**
141b Ovary 2-3-loculed
142a Stigmas 3 (rarely 2) .... **POLEMONIACEAE p. 328**
142b Stigma capitate, entire or slightly bilobed .... **SOLANACEAE p. 353**

Ferns and Fern Allies

OPHIOGLOSSACEAE
Adder's-tongue Family

Small fleshy perennials; leaves solitary or few, erect, simple or variously compound, consisting of a sterile foliaceous portion subtending a fertile segment; sporangia large, thick-walled; spores uniform.

Sterile blade entire .... **Ophioglossum**
Sterile blade lobed or compound... **Botrychium**

Botrychium

Rhizomes short; leaves 1-3, somewhat succulent to leathery, sterile blade ternate or 1-4 times pinnately compound, linear to fan-shaped, fertile portion 1-3-pinnate; sporangia sessile or nearly so; spores yellow.

Botrychium multifidum (Gmel.) Rupr. — Leather grape-fern

Plants 1-5 dm. in height, stout, leathery and evergreen; sterile blade stalked and attached near the ground level, 1-3-pinnate, nearly as wide or wider than long, ultimate segments more or less ovate, entire or crenate, fertile portion well separated from the sterile blade, much-branched, up to 2 dm. long.

(=*Botrychium silaifolium* Presl)

Sphagnum bogs along the coast, in meadows and along the margins of mountain lakes and streams; uncommon.

Ophioglossum

Roots usually bearing bulblets or plantlets; sterile blade simple, entire, more or less fleshy, fertile portion unbranched; sporangia in two rows, more or less embedded.

Ophioglossum pusillum Raf. — Northern Adder's-tongue

Sterile blade usually solitary, simple, entire, oblanceolate to ovate, 2.5-10 cm. long and 1-4 cm. wide, fertile portion 1-4 cm. long with 10-30 pair of sporangia.

Wet ground, bogs, meadows and wooded areas; uncommon.

PTERIDACEAE
Maidenhair Fern Family

Rhizomes bearing hairs and/or scales; stipes often dark-colored and usually with scales at least at the base; leaves all alike or dimorphic, leaf blades 1-6-pinnate; sori borne on the veins and often forming a continuous submarginal band or nearly covering the back of the leaf blade; a false indusium often formed by the reflexed or revolute leaf margins, true indusia lacking.

1a Leaves markedly dimorphic .... **Cryptogramma**
1b Leaves usually all alike, never markedly dimorphic
  2a Back of leaf blades covered with a yellow powder........... **Pentagramma**
  2b Back of leaf blades not covered with a yellow powder
    3a Stipe forked at the tip forming a fan-like array of 5-14 pinnae....... .... **Adiantum**
    3b Stipe not forked at the tip, pinnae not in a fan-like arrangement
      4a Leaf blades glabrous.. **Aspidotis**
      4b Leaf blades rusty-tomentose on the back............ **Cheilanthes**

Adiantum

Delicate ferns from chaffy rhizomes; stipes wiry, dark-colored and shiny; leaves once-pinnate to decompound; sori appearing marginal on the pinnules, covered by the reflexed margins (false indusia).

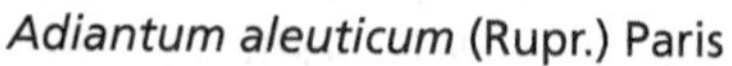

Adiantum aleuticum

Adiantum aleuticum (Rupr.) Paris
Western maidenhair

Rhizomes short-creeping, dark-colored; stipes about twice as long as the leaf blades; leaves up to 10 dm. tall, blades reniform to orbicular in outline; stipe forked at the top, then forming a fan-like array of 5-14 pinnae, pinnules numerous, oblong to obliquely triangular, with indentations on the upper margins; false indusia oblong to crescent-shaped.

(=*Adiantum pedatum* L. var. *aleuticum* Rupr.)

Shady areas along streams and on wet rocky cliffs.

Aspidotis

Rhizomes short-creeping, scaly; stipes dark reddish-brown; leaves crowded, glabrous, leaf blades 3-5-pinnate, ovate to triangular in outline; sporangia in marginal sori covered by the scarious revolute leaf margins.

Aspidotis densa (Brackenr.) Lellinger — Indian's dream

Rhizomes branched; leaves densely tufted, 5-30 cm. tall; stipes wiry, reddish-brown, shiny, several times longer than the blades; leaf blades 2-10 cm. long, 3-pinnate, ovate to deltoid in outline; glabrous, pinnae few, asymmetrical, pinnules linear, margins revolute; sori continuous along the length of the pinnules, except at the apex.

(=*Cheilanthes siliquosa* Maxon)

Rocky areas.

Cheilanthes

Small ferns, usually growing on rocks; leaves variously pubescent or rarely glabrous, leaf blades 1-4-pinnate, the segments usually small, orbicular to elongate or spatulate, the margins recurved; sori on or near the ends of veins, often concealed by the pubescence on the leaves or by the more or less modified revolute leaf margins forming the false indusia.

Cheilanthes gracillima D.C. Eat. Lace fern

Rhizomes branched, bearing brown scales; leaves numerous, 5-25 cm. tall; stipes dark brown, wiry, equal to the blades or much longer; leaf blades ovate to lanceolate in outline, 2-3-pinnate, pinnae numerous, pinnately divided, segments oblong to oval, densely rusty-tomentose on the lower surface, margins revolute, nearly concealing the numerous sporangia.

Dry rocky areas.

Cryptogramma

Small ferns usually growing on rocks; leaves markedly dimorphic, the fertile ones longer than the sterile, leaf blades 2-4-pinnate, ultimate segments of the sterile blades dentate to deeply cut, margins of the fertile blades reflexed to form false indusia; sori near the margins in a continuous line at the end of the veins.

Leaves herbaceous and thin, not persistent .... *C. cascadensis*
Leaves somewhat leathery, persistent.................................. *C. acrostichoides*

Cryptogramma acrostichoides R. Br. American parsley fern

Rhizomes much-branched; leaves densely tufted, dimorphic, somewhat leathery, 2-3-pinnate, fertile ones erect, up to 4 dm. tall, sterile leaves lower and more or less spreading; sporangia hidden by the revolute margins (false indusia) of the leaf segments.

Alpine cliffs and rock slides; also in the Columbia Gorge and near the coast.

Cryptogramma cascadensis E. Alverson Cascade parsley fern

Rhizomes much-branched; leaves tufted, dimorphic, herbaceous, sterile leaves spreading, 3-20 cm. long, fertile leaves erect, 5-25 cm. long, leaf blades 2-3-pinnate, ovate-lanceolate to deltoid in outline.

Rocky areas at high elevations in the mountains.

Pentagramma

Small ferns with scaly rhizomes; stipes dark-colored, wiry, shiny; leaves tufted, blades 1-3 pinnate, triangular in outline, lower surface covered by a dull white or yellow powder; sori borne along the veins.

Pentagramma triangularis (Kaulf.) Yatsk., Windham & E. Wollenw. Goldback fern

Rhizomes stout, covered with dark scales; stipes glabrous, about twice the length of the blades; leaves 6-40 cm. tall, leaf blades thin to leathery, broadly triangular in outline, 1-2-pinnate, basal pinnae largest, asymmetrical, lower surface covered with a yellow to golden-yellow powder.

[=*Pityrogramma triangularis* (Kaulf.) Maxon]

Rocky areas, often on lava outcrops.

DENNSTAEDTIACEAE
Bracken Family

Rhizomes often branching, bearing hairs or sometimes scales; stipe strongly grooved on the upper side; leaf blades once-pinnate to decompound; sori borne at or near the margin of the blade and usually covered by the reflexed margin.

Pteridium

Large coarse ferns with thick much-branched rhizomes; rhizomes and stipes with felt-like hairs, scales absent; leaves borne on long stout stipes, leaf blades large, pinnately decompound; sori marginal, more or less continuous, covered by the reflexed margins of the pinnules (false indusia); true indusia narrow, obscure.

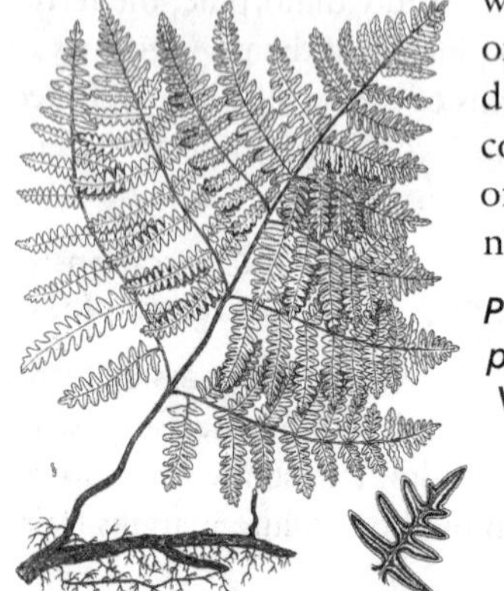

Pteridium aquilinum var. pubescens

Pteridium aquilinum (L.) Kuhn var. *pubescens* Underw.
Western bracken or Western brake fern

Rhizomes long-creeping; stipe yellow, about as long as the leaf blade; leaves 1.5-30 dm. tall, blades 3-pinnate, triangular to deltoid-ovate in outline; pinnules entire, toothed or lobed, tomentose on the lower surface, glabrous or nearly so above; false indusia villous.

A common, widespread, and extremely variable species.

THELYPTERIDACEAE
Marsh Fern Family

Rhizomes scaly at the tip; leaves all alike to slightly dimorphic; leaf blades pinnate to pinnate-pinnatifid; sori orbicular or oblong, borne along the veins; indusia, if present, orbicular to reniform.

Thelypteris

Leaf blades once-pinnate to pinnate-pinnatifid, or rarely 2-pinnate, pinnae entire to deeply pinnatifid, the lower pinnae usually greatly reduced; sori orbicular to elongate, borne along the veins; indusia usually present.

Thelypteris nevadensis (Baker) Clute ex C. V. Morton
Nevada marsh fern

Rhizomes slender, covered with bases of old leaves; stipes half the length of the blades or less, straw-colored; leaf blades 2-7 dm. long, lanceolate in outline, the apex long acuminate, pinnate-pinnatifid to 2- pinnate, pinnae sessile; sori small, submarginal; indusia present, but minute, often ciliate and glandular.

(=*Dryopteris oregana* C. Chr.)

Stream banks and seepage areas, mostly on the western slopes of the Cascades and westward.

BLECHNACEAE
Deer Fern Family

Rhizomes and stipes scaly; leaves all alike or dimorphic, the blades pinnatifid to 2-pinnate; sori elongate or linear, borne along the midrib; indusia usually present, but sometimes obscured by the dehisced sporangia.

Blechnum

Leaves often dimorphic, leaf blades pinnatifid to once pinnate, sterile ones numerous, spreading or ascending, fertile ones few, erect; sori in a continuous row parallel to the midrib; indusia attached to the margins and opening towards the midrib.

Blechnum spicant (L.) Smith — Deer fern

Rhizomes short-creeping, covered with chestnut-brown scales; stipes reddish-brown; leaves markedly dimorphic, sterile blades 1.5-10 dm. long, linear to linear-oblanceolate in outline, once-pinnate, pinnae linear-oblong, or the distant basal ones broader; fertile leaves erect in the center of the spreading or ascending sterile ones, 4-15 dm. tall, stipe long, blades of the fertile leaves with more narrowly linear and distant pinnae.

[=*Struthiopteris spicant* (L.) Weis]

Moist to wet areas in forests, near the coast and in the mountains.

ASPLENIACEAE
Spleenwort Family

Leaves crowded, erect to spreading, simple to 4-pinnate or several times pinnatifid; sori linear to oblong, borne on the veins; indusia usually present, lateral, usually entire and of the same shape as the sori.

Asplenium

Characters those of the family.

Asplenium trichomanes L. — Maidenhair spleenwort

Rhizomes short-creeping; stipes short, dark-colored; leaves densely tufted, 5-25 cm. long, blades once-pinnate, linear, tapering to both ends, pinnae oblong to nearly orbicular, usually asymmetrical, the margins shallowly crenate; indusia opaque, entire or nearly so.

Moist crevices or rock slides, usually on limestone.

DRYOPTERIDACEAE
Wood Fern Family

Rhizomes usually creeping; leaves all alike or rarely dimorphic, the blades simple to 1-5-pinnate; sori borne on the veins or the tips of the veins, rounded to elongate; indusia, when present, orbicular, elongate or sometimes hood-like, often peltate.

1a Pinnae auriculate on one side of the base; leaves once-pinnate; indusium attached at the center (peltate) without a sinus .... **Polystichum**
1b Pinnae not auriculate on one side of the base; leaves usually, at least in part, two or more times pinnate
  2a Stipes, at least the lower portion, dark-colored and shiny
    3a Margins of the pinnules bluntly toothed or shallowly lobed; indusia cup-like of many segmented hairs or scales............................. **Woodsia**
    3b Margins of the pinnules with small acute serrations; indusia hood-like, though often inconspicuous .... **Cystopteris**
  2b Stipes light-colored, or if at all dark, then not shiny
    4a Indusia orbicular to reniform .... **Dryopteris**
    4b Indusia absent or elongate and flap-like
      5a Leaves densely tufted, blades several times longer than wide .... **Athyrium**
      5b Leaves scattered, blades nearly as wide as long or wider; indusia absent.. **Gymnocarpium**

Athyrium

Leaves usually large, erect to spreading, 1-3-pinnate, the pinnules variously incised to pinnatifid; sori dorsal, oblong to crescent-, J- or horseshoe-shaped or rarely nearly orbicular; indusia usually present, laterally attached and flap-like.

Athyrium filix-femina (L.) Roth — Lady fern

Leaves up to 2 m. tall; densely tufted from stout, erect rhizomes; leaf blades broadly lanceolate in outline, tapering at both ends, 2-3-pinnate; sori oval to oblong or J-shaped to horseshoe-shaped; indusia usually lacerate-ciliate along the free margin.

This is a highly variable species with several named varieties.

Damp shaded areas in the forest and along streams.

Athyrium filix-femina

Cystopteris

Small ferns with scaly creeping rhizomes; leaf blades thin and delicate, 1-4-pinnate; sori nearly orbicular, borne on the veins; indusia membranous, attached by the base on the side toward the midrib and arching over the sori.

Cystopteris fragilis (L.) Bernh. Fragile fern or Brittle fern

Rhizomes creeping, densely scaly; leaves 5-40 cm. tall, usually clustered; stipes shorter than, or about equaling, the blades in length, scaly and dark-colored at the base; leaf blades 1-3-pinnate, mostly ovate to lanceolate in outline, but extremely variable; sori small; indusia hood-like, more or less fringed on the free margins.

Usually on cliff faces or in rocky areas and damp woods.

Dryopteris

Rhizomes creeping; leaves usually 1-3-pinnate, herbaceous to leathery; sori nearly orbicular; indusia orbicular or reniform, attached at the sinus.

1a Leaf blades essentially 3-pinnate ....*D. expansa*
1b Leaf blades pinnate-pinnatifid to 2-pinnate
  2a Stipes with two distinct types of scales, 1 hair-like and the other much broader; leaves not glandular.. *D. filix-mas*
  2b Stipes with broad to narrow scales, but all of one type; leaves often glandular along the rachis..*D. arguta*

Dryopteris arguta (Kaulf.) Maxon Coastal wood fern

Rhizomes stout, covered by bright chestnut-colored scales; stipes shorter than the blades; leaves to 10 dm. tall, blades herbaceous to slightly leathery, pinnate-pinnatifid to 2-pinnate at the base, lance-ovate to deltoid-lanceolate in outline, acuminate, pinnae sessile, pinnules oblong, the margins doubly serrate to incised; sori large, orbicular; indusia persistent.

Rocky outcrops and woods.

Dryopteris expansa (Presl) Fraser-Jenkins & Jermy

Spreading wood fern

Rhizomes stout, covered by light brown scales; stipes usually about as long as the blades; leaves 2.5-11 dm. tall, blades 3-pinnate, broadly triangular to ovate or oblong in outline, pinnae short-stalked, asymmetrical; sori small; indusia sub-persistent, glabrous, sometimes glandular.

[*Dryopteris dilatata* (Hoffm.) Gray misapplied]

Moist woods, usually in rocky areas, and along stream banks.

Dryopteris filix-mas (L.) Schott Male fern

Stipe with two distinct types of scales, 1 hair-like and the other much broader; leaf blades ovate-lanceolate in outline, pinnate-pinnatifid to 2-pinnate, margins of the pinnae serrate to lobed; sori borne in a row between the margin and the midrib; indusia not glandular.

Moist shady woods, often in rocky areas.

Gymnocarpium

Rhizomes long-creeping; leaves 2-3-pinnate-pinnatifid, the margins of the pinnae entire to crenate; sori borne between the midrib and the margin of the pinnae; indusia absent.

Gymnocarpium dryopteris (L.) Newman — Oak fern

Rhizomes slender, bearing a few light brown scales; stipes to 3 dm. long; leaves arising singly, leaf blades 3-28 cm. long, broadly triangular in outline, divided into 3 primary divisions, these 1-2 pinnate; sori small, submarginal.

(=*Dryopteris linnaeana* C. Chr.)

Moist ground in cool forests, on talus slopes or in swamps.

Polystichum

Rhizomes stout, densely scaly; leaves simple or 1-3-pinnate, blades usually firm in texture, scaly, pinnae usually auriculate and the margins commonly sharply serrate; sori rounded; indusia, if present, orbicular, centrally peltate.

1a Pinnae, at least the lower, deeply pinnatifid; rachis producing one or more scaly vegetative buds .... *P. andersonii*

1b Pinnae not deeply pinnatifid; rachis without vegetative buds

  2a Teeth of pinnae spreading-spinulose; stipe less than 1/6 the length of the blade.......... *P. lonchitis*

  2b Teeth of the pinnae incurved and ascending; stipe more than 1/5 the length of the blade

    3a Bases of the pinnae cuneate; indusia ciliate; common...... *P. munitum*

    3b Bases of the pinnae oblique; indusia entire or toothed, not ciliate...... .... *P. imbricans*

Polystichum andersonii M. Hopkins — Anderson's sword fern

Stipes 5-20 cm. long, chaffy; rachis producing one or more scaly vegetative bud(s); leaves 3-10 dm. tall, pinnae 2-10 cm. long, the lower ones deeply pinnatifid, all spinulose-toothed, the teeth ascending; indusia small, sparsely ciliate.

Moist forests.

Polystichum imbricans (D.C. Eat.) D. H. Wagner — Imbricated sword fern

Easily confused with *P. munitum*, but differing from it by crowded, usually over-lapping pinnae, smaller size (3-6 dm. tall) and scales of the upper portion of the stipe which are narrow and deciduous and by the never ciliate indusia.

[=*Polystichum munitum* var. *imbricans* (D.C. Eat.) Maxon]

In the high mountains and in eastern Washington and Oregon.

Polystichum lonchitis (L.) Roth — Holly fern

Rhizomes large, covered with old stipe bases; stipes short, covered with rusty-brown scales; leaves 1.5-6 dm. tall, blades once-pinnate, linear-oblanceolate in outline, taping to both ends, pinnae approximate, the lower ones reduced, others oblong to lanceolate, the base auriculate above, margins serrate with spreading-spinulose teeth; sori large; indusia entire or only minutely erose.

Rocky areas in the high mountains.

Juvenile plants of *P. munitum* are often confused with this species.

Polystichum munitum (Kaulf.) Presl — Sword fern

Rhizomes stout, covered with old stipe bases; stipes stout, densely covered with persistent large brown scales; leaves ascending or stiffly erect, 3-15 dm. tall, blades linear-lanceolate in outline, once-pinnate; pinnae linear-attenuate, the base auriculate above, the margins doubly serrate to incised with incurved teeth; sori large; indusia present, often long-ciliate.

Damp woods; common.

Woodsia

Small caespitose ferns; leaves 1-2-pinnate, often glandular or pubescent; sori orbicular, borne on the veins; indusia arising from below the sporangia, cup-shaped, becoming variously cleft or consisting of numerous stellate divisions.

Leaf blades without multicellular hairs along the midrib .... *W. oregana*
Leaf blades with flattened multicellular hairs, especially along the midrib .... *W. scopulina*

Woodsia oregana D.C. Eat. — Oregon Woodsia

Rhizomes short creeping, covered with pale brown scales; stipes half the length of the leaf blades or less; leaves 5-25 cm. tall, blades pinnate-pinnatifid or 2-pinnate, linear to lanceolate in outline, pinnae distant, triangular to oblong in outline, margins of the pinnules commonly reflexed and covering the sori; indusia minute, divided into a few beaded hairs.

Dry rocky areas.

Woodsia scopulina D. C. Eat. — Rocky Mountain Woodsia

Rhizomes creeping, covered with pale brown scales; stipes about half the length of the leaf blades; leaves 5-40 cm. tall, blades 1-2-pinnate, linear to oblong or lanceolate in outline, pinnae oblong to deltoid in outline, glandular-pubescent, especially on the lower surface, white hairs mixed with gland-tipped hairs, margins of the pinnules bluntly toothed to shallowly lobed; indusia deeply cleft.

Dry rocky areas.

POLYPODIACEAE
Polypody Family

Rhizomes creeping and scaly; leaves all alike or dimorphic, leaf blades simple, pinnatifid to pinnate; sori orbicular to oblong, borne on the veins; indusia absent.

Polypodium

Small often epiphytic ferns; leaves simple, variously pinnatifid to 1-3-pinnate, margins entire to serrate; sori large, orbicular to elliptic; indusia absent.

1a Leaves leathery, upper divisions of the veins joining .... *P. scouleri*
1b Leaves thin to only slightly leathery, upper divisions of the veins continuing free to the margins of the pinnae

2a Margins of the pinnae serrate, pinnae attenuate or acute at the tip .... *P. glycyrrhiza*
2b Margins of pinnae entire or crenate, pinnae obtuse to broadly acute at the tip
  3a Scales of the rhizome contorted and usually dentate......... *P. amorphum*
  3b Scales of the rhizome symmetrical, not contorted and usually entire or nearly so .... *P. hesperium*

Polypodium amorphum Suksd.
Rhizomes often whitish and glaucous, acrid in taste, scaly; leaves to 30 cm. long, the blades to 4 cm. wide, slightly leathery, oblong, pinnatifid, the margins entire to crenulate; sori orbicular.
(=*Polypodium montense* F. A. Lang)
Cliffs and rocky areas.

Polypodium glycyrrhiza D. C. Eaton — Licorice fern
Rhizomes with a licorice taste, creeping, covered with light brown scales; stipes usually short; leaves 1.5-7.5 dm. tall, blades lanceolate to oblong in outline, once-pinnate, divided nearly to the midrib, pinnae alternate, linear attenuate, abruptly broadened at the bases, margins of the pinnae serrate; sori orbicular to oval.
(=*Polypodium vulgare* L. var. *occidentale* Hook.)
Rocks, logs and mossy trees.

Polypodium hesperium Maxon — Western polypody
Rhizomes whitish and glaucous, acrid to sweet in taste, scaly; leaves differing from the preceding species by the shorter, oblong to oval, obtuse pinnae with entire to crenate margins; sori orbicular to elliptic.
(=*Polypodium vulgare* L. var. *columbianum* Gilbert)
Rocky areas, from the lowlands to the mountains.

Polypodium scouleri Hook. & Grev.
Leathery-leaf polypody or Coast polypody
Rhizomes covered with dark scales at first, but becoming naked; leaves few, 0.5-7 dm. tall, blades ovate in outline, leathery, pinnatifid, the pinnae oblong, obtuse, the margins crenate to obscurely crenate-serrulate; sori orbicular, crowded along the midrib, commonly confined to the upper pinnae.
Usually on trees or moss covered rocks; coastal.

AZOLLACEAE
Mosquito Fern Family

Often forming dense free-floating mats on the surface of quiet water; roots unbranched; stems repeatedly forked, bearing minute alternate leaves in 2 more or less overlapping rows, divided into a floating lobe and a submerged lobe, the latter serving as a float, upper lobe containing the nitrogen-fixing cyanobacteria, *Anabaena azollae*; sporocarps in pairs on the lower leaf lobe, unisexual, bearing either microsporangia containing numerous microspores or megasporangia containing a single megaspore, also containing glochidia with barbed tips.

Azolla

Characters those of the family.

Plants averaging over 2 cm. long, up to 6 cm. long; largest hairs of the upper leaf lobes unicellular .... *A. filiculoides*
Plants mostly less than 2 cm. long; largest hairs of the upper leaf lobes multicellular .... *A. mexicana*

Azolla filiculoides Lam. Duckweed fern
Plants 2-6 cm. long; upper lobe of the leaf about 1 mm. long, often tinged with red and with broad hyaline margins, lower leaf lobes about equal in size to the upper; glochidia not septate or rarely septate at the apex.
Free-floating, often in large masses, on still water, or stranded on mud.

Azolla filiculoides

Azolla mexicana Presl Mexican water-fern
Plants 1-2 cm. long; upper leaf lobes less than 1 mm. long, often tinged with purple or red, and with narrow hyaline margins, lower leaf lobes larger than the upper; glochidia septate.
Free-floating or stranded on mud.

MARSILEACEAE
Water-clover Family

Perennials from slender creeping rhizomes; leaves erect, the leaf blades lacking or with 2-4 palmately arranged leaflets; sporocarps hard, compressed ovate to globose, borne on the rhizome at the base of the petiole; sori enclosed in the sporocarp, each sporocarp bearing both microsporangia and megasporangia.

Marsilea

Plants aquatic or amphibious; leaves usually floating on the surface of the water, long-petioled, palmately 4-foliolate; sporocarps pubescent, ovoid to reniform; dehiscent by 2 valves; sori containing both megasporangia and microsporangia.

Upper tooth of the sporocarp 0.4-1.2 mm. long, acute and sometimes hooked .... *M. vestita*
Upper tooth of the sporocarp, if present, less than 0.4 mm. long, blunt........... .... *M. oligospora*

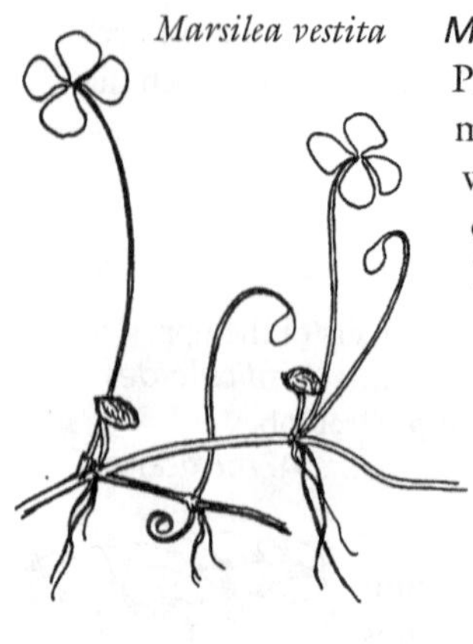

Marsilea oligospora Goodd.

Plants pubescent; petioles 3-6 cm. long; leaflets 5-15 mm. long, pilose; sporocarps nodding, pubescent, when young, but becoming glabrous, ovoid, with 1 or 2 teeth near the base, the lower tooth 0.2-6 mm. long, the upper tooth, if present, less than 0.4 mm. long, blunt.

In shallow ponds, marshes and river banks.

Marsilea vestita Hook. & Grev. Clover-fern

Plants usually pubescent; petioles 2-20 cm. long; leaflets cuneate, entire, 0.5-2 cm. long; sporocarps 1 to a peduncle, densely pubescent when young, but becoming glabrous, ovoid, with 2 teeth near the base, the upper tooth 0.4-1.2 mm. long, acute and sometimes hooked at the apex, the lower tooth 0.3-0.6 mm. long, obtuse.

Mainly in shallow pools and river banks.

EQUISETACEAE
Horsetail or Scouring rush Family

Perennial plants with creeping blackish rhizomes; aerial stems jointed, hollow except at the nodes, generally erect, simple or with whorled branches, often of two kinds, sterile and fertile, at least the sterile stems ridged horizontally; leaves scarious, united below to form sheaths, their tips generally free and persistent or deciduous; spores borne in terminal cones (strobili) consisting of whorls of peltate bracts closely fitted together about an axis, each bract bearing several sporangia on its under surface; spores greenish, minute, each spore bearing 4 strap-shaped bands; entire plant overlaid with silica, smoothly, or in orderly arrangements of tubercles or bands.

Equisetum

Characters those of the family.

1a Stems with regular whorls of branches, often sterile
  2a Teeth of the sheaths 5 to 12
    3a Teeth of the sheaths 1-3 mm. long; stems sterile .... *E. arvense*
    3b Teeth of the sheaths 3-7 mm. long; stems usually fertile.. *E. palustre*
  2b Teeth of the sheaths 14 or more
    4a Teeth of the sheaths 3-8 mm. long; grooves of the stem 20-40 .... *E. telmateia*
    4b Teeth of the sheaths 1.5-3.5 mm. long; grooves of the stem 9-25 .... *E. fluviatile*
1b Stems unbranched or with a few scattered branches, usually fertile
  5a Stems non-green, brown or flesh-colored
    6a Stems up to 8 mm. thick; sheaths, including the teeth, to 20 mm. long .... *E. arvense*

6b Stems 10-25 mm. thick; sheaths, including the teeth, 20-50 mm. long.. .... *E. telmateia*

5b Stems green

7a Stems evergreen and persistent throughout the winter

8a Strobilus apiculate; sheaths usually with 2 dark bands...... *E. hyemale*

8b Strobilus rounded at the tip; sheath usually with a single dark band... .... *E. laevigatum*

7b Stems annual

9a Teeth of the sheaths 3-7 mm. long, hyaline margined...... *E. palustre*

9b Teeth of the sheaths 1-3 mm. long

10a Teeth 1-2 mm. long, early deciduous leaving a dark rim on the sheath........ *E. laevigatum*

10b Teeth 1.5-3 mm. long, persistent........ *E. fluviatile*

Equisetum arvense L. — Common horsetail

Stems annual, dimorphic; sterile stems 1-6 dm. tall, 1.5-5 mm. thick, with 4-12 vertical grooves, green, sheaths about 12-toothed, dark at least at the apex, branches in dense whorls, solid, 3-4-angled, sheaths of the branches 4-toothed; fertile stems appearing before the sterile ones, unbranched, 5-30 cm. tall, and up to 8 mm. thick, non-green, sheaths 6-12-toothed; strobilus blunt, terminal on the stem.

A hybrid between this species and *E. fluviatile* is called *Equisetum* x*litorale* Kuehlew. ex Rupr.

In moist to wet, usually disturbed, habitats; often weedy.

Equisetum fluviatile L. — River or Water horsetail

Stems annual, all alike, up to 1 m. tall, 9-25-grooved; sheaths 4-9 mm. long, with approximately 18 (12-24) persistent black teeth 1.5-3 mm. long; branches, if present, 4-6-angled; strobilus rounded.

A hybrid between this species and *E. arvense* is called *Equisetum* x*litorale* Kuehlew. ex Rupr.

In shallow water of marshes, stream banks and bogs.

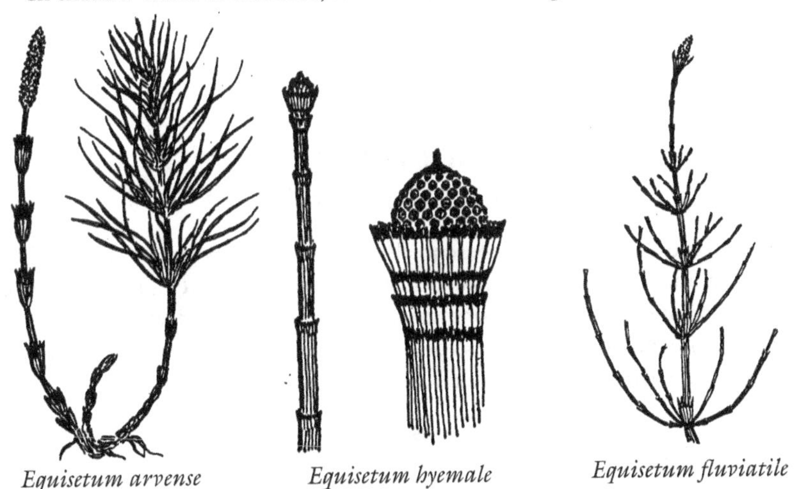

Equisetum arvense *Equisetum hyemale* *Equisetum fluviatile*

Equisetum hyemale L. subsp. *affine* (Engelm.) Calder & R. Taylor
Common scouring rush

Stems evergreen, all alike, up to 2 m. in height and 4-20 mm. thick, 16-48-grooved, ridges rough, irregular, stems usually simple, but occasionally with a few branches; sheaths black banded, one band at the base, the other at the tip, 14-50-toothed; strobilus apiculate.

A sterile hybrid between this species and *E. laevigatum* is called *Equisetum* x*ferrissii* Clute.

Wet to moist areas.

Equisetum laevigatum A. Braun — Smooth scouring rush

Stems annual or persistent, tufted, all alike, simple, or occasionally with a few branches, up to 1.5 m. tall, 2-8 mm. thick, usually with 14-30 grooves, the sheaths 10-28-toothed and dark banded apically, the teeth 1-2 mm. long and early deciduous leaving a dark rim on the sheath; strobilus usually more or less rounded.

A sterile hybrid between this species and *E. hymale* is called *Equisetum* x*ferrissii* Clute.

Moist, often gravelly, areas.

Equisetum palustre L. — Marsh horsetail

Stems annual, all alike, 2.5-9 dm. tall, deeply 5-10-grooved, sheaths loose, green, with persistent darker teeth, these with evident hyaline margins, branches 4-6-angled; strobilus blunt.

Usually in marshes and swamps.

Equisetum telmateia Ehrh. subsp. *braunii* (Milde) R. L. Hauke
Giant horsetail

Stems annual, dimorphic; sterile stems 0.5-30 dm. tall, 0.5-2 cm. thick, usually with 20-40 grooves, sheaths mostly 1-2.5 cm. long with 14-28 slender teeth, branches numerous, 4-6-grooved; fertile stems 1.5-6 dm. tall, 1-2.5 cm. thick, unbranched, tan to brown, sheaths 20-30-toothed; strobilus 4-10 cm. long, rounded at the apex.

Moist ground; often weedy.

ISOETACEAE
Quillwort Family

Amphibious or submerged perennials; the stems 2-3-lobed and corm-like, bearing fibrous roots below and a cluster of quill-like leaves above; leaf bases expanded and bearing the sporangia, these usually covered, at least partly so, by a translucent membrane; microspores gray or brown, small and numerous, megaspores white, large and variously sculptured.

Isoetes

Characters those of the family.

1a Stem 3-lobed; usually not submerged .... *I. nuttallii*
1b Stem 2-lobed; amphibious or submerged
  2a Peripheral strands of the leaves well developed .... *I. howellii*
  2b Peripheral strands of the leaves lacking .... *I. echinospora*

Isoetes echinospora Durieu — Bristle-like Quillwort

Plants typically submerged; stems 2-lobed; leaves up to 20 cm. in height, bright green, gradually tapering to a long slender tip, the bases white or brown; translucent membrane covering approximately 1/2 of the sporangium.

Lakes and persistent ponds.

Isoetes howellii Engelm. — Howell's Quillwort

Amphibious species; stems 2-lobed; leaves firm, gradually tapered to the tip, bright green, bases brown to black; translucent membrane covering 1/2 to 3/4 the sporangium.

Lake margins and vernal pools.

Isoetes nuttallii Engelm. — Nuttall's Quillwort

Plants nearly terrestrial; stems 3-lobed; leaves firm, 3-angled, light green to gray-green; translucent membrane covering the sporangium or nearly so.

Wet ground.

LYCOPODIACEAE
Club-moss Family

Trailing or erect plants, terrestrial or epiphytic; stems generally branching, the branches alternate or dichotomous; leaves linear to subulate, generally numerous, in 4 to many series, usually overlapping; sporangia solitary in the axils of leaves, or of bracts in cone-like structures (strobili); spores minute, uniform, numerous.

1a Sporangia borne in the axils of unmodified leaves; horizontal stems absent.. .... **Huperzia**
1b Sporangia borne in highly reduced bracts aggregated into strobili; horizontal stems present
  2a Strobili erect on leafy peduncles .... **Lycopodiella**
  2b Strobili sessile or borne on peduncles with remote reduced leaves
    3a Leaves 6-many-ranked; ultimate shoots, including the leaves, 5-12 mm. in diameter .... **Lycopodium**
    3b Leaves mostly 4-5-ranked; ultimate shoots, including the leaves, 2-6 mm. in diameter .... **Diphasiastrum**

Diphasiastrum

Horizontal stems long-creeping, erect shoots (including the leaves) 2-6 mm. in diameter; leaves on the horizontal stems distant, linear, lanceolate or scale-like, those of the erect shoots mostly imbricate in 4-5-ranks; strobili solitary and sessile or multiple and stalked.

Ultimate branches round in cross section; leaves 5-ranked .... *D. sitchense*
Ultimate branches flat in cross section; leaves 4-ranked........ *D. complanatum*

Diphasiastrum complanatum (L.) Holub Northern running-pine
Main stems prostrate and creeping, usually slightly below ground, bearing short, erect aerial branches, these conspicuously dichotomously branched, the branches flattened; leaves 4-ranked, linear to lanceolate; peduncles slender, distantly bracteate, forking once or twice, bearing 2-4 stalked cones.

(=*Lycopodium complanatum* L.)

Moist to dry woods and alpine slopes.

Diphasiastrum sitchense (Rupr.) Holub Sitka club-moss
Main stems prostrate and creeping along the ground or slightly below the surface, bearing appressed, lanceolate leaves; erect shoots clustered and branching from the base, branches dark green, the leaves 5-ranked, not overlapping; strobili solitary or rarely in pairs on upright shoots.

(=*Lycopodium sitchense* Rupr.)

Woods and alpine meadows.

Huperzia

Erect or decumbent; horizontal stems absent; shoots round in cross section, dichotomously branched; leaves appressed to spreading, triangular to oblanceolate; sporangia reniform, borne in the axils of leaves.

Huperzia occidentalis (Clute) Beitel
Shoots erect, becoming decumbent, 4-20 cm. long; leaves on the juvenile portion mostly larger than those of the mature portion; leaves all light green, the smallest 4-7 mm. long, triangular, the larger 6-10 mm., oblanceolate.

Often in wet areas in shaded forests.

Lycopodiella

Horizontal stems on the surface of the substrate; erect shoots scattered along the stems and unbranched; strobili solitary; sporangia subglobose.

Lycopodiella inundata (L.) Holub Northern bog club-moss
Creeping stems slender, leafy, rooting below, rarely forking; leaves linear, the margins entire; erect shoots arising from the creeping stems, each bearing an apical strobilus; leaves of the erect shoots closely appressed, those of the strobilus slightly shorter and somewhat spreading, widest at the base; strobili 1.5-6 cm. long.

(=*Lycopodium inundatum* L.)

Bogs near the coast and on the margins of lakes and ponds.

Lycopodium

Stems mainly trailing with some upright shoots scattered along them; leaves spirally arranged, mostly in 6 or more ranks, linear to narrowly lanceolate; strobili single and sessile or multiple and pedunculate; sporangia reniform.

Strobili solitary and sessile at the tips of leafy branches .... *L. annotinum*
Strobili 2-5, distinctly-stalked... *L. clavatum*

Lycopodium annotinum L. Bristly club-moss

Creeping stems sometimes reaching 3 m., sparsely leafy, giving rise to numerous erect or ascending densely leafy branches, these simple or somewhat forked; leaves stiff, spreading or reflexed, linear to lanceolate, sharp-pointed; strobili solitary at the ends of branches, sessile.

Moist woods and exposed rocky areas to timberline.

Lycopodium clavatum L.
Common club-moss, Ground-pine or Elk-moss

Horizontal stems bearing a few lateral branches plus clustered, erect shoots; leaves spreading or ascending, margins entire, bristle-tipped, at least at first; peduncles slender, bearing 2-5 strobili.

Moist woods and swampy areas to timberline.

SELAGINELLACEAE
Spike-moss Family

Herbaceous annuals or perennials; plants with leafy creeping or pendent freely branching stems, terrestrial or epiphytic; leaves of 2 kinds, arranged in 4 ranks, or all alike or nearly so, not arranged in distinct ranks; sporangia solitary in the axils of imbricated bracts arranged in clusters (strobili), sporangia of two sizes, the larger bearing 1-4 megaspores, the smaller bearing numerous microspores.

Selaginella

Characters those of the family.

1a Leaves 4-ranked and of 2 distinct sizes, giving the leafy stem a flattened appearance .... *S. douglasii*
1b Leaves several- to many-ranked, similar, the leafy stem cylindrical
 2a Plants weak-stemmed, bright green, usually pendent on mossy tree trunks or rocks; leaves long-decurrent .... *S. oregana*
 2b Plants with stiff stems, not bright green, terrestrial; leaves short-decurrent .... *S. wallacei*

Selaginella douglasii (Hook. & Grev.) Spring Douglas' spike-moss

Stems 1.5-3.5 dm. long, creeping and rooting, giving rise to alternate branches; leaves of 2 sizes, arranged in 4 series in such a manner as to give the leafy stem a flattened appearance; strobili paired, 4-angled, 5-15 mm. long, bracts closely imbricate, cordate-ovate, acuminate, keeled.

Moist shady rocky slopes.

Sellaginella oregana D.C. Eat. Oregon spike-moss

Stems 3-9 dm. long, freely arching, the branches alternate, recurved; leaves loosely imbricate, 2-3.5 mm. long, bright green, long-decurrent, entire or sparsely ciliate; strobili often paired, 1-6 cm. long.

Usually pendent on mossy tree trunks and moist shaded rocks.

Selaginella wallacei Hieron. Wallace's spike-moss

Stems prostrate, creeping and rooting, up to 2 dm. long, the branches numerous and loosely caespitose; leaves somewhat stiff, minutely bristle-tipped; strobili often paired, 1-9 cm. long, slender, curved, bracts slenderly bristle-tipped, somewhat ciliate.

This species is extremely variable depending on its habitat and especially on the amount of available moisture.

Rocky cliffs and bluffs, from sea level to alpine slopes.

Gymnosperms

TAXACEAE
Yew Family

Dioecious trees or shrubs; leaves (needles) linear, 2-ranked by a twist in their petioles; stamens united by filaments into a column; ovule solitary, terminal on a short axillary branch; seed bony, surrounded by a fleshy cup-like aril, the fruit thus appearing drupe-like.

Taxus

Evergreen trees or shrubs; leaves linear, flat, narrowly lance-shaped, and acute or sharp pointed, spirally arranged, but usually twisted so as to appear 2-ranked; stamens 3-14 in a scaly cluster; ovule in center of circular disk which later becomes red, cup-shaped and surrounds the bony seed.

Taxus brevifolia Nutt. Pacific or Western yew

Dioecious shrubs or small trees up to 25 m. high, trunk 2-12 dm. or more in diameter; bark thin, reddish, scaling off to a smooth surface; leaves linear, flat, acute at apex, 1.5-2.9 cm. long, deep yellow-green, pale beneath, midrib prominent; staminate clusters globose, small; seed borne on underside of a branch in a fleshy scarlet-red cup which is open at the apex.

Scattered in moist forests; Cascades to Coast Range.

Taxus brevifolia

PINACEAE
Pine Family

Trees or shrubs; leaves (needles) narrowly linear, alternate or in bundles of 1-5 in a sheath of bracts; stamens and ovules in different cones on the same tree; pollen cones spirally arranged each bearing 2 pollen sacs; ovulate cones spirally arranged, scales each bearing 2 naked ovules on upper side of the scale near the base, the scales becoming woody and each scale subtended by a bract; seeds usually winged.

1a Leaves in bundles of 2-5; cones maturing second or third year .... **Pinus**

1b Leaves borne singly; cones maturing in one season

  2a Cones erect, their scales deciduous .... **Abies**

  2b Cones pendulous, their scales persistent

    3a Bracts longer than cone-scales, 3-pointed; leaf-scar only slightly raised on the branchlets .... **Pseudotsuga**

3b Bracts shorter than the scales; branchlets roughened by persistent woody leaf-bases
  4a Leaves sessile on woody peg-like bases; leaf apex acute or sharp-pointed. ..... **Picea**
  4b Leaves petioled on woody bases; leaf apex blunt ..... **Tsuga**

Abies

Tall evergreen tree with conspicuous resin-pockets on trunk bark; leaves linear, sessile, spirally arranged, but spreading to give a 2-ranked appearance, often curved upward on upper branches, leaving a smooth circular scar on falling; pollen and ovulate cones borne on branchlets of previous year's growth; the pollen cones numerous on under side of branchlets; the seed cones on upper side of branchlets, erect, maturing first autumn, scales individually shed leaving the persistent cone axis on the tree; scales thin, incurved at the rounded apex; seeds winged.

1a Cones with exserted reflexed bracts, leaves 4-angled, at least on mid and upper branches ..... *A. procera*
1b Cones with concealed bracts, leaves not 4-angled
  2a Resin ducts of the leaves medial; leaves blue-green with stomata on both sides ..... *A. lasiocarpa*
  2b Resin ducts of the leaves marginal and near the lower surface
    3a Leaves 2-ranked, forming a flat spray and with alternating short and long leaves ..... *A. grandis*
    3b Leaves, at least of the mid and upper branches, more or less erect, leaves of nearly equal length
      4a Leaves bluish- or whitish-tinged, 1.5-7 cm. long ..... *A. concolor*
      4b Leaves deep lustrous green above, 3 cm. or less in length ..... *A. amabilis*

Abies amabilis (Dougl.) Forbes — Pacific silver or Red fir

Large trees as much as 70-80 m. high, trunks to 2.5 m. in diameter; bark ashy-gray, conspicuously marked with large chalky-white areas, thin, smooth, or in old trees divided into small plates near the base; leaves spreading, those on the underside of branch curved upward, flat, prominently grooved on the upper side, silvery white beneath, deep lustrous green above, midrib prominent, apex blunt or obtuse, 1-3 cm. long; pollen cones bright red; seed cones dark purple, oblong, 1-1.5 dm. long; bracts slightly toothed, about 12 mm., as long as scales; seeds about 12 mm. long, dull yellowish-brown, with light brownish wings.

Usually between 1,000 and 4,000 feet in the Cascades and Coast Mountains.

Abies concolor (Gord. & Glend.) Lindl. ex Hildebr. — White fir

Large trees up to 70-80 m. high; trunks 1.5-2.5 m. in diameter; bark smooth and silvery in young trees, thickening with age and becoming deeply ridged and ash-gray to brownish in color; leaves flat, 1.5-7 cm. long, erect on mid to upper branches, bluish- or whitish-tinged, apex usually obtuse but sometimes acute or notched; pollen cones rose-colored to red; seed cones oblong, 7-12 cm. long, olive green to purplish-brown; bracts concealed.

It appears that most of our trees that key to *Abies concolor* are, in fact, hybrids of *Abies concolor* and *Abies grandis.*

From 2,000 to 5,000 feet in the mountains.

Abies grandis (Dougl.) Lindl. Grand or Lowland white fir

Large trees reaching 80-90 m., trunks reaching 1.5 m. in diameter; bark ashy-brown, hard, smooth at first, but becoming shallowly-furrowed; leaves spreading, somewhat 2-ranked, 1.5-6 cm. long, alternating short and long leaves, flat, deeply grooved, apex blunt or notched, dark yellow-green above, silvery beneath; pollen cones light yellow; seed cones cylindrical, 5-12 cm. long, dark- or yellow-green to purplish; scales broadly fan-shaped, entire; bracts small, short-pointed, concealed.

Common; from the coast to 5,000 feet in the Cascades.

Abies lasiocarpa (Hook.) Nutt. Alpine or Subalpine fir

Small trees up to 20-30 m. high, trunks to 15 dm. in diameter; bark thin, hard, flinty, chalky-white to pale-brown; twigs covered with minute, rusty hairs for 2-3 years; leaves 1.8-4 cm. long, thickened in the center, curved upward, shiny blue-green with silvery tinge due to the stomata on both surfaces; pollen cones bluish; seed cones dark purple, oblong-cylindric, 6-10 cm. long; scales about 2.5 cm. long, fan-shaped; bracts 1/3 as long as the scale, red-brown; seeds ivory-brown with large, shiny purplish wings.

Subalpine to alpine forests.

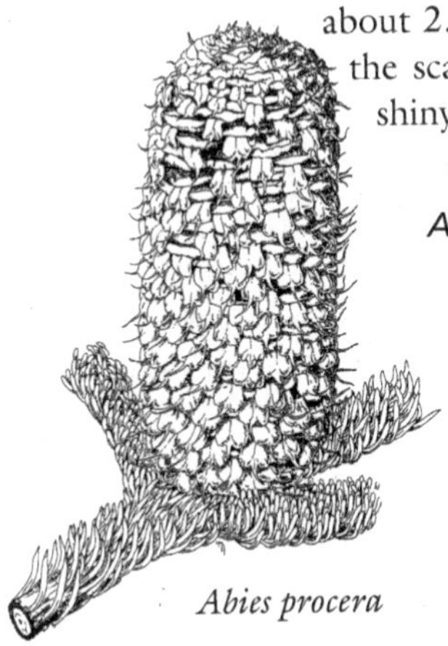

Abies procera Rehd. Noble fir

Trees to 80 m. tall, trunks 2.5 m. or less in diameter; bark rather thin, grayish-brown, becoming ridged and broken into long irregular plates; leaves 1-3.5 cm. long, more or less flat on lower branches, but conspicuously 4-angled and incurved on mid and upper branches, blue-green; pollen cones reddish or purple; seed cones light yellow-green or brown, oblong, 10-15 cm. long, with fan-shaped scales which are nearly covered by large reflexed, fringe-margined bracts; seeds dull red-brown with pale brown wings.

[=*Abies nobilis* (Dougl.) Lindl.]

Western slope of Cascades and high peaks of Coast Range.

Abies procera

Picea

Tall evergreen trees with straight tapering trunks; bark thin and scaly, becoming thick and furrowed; leaves linear, borne singly, spreading in all directions around the twigs, usually sharp-pointed, somewhat flattened to triangular in cross section or more often 4-angled, falling away in drying from woody "peg-like" projections; pollen cones nearly sessile; seed cones pendulous, maturing first season, usually scattered over upper half of tree, scales thin, persistent, longer than bracts.

Leaves prominently 4-angled; occurring in the mountains ...... *P. engelmannii*
Leaves flattened or broadly triangular in cross section; coastal and lowland trees .... *P. sitchensis*

Picea engelmannii Engelm. Engelmann spruce
Trees to 60 m. high, trunks to 1.5 m. in diameter; bark from gray-brown to russet-red, small scales loosely attached; leaves stiff, 4-angled, 1-2.5 cm. or more in length, deep blue-green, sometimes with a silvery or whitish tinge; pollen cones purplish; seed cones from a light brown to a dark cinnamon-brown, oblong, 2.5-7 cm. long; scales ovate, truncate, retuse or rounded; seeds blackish-brown with small wings.

In the Cascades, in our area.

Picea sitchensis (Bong.) Carr. Sitka spruce
Trees reaching 80 m., trunks to 2 m. in diameter; bark gray to reddish-brown; leaves flattened to triangular in cross section, 1.5-3 cm. long, sharp-pointed, stiff, blue-green to bright yellow-green; pollen cones purple, 2-2.5 cm. long, borne on a short peduncle; seed cones dull brown, oblong, 5-10 cm. long; scales oblong-ovate, papery, denticulate; seeds small, winged.

Coastal.

Picea sitchensis
a. scale
b. winged seed

Pinus

Evergreen trees with 2 kinds of leaves, the scale-like with deciduous tips forming a sheath in the axils of which arise the needle-like foliage leaves in fascicles of 2 to 5, rarely single; pollen cones arranged in whorls at the base of new shoots; seed cones becoming woody upon maturing, usually in second year.

1a Needles consistently in fascicles of 5's; cone scales unarmed
  2a Cones 3.5-8 cm. long, short-stalked .... *P. albicaulis*
  2b Cones over 8 cm. long, long-stalked, slender
    3a Cones 10-25 cm. long .... *P. monticola*
    3b Cones 25-55 cm. long... *P. lambertiana*
1b Needles in fascicles of 2 or 3
  4a Needles typically in 2's, 2-7 cm. long.................................. *P. contorta*
  4b Needles typically in 3's, 6-30 cm. long
    5a Cones strongly asymmetrical, persistent, buff-colored; needles 6-20 cm. long; small trees.. *P. attenuata*
    5b Cones nearly symmetrical, not persistent, reddish-brown; needles 7-30 cm. long; large trees... *P. ponderosa*

Pinus albicaulis Engelm. Whitebark pine
Scraggly trees, 5-20 m. high, sometimes nearly prostrate, trunks 0.5-1.5 m. in diameter; bark thin, whitish and smooth, or sometimes scaly on main trunk; needles 5 per fascicle, 3-7 cm. long, more or less curved, dark yellow-green; pollen cones pink to red, turning golden brown; seed cones yellowish-brown

to purplish, oval, 3.5-8 cm. long, nearly as thick, scales thick, broad, rounded at apex, with a short incurved protuberance.

At or near timberline.

Pinus attenuata Lemmon — Knobcone pine

Small trees to 24 m. high, trunks rarely to 1 m. in diameter; bark thin, dull brown, ridged and scaly on mature trees; needles 3 per fascicle, 6-20 cm. long, light yellow-green, the margins finely serrulate; pollen cones brownish-purple to orange; seed cones remaining closed many years, opening upon burning, buff-colored, ovoid, oblique, acutely or bluntly pointed and somewhat recurved at tip, 8-15 cm. long, scales on convex side of the cone thickened and knob-like, with a terminal umbo and prickle; seeds blackish, ovoid, about 5-7 mm. long, the surface slightly roughened, the wings long and narrow.

Along McKenzie River and southward.

Pinus contorta Dougl. ex Loud. — Lodgepole, Shore or Coast pine

Scrubby trees, usually less than 20 m., but some up to 50 m., trunk rarely more than 4-5 dm. in diameter; bark dark purplish red-brown and cross-checked; needles in fascicles of 2, rarely some in 3's, 2-7 cm. long, dark yellow-green to dark green, the margins finely serrulate; pollen cones reddish- or yellowish-brown; seed cones maturing the second year, variably persistent, small, ovoid, 2-6 cm. long, scales somewhat thickened, with exposed part raised and bearing a stubby or slender prickle; seeds ovoid, winged.

Two varieties occur in our range: *Pinus contorta* Dougl. ex Loud. var. *contorta* which is common along the coast and *Pinus contorta* Dougl. ex Loud. var. *murrayana* (Grev. & Balf.) Engelm. which is the tallest and best formed variety of the species and occurs inland.

Pinus lambertiana Dougl. — Sugar pine

Trees 50-75 m. tall, trunk to 2.5 m. or more in diameter; bark gray to reddish-brown, 4-8 cm. thick, closely and deeply fissured and split into small scales; needles 5 per fascicle, 5-10 cm. long, deep blue-green with a whitish tinge; pollen cones yellowish-brown; seed cones pale reddish-brown to deep yellowish-brown, long-oblong, 25-55 cm. long, scales broad, slightly thickened, rounded and thin at apex; seeds smooth, chocolate to black-brown, 6-20 mm. long with broad wings.

Cascade slopes, principally south of our area.

Pinus monticola Dougl. — Western white pine

Trees to 40 m. high, rarely taller, trunks to 2.4 m. in diameter; bark gray, to cinnamon color, longitudinally cracked, more or less forming squares; needles 5 per fascicle, slender, 4-10 cm. long, bluish-green with whitish tinge; pollen cones yellow or cinnamon-brown; seed cones black-purple or green when young, buff-brown at maturity, pendulous, slender, 10-25 cm. long, resinous, scales smooth, with rounded apex; seeds 5-8 mm. long, with wings about 3 times as long.

Cascades, with scattered trees in Coast Range.

Pinus ponderosa Dougl. Ponderosa, Western yellow or Bull pine

Large forest trees up to 75 m. high, trunks 0.5-2.5 m. in diameter; bark thick, yellowish- to reddish-brown, deeply furrowed and cross checked into scaly plates; needles in fascicles of 3, rarely 2 or 5, 7-30 cm. long, yellow-green, slightly twisted; pollen cones arranged in rosette-like clusters; seed cones reddish-brown, ovoid, 5-15 cm. long, basal scales left on tree when cone breaks away, scales thickened with short prickle at apex; seeds ovate, 4-12 mm. long, yellowish, often mottled with purple, winged.

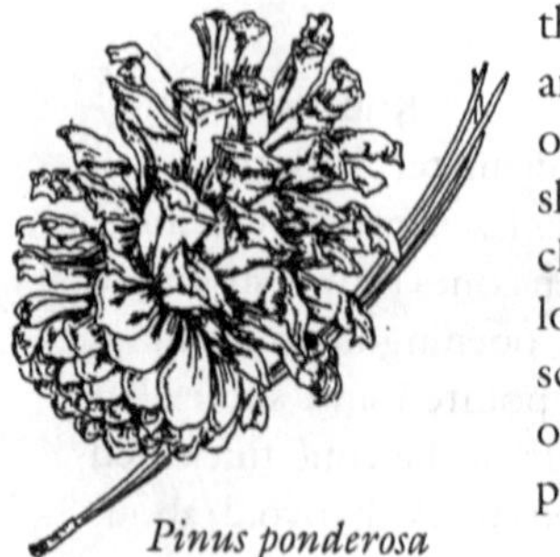

Pinus ponderosa

Cascades, sometimes in the Coast Range and scattered through the Willamette Valley.

Pseudotsuga

Tall evergreen trees with broadly pyramidal crowns; leaves (needles) linear, flat, spirally arranged, sometimes giving a 2-ranked appearance; pollen cones scaly, axillary, usually growing back from the end of twig; ovulate cones erect, terminal or axillary, bristly and scaly, becoming pendulous, maturing the first autumn, bearing prominent 3-pointed bracts protruding beyond the cone-scales; seeds winged.

Pseudotsuga menziesii

Pseudotsuga menziesii (Mirb.) Franco Douglas-fir

Large forest trees to 100 m. tall, trunks reaching 3 m. or more in diameter; bark ashy brown, in young trees thin and smooth, becoming thick, rough, and deeply furrowed in age; leaves generally spreading all around the twig, linear, flat, slightly grooved above and ridged below, 1.5-4 cm. long, blue-green to yellow-green, somewhat paler on under side; pollen cones orange to pale brown, enclosed in reddish-brown bracts; seed cones cinnamon or reddish-brown, long oval to somewhat pointed, 4-10 cm. long, with thin tridentate bracts protruding beyond the broad, thin, rounded cone scales; seeds dull russet-brown, 5-6 mm. long, with longer wings.

Common.

This is the state tree of Oregon.

Tsuga

Evergreen trees, the leading shoot usually drooping; leaves (needles) short, linear, flat to more or less angular, usually of unequal length, extending from jointed petioles on rough, drooping branchlets; pollen cones globose, pendulous, pollen cones arising from axillary winter buds, anthers tipped with a short spur or knob; seed cones erect, terminal on year-old branchlets, composed of thin overlapping scales, maturing in one season.

Leaves in flat sprays; cones 1.5-2.5 cm. long.......................... *T. heterophylla*
Leaves spreading around the branchlets; cones 3-7.5 cm. long ....
.... *T. mertensiana*

Tsuga heterophylla (Raf.) Sarg.
Western hemlock

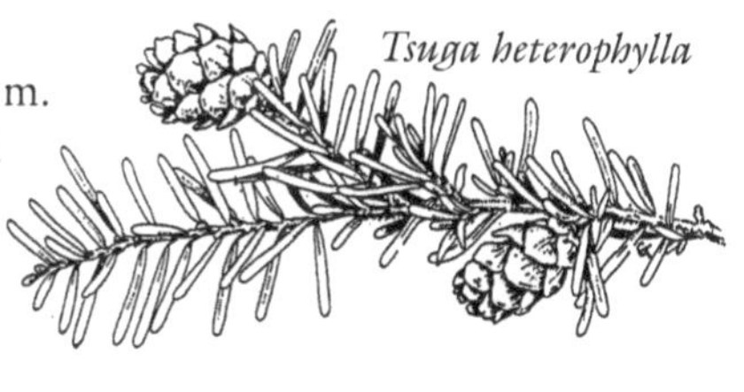
Tsuga heterophylla

Trees reaching 70 m. high, trunks to 2 m. in diameter; bark thick, deeply-furrowed and narrow cross-ridged, russet-brown, tinged with red, bark of young branches thin, scaly, and russet-brown; branches and branchlets slender and drooping, leaf- bearing branchlets finely pubescent; leaves petioled, 5-20 mm. long, rarely longer, appearing imperfectly 2-ranked on the branchlet, glossy yellow-green; pollen cones borne on thread-like stalks, either solitary or in clusters; seed cones reddish-brown, oblong and conical, 1.5-2.5 cm. long, cone scales longer than broad, rounded at apex; seeds light brown, 3 mm. long, with wings twice as long.

West slopes of the Cascades, in the Coast Range, and along the coast. This is the state tree of Washington.

Tsuga mertensiana (Bong.) Carr. Mountain hemlock

Trees to 40 m. high; trunks to 1.2 m. in diameter, branches drooping, giving a conical or pyramidal appearance; bark deeply-fissured, dull gray to dark reddish-brown; leaves standing out all around the branchlets, 1.5-2 cm. long, pale to dark blue-green, bearing stomata on both surfaces; branchlets puberulent; pollen cones violet-purple, on slender stalks; seed cones yellowish-green to bluish-purple, oblong, 3-7.5 cm. long, scales thin, spreading or even recurving; seeds brown, winged.

Mostly above 4,000 feet in the Cascades; when growing near timberline these trees may become dwarfed and contorted.

Tsuga mertensiana

CUPRESSACEAE
Cypress Family

Monoecious or dioecious trees or shrubs; leaves opposite or whorled, scale-like, rarely linear, thickly clothing the branchlets; pollen cones small, stamens shield-like, bearing 2-6 pollen sacs; seed cones with several opposite or whorled scales each bearing 1 to several erect ovules at the base, becoming woody or berry-like, scales shield-shaped or imbricated; seeds angled or winged.

1a Fruit berry-like; monoecious or dioecious species .... **Juniperus**
1b Fruit a woody cone; monoecious species
  2a Cones subglobose, scales shield- or wedge-shaped .... **Chamaecyparis**
  2b Cones and scales oblong, scales imbricate
    3a Cone scales 6, middle pair fertile; leaves appearing to be in whorls of 4 .... **Calocedrus**
    3b Cone scales 8-12; leaves in pairs, 4-ranked .... **Thuja**

Calocedrus

Monoecious aromatic forest trees, with naked buds; leaves scale-like, 4-ranked, compressed and keeled; cones terminal, the pollen cones oblong with many stamens; seed cones oblong, of 6 acuminate scales, the lower pair much-reduced, cones mature first year, with 2 seeds to a scale.

Calocedrus decurrens (Torr.) Florin — Incense cedar

Forest trees to 50 m. high, trunks to 2 m. in diameter; bark 5-7.5 cm. thick, fibrous, cinnamon brown, furrowed and ridged; leaves scale-like, decurrent, opposite, appearing to be in whorls of 4, light green; pollen cones yellow; seed cones yellow-green at first, becoming woody, reddish-brown, pendulous, 2-2.5 cm. long, scales 6, but only the middle pair seed bearing, the lower pair small; seeds oblong, yellowish-brown, winged.

(=*Libocedrus decurrens* Torr.)

West slopes of Cascades, Lane Co., Oregon, south.

Chamaecyparis

Evergreen trees with 2-ranked branchlets; leaves small, scale-like, opposite, in 4-ranks; pollen cones oblong with numerous stamens; seed cones small, globose, composed of 4 to 10, rarely 12, thick, peltate scales, each bearing 2 to 5 ovules; seeds 2-winged, maturing the first year.

Cone scales 4-6, each with a prominent point; glands of the leaves, if present, indistinct and not linear... *C. nootkatensis*
Cone scales 6-10, without prominent points; leaves typically with linear glands .... *C. lawsoniana*

Chamaecyparis lawsoniana (Murr.) Parl. — Port-Orford-cedar

Trees to 50 m. tall, trunks to 2 m. in diameter; bark thick and reddish-brown; leaves scale-like, arranged in opposite pairs, appressed to the twigs, usually blue-green, facial leaves typically bearing linear glands; pollen cones dark brown to red; seed cones subglobose with 6-10 peltate scales; seeds 2-5 per scale, winged.

This species is found in only a few scattered localities within our range; it is much more common in southwestern Oregon and is extensively used as an ornamental.

Chamaecyparis nootkatensis (Lamb.) Spach
Alaska-cedar or Yellow-cypress

Tree to 40 m. high or dwarfed at high elevations; trunks up to 1.5 m. in diameter; bark thin, furrowed, gray-brown on outside, reddish- to cinnamon-brown when broken; leaves scale-like, young leaves closely appressed and long-pointed, sharply acute, blue-green; pollen cones oblong, light yellow; young seed cones light reddish in color becoming deep russet-brown, with a conspicuous bloom, globose, usually with 4-6 peltate scales; seeds 1-5 per scale, ovate, wing-margined.

In the Olympic and Cascade Mountains.

Juniperus

Dioecious or rarely monoecious aromatic evergreen small trees or shrubs; leaves scale- or needle-like, opposite and 4-ranked or arranged three in a whorl and 6-ranked; pollen cones yellow, stamens numerous; seed cones of fleshy scales which unite and become berry-like; seeds wingless.

Juniperus communis L. var. *montana* Ait. Common or Dwarf juniper
Low prostrate shrubs, the spreading branches stout, 0.5-2 m. long, and forming patches several meters in diameter; bark of thin, loosely attached scales, deep brown, tinged with red; leaves mostly in whorls of three, rigid, linear-lanceolate, acute, 6-12 mm. long, convex on the lower side, shining dark green to glaucous; pollen cones 3-6 mm. long; fruit a berry-like cone 6-10 mm. in diameter, globose, bluish, but covered with a white bloom; seeds bony, 1-3.

(=*Juniperus communis* L. var. *saxatilis* Pall.)

Open woods to rocky slopes; from near sea level to the summit of Cascades.

Thuja

Evergreen trees with flattened, spay-like branches; leaves scale-like, opposite, 4-ranked, the lateral pairs keeled, the facial pairs flat; seed cones maturing the first year, scales woody but thin and flexible, 4-6 pairs, the lowermost pair the shortest; seeds 2-winged.

Thuja plicata D. Don Western red cedar
Large trees reaching 50-70 m., trunk diameter has been known to be as much as 5 m.; bark thin, longitudinally fissured, reddish- to cinnamon-brown, peeling off in long fibrous strips; leaves ovate, sharp pointed, 3-6 mm. long, bright green; seed cones crowded near tips of branchlets, oblong 8-14 mm. long, with about 6 pairs of leathery brown scales, each scale bearing 2 or 3 seeds; seeds 2-winged.

Usually in moist sites; Coast Range, Cascades, and along the Columbia River.

Monocots

TYPHACEAE
Cat-tail Family

Perennial marsh or aquatic herbs with creeping rhizomes and glabrous cylindrical stems; leaves linear, alternate, sheathing at the base; flowers unisexual, borne in crowded spikes, the staminate flowers borne above the pistillate ones; staminate flowers of several stamens united by the filaments, surrounded by hairs or scales; pistillate flowers of a single 1-loculed, 1-ovuled pistil surrounded by hairs; fruit fusiform, minute, thin-walled.

Typha

Characters those of the family.

Leaf blades flat, 1-3 cm. wide; pistillate spikes often contiguous with the staminate spikes; pistillate spikes over 2 cm. in diameter in fruit .... *T. latifolia*

Leaf blades convex on the back, 12 mm. wide or less; staminate and pistillate spikes separated; pistillate spikes less than 2.2 cm. in diameter in fruit .... *T. angustifolia*

Typha angustifolia L. — Narrow-leaved cat-tail

Stems 1-3 m. tall from slender creeping rhizomes; leaves exceeding the inflorescence, blades 3-12 mm. wide, convex on the back, the leaf sheath with scarious margined auricles; staminate and pistillate portions of the inflorescence separated by 0.5-8 cm.; pistillate portion 1.5-2.2 cm. thick.

(Hybrids with *T. latifolia* L. are called *T.* x*glauca* Godron)

Shallow or slow-moving water and marshes; uncommon in our area.

Typha latifolia L. — Common cat-tail

Stems stout, 1-3 m. long; leaf blades nearly flat, 1-3 cm. broad, sheathing at the base, the margins scarious; spike dark brown, sometimes reaching more than 2.5 dm. in length, and 2-3 cm. in diameter; the staminate and pistillate portions usually contiguous or nearly so.

(Hybrids with *T. angustifolia* L. are called *T.* x*glauca* Godron)

Common in marshy places and in shallow quiet or slow-moving water.

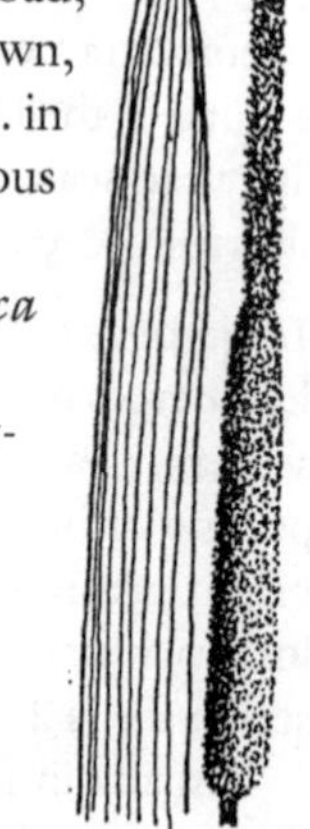

Typha latifolia

SPARGANIACEAE
Bur-reed Family

Rhizomatous perennials; leaves alternate, erect or floating, and basally sheathing; flowers unisexual, borne in capitate-globose clusters, the staminate usually above the pistillate; staminate flowers of 2-8 stamens and 3-5 minute chaffy bracts; pistillate flowers with a single pistil and 3-6 broad bracts; fruit hard and nut-like, indehiscent, beaked or not.

Sparganium

Characters those of the family.

1a Stigmas 2 on over 50 percent of the pistillate flowers; fruit not constricted near the middle; plants usually emergent .... *S. eurycarpum*
1b Stigma 1; fruit usually constricted near the middle
  2a Leaves keeled on the back, at least at the base (sometimes flat on the upper leaves, if the plants are submerged)................................*S. emersum*
  2b Leaves flat on the back; plants usually submerged
    3a Beak of the fruit 0.5-1.5 mm. long, curved........................*S. natans*
    3b Beak of the fruit 1.5-2 mm. long, straight.............*S. angustifolium*

Sparganium angustifolium Michx. — Floating bur-reed

Plants typically submerged; stems slender, up to 2 m. or more tall; leaves flat, 2-5 (10) mm. wide, floating; inflorescence unbranched, usually erect at the surface of the water; staminate heads 1-4, contiguous; pistillate heads 2-5, below the staminate heads; fruit fusiform, constricted near the middle, tapering to the straight beak, borne in heads to 1.5 cm. across.

Usually in shallow water, up to 2.5 m. in depth; ponds, lakes and streams.

Sparganium emersum Rehmann — Simple-stem bur-reed

Stems simple, erect, 1.5-15 dm. tall, rarely taller; leaves 4-15 mm. wide, greatly exceeding the inflorescence, leaves keeled at the base, at least on the lower leaves, upper leaves sometimes flat if submerged; staminate heads 3-10; pistillate ones 1-6, the lower on separate peduncles; fruit fusiform, long-beaked, in heads 1-3 cm. across.

In and around lakes, ponds and marshes.

Sparganium emersum

Sparganium eurycarpum Engelm. — Broad-fruited bur-reed

Stems erect, 5-12 (25) dm. tall, branched; leaves slightly keeled on the back, at least at the base, 6-20 mm. wide, shorter than or equal to the inflorescence; inflorescence branched; staminate heads 5-12; pistillate heads 1-2, the stigmas usually 2; fruit wedge-shaped, sessile, abruptly narrowed to the short beak, in heads 2-3.5 cm. across.

Margins of lakes, ponds and in wet meadows and marshes; tolerant of brackish water.

Sparganium natans L. Small bur-reed

Stems simple, 3-10 dm. long when submerged and 1-3 dm. tall when emergent; leaves flat, 2-10 mm. wide, about equalling the inflorescence; inflorescence unbranched with 1-3 heads, often with staminate and pistillate flowers intermixed; fruit more or less fusiform, with a curved beak, in heads 8-12 mm. across.

(=*Sparganium minimum* Fries)

In shallow water of bays, bogs, ditches and marshes.

NAJADACEAE
Water-nymph Family

Submerged monoecious or dioecious annuals or rarely perennials; leaves linear, subopposite or appearing whorled, dilated and sheathing at the base; flowers axillary, minute, unisexual, the staminate of a single stamen enclosed by a membranous bract and surrounded by an entire to 4-lobed perianth-like structure; pistillate flowers usually naked, consisting of a single pistil with 2-4 stigmas; fruit a fusiform achene.

Najas

Characters those of the family.

Seeds widest above the middle, smooth and shiny ....*N. flexilis*

Seeds widest near the middle and tapered to both ends, dull, pitted................. .... *N. guadalupensis*

Najas flexilis (Willd.) Rostkov & Schmidt Wavy or Slender water-nymph

Monoecious; stems submerged, much-branched; leaves linear, 1-3 cm. long and 0.5-2 mm. wide with an expanded base, tapering from the base to a long slender tip, the margins minutely toothed; seeds widest above the middle, shiny and smooth.

In fresh or brackish quiet water.

Najas guadalupensis (Spreng.) Magnus Guadalupe water-nymph

Similar to *Najas flexilis* but with dull, pitted seeds widest near the middle and tapered to both ends

In fresh or brackish quiet or slow-moving water; northern Oregon and Washington coastal areas.

ZANNICHELLIACEAE
Horned Pondweed Family

Submerged monoecious or dioecious annuals or perennials; leaves opposite, alternate or appearing whorled with membranous sheathing stipules; flowers small, unisexual, naked or the staminate ones with 3 minute membranous scales, stamens 1-3; pistillate flowers with 1-8 distinct carpels subtended by a small bract; fruit drupe-like.

Zannichellia

Monoecious; leaves entire; flowers naked, staminate and pistillate flowers often in the same leaf axil; staminate flowers composed of a single stamen; pistillate flowers usually with 4-5 pistils together in a cup-shaped bract; stigma peltate; fruit often serrulate-margined and beaked.

Zannichellia palustris L. Horned pondweed

Stems submerged, much-branched, 3-10 dm. long; leaves opposite, 2-10 cm. long and less than 1 mm. wide, 1-veined, acute at the apex; pistillate flowers with 4-5 pistils; fruit usually short-stipitate, often serrulate-margined and with a slender short beak.

Fresh or brackish water.

RUPPIACEAE
Ditch-grass Family

Submerged annuals or perennials; stems filiform, much-branched; leaves alternate, linear, with sheathing adnate stipules, these with free, ligule-like tips; flowers perfect, minute, sessile on a short spadix enclosed in the leaf bases, perianth absent, stamens typically 2, one above the 4-8 (16) pistils and one below them, stigma sessile and peltate; fruit small and drupe-like, borne on a long slender stipe that continues to elongate and often becomes spirally coiled when fruit matures.

Ruppia

Characters those of the family.

Ruppia maritima L. Ditch-grass

Stems from long creeping rhizomes, olive-green to brownish up to 8 dm. long, completely submerged; leaves 1-12 cm. long and up to 0.5 mm. wide, the sheathing stipules adnate, membranous, often with a free, ligule-like tip; peduncles axillary, as much as 3 dm. long when the fruit is mature and often becoming spirally coiled.

Saline, alkaline or brackish water.

POTAMOGETONACEAE
Pondweed Family

Perennial aquatics, often rhizomatous; leaves alternate or rarely opposite, often dimorphic, the submerged leaves thin and the floating leaves leathery, with a longer petiole and a more expanded blade; flowers perfect, with 4 clawed perianth parts; stamens 4, each attached to a perianth segment; pistils 1 or 4, sessile or nearly so; fruit an achene.

Potamogeton

Characters those of the family.

1a Submerged leaves with the stipules fused to the petioles (the leaves, therefore, not appearing attached at the nodes)
  2a Leaves of two types, floating and submerged.................. *P. diversifolius*
  2b Leaves all submerged
    3a Leaves 3-8 mm. broad, very finely 20-60-veined, auricled at the base ....*P. robbinsii*
    3b Leaves up to 1 mm. broad, 1-3-veined ....*P. pectinatus*
1b Submerged leaves with stipules free from the rest of the leaf, the leaf bases or petioles attached directly at the nodes
  4a Leaf margins crisped and serrulate; leaf blades narrowly oblong, 2-8 cm. long, 3-12 mm. wide, all submerged ....*P. crispus*
  4b Leaf margins of submerged leaves usually not crisped, but if so, then entire
    5a Stems flattened, winged, more than half as broad as the leaves; leaves 2-5 mm. wide, all submerged .... *P. zosteriformis*
    5b Stems terete, or if somewhat flattened, not at all winged
      6a Leaves all submerged and 2.5 mm. or less in width
        7a Nodes with a pair of yellowish globose glands..........*P. pusillus*
        7b Nodes without glands... *P. foliosus*
      6b Some of the leaves either floating or, if all submerged, then averaging over 2.5 mm. in width
        8a Leaves more or less cordate and clasping the stem, all submerged
          9a Stipules 3-10 cm. long, persistent; stems more or less zigzag in appearance .... *P. praelongus*
          9b Stipules 1-2 cm. long, soon shredding into numerous white fibers; stems straight..*P. richardsonii*
        8b Leaves petiolate or tapering to the base, or linear and sessile; floating leaves often present
          10a Submerged leaves leathery, subterete and less than 2.5 mm. wide..*P. natans*
          10b Submerged leaves without this combination of characters
            11a Submerged leaves petioled
              12a Submerged leaves folded along the midrib and falcate, 15-75 mm. wide, 19-50-veined....*P. amplifolius*
              12b Submerged leaves usually flat, never falcate, less than 20-veined
                13a Fruits reddish; petioles of the floating leaves usually longer than the blades, petioles of submerged leaves 2-13 cm. long.............................. *P. nodosus*
                13b Fruits green to gray-green; petioles of the floating leaves usually shorter than the blade; submerged leaves subsessile or short-petioled ....*P. illinoensis*
            11b Submerged leaves sessile
              14a Submerged leaves ribbon-like and with a median stripe with reticulate veining on each side of the midrib . ...*P. epihydrus*

14b Submerged leaves not ribbon-like and without a median stripe
15a Plants reddish-tinged; floating leaves, if any, not markedly differentiated from the submerged ones; fruit not compressed, not sessile .... *P. alpinus*
15b Plants green; fruit laterally compressed, sessile
16a Submerged leaves 1-12 mm. wide, the apex acuminate, 3-9-veined; stipules 0.5-3 cm. long .... *P. gramineus*
16b Submerged leaves 10-50 mm. wide, the apex acute-mucronate, 7-19-veined; stipules 1-8 cm. long.................................... *P. illinoensis*

Potamogeton alpinus Balbis — Northern or Reddish pondweed

Stems slender, up to 1 m. in length, only sparingly branched, if at all; entire plant reddish-tinged; submerged leaves lanceolate to oblong-linear, 2.5-25 cm. long, 7-18 mm. wide, sessile, stipules 1.5-4 cm. long, membranous and early deciduous; floating leaves, when present, not markedly differentiated from the submerged ones, but short-petioled, 2.5-10 cm. long and 1-2.5 cm. wide; spikes up to 3.5 cm. long, the flowers crowded; fruit 2.5-3.5 mm. long with a short curved beak and a prominent dorsal keel.

Quiet water.

Potamogeton amplifolius Tuckerman — Large-leaved pondweed

Stems stout, up to 1 m. long, branched above, if at all; submerged leaves diverse, the lower narrowly lanceolate and short-lived, the upper broadly lanceolate to ovate, folded along the midrib, thus appearing falcate, 8-20 cm. long, 1.5-7.5 cm. wide, 19-50-veined, stipules 1.5-12 cm. long; floating leaves leathery, lanceolate to ovate, apex acute, rounded to cordate at the base, 5-10 cm. or more long and 3-5 cm. wide, 25-50-veined and long-petioled; spikes up to 8 cm. long; fruit 3.5-5 mm. long, prominently beaked and 3-keeled.

Usually in deep water.

Potamogeton crispus L. — Curled-leaved pondweed

Stems stout, more or less flattened, much-branched, up to 9 dm. in length; leaves all submerged, sessile, oblong, 2-8 cm. long and 3-12 mm. wide, the margins strongly crisped and finely serrulate, stipules 3-8 mm. long, scarious, deciduous; spikes 1-2 cm. long; fruit about 3-6 mm. long, with 3 obtuse keels and a prominent beak; plants also reproducing vegetatively from buds in the leaf axils.

Ponds and streams. Introduced from Europe.

Potamogeton diversifolius Raf. — Diverse-leaved pondweed

Stems slender, terete, much-branched, 2-5 dm. long; leaves dimorphic, the submerged leaves linear, 2-4 cm. long, 0.5-1.5 mm. wide, sessile, stipules 4-20 mm. long, fused for about 1/2 their length; floating leaves elliptic, 1-3 cm. long and 3-8 mm. wide, short-petioled; spikes about 1.5 cm. long; fruit suborbicular, 1-2 mm., prominently dorsally keeled, the keel wavy to finely toothed and with 2 low, lateral keels.

Quiet and low-moving water.

Potamogeton crispus

Potamogeton epihydrus Raf. Ribbon-leaf pondweed
Stems more or less flattened, usually sparingly branched, 4-18 dm. long; leaves dimorphic, submerged leaves sessile, linear, 5-25 cm. long, 1-10 mm. wide, with a reticulate median stripe on both sides of the midrib; the ligule-like stipules 1-3 cm. long; floating leaves long-petioled, elliptic to oblong, 2-8 cm. long, 0.5-3 cm. wide, leathery; spikes 1.5-4 cm. long; fruit obovate, 2-4.5 mm. long, 3-keeled.
Shallow to deep water.

Potamogeton foliosus Raf. Leafy pondweed
Stems much-branched, obscurely winged, 2-10 dm. long; leaves all submerged, sessile, narrowly linear, 1-10 cm. long, 1-2.5 mm. wide, stipules 2-20 mm. long; spikes 1-5 mm. long; flowers few, minute; fruit 1.5-2.5 mm. long, prominently dorsally keeled, the keel wavy to toothed.
Shallow water.

Potamogeton gramineus L. Grass-leaved pondweed
Stems much-branched, 3-8 (15) dm. long; leaves dimorphic, submerged leaves sessile, 2-11 cm. long and 1-12 mm. wide, the ligule-like stipules 0.5-3 cm. long; floating leaves leathery, long-petioled, elliptic, 2-7 cm. long, 1-2.5 cm. wide; spikes 1-4 cm. long; fruit 2-3 mm. long; 3-keeled.
Quiet or slow-moving water.

Potamogeton illinoensis Morong Shining pondweed
Stems simple or branched, terete, up to 1.5 m. long; leaves dimorphic, submerged leaves subsessile or short-petioled, elliptic to oblanceolate, 5-20 cm. long, 1-5 cm. wide, stipules 1-8 cm. long; floating leaves more or less leathery, elliptic to oblong-elliptic, 4-18 cm. long, 2-6.5 cm. wide; spike 3-6 cm. long; fruit obovoid, 2.5-3.5 mm. long, prominently 3-keeled, short-beaked.
Usually in deep quiet or slow-moving water.

Potamogeton natans L. Common floating pondweed
Stems subterete, 3-16 dm. long; leaves dimorphic, submerged leaves leathery, subterete, 10-30 cm. long, 1-2.5 mm. wide, stipules 4-11 cm. long, persistent; floating leaves leathery, long-petioled, ovate-lanceolate, 3-11 cm. long, 1.5-6 cm. wide, cordate at the base; spikes 3-5 cm. long, flowers numerous; fruit obovoid, 3-5 mm. long, keeled or not.
Shallow quiet fresh or brackish water.

Potamogeton nodosus Poir. Long-leaved pondweed
Stems subterete, simple or branched above, 4-25 dm. long; leaves dimorphic, submerged leaves petioled, lanceolate, 5-20 cm. long, 1-3.5 cm. wide, stipules 3-9 cm. long; floating leaves long-petioled, leathery, elliptic, 3-12 cm. long, 1.5-4 cm. wide; spikes 2-6 cm. long; fruit 3-4.5 mm. long, 3-keeled, and with a short straight beak.
Shallow to deep water of ponds and lakes or slow-moving streams.

Potamogeton pectinatus L. Fennel-leaf pondweed

Stems from slender, creeping, tuber bearing rhizomes, filiform, dichotomously much-branched, up to 8 dm. long; leaves all submerged, filiform, 2-15 cm. long, rarely longer, 0.2-1 mm. wide, apex sharp-pointed; stipules 1-5 cm. long, fused for 2/3 their length; spikes 1-3 cm. long; fruit obovoid, 3-5 mm. long, the dorsal keel rounded, short-beaked.

[=*Stuckenia pectinata* (L.) Borner]

Shallow quiet or slow-moving water.

Potamogeton praelongus Wulf. White-stemmed pondweed

Stems more or less zigzag in appearance, subterete, whitish, 1-3 m. long; leaves all submerged, sessile, cordate to clasping the stems, oblong-lanceolate, 5-35 cm. long, 1-4 cm. wide, the tip boat-shaped, stipules 3-10 cm. long; spikes 2.5-5 cm. long; fruit obovoid, 4-5 mm. long, prominently keeled and short-beaked.

Deep water.

Potamogeton pusillus L. Small pondweed

Stems terete, much-branched, up to 1 m. long; leaves all submerged, sessile and with a pair of small glands just below the point of attachment of the leaf; narrowly linear, 1-7 cm. long, 0.5-2.5 mm. wide, stipules 3-15 mm. long, membranous but becoming fibrous; spikes up to 6 cm. long; fruit obovoid, 2-2.5 mm. long, obscurely keeled, short-beaked.

Shallow water.

Potamogeton richardsonii (Bennett) Rydb. Richardsons' pondweed

Stems stout, terete, sparsely branched, 3-6 dm. long; leaves all submerged, sessile and clasping at the base, lanceolate, 1.5-10 cm. long, 5-28 mm. wide, more or less crisp-margined, stipules 1-2 cm. long, fibrous, soon shredding; spikes 1.5-4 cm. long; fruit obovate, 2.5-4 mm. long, rounded or faintly short-keeled.

Shallow to deep, usually alkaline, water.

Potamogeton richardsonii

Potamogeton robbinsii Oakes Robbins' pondweed

Stems stout, terete, usually much-branched, 1-5 dm. long, or occasionally longer; leaves all submerged, stiffly 2-ranked, sessile and more or less auricled at the base, linear to narrowly lanceolate, 3-12 cm. long, 3-4 (8) mm. wide, stipules 1-3 cm. long, fused for about 1/4 their length, shredding at the tip; inflorescence often branched, spikes 7-20 mm. long, few-flowered; fruit 3-5 mm. long, keeled, inconspicuously beaked.

Quiet, often deep, water.

Potamogeton zosteriformis Fern. Eel-grass pondweed

Stems flattened and winged, 3-8 dm. long, 0.7-4 mm. wide, usually branched; leaves all submerged, sessile, 5-20 cm. long, 2-5 mm. wide, stipules 1-4 cm.

Potamogeton zosteriformis

long, becoming shredded; spikes 1-3 cm. long; fruit 4-5 mm. long, the dorsal keel sharp and wavy to toothed, beak slightly curved.

Quiet and slow-moving water.

ZOSTERACEAE
Eel-grass Family

Monoecious or dioecious marine aquatics; leaves submerged, alternate, 2-ranked or tufted, sheathing at the base and with a ligule between the sheath and the blade; flowers sessile borne on a one-sided spadix, surrounded by a membranous spathe; perianth absent; staminate flowers with a solitary sessile stamen; pistillate flowers of a single compound pistil, stigmas 2; fruit 1-seeded, achene-like or drupaceous.

Rhizomes short, older portion covered with fibers; dioecious species .... **Phyllospadix**
Rhizomes long, not covered with fibers; monoecious species .... **Zostera**

Phyllospadix

Dioecious perennials from short stout rhizomes; leaves submerged at least at high tide; staminate flowers sessile, each enclosed by membranous scale-like bracts, and the entire inflorescence enclosed, at first by a membranous spathe, this rupturing at maturity; pistillate inflorescence similarly enclosed, and the pistillate spadix bordered on each side by a series of flap-like bracts which cover the flowers.

1a Leaves 5-7-veined; rhizome nodes with 2 roots .... ***P. serrulatus***
1b Leaves 3-veined; rhizome nodes with 2 clusters of 3-5 roots
 2a Leaves 0.5-1.5 mm. wide; pistillate inflorescences several per stem .... ***P. torreyi***
 2b Leaves 1-4 mm. wide; pistillate inflorescence solitary .... ***P. scouleri***

Phyllospadix scouleri **Hook.** Scouler's surf-grass

Rhizomes short, stout, each internode with 2 clusters of 3-5 roots, older portion of the rhizome covered with yellow to gray fibers; leaf sheaths 4-35 cm. long and 6-7 mm. wide, the blades up to 2 m. long, 1-4 mm. wide, 3-veined, tip obtuse to slightly emarginate; pistillate spadix solitary, 14-26-flowered.

From the lower intertidal zone to the shallow subtidal zone.

Phyllospadix serrulatus **Rupr. ex Ascher.**

Rhizomes short, stout, internodes each with 2 roots, older portion of rhizome covered with brownish fibers; leaf sheaths 3.5-20 cm. long, the blades up to 1 m. in length and 2-8.5 mm. wide, 5-7-veined, margins minutely toothed near the apex, the tip truncate, obtuse or with a small notch; pistillate spadix solitary, 8-10-flowered.

Usually on rocks in the ocean, but occasionally on muddy sediment; Cape Arago, Oregon to Alaska.

Phyllospadix torreyi Wats. Torrey's surf-grass

Rhizome internodes each with 2 clusters of 3-5 roots, older portion of the rhizome covered with yellow to gray fibers; leaf sheaths 7-55 cm. long, leaf blades to 2 m. in length and 0.5-1.5 mm. wide, 3-veined, tip obtuse to slightly emarginate; pistillate spadix 14-20-flowered, several on a branched stem.

Rocky areas in the ocean.

Zostera

Monoecious submerged marine species from long rhizomes; leaves alternate, cauline, linear, basally sheathing, 3-11-veined; inflorescence enclosed by the leaf sheath; staminate flowers alternating with the pistillate ones.

Leaf sheath open with overlapping margins; leaf blades 3-veined, 0.75-1.5 mm. wide.......................*Z. japonica*

Leaf sheath closed, rupturing with age; leaf blades 5-11-veined, 1.5-12 mm. wide.......................*Z. marina*

Zostera japonica Ascher. & Graeb.

Perennials, rhizomes to 1.5 mm. thick; leaf sheaths 1.2-6 cm. long, open with overlapping margins; leaf blades 3-30 cm. long, 0.75-1.5 mm. wide, 3-veined, the tip obtuse to slightly emarginate; reproductive shoots short; spadix 8-15-flowered, the spathe 1-4 cm. long; fruit about 2-4 mm. long.

In tidal flats and estuaries with brackish water.

Zostera marina

Zostera marina L. Eel-grass

Rhizomes long and creeping, 2-6 mm. thick; stems branched, up to 2.5 m. in length; leaf sheaths 5-20 cm. long, closed, but rupturing with age; leaf blades up to 2 m. in length, 1.5-12 mm. wide, 5-11-veined, apex obtuse to slightly mucronate; reproductive shoots up to 1.5 m. long, much-branched; spadix with as many as 40 flowers, the spathe 4-8.5 dm. long; fruit 2-5 mm. long, often beaked.

Mostly subtidal, often forming large "meadows", sometimes in water as much as 10 m. in depth; common.

JUNCAGINACEAE Arrow-grass Family

Annuals or perennials with basally sheathing terete leaves; flowers perfect or unisexual with the staminate and pistillate flowers on the same plant; perianth parts, if present, usually 6, in 2 similar whorls; stamens 1, 4 or 6; fruits several and simple or single and compound.

Triglochin

Rhizomatous perennials; leaves basally tufted, sheathing at the base with a ligule between the sheath and the blade, terete to somewhat flattened; flowers perfect; sepals 3; petals 3 or absent; stamens 3-6; pistils 3 or 6, rarely more.

1a Apex of the ligule 2-lobed; fertile pistils 6 .... *T. maritima*
1b Apex of the ligule entire; fertile pistils 3, sterile pistils 3
  2a Stamens 6; fruit linear to clavate, 5-8 mm. long, weakly ridged ....
  .... *T. palustris*
  2b Stamens 1-3; fruit subglobose, less than 3 mm. long, strongly 3-keeled .
  .... *T. striata*

Triglochin maritima L. — Seaside arrow-grass

Rhizomatous perennials 0.5-12 dm. tall; leaves linear, subterete to more or less flattened, 1-8 dm. long, 0.5-5 mm. wide; ligule 0.5-5 mm. long, the tip 2-lobed; flowers in a dense raceme; perianth parts 6, greenish-yellow, 1.5-2 mm. long; stamens usually 6; pistils 6, all fertile; fruit 2-5 mm. long.

(Includes *Triglochin concinna* Davy)

Saline marshes along the coast and alkaline meadows inland.

Triglochin palustris L. — Marsh arrow-grass

Perennials from a short, stout rhizome, 1.5-6 dm. tall; leaves linear, more or less flattened, sharp-pointed, 5-30 cm. long and 1-2 mm. wide, ligule 0.5-1.5 mm. long, entire; perianth segments usually 6, greenish-yellow, 1-2 mm. long; stamens 6; pistils 6, 3 sterile and 3 fertile; fruit linear to clavate, 5-8 mm. long, separating at the base and remaining suspended from the tip until falling.

Mud flats, lake shores, along streams, often where brackish or alkaline.

Triglochin striata Ruiz Lopez & Pav. — Three-ribbed arrow-grass

Perennials from slender rhizomes, 1-2 (4.5) dm. tall; leaves flattened, 5-35 cm. long, 1-2 mm. wide, ligule 1-2.5 mm. long, entire; perianth parts 3-6, greenish-yellow, 0.5-1 mm. long; stamens 1-3; pistils 6, 3 sterile and 3 fertile; fruit subglobose, 1-3 mm. long.

Saline or brackish marshes or bogs; coastal.

ALISMATACEAE
Water-plantain Family

Aquatic or marsh herbs with basal leaves; flowers perfect or unisexual; sepals 3, usually green; petals usually present, white, pink or purplish, deciduous; stamens 3-many; pistils 6-many; fruit usually an achene, but sometimes dehiscent.

Flowers usually unisexual; blades of the leaves (unless submerged) sagittate;
  pistils in a spiral arrangement .... **Sagittaria**
Flowers perfect; blades of the leaves not sagittate; pistils arranged in a ring......
  .... **Alisma**

Alisma

Perennials from a lactiferous corm-like caudex; leaves all basal, linear to ovate; flowers in compound whorled panicles; sepals 3, green, petals 3, white to pink; stamens 6-9, opposite the petals; pistils 15-25 borne in a single whorl around the margin of the receptacle; fruit an achene.

1a Petals acute to acuminate at the apex, pinkish-purple .... *A. lanceolatum*
1b Petals obtuse at the apex, white, but often becoming pinkish or purplish in age
  2a Leaf blades ovate to oblong-lanceolate, 3-15 cm. wide; inflorescence usually greatly surpassing the leaves .... *A. triviale*
  2b Leaf blades linear to narrowly lanceolate, 0.2-2 (3) cm. wide; inflorescence not much, if at all, surpassing the leaves .... *A. gramineum*

Alisma gramineum Lejeune — Grass-leaved water-plantain

Plants 5-30 cm. tall, rarely taller; leaves usually equalling or taller than the inflorescence; leaf blades linear to narrowly lanceolate, often floating, 3-10 cm. long, 0.8-2 (3) cm. wide, gradually tapering to the sheathing petiole; stamens 6-9; petals white but becoming pinkish or purplish tinged in age, apex obtuse; achenes suborbicular, 2-2.7 mm. long, with 2 shallow grooves and a central ridge, the beak erect or becoming curved.

Marshy areas, ponds and ditches, sometimes nearly submerged.

Alisma lanceolatum With.

Leaves lanceolate, usually cuneate or only slightly rounded at the base, leaf blades 6-20 cm. long, 1-4 cm. wide; inflorescence usually much exceeding the leaves; petals pinkish-purple, apex acute to acuminate; fruit with an erect beak.

Slow-moving or quiet water.

Alisma triviale Pursh — Northern water-plantain

Plants up to 1 m. in height; leaves shorter than the inflorescence, leaf blades ovate to oblong-lanceolate, 5-30 cm. long and 3-15 cm. wide, rounded or cordate at the base, long petioled; panicle large, many-flowered; petals white, 4-6 mm. long, apex obtuse; achenes obovate, 2-3 mm. long usually centrally grooved, at least near the tip, often 3-ribbed, the beak erect.

(*Alisma plantago-aquatica* L. misapplied)

Fresh water marshes, stream banks, ditches and ponds.

Alisma triviale

Sagittaria

Monoecious, rarely dioecious and occasionally with some perfect flowers; annuals or perennials, often rhizomatous and often producing tubers; leaves all basal, usually sagittate; flowers borne in whorls, the staminate above the pistillate; petals white; stamens 7-30; pistils many, spirally arranged on a globose receptacle; fruit a compressed achene with a short beak.

Beaks of the achenes 1-2 mm. long, horizontal .... *S. latifolia*
Beaks of the achenes 0.1-0.4 mm. long, erect................................. *S. cuneata*

Sagittaria cuneata Sheld. — Northern arrowhead

Similar to *Sagittaria latifolia* but the achenes with shorter beaks, these 0.1-0.4 mm. long and erect as opposed to the 1-2 mm. horizontal beaks of *Sagittaria latifolia*. Basal lobes of the leaves are also much shorter than the terminal lobe.

Muddy banks and shallow water of rivers and lakes as well as ditches and other wet areas.

Sagittaria latifolia Willd. — Broad-leaved arrowhead or Wapato

Sagittaria latifolia

Monoecious or dioecious, occasionally with perfect flowers; rhizomatous tuber producing perennials 2-11 dm. tall; leaves 6-30 cm. long, the blades usually sagittate, the lateral lobes only slightly shorter than the terminal lobe in length, 2.5-20 cm. long and about as broad, submerged leaves, if present, usually ribbon-like; scapes with 2-8 whorls of flowers (flowers usually 3 per whorl); petals 1-2 cm. long, white; fruiting heads 1.5 cm. or more in diameter.

Slow moving streams, ditches, ponds, lakes and swampy areas.

The tubers formerly were an important article of the Native American diet.

HYDROCHARITACEAE
Frog's-bit Family

Submerged perennials; leaves simple, basal, alternate, opposite or whorled; flowers mostly unisexual, the staminate and pistillate flowers on the same plant or on separate plants or the flowers rarely perfect, the flowers enclosed in a papery spathe or a pair of opposite bracts; perianth in 1 or 2 series of 3, free; stamens 2-many; ovary inferior; fruit indehiscent, many-seeded.

1a Leaves few, long and ribbon-like, basal on very short stems....... **Vallisneria**
1b Leaves many, short, opposite or whorled on well developed stems
  2a Main leaves in whorls of 4-6; averaging at least 2.5 cm. long....... **Egeria**
  2b Main leaves 2-3 at a node, rarely as much as 2.5 cm. long.......... **Elodea**

Egeria

Dioecious submerged perennials; stems simple or branched; leaves sessile, crowded and in whorls of 4-6, but typically 5 or sometimes the lowermost leaves opposite; staminate flowers 2-3, stamens 9, the spathe slender, the perianth tube elongated; pistillate flowers solitary and axillary, the spathe slender, the perianth tube elongated, ovary inferior, styles 3.

Egeria densa Planch. Brazilian waterweed

Stems usually dichotomously branched; the leaves crowded, 2-4 cm. long and 2-5 mm. wide, oblong, acuminate, the margin finely serrulate, mostly in whorls of 4 or 6. Our plants all staminate; reproducing entirely vegetatively.

[=*Elodea densa* (Planch.) Casp. and *Anacharis densa* Marie-Victorin]

Streams, ponds, lakes and ditches; native of Brazil and widely sold for use in aquaria.

Egeria densa

Elodea

Dioecious or monoecious or occasionally with some perfect flowers; submerged perennials; stems often branched; leaves sessile, opposite and/or in whorls of 3, margins finely serrulate; flowers solitary, axillary, the spathe usually 2-lobed; staminate flowers floating, stamens 3-9; pistillate flowers floating, ovary inferior, stigmas 3, often 2-lobed.

1a Leaves averaging less than 1.5 mm. wide, tapering to a slender point .... *E. nuttallii*

1b Leaves averaging over 1.5 mm. wide, obtuse or abruptly pointed at the tip

  2a Leaves mostly in 3's; anthers 3 mm. long or less .... *E. canadensis*

  2b Leaves predominately opposite, but some in whorls of 3; anthers 3-4.5 mm. long .... *E. bifoliata*

Elodea bifoliata St. John Long-sheath waterweed

Leaves predominately opposite, but some in whorls of 3, linear to narrowly oblong, minuted toothed, 4.5-25 mm. long, 1.5-4 mm. wide; staminate flowers solitary, stamens 7-9; pistillate flowers with a stipe-like floral tube up to 3 dm. long, aborted stamens often present.

(=*Elodea longivaginata* St. John)

In quiet or slow-moving water.

Elodea canadensis Richard in Michx. Common waterweed

Usually a dioecious species, but occasionally with a few perfect flowers; stems dichotomously branched, leaves linear to narrowly ovate, finely serrulate, obtuse or abruptly pointed, mostly opposite below, but in whorls of 3 in the middle and above, crowded, 5-17 mm. long, 1.5-3 mm. wide; staminate flowers rare in our area, solitary in an inflated spathe; sepals and petals about 3.5-5 mm. long, the petals slightly longer than the sepals, stamens 7-9; pistillate flowers with a floral tube up to 15 cm. long, sepals and petals nearly equal, 2-2.5 mm. long, stigmas 3, 2-lobed; fruit a cylindrical capsule 5-6 mm. long.

Ditches, ponds, lakes and streams.

Elodea nuttallii (Planch.) St. John Nuttall's waterweed

Dioecious species; stems dichotomously branched; leaves linear to narrowly lanceolate, finely serrulate, tapered to a slender acute tip, mostly in whorls of

3, 4-16 mm. long, 0.5-1.7 mm. wide; staminate flowers sessile and solitary, sepals 2 mm. long, petals, if present, about 0.5 mm. long, stamens 9; pistillate flowers with a long floral tube, sepals and petals 1-1.5 mm. long, styles 3, obscurely 2-lobed.

Shallow water in streams, lakes, ponds and ditches.

Vallisneria

Dioecious perennials; plants submerged; stems very short with tufted ribbon-like leaves; staminate flowers crowded on a short spadix borne in a 2-3-parted spathe, becoming deciduous and free-floating at anthesis; sepals 3, unequal; stamens 1-3; pistillate flowers solitary, borne on long scapes at the surface of the water; sepals and petals 3; fruit indehiscent.

Vallisneria americana Michx. **Tapegrass**

Rhizomatous perennials with fibrous roots and fleshy propagating buds; leaves in basal clusters, ribbon-like, thin, linear, obtuse, 1-10 (20) dm. long, 2-20 mm. wide, the veins prominent; staminate flowers numerous, crowded on the spadix, enclosed in the spathe, detached and free-floating at anthesis; pistillate flowers usually solitary, borne on scapes up to 1-2 m. long, floating on the water surface, but retracting and coiling after pollination, sepals 3, 5-6 mm. long, petals 3, 2-3 mm. long, white; fruit cylindrical, 5-18 cm. long.

In lakes, streams and bays.

POACEAE
Grass Family

Annual or perennial herbs, rarely woody; stems (culms) usually hollow at the internodes, solid at the nodes; leaves 2-ranked, composed of the sheath and the blade, ligule usually present at the collar; flowers (florets) without a distinct perianth, arranged in spikelets consisting of a rachilla and 2 to many 2-ranked bracts, the lower pair empty (glumes); stamens 1-6, usually 3; pistil 1, styles typically 2; fruit a caryopsis.

The alternate name for this family is **Gramineae.**

For keys and descriptions for members of the Poaceae refer to other floras, for example: Hitchcock, C. Leo & A. Cronquist. 1973. Flora of the Pacific Northwest. University of Washington Press, Seattle, WA.

CYPERACEAE
Sedge Family

Annual or perennial herbs, the latter rhizomatous; aerial stems (culms) generally 3-angled, or terete, typically solid, naked or leafy; leaves 3-ranked, consisting of narrow parallel veined blades and closed sheaths, or sometimes reduced to sheaths alone; flowers (florets) minute, borne in spikelets in simple or compound arrangements, each floret typically subtended by a reduced bractlet (scale), and the inflorescence often subtended by 1 or more involucral leaves (bracts); florets perfect or unisexual; perianth none or consisting of bristles; styles 2 or 3; ovary 1-loculed, 1-ovuled, becoming a biconvex or 3-angled achene.

For keys and descriptions for the members of Cyperaceae refer to other floras, for example: Hitchcock, C. Leo & A. Cronquist. 1973. Flora of the Pacific Northwest. University of Washington Press, Seattle, WA.

ARACEAE
Arum Family

Perennial herbs with large leaves; flowers small, often unisexual, many, on a fleshy spadix often surrounded by a showy, generally colored spathe; fruit a berry (in ours).

Lysichiton

Perennials from a stout caudex with conspicuous leaf scars; leaves basal, very large and thin, with thick spongy midribs; spathe white or yellow, sheathing at the base; flowers perfect; calyx lobes 4; stamens 4, opposite the calyx lobes; ovary 2-loculed; berry 1-2-seeded, embedded in the spadix.

Lysichiton americanum Hult. & St. John
Yellow skunk cabbage

Leaves 3-15 dm. long, oblong-elliptic; spadix thick, becoming 10-13 cm. long and 2.5 cm. thick at maturity; spathe to 4 dm. long, somewhat hood-shaped, bright yellow, enclosing the spadix at first, but the spadix becoming exserted at maturity; stems and leaves, especially when bruised, producing a skunk-like odor.

Lysichiton americanum

Low areas in pastures, swamps, marshes, stream banks and bogs.

LEMNACEAE
Duckweed Family

Very small free-floating or submerged plants without definite leaves or stems (i.e. a frond); frond flattened or spherical; reproduction mainly vegetative by budding; flowers rarely present, minute; perianth lacking; stamens 1 or rarely 2; pistil 1; fruits variously interpreted as achene-like, utricules or follicles.

1a Rootlets absent; fronds veinless
  2a Fronds globose, ovoid or boat-shaped, 0.3-1.5 mm. long .... **Wolffia**
  2b Fronds flat, 3-9 mm. long (ours) .... **Wolffiella**

1b Rootlets usually present, but if lacking, then the fronds 1 to many-veined, although sometimes obscurely so
3a Plants with only one rootlet per frond (sometimes absent in *L. trisulca*), without a scale at the point of attachment of the rootlet .... **Lemna**
3b Plants with more than one rootlet per frond and with a scale at the point of attachment of the rootlets .... **Spirodela**

Lemna

Fronds solitary or more often forming clusters, flattened, 1-7-veined, bearing 1 simple rootlet per frond; marginal pouches with flowers or vegetative buds.

1a Fronds narrowed to a long slender stipe, 2-20 mm. long, frequently remaining attached in chain-like colonies; submerged .... *L. trisulca*
1b Fronds not narrowed to a long-stalk; individual fronds not over 8 mm. long; floating
2a Fronds 1-veined
3a Base of the fronds symmetrical, fronds thick .... *L. minuta*
3b Base of the fronds asymmetrical, fronds thin .... *L. valdiviana*
2b Fronds 3-7-veined, sometimes inconspicuously so
4a Upper surface usually mottled, the apex asymmetrical; air spaces of the under surface prominent and usually inflated, both surfaces often tinged with red .... *L. gibba*
4b Upper surface dark green, not mottled, the apex symmetrical, under surface without inflated air spaces
5a Fronds with papillae of nearly equal size along the midline, usually reddish-tinged on the under surface; forming rootless turions ....
.... *L. turionifera*
5b Frond without prominent papillae or with only 1, not red underneath; not forming turions .... *L. minor*

Lemna gibba L. — Inflated duckweed or Wind bags

Fronds broadly elliptic to nearly orbicular, asymmetrical at the apex, 1-8 mm. long, thick, green or yellow-green and red mottled, lower surface usually with inflated air spaces, these commonly bordered in red, usually 4-5-veined.

Fresh or brackish water.

Lemna minor L. — Lesser duckweed or Water lentil

Fronds obovate-elliptic to suborbicular, 2-8 mm. long, shiny green and smooth, generally symmetrical, usually obscurely 3-veined, but if 5-veined, the outer ones branching from the inner ones.

Common in fresh water.

Lemna minuta Kunth

Fronds elliptic to obovate, 1-4 mm. long, 0.7-2 mm. wide, more or less symmetrical, pale green, inconspicuously 1-veined, typically thicker in the middle and tapering to the thin outer margins.

(=*Lemna minuscula* Herter)

Quiet water.

Lemna trisulca L. — Star or Ivy-leaf duckweed

Fronds lanceolate to oblong-elliptic, tending to form large connected mats, usually just under the surface of the water, base of the frond tapered and stalk-like, fronds 5-35 mm. long, including the stalk, obscurely 1-3-veined, minutely serrulate toward the apex, translucent green; new fronds arising at right angles to the mature frond; rootlets sometimes lacking.

Still and slow-moving water.

Lemna turionifera Landolt

Fronds obovate, 2-5 mm. long, shiny green above and often reddish-tinged on the lower surface, symmetrical, upper surface with a row of minute papillae along the midline, 3-veined. This species forms turions (i.e. winter buds which are dense rootless daughter plants that sink to the bottom and over winter there).

Still and slow-moving fresh water.

Lemna valdiviana Philippi

Fronds lanceolate to elliptic, 2-5 mm. long, 0.5-2 mm. wide, thin, transparent green, the upper surface smooth, 1-veined; often several fronds remaining clustered together.

Still and slow-moving water.

Spirodela

Fronds solitary or more often in clusters, 3-21-veined, bearing up to 21 rootlets per frond, flattened, orbicular to oblong-obovate, often reddish-purple on the under surface; flowers borne in pouches, one pouch on either side of the basal end.

Fronds nearly orbicular, 7-12-veined, 3-10 mm. long; rootlets 7-16 (21) .... *S. polyrrhiza*

Fronds oblong-obovate, obscurely 3-5-veined, 1.5-8 mm. long; rootlets 2-7 (12) .... *S. punctata*

Spirodela polyrrhiza (L.) Schleiden — Great duckweed

Fronds nearly orbicular, symmetrical, 3-10 mm. long, usually several remaining clustered together, dark green on the upper surface and reddish-purple beneath, conspicuously 7-12-veined; rootlets 7-16 (21).

Quiet or slow-moving water.

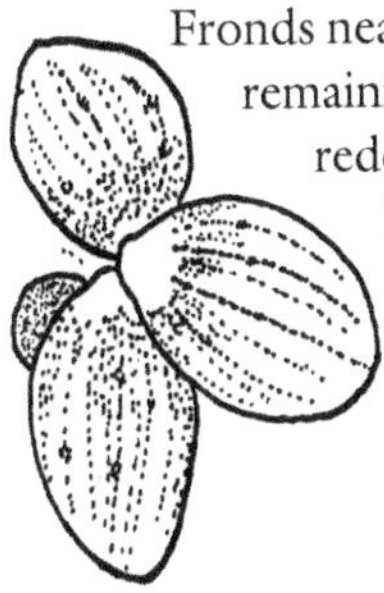

Spirodela polyrrhiza

Spirodela punctata (G. Meyer) C. Thompson

Fronds oblong-obovate, 1.5-8 mm. long, 1.5-4 mm. wide, asymmetrical, obscurely 3-5-veined, often with a row of minute papillae along the midline; rootlets 2-7 (12).

Quiet water.

Wolffia

Fronds minute, 0.3-1.6 mm. long, rootless and without veins; floating or submerged, globose, ovoid or boat-shaped, usually occurring in unequal pairs; reproducing vegetatively from buds arising from a cleft at one end of the frond; flower solitary, minute, produced on the upper surface.

1a Fronds spherical or ovoid, without brown pigment cells
  2a Fronds ovoid, 0.4-0.8 mm. long, 1.3-2 times as long as wide................ *W. globosa*
  2b Fronds spherical, 0.4-1.4 mm. long, 1-1.3 times as long as wide........... *W. columbiana*
1b Fronds boat-shaped, with brown pigment cells present, but visible only in dried material
  3a Tip of the fronds clearly pointed and slightly turned upwards; upper surface distinctly flattened .... *W. borealis*
  3b Tip of the fronds rounded, upper surface more or less rounded, and with a prominent central protuberance...................................... *W. brasiliensis*

Wolffia borealis (Hegelm.) Landolt & Wildi
Fronds boat-shaped, 0.5-1.5 mm. long, upper surface definitely flattened, the apex acutely pointed and appearing upturned in side view; brown pigment cells present, but visible only in dried material.
  Fresh water.

Wolffia brasiliensis Wedd.
Fronds boat-shaped, 0.7-1.5 mm. long, upper surface with a prominent central conical protuberance, apex of the frond rounded; brown pigment cells present, but visible only in dried material.
  Quiet water.

Wolffia columbiana Karsten
Fronds solitary or in pairs, spherical, translucent green, 0.4-1.4 mm. long, the apex rounded.
  Quiet or slow moving fresh water.

Wolffia globosa (Roxb.) Hartog & Plas
Fronds ovoid, 0.4-0.8 mm. long, 0.3-0.5 mm. wide, the apex usually rounded, but occasionally slightly pointed.
  Quiet water.

Wolffiella

Fronds submerged, except when flowering, typically several remaining together, variously shaped, but longer than wide, flat, margins entire, rootless and without veins; flowers minute, produced in a pouch on the upper surface.

Wolffiella gladiata (Hegelm.) Hegelm.
Fronds usually clustered together, 3-9 (12) mm. long, about 0.5 mm. wide. sickle-shaped, tapered at both ends; flowering fronds wider at the base than the vegetative ones; air spaces present throughout most of the frond.
  [=*Wolffiella floridana* (D. Smith) C. Thompson]
  Quiet water; in our area known only from Washington.

JUNCACEAE
Rush Family

Annual or perennial herbs with simple pithy cylindrical stems; leaves usually consisting of a blade and a generally open sheath (closed in *Luzula*), the blades terete, channeled, gladiate or flattened, or the leaves sometimes reduced to sheaths; flowers very small, consisting of 6 chaffy perianth segments, 3-6 stamens and 3 carpels united to form a capsule.

Two genera, *Juncus* and *Luzula*, occur in our area and can be distinguished on the basis of the closed sheaths and many seeded capsules in *Juncus*, as opposed to open sheaths and 3-seeded capsules in *Luzula*.

For keys and descriptions for the members of Juncaceae refer to other floras, for example: Hitchcock, C. Leo & A. Cronquist. 1973. *Flora of the Pacific Northwest.* University of Washington Press, Seattle, WA. or Flora North America Editorial Committee, eds. 2000. *Flora of North America North of Mexico.* Oxford University Press, New York, NY. vol. 22.

LILIACEAE
Lily Family

Herbs (ours) with perennial bulbs, corms, or rhizomes, the peduncles and leaves dying and withering after blossoms and seeds are produced; flower parts generally in 3's, rarely in 2's; sepals and petals often similar in appearance; stamens six or fewer; ovary superior, generally with three compartments; fruit a capsule or berry.

1a Styles separate
  2a Leaves in a single whorl of 3 .... **Trillium**
  2b Leaves not in a single whorl of 3
    3a Stems very leafy
      4a Leaves broad, large, many-veined .... **Veratrum**
      4b Leaves narrow, grass-like .... **Xerophyllum**
    3b Leaves mostly basal, cauline leaves absent or few
      5a Flowers (in ours) nodding, greenish-yellow to purplish .... **Stenanthium**
      5b Flowers not nodding, white or yellowish
        6a Plants from a bulb; perianth parts with basal glands .... **Zigadenus**
        6b Plants rhizomatous; perianth parts without glands although the stems very glandular .... **Triantha**
1b Styles united to the middle or above; stigmas often 3
  7a Sepals and petals distinctly different
    8a Leaves 3, in a single whorl .... **Trillium**
    8b Leaves not in a single whorl of 3
      9a Leaves 2, broad, paired, basal; flowers on long slender petioles from a subsessile umbel .... **Scoliopus**

9b Basal leaf 1; cauline leaves present, narrow; flowers solitary or inflorescence umbel-like .... **Calochortus**
7b Sepals and petals quite similar, all petal-like
10a Flowers small, many, in a scapose, bracted umbel or a bracted subcapitate cluster
11a Plant with an onion or garlic odor; perianth parts free or nearly so .. .. **Allium**
11b Plant without onion odor; perianth parts united below into a tube
12a Fertile stamens 6.. **Triteleia**
12b Fertile stamens 3 (in ours)
13a Inflorescence an open umbel .... **Brodiaea**
13b Inflorescence a dense umbel or a subcaptiate cluster .... **Dichelostemma**
10b Flowers not in scapose, bracted umbels nor in a bracted subcapitate cluster
14a Leaves all basal or nearly so
15a Leaves numerous; flowers many, in long racemes........ **Camassia**
15b Leaves typically 2, rarely 3; flowers 1-6
16a Plants from a deep seated corm; flowers nodding, the perianth segments reflexed or spreading............................... **Erythronium**
16b Plants rhizomatous; perianth segments erect .... **Clintonia**
14b Leaves not all basal
17a Only a few of the leaves alternate, the others basal......... **Lloydia**
17b Leaves not as above
18a Plants from a bulb; some leaves typically in whorls, (in ours), occasional plants with leaves alternate and/or opposite; flowers large, showy
19a Flowers broadly bell-shaped, greenish-brown, mottled with purple (ours); nectaries of the perianth conspicuous................ .... **Fritillaria**
19b Flowers trumpet-shaped, white, pinkish or orange (in ours); nectaries small, obscure .... **Lilium**
18b Plants rhizomatous; leaves definitely alternate
20a Flowers small, in terminal racemes or panicles; main stems unbranched.. **Maianthemum**
20b Flowers 1-3 per axil or at the tip of the branches, more or less hidden by the leaves
21a Flowers borne at the tip of the branches....... **Prosartes**
21b Flowers axillary .... **Streptopus**

Allium

Leaves basal, sheathing the base of the peduncle, generally few and slender, sometimes broad, arising with the peduncle from a bulb; flowers borne in an umbel or head-like inflorescence subtended by several tissue-like bracts; sepals and petals (collectively) 6, not at all or only very slightly united; leaves and bulbs with characteristic onion or garlic taste and odor; fruit a capsule, often crested at summit.

1a Alpine or sub-alpine plants; peduncle flattened at least beneath the umbel
  2a Peduncle 2-9 dm. tall, cylindrical below .... *A. validum*
  2b Peduncle 0.5-1.5 dm. tall, flattened the entire length .... *A. crenulatum*
1b Not alpine plants; peduncle cylindrical
  3a Flowers, all, or in part replaced by bulbils; plants with an odor of garlic.. .... *A. vineale*
  3b Flowers not replaced with bulbils; plants with an odor of onion
    4a Perianth segments long-acuminate; flowers pink to deep rose-colored .... *A. acuminatum*
    4b Perianth segments not long-acuminate; flowers white to pale pink
      5a Umbels nodding; bulb oblong; stamens longer than the perianth . ... *A. cernuum*
      5b Umbels erect; bulb ovoid to subglobose; stamens not longer than the perianth .... *A. amplectens*

Allium acuminatum Hook. Wild onion

Bulbs nearly globose to ovoid; leaves narrow, generally shorter than the flowering stem; flowering stem 1-2.5 dm., sometimes taller, stout; umbel 10-30-flowered, pedicels 10-25 mm. long; flowers pink to deep rose-colored, the segments 12 mm. or slightly less in length, narrowly acuminate, the tips recurved, minutely toothed; stamens shorter than perianth segments, their filaments widened below and attached to the perianth; ovary with 6 crests near apex; seed surface minutely granular.

In gravelly open places and exposed dry hillsides.

Allium amplectens Torr. Slim-leaf onion

Peduncle 1.5-4.5 dm. tall, longer than the leaves; leaves several grass-like, often withered at the time of flowering; umbel head-like, erect, subtended by 2 broad, tissue-like bracts; flowers white or pale pink; stamens not extending beyond the perianth.

Open or wooded areas and sometimes a weed in grain fields.

Allium cernuum Roth Nodding onion

Peduncle 3-4.5 dm. tall, longer than the leaves; leaves several, slender, grass-like; umbel head-like, generally nodding, subtended by 2 tissue-like bracts, pedicels 12 mm. long; flowers pink; stamens extending beyond perianth segments.

In moist, often rocky places.

Allium crenulatum Weig. Mountain onion

Peduncle 0.5-1.5 dm. tall, stout, exceeded by the leaves; leaves 2, broad, curved, 1-1.5 dm. long; umbel compact, subtended by 2 or 3 broad bracts narrowing to an acute tip, pedicels 6-12 mm. long; flowers 6-12, deep red, 6-10 mm. long; capsule 6 crested at the summit.

Found at altitude of 1,000 m. and above. Common on Mary's Peak.

Allium crenulatum

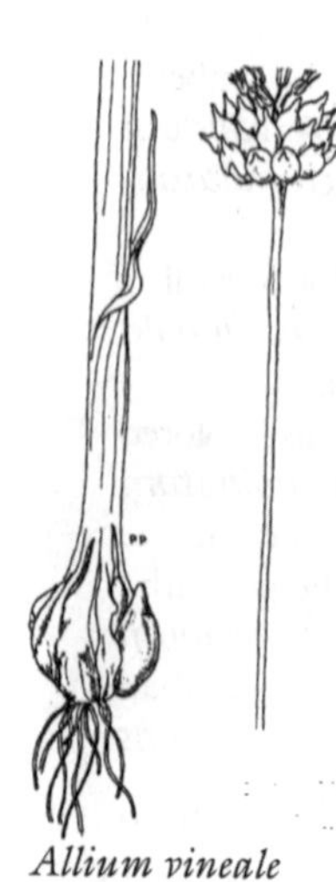
Allium vineale

Allium validum Wats. Tall mountain onion
Bulb crowning a stout horizontal rhizome; flowering stem 2-9 dm. tall, cylindrical below, 2-edged beneath umbel; leaves nearly flat, 6-12 mm. wide, more or less obtuse, nearly as long as scape; umbel dense, bracts broad, abruptly pointed, pedicels about 12 mm. long; flowers deep-rose or paler, segments about 6 mm. long, narrow acuminate; stamens generally exserted; capsule not crested; seeds purplish-black, somewhat wrinkled, minutely but indistinctly granular.

Common at high altitudes in wet meadows.

Allium vineale L. Wild or Field garlic
Entire plant with a garlic odor; bulbs clustered, the outer coat papery and brittle, individual bulblets asymmetrical; stems to 1 m. tall; leaves subcylindrical, hollow, sheathing the lower part of the stem; the flowers all or in part replaced by small bulbils, these fall and form new plants or sprout in place to form a bushy ball of green seedlings.

Weedy in pastures, cultivated fields, along roadsides and in flower beds; native of Europe.

Brodiaea

Leaves all basal, mostly few and slender, arising with the peduncle from a scaly corm; flowers borne in an open umbel; perianth segments similar, partly united; fertile stamens 3 often alternating with 3 sterile stamens (staminodia), filaments slender or winged; style 1, stigmas 3; fruit a capsule.

Perianth 2-3.5 cm. long, the tube bell-shaped ....*B. coronaria*
Perianth 3-5 cm. long, the tube funnel-shaped ....*B. elegans*

Brodiaea coronaria (Salisb.) Jepson Large cluster-lily or Harvest lily
Scapes 0.5-3 dm. tall; umbel 2-10-flowered, the pedicels 1-9 cm. long; flowers 2-3.5 cm. long, narrowly bell-shaped, the segments longer than the tube, violet-purple, with darker mid-ribs; anther-bearing stamens 3, about equaling or shorter than the 3 lanceolate staminodia; ovary angled; capsule sessile.

Common on dry hillsides in early summer.

Brodiaea elegans Hoover Elegant cluster-lily
Scapes up to 4 dm. tall; pedicels 5-10 cm. long; perianth blue-purple to violet, 3-5 cm. long, the tube funnel-shaped and shorter than the lobes; anthers longer than the flattened filaments; the 3 staminodia flat or the margins incurved, subequal to the fertile stamens; capsule sessile.

Dry hillsides and open meadows.

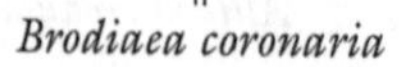
Brodiaea coronaria

Calochortus

Stem arising from a corm with 1 narrow basal leaf (or rarely 2); cauline leaves shorter and narrower than the basal leaf; flowers few and generally showy; sepals and petals unlike, sepals generally more slender, colored like the petals or greenish; petals generally broad, with a gland on the surface or in pocket at the base, glands sometimes present at base of sepals also; stamens 6; capsule 3-angled or 3-winged.

Subalpine; sepals with a dark purple spot inside near base .... *C. subalpinus*
Plants of the valleys and low hills; sepals without purple spot........... *C. tolmiei*

Calochortus subalpinus Piper — Alpine cat's ear

Stems somewhat weakly erect, 7.5-20 cm. tall; sepals broadly ovate or lanceolate, acute, with a pit near the base marked inside by a dark purple spot; petals whitish to cream, purplish at the base, 1.8-3.2 cm. long, obovate, somewhat pubescent, gland covered by a narrow scale, this covered by long hairs; anther about as long as the filament, beaked; capsule nodding, narrowly elliptic.

Open forests at high elevations in the mountains.

Calochortus tolmiei Hook. and Arn. — Cat's ear

Plants 1.5-3 dm. tall, stems generally exceeded in height by basal leaf; basal leaf 6-12 mm. wide, conspicuously parallel-veined, bluish green, somewhat curved, cauline leaves much shorter and narrower, few, alternate except for the upper pair; flowers 2.5-4 cm. in diameter, 1 or several borne in the angle between the upper pair of leaves; sepals slender, pointed, greenish, or often dark purplish on back, without glands; petals broadly obovate, mucronate, deep lavender to nearly white, pubescent on inner surface down to the basal gland which is smooth; capsule broadly winged.

Common in somewhat moist fields and road sides, where shaded by grasses and shrubs.

Calochortus tolmiei

Camassia

Perennials from bulbs; leaves several, linear-lanceolate, all basal; flowers 3-5 cm. in diameter, blue or white, in a simple raceme, each flower subtended by a slender bract; sepals and petals (collectively) 6, not at all united; stamens 6, shorter than the sepals and petals; style 3-cleft at apex; capsule 3-lobed.

Perianth segments spreading wheel-like, twisting together upon withering .... *C. leichtlinii*
One perianth segment turned downward, the other 5 erect to horizontal, each twisting separately upon withering.. *C. quamash*

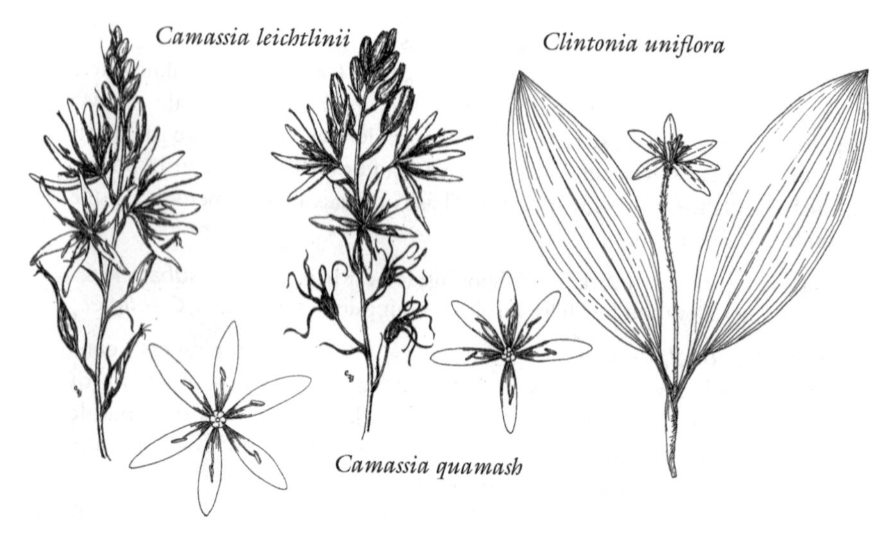

Camassia leichtlinii (Baker) Wats. Large camas

Scapes 3-7 dm. tall, somewhat stout; flowers dark purplish-blue or sometimes white or cream-colored; sepals and petals spreading equally from the center like spokes of a wheel, twisting together above the capsule upon withering, later falling together.

Common in wet meadows, moist hillsides and road sides.

Camassia quamash (Pursh) Greene Small camas

Scapes 1-7 dm. tall, generally slender; flowers dark blue or often lighter, generally more delicate than those of *C. leichtlinii;* sepals and petals not spreading equally from center, but one turned downward, the other 5 erect to horizontal, each withering separately after flowering, and falling separately.

Common in moist meadows.

Clintonia

Peduncle solitary, arising with 2 to several leaves from a creeping rhizome; flowers solitary or borne in an umbel; sepals and petals collectively 6, similar, not at all united; stamens 6, attached to bases of sepals and petals; ovary 2-3-loculed; fruit a berry.

Clintonia uniflora (Schult.) Kunth Queen's cup

Leaves 2 or sometimes 3, oblong to narrowly or broadly elliptic, conspicuously parallel-veined, mucronate, narrowed to a short petiole-like base; peduncle 10 cm. or less long, shorter than the leaves, 1-flowered; flower clear white, bell-shaped; stamens shorter than sepals and petals; berry 6-10 mm. long, dark blue.

Coniferous woods of the Coast Range and Cascades.

Dichelostemma

Dichelostemma congestum

Perennials from scaly corms; leaves 2-5, all basal; inflorescence subcapitate to umbel-like, subtended by 2-5 papery bracts; perianth 6-parted, the tube bell-shaped to cylindrical; stamens 3 or rarely 6, staminodia often reduced; capsule usually sessile.

Dichelostemma congestum (Sm.) Kunth Field cluster-lily
Scapes 6 dm. or more tall, sometimes bent or twisted; flowers in a compact cluster, few-20-flowered, subtended by several ovate bracts; flowers light to dark purplish-blue, 12-20 mm. long; stamens with anthers 3, the short filaments united to the perianth, deeply-lobed, staminodia 3.

(=*Brodiaea congesta* Sm.)

Fields and hillsides.

Erythronium

Stem short, from a corm; leaves 2, appearing basal, the peduncle arising between them; flowers large, solitary to several; perianth segments 6, nearly alike, all or only the 3 inner segments provided with nectar glands and crests; stamens 6, shorter than the perianth segments and free from them; style 3-cleft or entire; capsule 3-loculed, somewhat 3-angled.

1a Leaves mottled (see also *E. elegans* if plants from the Coast Range)
  2a Perianth rose-pink, center paler with broken yellow bands .. *E. revoltum*
  2b Perianth cream-colored to nearly white .... *E. oregonum*
1b Leaves not mottled; montane species
  3a Perianth golden yellow .... *E. grandiflorum*
  3b Perianth white to pale pink
    4a Leaves ovate .... *E. montanum*
    4b Leaves narrowly lanceolate .... *E. elegans*

Erythronium elegans Hammond & Chambers
Leaves narrowly lanceolate, 7-15 cm. long, gradually narrow to the winged petiole, usually not mottled; scape 15-30 cm. tall, 1-4-flowered; perianth white to pale pink with basal yellow stripes, often reddish on the under surface; anthers golden yellow; stigma deeply 3-lobed.

Coast Range, Tillamook, Lincoln and Polk Counties, Oregon.

Erythronium grandiflorum Pursh Yellow fawn lily
Leaves oblong to lanceolate, not mottled; scape 1-3-flowered; perianth parts golden yellow, the base paler; filaments dilated at the base; style filiform, stigma deeply 3-lobed.

Cascades and Coast Range.

Erythronium montanum Wats. Alpine fawn lily
Leaves ovate, not mottled; scapes 1-5-flowered; perianth parts white or sometimes streaked with pink or yellow at the base; filaments flattened, slightly dilated below, anthers yellow; style filiform, stigma deeply lobed.

High elevations in the Cascades.

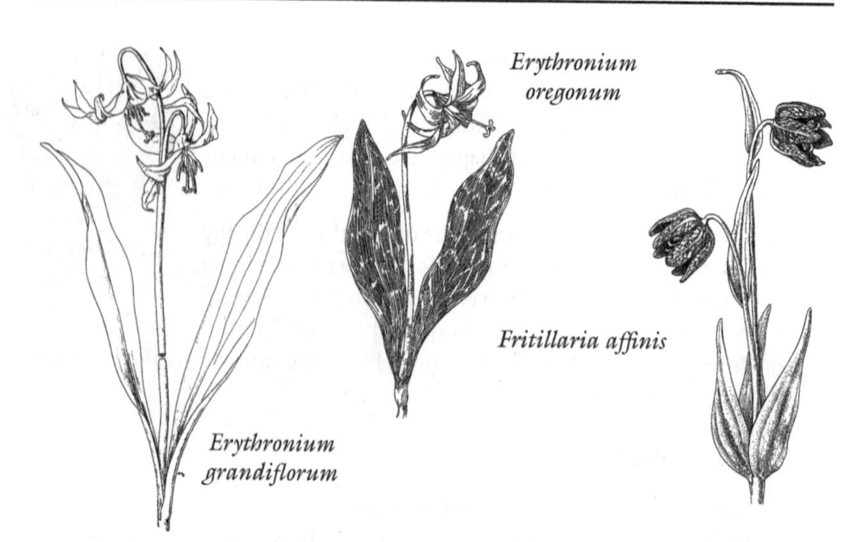

Erythronium oregonum Appleg. Giant fawn lily or Lamb's-tongue

Leaves mostly 2, but sometimes 3, broadly oblanceolate, mottled; scapes 1-6-flowered; perianth parts cream-colored with a golden base; filaments dilated at the base; style clavate, stigma deeply 3-lobed.

Open woods and fields, western Oregon northward.

Erythronium revolutum G. J. Sm. Coast fawn lily

Leaves lanceolate to oblong, mottled; scape usually with a single flower; perianth parts rose-pink; filaments dilated at the base, anthers yellow; style filiform, stigma deeply lobed.

Coastal.

Fritillaria

Stems unbranched, leafy, arising from a fleshy-scaled bulb; leaves alternate or whorled or both; flowers generally showy, solitary or in racemes; sepals and petals similar, each with nectar gland near base; stamens 6; capsule 6-angled or winged.

Fritillaria affinis (Schultes) Sealy Rice-root lily

Stems 3-8 dm. tall; bulb covered with minute bulblets like grains of rice, hence common name of plant; leaves ovate-lanceolate, some in several whorls, the rest alternate; flowers 2.5-4 cm. in diameter, broadly bell-shaped, greenish-brown, mottled with purple in a checker-board effect, each petal and sepal with a deep nectar gland; style 3-cleft; capsule broadly winged.

(=*Fritillaria lanceolata* Pursh)

Shaded spots in clearings and shady moist road sides.

Lilium

Stems leafy, arising from scaly bulbs or rhizomes; flowers showy, erect or nodding, solitary or several in a raceme; perianth segments similar, 6, each with a nectar gland at the base; stamens 6, anthers attached at center to filament; style 1, stigmas 3.

Flowers orange, dark-spotted .... *L. columbianum*
Flowers white (later pinkish) with minute dots ....*L. washingtonianum*

Lilium columbianum Hanson
Tiger or Oregon lily

Stems 4.5-18 dm. tall, leafy; leaves mostly in whorls, sometimes lowermost and uppermost alternate, lanceolate or oblanceolate, generally acute at apex, 2.5-10 cm. long; flowers solitary or several, deep orange or yellow with dark purple spots, the sepals and petals strongly turned backward.

Common in open woods and burned-over areas.

Considerable variation occurs in this species, particularly in the shape and arrangement of leaves, color of flower, and character of spotting.

Lilium columbianum

Lilium washingtonianum Kell. Cascade lily

Stems 9-18 dm. tall; leaves mostly in whorls, some alternate, long-elliptic or oblanceolate, acute at apex; flowers solitary or more often several in a raceme, 6-10 cm. long, tubular spreading to bell-shape, white at first, sometimes with minute dots, becoming pinkish to purplish with age; capsule angled, sometimes narrowly winged.

Common in the Cascades, but also found in fields of the Willamette Valley.

This fragrant lily is one of the most beautiful of those native to the Pacific Coast.

Lloydia

Low plants with grass-like basal leaves and shorter cauline leaves; flowers white, solitary or in a raceme; perianth segments with a narrow transverse gland near the base; stamens 6, the anthers attached by the base to the filaments; ovary 3-angled, 3-loculed, with many seeds; style 1, stigma 3-lobed.

Lilium washingtonianum

Lloydia serotina (L.) Sweet Alpine lily

Bulb crowning a slender creeping rhizome; stems 0.5-2 dm. tall, arising from a cluster of narrow basal leaves, and bearing several reduced leaves; flowers generally solitary, about 12 mm. long, broadly top-shaped, white with purple veins, and sometimes rose-tinged on back; capsule obovoid.

Alpine, uncommon.

Maianthemum

Stem simple from a horizontal rhizome, bearing several leaves and a terminal inflorescence of small white flowers; perianth segments 4-6, similar; ovary 2-3-loculed; fruit a globose berry.

1a Perianth parts 4, in 2 petal-like pairs .... *M. dilatatum*
1b Perianth parts 6
  2a Flowers few, in racemes.. *M. stellatum*
  2b Flowers many, in panicles .... *M. racemosus*

Maianthemum dilatatum **(Wood) Nels. & Macbr.**
Wild lily-of-the-valley

Stem erect, bending at a slight angle at each node, 7-35 cm. tall; cauline leaves triangular, heart-shaped at base, 5-13 cm. long, often long-tapering at the apex; basal leaf long-petioled, broadly heart-shaped, short pointed; raceme terminal, 5 cm. or less long, bearing many small white to cream-colored flowers, attached singly or several in each cluster; perianth segments 4, in 2 petal-like pairs; berry red, 6 mm. in diameter.

[=*Maianthemum bifolium* DC. var. *kamtschaticum* (Gmel.) Jepson]

Shady moist woods, often along streams.

Maianthemum dilatatum

Maianthemum racemosus **(L.) Link subsp. *amplexicaule*** **(Nutt.) LaFrankie**
Large false Solomon's seal

Rhizomes creeping, stout, with conspicuous scars; stems 3-9 dm. tall, leaves flat, broadly elliptic to lanceolate; flowers cream-colored or white, many, in a panicle; berries red, dark-dotted.

[=*Smilacina racemosa* (L.) Desf.]

Common in moist woods.

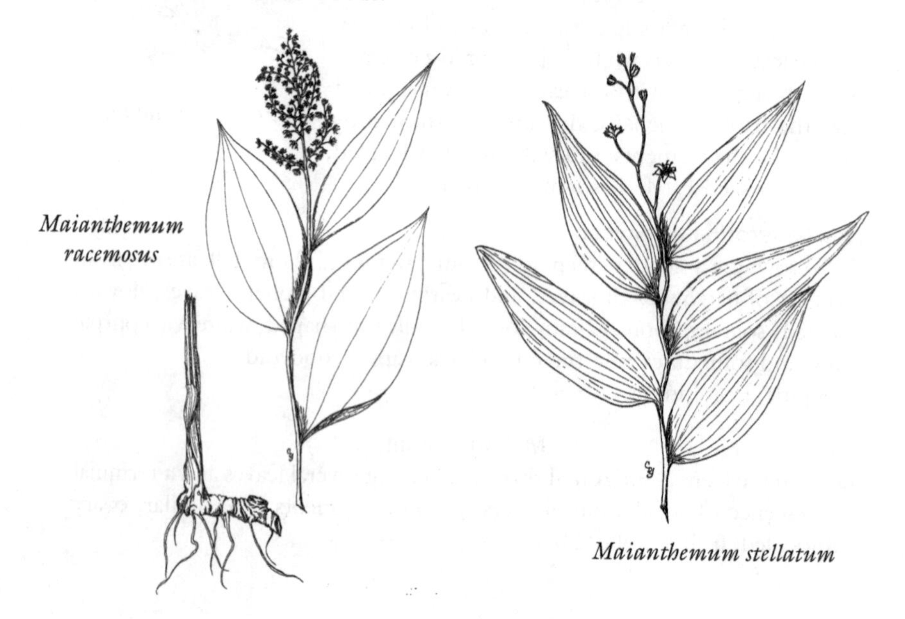

Maianthemum racemosus

Maianthemum stellatum

Maianthemum stellatum (L.) Link — Small false Solomon's seal

Rhizomes slender, creeping; stems 3-6 dm. tall, leafy, the leaves appearing 2-ranked; leaves mostly flat, broadly lanceolate or ovate, long-acuminate; raceme few-flowered; flowers cream-colored to white; berry red.

[=*Smilacina stellata* (L.) Desf. and *Smilacina sessilifolia* Nutt.]

Common in moist woods.

Prosartes

Stem branched, erect from a slender rhizome; lowermost leaves mostly below ground, scale-like sheathing; foliage leaves ovate to oblong-elliptic, thin; flowers drooping, cream-colored or greenish, few or solitary, borne terminally; fruit a berry.

Perianth segments flared only at the tips, 15-28 mm. long; stamens much shorter than the perianth segments and concealed by them.......... *P. smithii*

Perianth segments flared outward, 8-18 mm. long; stamens equal to or longer than the perianth segments.. *P. hookeri*

Prosartes hookeri Torr. — Fairy bells

Stems 3-6 dm. tall; leaves and stems minutely downy or smooth; foliage leaves ovate, tapering to a long point at apex, cordate at base, upper leaves sometimes oblique; flowers greenish-white, narrow at base, 8-18 mm. long, borne on pubescent pedicels, perianth segments spreading; stamens exserted, equal to or longer than the perianth; style entire or nearly so; berry scarlet, 15 mm. long or more, short-pubescent, generally beaked, beak not wrinkled.

[=*Disporum hookeri* (Torr.) Britt. var. *oreganum* (Wats.) Q. Jones]

Found with *P. smithii* in moist shady woods.

Prosartes hookeri

Prosartes smithii (Hook.) F. H. Utech, Z.K. Shinwari & Kawano — Fairy lanterns

Stems 3-9 dm. tall, leaves and stems minutely downy or glabrous; foliage leaves broadly ovate or long ovate, acute or tapering to a slender point, obtuse or somewhat heart-shaped at base (sometimes one side of the base oblique); flowers cream colored or white, the base broad; perianth segments 15-28 mm. long, not spreading except slightly at apex; stamens shorter than the perianth and concealed by it; style 3-cleft at apex; berry yellow to orange, short beaked somewhat wrinkled.

[=*Disporum smithii* (Hook.) Piper]

Common in moist shady woods.

Prosartes smithii

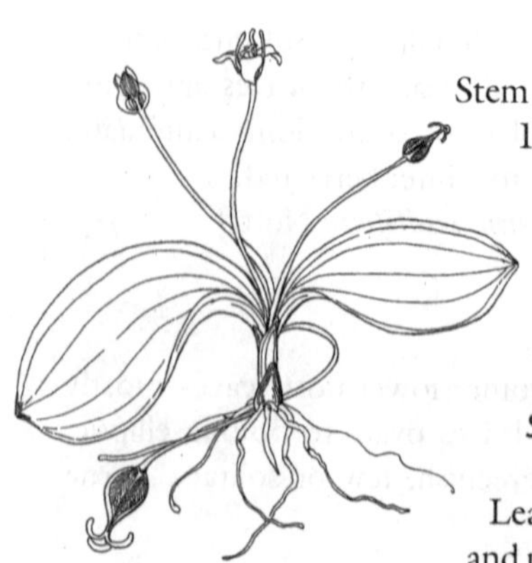
Scoliopus hallii

Scoliopus

Stem very short, entirely below ground, sending up 1 to several pedicels above the 2 basal leaves, all from a short underground rhizome sheathed with leaf bases; flowers small, 6-18 mm. in diameter, purplish and green, inconspicuous; stamens 3, opposite the sepals; ovary 3-angled, 1-loculed, bursting irregularly in fruit.

Scoliopus hallii Wats.

Fetid adder's-tongue or Slink-pod

Leaves 2, oblong-elliptic, 6-20 cm. long, obtuse and mucronate at apex, sheathing at base below the surface of the ground, mottled with purple; pedicels 1 to several, a little shorter than the leaves; sepals greenish, striped or mottled with purplish-brown, strongly turned backward at about the middle; petals narrow, incurved over pistil; style-branches 3, recurved; capsule sharply 3-angled, erect in flower but carried downward after flowering and more or less buried under leaves of the forest floor.

Odor of the flower faint, not very disagreeable. This flower is one of the earliest to blossom in the spring, having already begun to bury its capsules by the time most of the early flowers are blooming.

Coniferous woods.

Stenanthium

Erect glabrous herbs, perennial by bulbs, with narrow and mostly basal leaves; flowers narrowly bell-shaped, greenish, brownish or purplish, several to many in a narrow raceme or panicle; perianth segments united to base of ovary; stamens short.

Stenanthium occidentale Gray

Stems slender 1.5-3 dm. tall; leaves narrow, shorter than the stem; flowers narrowly bell-shaped, greenish-yellow to purplish, about 12 mm. long, in a narrow raceme.

Moist rock crevices and stream banks in the mountains.

Streptopus

Stems simple or more often branched, from creeping rhizomes; leaves alternate, ovate, pointed at apex, sessile or clasping; flowers white or greenish, borne singly or in pairs in the leaf axils; peduncle slender, bent or twisted at the middle; stamens 6; fruit a berry.

Plants 3-12 dm. tall; flowers whitish with green tinge; fruit elongated .... *S. amplexifolius*

Plants 1.5-4 dm. tall; flowers pink to purplish or sometimes whitish with reddish-purple splotches; fruit nearly globose .... *S. lanceolatus*

Streptopus amplexifolius (L.) DC. Twisted-stalk

Stems 3-12 dm. tall, branching; leaves broad to long-ovate, acuminate, clasping at base, green above, glaucous beneath, 5-12 cm. long; flowers 15 mm. or less long, whitish, greenish-tinged, hidden by the leaves; peduncle bearing small gland at bend; anthers unequal; fruit a yellow or red, translucent, elongated berry.

Shady woods and stream borders.

Streptopus lanceolatus (Ait.) Reveal var. *curvipes* (Vail) Reveal

Stems simple or less often branched, 1.5-4 dm. tall; leaves long, ovate to lanceolate, sessile, acuminate, glabrous or slightly pubescent; flowers pink to purplish or sometimes whitish with reddish-purple splotches; berry red, translucent, nearly globose, slightly 3-angled.

(=*Streptopus curvipes* Vail)

In moist places at low to middle elevations in the mountains.

Triantha

Perennial herbs with short erect or horizontal rhizomes and slender stems; leaves narrow, clustered at the base, a few 2-ranked folded leaves on lower part of stem; flowers small, white or greenish, in a short panicle or raceme, with bracts at base of pedicels, and each flower subtended at base by 3 papery bractlets which are united for at least 1/2 their length; stamens 6; ovary 3-lobed above, styles 3, short, recurved; capsule 3-beaked; seeds with a tail at both ends.

Triantha occidentalis (Wats.) R.R. Gates

Stems to 3 dm. or more tall, sticky-pubescent with small reddish glands; leaves shorter than the stem, 3-6 mm. or more wide; raceme or panicle dense, generally about 2.5 cm. long.

(=*Tofieldia occidentalis* Wats.)

In mountain bogs and along the coast.

Triantha occidentalis

Trillium

Stem unbranched from a short stout vertical rhizome; leaves 3, borne in a whorl at apex of stem; flower solitary at center of leaf whorl; sepals 3, green; petals 3, larger than the sepals, white, yellow, greenish, pink or purple; stamens 6, shorter than the perianth segments; ovary 3-6-angled, 3-loculed at base, 1-loculed with 3 projecting placentae at summit; fruit a fleshy capsule.

Flower stalked above the whorl of leaves .... *T. ovatum*
Flower sessile within the whorl of leaves .... *T. albidum*

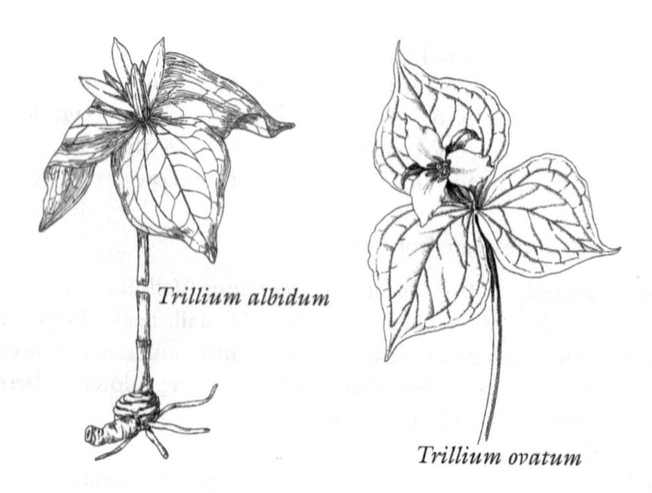

Trillium albidum Freeman — Sessile or Giant Trillium

Stems 1.5-6 dm. tall; leaves broadly ovate to nearly orbicular, acute, usually mottled; petals slender, oblanceolate, white to yellowish or greenish or pinkish or even purplish near the base; flower sessile in center of leaf whorl; capsule more or less 6-angled.

[*Trillium chloropetalum* (Torr.) How. misapplied]

Moist shaded woods.

Trillium ovatum Pursh — White or Western Trillium or Wood lily

Stems 1-5 dm. tall; leaves broadly ovate, acute or tapering to a slender point at apex, abruptly narrowed at base; petals broad, obtuse or acute, pure white at first, changing to deep pink with age; capsule broad with narrow wings.

Moist shaded woods.

Triteleia

Perennials from scaly corms; leaves 1-3, all basal; inflorescence umbel-like, subtended by papery bracts; perianth tube usually funnel-form; stamens 6; style 1, stigmas slightly 3-lobed; fruit a capsule, usually stalked.

Triteleia hyacinthina (Lindl.) Greene — White cluster-lily

Scapes 2.5-7 dm. tall; leaves keeled on the under side; umbel somewhat loose, 8-40-flowered, subtended by several slender bracts; pedicels 0.5-5 cm. long; perianth segments 6-16 mm. long, white or whitish or tinged with purple with a central green stripe; fertile stamens 6, filaments dilated at base, narrowed above; ovary short-stalked, bearing glands near base of style.

[=*Brodiaea hyacinthina* (Lindl.) Baker]

Fields.

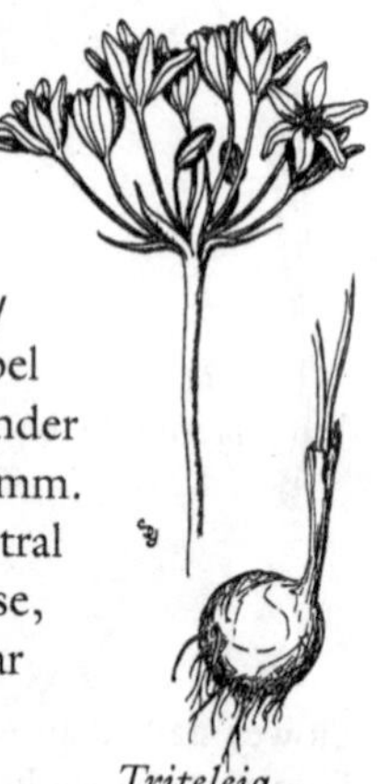
Triteleia hyacinthina

Veratrum

Rhizomes stout; stems tall with large coarse parallel-veined leaves which are often more or less folded; stem and leaves covered with fine hairs; flowers many, greenish or cream-colored, borne in a large terminal panicle; perianth segments 6, attached to the base of the ovary; stamens 6, free from the perianth; styles 3; fruit a 3-loculed capsule.

1a Panicle branches slender, generally nodding; flowers green to yellow-green. .... *V. viride*

1b Panicle erect; flowers white or whitish-yellow, or sometimes greenish at the base

2a Flowers short-pediceled; ovary not woolly .... *V. californicum*

2b Flowers with long slender pedicels (6-18 mm.); ovary densely woolly .... *V. insolitum*

Veratrum californicum Durand California false hellebore

Stems stout, 9-25 dm. tall, very leafy; lower leaves strongly sheathing at the base, broadly elliptic, reaching 3-4 dm. in length, 2-3 dm. in width, upper leaves narrower; panicle often dense, ending above in a greatly elongated raceme in *var. caudatum* (Heller) C. L. Hitchc. and without the elongated terminal raceme in *var. californicum;* flowers whitish with green base, segments 8-17 mm. long, narrow, with 1-2 basal glands; capsule narrow, 3 cm. or less long.

[*Veratrum caudatum* Heller = *V. californicum* Durand var. *caudatum* (Heller) C.L. Hitchc.]

Moist woods, marshy meadows and stream banks.

Veratrum insolitum Jepson
Siskiyou or Woolly false hellebore

Stems to 9-17 dm. tall; stem, axis of the panicle, pedicels and ovary (sometimes also the under side of the leaves and perianth segments) densely gray-woolly; perianth segments 6-9 mm. long bearing 2 basal glands; capsule 2-3 cm. long, more or less woolly.

Woods or open areas. More common south of our range.

Veratrum viride Ait. Green false hellebore

Stems 7.5-18 dm. tall; panicle open, the lateral branches generally nodding; flowers green; perianth segments 6-10 mm. long; capsule 2-3 cm. long.

Wet meadows in the Cascades at approximately 1,000 m. and above, and eastward.

Veratrum californicum

Xerophyllum

Stems unbranched, stout, with a dense basal tuft of narrow leaves which continue on up the stem, decreasing in length upward to the flowers; leaves stiff, needle-like or grass-like with somewhat enlarged bases and rough edges; flowers cream-colored, borne in a dense raceme; sepals and petals similar, remaining on the pedicels after withering; ovary 3-lobed.

Xerophyllum tenax (Pursh) Nutt.

Bear-grass or Indian basket-grass

Stems 4.5-18 dm. tall; basal leaves 3-9 dm. long, 3-10 mm. wide; raceme 1.5-4.5 dm. long, densely covered with minute flowers on long pedicels, forming a pyramid of blossoms until all buds are opened, each flower subtended by a narrow bract longer than the bud and nearly equalling the pedicel of the flower; stamens slightly longer than the sepals and petals.

Open woods in the Coast Range and Cascades.

Xerophyllum tenax

Zigadenus

Leaves slender, mostly basal, arising with the peduncle from a bulb; outer scales of the bulb generally dark; flowers small, yellowish-white, in a raceme or panicle, each flower subtended by a cream-colored tissue-like bract, and the peduncle generally bearing 2 or more larger bracts below the flower cluster; sepals and petals each bearing a greenish gland at the base; styles 3; capsule 3-lobed.

Zigadenus venenosus Wats. Death camas

Peduncle 3-6 dm. tall, bearing a raceme of small cream-colored flowers; leaves slender, several, often folded, shorter than the peduncle; bract at base of flowers slender and nearly as long as the pedicels; sepals and petals nearly alike, each with a conspicuous gland in a depression at the base; stamens about equalling the sepals and petals in length.

Moist and often shady places in fields and on hillsides.

Z. elegans Pursh with perianth segments 12 mm. long, and *Z. paniculatus* (Nutt.) Wats. with flowers in panicles, may rarely be found in our limits.

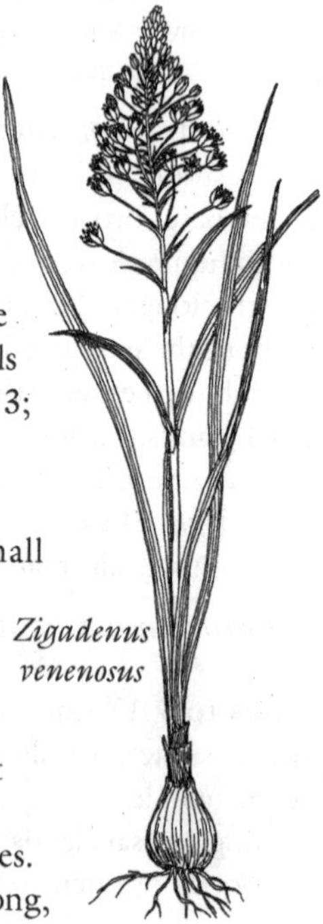

Zigadenus venenosus

IRIDACEAE
Iris Family

Perennial herbs; leaves 2-ranked and more or less equitant, grass-like, sheathing; flowers one to several, generally showy; sepals and petals colored; stamens three; ovary inferior; stigmas 3; capsule 3-loculed.

1a Perianth without a distinct tube .... **Sisyrinchium**
1b Perianth with a distinct tube
  2a Perianth segments in 2 unlike series .... **Iris**
  2b Perianth segments more or less alike, orange .... **Crocosmia**

Crocosmia

Perennials from scaly corms and/or rhizomes; inflorescence usually a panicle of spikes; perianth funnel-form, the tube usually slightly curved, lobes nearly equal, spreading; stamens exserted; capsule longer than broad.

Crocosmia xcrocosmiiflora (Burb. & Dean) N. E. Br.
Stems less than 1 m. in height; leaves 3-6 dm. long, 1-2.5 cm. wide, broadly linear to sword-shaped; perianth 2-5 cm. long, orange to crimson-orange.

This is a hybrid between *C. pottsii* (Baker) N. E. Br. and *C. aurea* (Pappe) Planch. It has escaped from cultivation and become established, especially along coastal road sides.

Iris

Stems erect from stout, often creeping, rhizomes; leaves mostly basal, those on the stem generally reduced in length; flowers large, showy, solitary or several, borne in the axils of sheath-like bracts; sepals colored (not green), spreading or turned backward; petals erect; style divided into 3 petal-like spreading branches, each 2-lobed at apex; stigma a small triangular flap at base of lobes; stamens 3, each closely appressed beneath a style branch; capsule oblong, 3-angled.

1a Flowers normally white with purple markings; petals notched; perianth tube 3-6 mm. long .... *I. tenuis*
1b Flowers purple or yellow or, if white, the perianth tube over 6 mm. long
  2a Leaves 1-2 cm. wide; flowering stems well over 5 dm. in height; flowers yellow .... *I. pseudoacorus*
  2b Leaves less than 1 cm. wide; flowering stems less than 5 dm. tall
    3a Perianth tube 4-12 cm. long .... *I. chrysophylla*
    3b Perianth tube 1 cm. or less in length .... *I. tenax*

Iris chrysophylla How. — Yellow-flowered Iris
Plants often forming clumps from slender rhizomes; flowering stems usually less than 2 dm. tall; leaves 2-4 dm. long and 3-7 mm. wide; perianth tube slender, 4-12 cm. long, perianth pale yellow to cream-colored, sometimes bluish-tinged with darker lines; sepals 4.5-6.5 cm. long; petals much shorter than the sepals.

Open coniferous forest, mainly south of our range.

Iris pseudoacorus L. — Water flag
Plants often forming dense clumps from short thick rhizomes; stems approximately 1-1.5 m. in height; lower leaves 5-9 dm. long, 1-2 cm. wide; flowers pale to bright yellow with purple to brown markings; perianth tube 12-13 mm. long; sepals about 6 cm. long; the petals shorter and more narrow; capsule 5-8 cm. long.

Introduced from Europe and escaped from cultivation and well established along streams, ponds, lakes and in marshy areas.

Iris tenax Dougl. Purple Iris or Flag

Stems slender or rather stout, arising with many leaves from a fibrous-covered rhizome; basal leaves grass-like, 1-5 dm. long, 2-6 mm. wide, acuminate, conspicuously parallel-veined; perianth tube 6-10 mm. long; sepals reaching 7.5 cm. in length, broad at summit, purple with yellow center, the whole sepal conspicuously veined with darker purple; petals narrower, erect, uniformly purple or lavender; or flowers rarely yellow; style branches purple or lavender.

Iris tenax

Common and very showy in fields and roadsides. An albino form, having white flowers, is occasionally found.

Iris tenuis Wats. Clackamas Iris

Stems slender, 2-4 dm. tall, arising with broad leaves from a slender rhizome; basal leaves equalling or slightly longer than the stems, 1-3.5 dm. long and 7-18 mm. wide, broadest about middle, often somewhat curved, long-acuminate, conspicuously veined, cauline leaves several, scarious, bract-like; bracts 2, usually arising together, scarious; flowers generally 2, on short pedicels scarcely exceeding the bracts, white, somewhat veined with purple and yellow; perianth tube 3-6 mm. long.

Known only from the Clackamas River in Clackamas County, Oregon.

Sisyrinchium

Tufted perennials; stems often flattened and winged, sometimes bent at the nodes; leaves grass-like; flowers quickly wilting, produced in a few-flowered umbel from between 2 sheathing bracts; sepals and petals similar, spreading; filaments of stamens usually united into a tube; ovary 3-loculed, stigmas 3; fruit a capsule.

1a Flowers yellow .... *S. californicum*
1b Flowers blue or purple
  2a Floral bracts of about equal length; peduncle branching......... *S. bellum*
  2b Outer floral bract generally much longer than inner
    3a Stems flattened, basal leaves not bract-like.............. *S. sarmentosum*
    3b Stems cylindrical, basal leaves bract-like ....*S. douglasii*

Sisyrinchium bellum Wats. Purple-eyed-grass

Stems flattened, 2-5 dm. tall, tufted, erect, often branching near apex; floral bracts generally nearly equal, often reddish, the outer broadly overlapping the inner, margins widely hyaline, apex generally abruptly acute; flowers 10-17 mm. long, dark purple to bluish-violet with yellow centers.

Occasional in moist areas, becoming more abundant south of our range.

Sisyrinchium californicum (Ker Gawl.) Dry. Golden-eyed-grass

Stems stoutish, broadly winged, 1-3 dm. or more tall, longer than the leaves; floral bracts broad, somewhat unequal; flowers 8-18 mm. long, golden yellow, lined with brown; anthers about equalling the filaments; capsule obovoid, nearly black at maturity.

In moist areas, especially in marshes along the coast.

Sisyrinchium douglasii Dietr.
Grass widows

Stems 1.5-3 dm. tall, not flattened; basal leaves bract-like, cauline leaves broadly sheathing; floral bracts unequal; flowers 1.5-2.5 cm. long, reddish-purple, showy, filaments fused for half their length.

In moist open areas; more common south of our range.

This species is sometimes split out of the genus *Sisyrinchium* and is then called *Olsynium douglasii* (Dietr.) E. P. Bicknell.

Sisyrinchium sarmentosum Suksd.
Blue-eyed-grass

Stems tufted, erect 2-4.5 dm. tall, longer than the leaves; leaves 3-6 mm. wide; bracts enclosing the flowers unequal in length, one often twice the length of the other; sepals and petals from pale to dark blue or purplish, yellow at the base, 1-3-toothed at apex, the middle tooth conspicuously slender; capsule globose.

Common in meadows.

Sisyrinchium bellum

ORCHIDACEAE
Orchid Family

Perennial herbs with sheathing leaves, these sometimes reduced to scales; flowers irregular; sepals 3; petals 3, 2 alike, the third forming a lip, sometimes slipper-like; stamens and style united to form a "column"; ovary inferior, sometimes twisted, 1-loculed with 3 placentae; fruit a capsule, opening at maturity by 3 valves; seeds numerous, minute.

1a Green foliage leaves not present
  2a Plant white, becoming yellow-brown in age .... **Cephalanthera**
  2b Plant reddish or purplish .... **Corallorhiza**
1b Green foliage leaves present
  3a Flowers solitary or few, showy
    4a Flowers pink or "orchid" color .... **Calypso**
    4b Flower brownish or yellowish, with a white or yellow sac-like lower lip .... **Cypripedium**
  3b Flowers many, in a spike or raceme
    5a Leaves 2, near middle of stem .... **Listera**
    5b Leaves not as above
      6a Flowers spurred
        7a Leaves nearly all basal, but not suborbicular................. **Piperia**
        7b Stems leafy or, if leaves all basal, then suborbicular .... **Platanthera**

6b Flowers not spurred
8a Flowers greenish and reddish or purplish, in a loose raceme .... **Epipactis**
8b Flowers white or greenish, in a spike
9a Leaves mottled; spike not twisted .... **Goodyera**
9b Leaves not mottled; spike twisted .... **Spiranthes**

Calypso

Stem and single foliage leaf arising from a bulbous corm with thick roots; stem short with several sheathing scale-like leaves, bearing a single showy flower; the 3 sepals and 2 upper petals similar; lower petal forming a sac-like or slipper-like lip; column broad, petal-like.

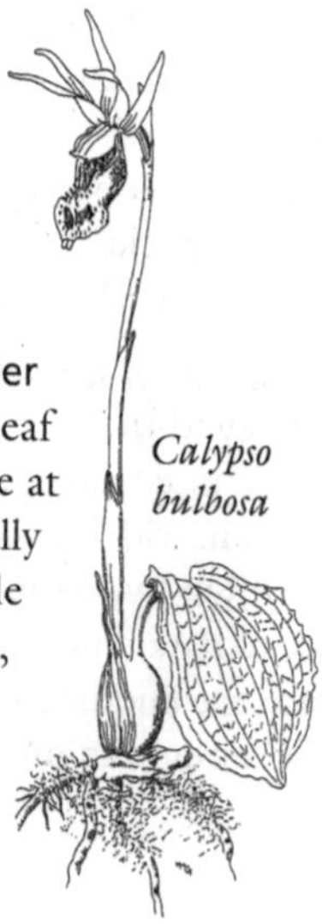

Calypso bulbosa (L.) Oakes

Angel-slipper or Fairy-slipper

Plants 1-2 dm. tall, the blade of the single foliage leaf broadly ovate, 3-8 cm. long, obtuse or sometimes acute at apex, heart-shaped or square-cut at base or occasionally tapering into the petiole, conspicuously veined, petiole long, slender; sepals and petals deep reddish-purple, becoming paler with age, rarely salmon-colored, 1.5 to more than 2.5 cm. long, lip generally slightly longer, pointed at apex, with 2 short lower spurs, purple striped and mottled.

The flower has a lovely delicate fragrance.

In leaf mold of coniferous woods.

Calypso bulbosa

Cephalanthera

Stems erect, bearing a terminal raceme of flowers; leaves (in ours) reduced to sheathing scales; sepals and 2 petals similar, nearly equal in length, third petal forming the lower lip, 3-lobed; pollen masses separate, borne above the stigma.

Cephalanthera austiniae (Gray) Heller — Phantom orchid

Whole plant white, but becoming yellow-brown in age, saprophytic, stems 2-5 dm. tall; leaves 3 or more, reduced to wide-mouthed sheaths; flowers generally many; lip shorter than the other perianth segments, its middle lobe much broader than the lateral lobes, 3- 5-ridged; upper sepal and petals somewhat hooded over column.

[=*Eburophyton austiniae* (Gray) Heller]

Coniferous woods.

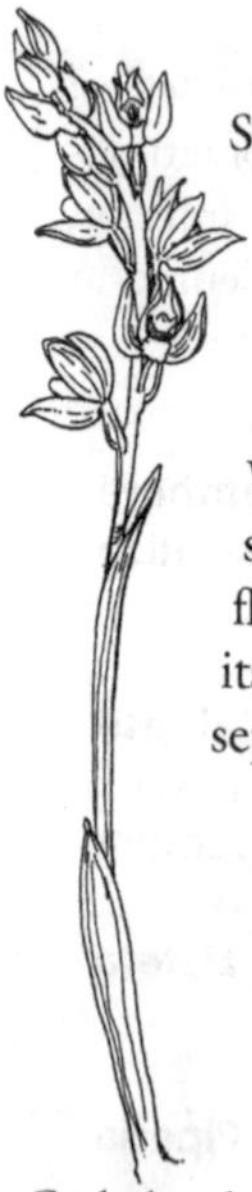

Cephalanthera austiniae

Corallorhiza

Purplish, reddish, brownish, or yellow saprophytes (i.e., securing their food from dead organic material) or parasites; roots branched, coral-like; leaves scale-like, sheathing the stem; flowers borne in a terminal raceme; perianth segments nearly equal in length, 2 petals and the sepals similar, the third petal broader and forming the lower lip which is 1-3-ridged; flowers in some species spurred on lower side at base of the perianth segments; capsules reflexed after flowering.

1a Spur none; lower lip prominently striped with purple.................. *C. striata*
1b Spur often short, but present; lower lip not prominently striped
  2a Lower lip with 4 or more red or purple spots.................... *C. maculata*
  2b Lower lip with 0-2 red or purple spots.......................... *C. mertensiana*

Corallorhiza maculata Raf. — Spotted coral-root

Stems 2-5 dm. tall; raceme 5-15 cm. long; 2 petals and the sepals purplish-brown like the stem, lateral sepals forming a short spur with base of column, lower lip broad, with one wide central lobe and 2 narrow lateral ones, red to purple-spotted; capsule 1.5-2 cm. long.

Coniferous woods, growing in decomposing leaf litter.

Corallorhiza mertensiana Bong. — Western coral-root

Stems reddish, 1.5-5 dm. tall; raceme 1-2 dm. long; peduncle deep purple; flowers red or purplish, spur short with free tip, lower lip broad, generally entire, dark red or purple, usually not spotted; capsule 1.5-2.5 cm. long.

Coniferous or deciduous woods, growing in decomposing leaf litter.

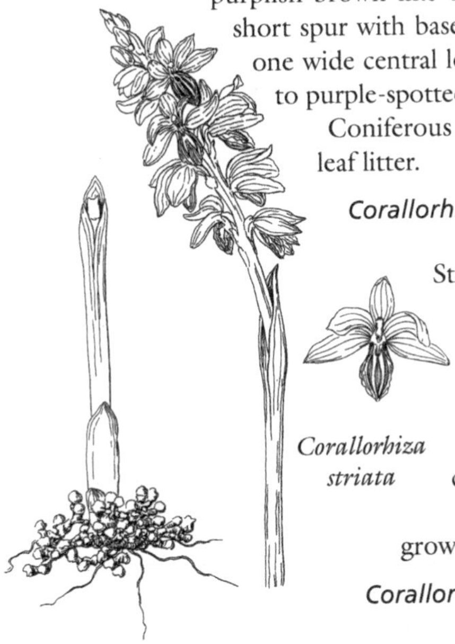

Corallorhiza striata

Corallorhiza striata Lindl. — Striped coral-root

Stems 1.5-5 dm. tall, often several in a cluster together with remnants of last year's stems; raceme 3-20 cm. long; sepals and upper petals cream or flesh-colored with slender reddish-purplish stripes, spur none, lower lip broad, entire, with edges incurved, and with 3 broad purple stripes; capsule 1-2.5 cm. long.

Coniferous woods, growing in decomposing leaf litter.

Cypripedium

Stems erect, leafy, from short horizontal rhizomes bearing cord-like roots; leaves 2 to many, the base of the stem enclosed in sheathing scales; flowers 1-several, generally large and showy, subtended by leafy or scale-like bracts; sepals spreading, the 2 lateral nearly or quite united; upper 2 petals similar to sepals, lower petal large and sac-like; fertile stamens 2, sterile stamen 1, arched over stigma.

Lip yellow often dotted with purple; flowers usually solitary (rarely 2)... *C. calceolus*
Lip white to purplish-tinged; flowers usually 2 (rarely 1 or 3).. *C. montanum*

Cypripedium calceolus L. Yellow Lady's-slipper

Stems leafy, 1.5-4 dm. tall; herbage pubescent and glandular; leaves mostly clasping or sheathing, elliptic to broadly lanceolate, 6-17 cm. long and up to 7 cm. wide; flowers solitary, or rarely 2; sepals and petals greenish-yellow or purplish-brown or mottled, usually slightly twisted, the lip 2-3 cm. long, pouch-like, yellow but sometimes dotted with purple.

(=*Cypripedium parviflorum* Salisb.)

Damp woods and bogs; uncommon.

Cypripedium montanum Dougl. ex Lindl. Lady's-slipper

Stems 1.5-6 dm. tall, rough, as are also the leaves, with short glandular hairs; leaves broadly elliptic to long-ovate, the lowermost sometimes 15 cm. long; flowers usually 2, rarely 1 or 3, on very short pedicels; the 2 upper petals somewhat narrower than the sepals, wavy, often twisted, sepals and upper petals all longer than lip, spreading, brown or purplish; lip large, white to purplish or purple-veined; ovary covered with short glandular hairs.

Coniferous woods; uncommon.

Epipactis

Plants leafy-stemmed from creeping rhizomes; flowers in racemes, the lower bracts sometimes leaf-like; flowers with conspicuous lower lips narrowed at the middle; capsules turned downward at maturity.

Epipactis gigantea Dougl. ex Hook. Stream orchid

Stems generally stout, 2-10 dm. or more tall; leaves lanceolate to elliptic, the upper, narrower, mostly smooth, 5-15 cm. long, the numerous parallel veins conspicuous; flowers generally few, distinct, in a raceme, each on a short pedicel; calyx greenish; corolla purplish, the lip darker and conspicuously veined, basal portion of lip concave, winged, apical portion spreading, ridged.

Stream banks, seepage areas, wet meadows and lake margins; uncommon.

Epipactis gigantea

Goodyera

Stems from a creeping rhizome; leaves white-mottled, in a basal cluster, the cauline leaves much-reduced and bract-like; flowers white, small, borne in a spike; upper sepal and lateral petals more or less concealing the saccate lip.

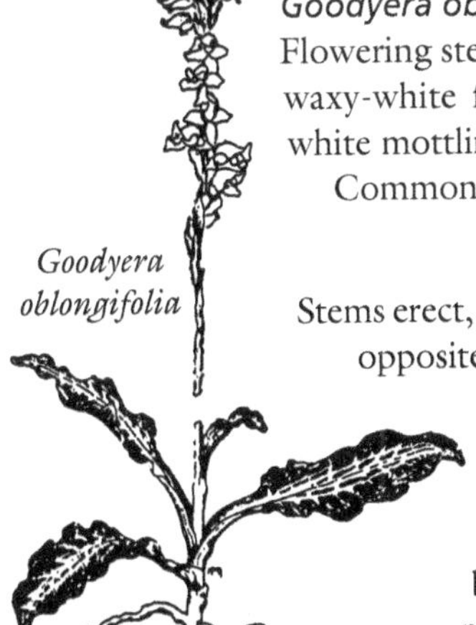
Goodyera oblongifolia

Goodyera oblongifolia Raf. — Rattlesnake plantain

Flowering stems about 3 dm. tall, bearing a spike of small waxy-white flowers; basal leaves clustered, green with white mottlings along the larger veins.

Common in deep woods in decomposing leaf litter.

Listera

Stems erect, slender, bearing a single pair of broad, sessile, opposite foliage leaves, with a sheathing scale-leaf at base of stem above rhizome; flowers small, inconspicuous, in an open terminal raceme; sepals and the 2 upper petals similar, slender; lower petal forming the lip, longer and broader, 2-toothed or cleft at apex; pollen masses 2; capsule ovoid.

1a Leaves heart-shaped at base; lower lip of the corolla 2-cleft to near the middle .... *L. cordata*

1b Leaves not heart-shaped at base; lower lip of the corolla not cleft to the middle

2a Ovary glandular-pubescent; sepals sharply recurved at maturity; lower corolla lip 6-12 mm. long .... *L. convallarioides*

2b Ovary glabrous; sepals not sharply recurved at maturity; lower corolla lip 4.5-7mm. long .... *L. caurina*

Listera caurina Piper — Northwest twayblade

Plants 1-3 dm. tall; leaf blades 2-7 cm. long, abruptly tapered to the base; sepals 3-4 mm. long, lanceolate; lower lip of the corolla 4.5-7 mm., shallowly 2-lobed with 2 minute teeth, 1 on each side of the base; ovary glabrous.

Woods, principally in the mountains.

Listera convallarioides (Sw.) Nutt. — Broad-leaved twayblade

Stems 8-30 cm. tall, pubescent above the leaves; leaves broadly ovate to round, generally obtuse or rarely acute at apex, 1.5-8 cm. long, flowers greenish, 6-15, on rather long pedicels; sepals turned sharply back against pedicels at maturity; lower lip of the corolla 6-12 mm. long, 2-lobed, toothed on each side of base; ovary glandular-pubescent.

Woods in the mountains.

Listera cordata

Listera cordata (L.) R. Br. — Heart-leaved twayblade

Stems 7.5-20 cm. tall, pubescent above the leaves; leaves broadly ovate, heart-shaped at base, 1.8-5 cm. long; flowers greenish, yellowish, or purplish, very small; lower lip of the corolla 5-10 mm. long, cleft to the middle into 2 very slender lobes, toothed on each side of base.

Coniferous woods.

Piperia

Stems from a tuber-like or bulb-like base; basal leaves 2-5, cauline leaves bract-like; flowers in a bracteate raceme or a spike; flowers spurred, white or greenish; sepals and lateral petals more or less equal; capsule erect to spreading.

Spur 6-18 mm. long .... *P. elegans*
Spur 1.5-5 mm. long .... *P. unalascensis*

Piperia elegans (Lindl.) Rydb. — Elegant rein-orchid

Stems 1.5-7 dm. tall; basal leaves lanceolate or oblanceolate, 7-30 cm. long, cauline leaves bract-like, few; spike 0.5-4 dm. long, generally dense; flowers greenish-white or white, less than 12 mm. long; lower lip of the corolla lanceolate, slightly lobed at base, spur 6-18 mm. long, straight or strongly curved.

[=*Habenaria elegans* (Lindl.) Boland.]
Dry open woods.

Piperia unalascensis (Spreng.) Rydb. — Alaska rein-orchid

Stems 3-4.5 dm. tall, from fleshy tubers; lower leaves oblanceolate or long-elliptic, more or less obtuse, 12-15 cm. long, about 2.5 cm. wide; cauline leaves few, bract-like; spike very slender, 1-3 dm. long, the bracts shorter than the flowers; flowers greenish, less than 12 mm. long, lower lip of the corolla obtuse, slightly lobed at base, shorter than the ovary, spur 1.5-5 mm. long, slender, slightly longer than lip.

[=*Habenaria unalascensis* (Spreng.) Wats.].
Dry or moist woods.

Platanthera

Leaves basal and cauline, gradually reduced upwards; inflorescence a bracteate spike, flowers generally many, white to greenish or yellowish-green, spurred.

1a Leaves 2 or 3, nearly basal, suborbicular .... *P. orbiculata*
1b Stems leafy and leaves not suborbicular
  2a Flowers white to cream-colored; lip broadened just above the base
    3a Spur usually strongly curved, 1 to 2 times the length of the lip .... *P. leucostachys*
    3b Spur only slightly curved, about equal to the lip in length .... *P. dilatata*
  2b Flowers greenish or purplish
    4a Spur saccate or club-shaped, straight .... *P. stricta*
    4b Spur wider at the base than at the tip, usually curved
      5a Tip of the spur acute; flowers of the spike rarely over-lapping .... *P. sparsiflora*
      5b Tip of the spur blunt; spike densely many-flowered .... *P. hyperborea*

Platanthera dilatata (Pursh) Lindl. — White orchis or Bog-candle

Stems stout, up to 12 dm. tall; leaves lanceolate; spike usually dense, 1-2 dm. long; flowers white, spur about equalling the lip in length, only slightly curved.

[=*Habenaria dilatata* (Pursh) Hook.]
Mountain bogs.

Platanthera hyperborea (L.) Lindl.
Green-flowered bog-orchid
Stems 1.5-10 dm. tall; cauline leaves 4-12 cm. long and 6-30 mm. wide; flowers in a dense bracteate spike; perianth green to yellowish-green, spur cylindric, slightly curved, the tip blunt.
[=*Habenaria hyperborea* (L.) R. Br.]
Wet coniferous forest.

Platanthera leucostachys Lindl. White-flowered bog-orchid
Stems 1.5-10 dm. tall; cauline leaves 5-25 cm. long, 9-30 mm. wide; spike generally dense; flowers white to cream-colored, spur 5-15 mm. long, slender, cylindric to clavate, usually strongly curved, once or twice as long as the lip.
[=*Habenaria dilatata* (Pursh) Hook. var. *leucostachys* (Lindl.) Ames]
Wet meadows to boggy ground.

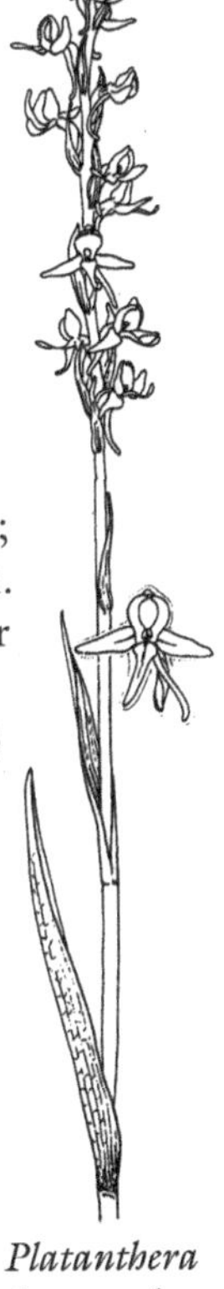
Platanthera leucostachys

Platanthera orbiculata (Pursh) Lindl.
Round-leaved orchid
Stems 2-6 dm. tall; basal leaves usually 2, orbicular or nearly so, 9-20 cm. long, shining, cauline leaves bract-like; spike loose; lower lip of the corolla narrow, white, about 12 mm. long, spur generally much longer than the ovary.
[=*Habenaria orbiculata* (Pursh) Torr.]
Moist woods; uncommon.

Platanthera sparsiflora (S. Wats.) Schltr.
Sparse-flowered bog-orchid
Plants 2.5-5.5 dm. in height; cauline leaves 4-15 cm. long and 5-30 mm. wide; inflorescence more or less open; flowers green, spur cylindric, slightly curved, more or less equal to the lip, the tip acute.
(=*Habenaria sparsiflora* Wats.)
Wet meadows and stream banks.

Platanthera stricta Lindl. Slender bog-orchid
Stems 2-10 dm. tall; lowest 2 to 3 cauline leaves often bladeless and sheathing, middle cauline leaves oblong-elliptic, 3-15 cm. long and 5-25 mm. wide, upper leaves becoming more narrow; flowers green or purplish-tinged, spur saccate or club-shaped, straight, half as long or equal to the lip.
(=*Habenaria saccata* Greene)
Wet meadows or boggy areas.

Spiranthes

Stems from enlarged fleshy, clustered roots; stems leafy below, the leaves reduced upwards to sheathing bracts, flowers in a spirally twisted spike; upper sepal and lateral petals often fused and forming a hood, lower sepals nearly free and about equalling the lip, lip deeply grooved, spur absent.

Spiranthes romanzoffiana

Spiranthes romanzoffiana Cham.

Hooded Ladies' Tresses or Twisted orchid

Plants 1-6 dm. in height; herbage glabrous; leaves mainly near the base of the stem, linear to oblong, 5-25 cm. long, 5-12 mm. wide, abruptly reduced to lanceolate sheathing bracts; flowers white to cream or greenish-white, arranged in a spirally twisted spike. Two varieties, sometimes treated as separate species occur in our area. They may be separated as follows: *var. romanzoffiana* has the lip constricted below the tip and lacks prominent basal callosites and terminal puberulence, while *var. porrifolia* (Lindl.) Ames & Correll (=*S. porrifolia* Lindl.) has a triangular to lanceolate lip, puberulent tip and prominent callosities.

Wet meadows, marshy areas and seeps.

Dicots

SALICACEAE
Willow Family

Deciduous trees or shrubs with characteristically bitter bark; leaves alternate; staminate and pistillate flowers in catkins, borne on separate plants; perianth reduced to a glandular disk or glands; flower subtended by a bract; stamens rarely 1, generally 2 to many; ovary 1-loculed, becoming a 2-4-valved capsule containing many seeds each with a tuft of long silky or cottony hairs; style branches 2-4.

Flower parts on a cup-like disk; bud-scales more than one; stamens usually many; leaves long-petioled .... **Populus**
Flower parts not on a disk, but gland(s) present; bud-scale one; stamens 1-8, usually 2; leaves short-petioled .... **Salix**

Populus

Soft-wooded trees with scaly, usually resinous buds and early deciduous stipules; leaves generally long-petioled; both staminate and pistillate catkins pendulous, appearing before the leaves; bracts fringed; stamens 8-60, borne on a disk; pistil borne on a disk; ovary 1-loculed; stigmas 2-4; seeds many; capsule 2-4-valved, globose at maturity.

Leaves long-ovate; petioles not flattened .... *P. balsamifera*
Leaves as broad as long; petioles flattened .... *P. tremuloides*

Populus balsamifera L. subsp. *trichocarpa* (Torr. & Gray) Brayshaw
Black cottonwood

Trees 15-60 m. tall with erect branches and yellowish-gray bark which is deeply fissured on old trunks; winter buds large, pointed, resinous; leaves long-petioled, the petioles not flattened, the blades pinnately veined, ovate, slightly serrate or finely scalloped, the teeth often gland tipped, truncate or cordate at the base, acute or abruptly acuminate at apex, shining and fragrantly resinous, whitish beneath, often stained with resin at maturity; bracts with long hairs; capsule 5-8 mm. long, nearly round, often pubescent; seeds very cottony.

(=*Populus trichocarpa* Torr. & Gray)

Low ground and along streams and lake shores.

Populus balsamifera

Populus tremuloides Michx. Quaking aspen

Small slender trees 4-20 m. tall with somewhat drooping branches and smooth greenish bark which becomes more or less fissured with age; petioles 2/3 as long to equal to the blade, flattened; leaf blades pinnately veined, broadly ovate to

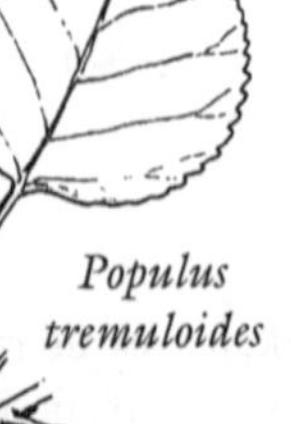

Populus tremuloides

nearly orbicular, more or less serrate to finely scalloped or wavy and white ciliate, except at the rounded or somewhat cordate base, abruptly acute or acuminate at apex, dark green above, yellowish beneath; bracts long-fringed; capsules 4-6 mm. long, longer than broad.

Stream banks and in drier locations; only rarely west of the Cascade Mountains.

Salix

Shrubs or occasionally trees; buds covered by a single nonresinous scale; pistillate and staminate flowers usually with an entire floral bract and a ventral gland (nectary), staminate flowers sometimes also with a dorsal gland (nectary); stamens typically 2, but 1-8 in some species; ovary 2-loculed; stigmas 2; capsule 2-valved, elongated at maturity.

Key to pistillate plants

1a Pistillate floral bracts deciduous
  2a Ovary glabrous
    3a Petioles with 2 or more glands near the base of the blades. ..*S. lucida*
    3b Petioles without glands.. *S. melanopsis*
  2b Ovary pubescent
    4a Under surface of leaves not glaucous, leaves silky pubescent when fully expanded; common.. *S. sessilifolia*
    4b Under surface of leaves glaucous, leaves typically becoming glabrous or nearly so by the time they are fully expanded..................*S. fluviatilis*
1b Pistillate floral bracts persistent
  5a Ovary pubescent
    6a Leaf margins strongly rolled under
      7a Leaf blades dull above; capsules 3.5-5.5 mm. long; plants less than 8 m. tall.. *S. sitchensis*
      7b Leaf blades shiny above; capsules 4.5-11 mm. long; plants 3-20 m. tall...*S. scouleriana*
    6b Leaf margins not rolled under or only slightly so
      8a Floral bracts tawny to light brown........................... *S. geyeriana*
      8b Floral bracts brown to black or bicolored
        9a Subalpine species; young leaves only sparingly pubescent......... ....*S. planifolia*
        9b Lowland species; young leaves densely pubescent.................. ....*S. hookeriana*
  5b Ovary glabrous
    10a Leaves not glaucous on the under surface
      11a Upper surface of leaves dull; floral bracts tawny or bicolored...... ....*S. commutata*
      11b Upper surface of leaves shiny or glossy; floral bracts dark brown. ....*S. boothii*
    10b Leaves glaucous on the under surface
      12a Floral bracts tawny or rose-colored...................... *S. pedicellaris*
      12b Floral bracts dark brown to black or bicolored

13a Upper surface of leaves dull; flowers appearing with the leaves .. *S. prolixa*
13b Upper surface of leaves shiny, flowers usually appearing before the leaves
14a Mature leaf blades broadly ovate to elliptic or obovate, 2-6.5 cm. wide; styles 0.6-2.3 mm. long...*S. hookeriana*
14b Mature leaves narrowly oblong to oblanceolate, 0.5-3.2 cm. wide; styles 0.1-0.6 mm. long...*S. lasiolepis*

Key to staminate plants

1a Stamens 4-5; petioles with 2 or more glands near the base of the blades...... ....*S. lucida*
1b Stamens 1 or 2; petioles without glands
2a Stamen 1 ....*S. sitchensis*
2b Stamens 2
3a Flowers with 2 glands (nectaries)
4a Apex of leaf acute; petioles not silky; anthers 0.6-0.9 mm. long ....*S. melanopsis*
4b Apex of leaf acuminate; petioles silky; anthers 0.8-1.3 mm. long
5a Catkins slender, 2-3.5 cm. long ....*S. sessilifolia*
5b Catkins 3-7 cm. long ....*S. fluviatilis*
3b Flowers with a single gland (nectary)
6a Floral bracts tawny, rose-colored or light brown
7a Leaves not glaucous on the under surface; catkins 1-8 cm. long ....*S. commutata*
7b Leaves glaucous on the undersurface; catkins 0.5-2 cm. long
8a Plants forming colonies by layering.............*S. pedicellaris*
8b Plants not colonial ....*S. geyeriana*
6b Floral bracts dark brown to black
9a Catkins appearing with the leaves on distinct leafy shoots......... ....*S. prolixa*
9b Catkins usually appearing before the leaves
10a Leaves not glaucous on the under surface ....*S. boothii*
10b Leaves glaucous on the under surface
11a Subalpine species with young leaves glabrous or nearly so...*S. planifolia*
11b Not subalpine species; young leaves pubescent; common species
12a Leaf margins strongly rolled under ...*S. scouleriana*
12b Leaf margins not rolled under or only slightly so
13a Mature leaf blades broadly ovate, elliptic, to obovate, 2-6.5 cm. wide; filaments often pubescent on the lower half...................................*S. hookeriana*
13b Mature leaves narrowly oblong to oblanceolate, 0.5-3.2 cm. wide; filaments glabrous......*S. lasiolepis*

Salix boothii Dorn — Booth's willow

Erect shrubs 0.25-6 m. in height with brown bark; leaf blades 2.5-10 cm. long, narrowly oblong to elliptic, the margins entire to finely serrate, the apex short-acuminate, pubescent at first but usually becoming nearly glabrous; stipules foliaceous; catkins usually appearing before the leaves; bracts dark brown, pubescent; stamens 2; pistillate flowers with a glabrous ovary.

Wet subalpine areas.

Salix commutata Bebb — Under-green willow

Erect shrubs 0.2-3 m. in height; new growth with villous-tomentose twigs; leaf blades 1-10 cm. long, narrowly oblong to elliptic-obobvate, entire to serrulate, obtuse, acute or acuminate at the apex and rounded to cordate at the base, woolly-villous on both surfaces, at least when young; stipules up to 1 cm. long, glandular-serrate, foliaceous; catkins appearing at the same time as the leaves; bracts brown or tawny, woolly-villous; stamens 2; ovary glabrous.

Open woods and along streams; moderate to high elevations in the mountains.

Salix fluviatilis Nutt. — Columbia River willow

Erect shrubs or small trees 2-6.5 m. in height; young twigs and young leaves with a silvery pubescence, but often becoming nearly glabrous; stipules usually small and early deciduous; leaf blades nearly sessile, 5-10 cm. long, linear to narrowly elliptic, nearly entire, but often with a few scattered teeth or the margins wavy, apex acuminate; catkins appearing with or after the leaves; bracts tawny, pubescent; stamens 2; ovary villous.

(=*Salix exigua* Nutt. var. *columbiana* Dorn)

Along the Columbia River.

Salix geyeriana Anderss. — Geyer's willow

Erect shrubs 1-5 m. in height; twigs tomentose at first, but becoming glabrous and usually glaucous; leaves short-petioled, leaf blades 3-7.5 cm. long, lanceolate to oblanceolate, entire or nearly so, pubescence silky, lower surface glaucous, cuneate at the base and acuminate at the apex; catkins appearing before or with the leaves; bracts tawny to light brown, pubescent; stamens 2; ovary pubescent.

Wet areas, especially in meadows and along streams.

Salix hookeriana Hook. — Coastal or Hooker's willow

Erect shrubs or small trees 0.6-8 m. in height; young twigs and young leaves usually pubescent; leaf blades 3.5-12 cm. long, broadly ovate to elliptic or obovate, entire to remotely serrate or wavy-margined, under surface glaucous, pubescent or rarely nearly glabrous, upper surface pubescent or rarely glabrous; catkins usually appearing before the leaves; bracts dark brown to nearly black; stamens 2; ovary glabrous or pubescent.

(Includes *Salix piperi* Bebb)

Wet or moist areas, especially common along the coast, but also inland.

Salix hookeriana

Salix lasiolepis Benth. — Arroyo willow

Erect shrubs or small trees 1.5-10 m. in height; young twigs pubescent, usually becoming glabrous; leaf blades 3-12.5 cm. long, narrowly oblong to oblanceolate or obovate, white tomentose at first, but becoming glabrous at least above, under surface glaucous, upper surface shiny, base cuneate, apex acute or obtuse, the margins entire to shallowly toothed; catkins appearing before the leaves; bracts dark brown; stamens 2; ovary glabrous.

Wet areas or hillsides to meadows and coastal headlands.

Salix lucida Muhl. — Shining willow

Erect shrubs or trees 1-11 m. in height; twigs glabrous or pubescent; stipules foliaceous, gland-toothed; petioles bearing 2 or more large glands; leaf blades 5-17 cm. long, linear to lanceolate or oblanceolate, serrulate, white pubescent at first, usually becoming glabrous, apex long-acuminate, base cuneate to rounded; catkins appearing with the leaves; bracts tawny; stamens 4-5; ovary glabrous. Two subspecies occur in our area: subsp. ***caudata*** (Nutt.) E. Murray is not glaucous on the under surface of the leaves and subsp. ***lasiandra*** (Benth.) E. Murray is glaucous on the under surface.

(=*Salix lasiandra* Benth.)

Wet to fairly dry sites, often along streams and in meadows.

Salix melanopsis Nutt. — Dusky willow

Erect shrubs less than 4 m. in height, usually forming large colonies; twigs often silky pubescent at first, but becoming glabrous; leaves short-petioled, leaf blades 3-8.5 cm. long, narrowly oblong to elliptic or oblanceolate, acute at both ends, entire to serrulate, villous at first, but often becoming nearly glabrous; catkins appearing with or after the leaves; bracts tawny to light brown, villous; stamens 2; ovary glabrous.

River and stream banks.

Salix pedicellaris Pursh — Bog willow

Erect to decumbent shrubs 2-14 dm. in height, forming colonies by layering; twigs dark-colored, glabrous or nearly so; leaves short-petioled, leaf blades 2-7 cm. long, narrowly oblong, elliptic to oblanceolate, entire, silky pubescent at first, but soon becoming glabrous; catkins developing with the leaves; bracts tawny or sometimes pinkish-tinged, villous; stamens 2; ovary glabrous.

Bogs.

Salix planifolia Cham. — Tea-leaf willow

Decumbent to nearly erect shrubs 0.15-4 m. tall with dark, glabrous, stout, divaricate branches and large chestnut-colored buds; leaf blades 2-6.5 cm. long, narrowly oblong, elliptic to oblanceolate, entire to serrulate or crenate, acute at both ends, glabrous or silky pubescent, the under surface glaucous, the upper surface shiny; catkins appearing before the leaves; bracts dark; stamens 2; ovary pubescent.

Subalpine meadows, lake margins and stream banks.

Salix prolixa

Salix prolixa Anderss. — Mackenzie's willow

Shrubs less than 5 m. in height; stipules ovate; leaf blades 5-15 cm. long, lanceolate to ovate, margin gland bearing and/or serrate, acuminate at the apex, sometimes pubescent when young, but usually glabrous, under surface glaucous; catkins appearing with the leaves; bracts brown, pubescent; stamens 2; ovary glabrous.

[=*Salix mackenzieana* (Hook.) Anderss.]

Along streams.

Salix scouleriana Hook. Scouler's willow

Shrubs or trees 2-12 m. in height; young branches velvety; stipules large and foliaceous on young vigorous shoots, but usually inconspicuous on older branches; petioles usually velvety; leaf blades 3-10 cm. long, elliptic to oblanceolate or obovate, entire to irregularly serrate, the margins inrolled, more or less tomentose on the glaucous under surface, upper surface usually becoming glabrous and shiny, the apex rounded to acute, the base cuneate to rounded; catkins usually appearing before the leaves; bracts dark brown; stamens 2; ovary densely long silky.

Moist to dry forests, river banks and around lakes.

Salix sessilifolia Nutt. Northwest sandbar willow

Erect shrubs 2-8 m. in height; twigs silky pubescent, sometimes becoming glabrous; leaves short-petioled; leaf blades 3-12 cm. long, linear to narrowly oblanceolate, margins with a few scattered teeth, acuminate at the apex, cuneate to rounded at the base, silky pubescent; catkins appearing with or after the leaves; bracts tawny or brown; stamens 2; ovary pubescent.

Along streams.

Salix sitchensis Bong. Sitka willow

Shrubs or small trees, 1-8 m. in height; young twigs silky pubescent; leaf blades 3-12 cm. long, elliptic to oblanceolate or obovate, acute to obtuse at the apex, cuneate at the base, entire or with a few scattered glandular teeth, the margins inrolled at least near the base, under surface pubescent, upper surface pubescent at first, but sometimes becoming nearly glabrous; catkins appearing before or with the leaves; bracts tawny to brown; stamen 1; ovary pubescent.

Stream banks, moist woods and other wet areas.

MYRICACEAE
Sweet-gale Family

Fragrant shrubs or small trees with alternate, simple, resinous-dotted leaves; monoecious or dioecious or rarely with both perfect and imperfect flowers; flowers in axillary catkin-like spikes; sepals and petals absent; staminate flowers subtended by a bract and sometimes also bractlets; pistillate flowers subtended by a bract and usually 2-4 bractlets; ovary superior, 1-loculed, stigmas 2; fruit drupaceous or nut-like.

Myrica

Deciduous or evergreen shrubs or small trees; leaves resinous-dotted and aromatic, generally toothed or lobed; flowers mostly unisexual; staminate flowers with 2-22 stamens; fruit smooth or warty, usually waxy.

Monoecious evergreen shrubs .... *M. californica*
Dioecious deciduous shrubs .... *M. gale*

Myrica californica Cham. — Wax myrtle

Monoecious shrubs or small trees, much-branched, 1-10 m. tall; leaves evergreen, leathery, 4-13 cm. long, narrow, elliptic to oblanceolate, tapering at both ends, mostly toothed or sometimes nearly entire, dotted with black or yellow glands; pistillate spikes above the staminate or some spikes with both staminate and pistillate flowers intermixed; stamens 3-16, united by the filaments; fruit warty and usually with a white waxy coating.

Common on and near the coast.

Myrica californica

Myrica gale L. — Sweet-gale

Dioecious low shrubs up to 2 m. tall; leaves deciduous, 3-6 cm. long, oblanceolate, obtuse, usually serrate near the apex, generally puberulent at least below, dotted with yellow waxy glands; flowers appearing before the leaves; staminate flowers mostly with 3-5 stamens, the filaments free; fruit smooth except for the wax glands.

Bogs or swamps and along lake margins; coastal and inland.

BETULACEAE
Birch Family

Monoecious trees or shrubs; bark more or less smooth but with lenticels; leaves deciduous, alternate, simple, usually serrate, stipules early deciduous; staminate catkins long, slender, drooping, flowers bracteate, stamens 1-6; pistillate catkins shorter, or the pistillate flowers clustered in a scaly bud; ovary 1-2-loculed; fruit 1-seeded nuts, nutlets or samaras.

1a Fruit a nut enclosed in an involucre; leaf bases cordate .... **Corylus**
1b Fruit a nutlet or a samara, not enclosed in an involucre; leaf bases typically not cordate
  2a Pistillate catkins forming woody cone-like structures, falling whole .... **Alnus**
  2b Pistillate catkins not becoming woody, disintegrating before falling .... **Betula**

Alnus

Deciduous trees or shrubs; winter buds stalked; staminate catkins drooping and falling after flowering, the pistillate catkins erect, becoming woody at maturity, and persisting until the next season's flowering; staminate flowers 3 per bract, stamens 1-6; pistillate catkins cone-like the flowers usually 2 per bract; fruit a samara or a wingless nutlet.

1a Flowers appearing with the leaves on new twigs; leaves not at all revolute .... *A. viridis*

1b Flowers appearing before the leaves, on the previous year's twigs

  2a Leaves strongly revolute, coarsely doubly serrate or rarely pinnatifid; fruits with winged margins .... *A. rubra*

  2b Leaf margins more or less flat, sharply denticulate; margins of the fruits not truly winged .... *A. rhombifolia*

Alnus rhombifolia Nutt. White alder

Large shrubs or trees reaching a height of 25 m. with smooth pale bark; leaves 4-10 cm. long, more or less elliptic with sharply denticulate margins, mostly finely rusty-pubescent beneath; fruit with a thin margin but not truly winged.

Along streams, often with *A. rubra*.

Alnus rhombifolia

Alnus rubra Bong. Red alder

Trees 10-30 m. tall with smooth gray bark; twigs reddish; winter leaf buds gummy; leaves 5-30 cm. long, broadly ovate, elliptical, or obovate, doubly serrate, mostly smooth, the under surface often pubescent and lighter than the upper surface; pistillate catkins becoming woody, dark brown or black, 1.5-2.5 cm. long; body of fruit much broader than the wings.

Common along streams.

Alnus rubra
a. old pistillate catkins
b. young pistillate catkins
c. staminate catkins
d. fruit

Alnus viridis (Villars) DC. Wavy-leaved alder

Large shrubs or small trees to 10 m. in height; bark reddish-brown but becoming gray to nearly black; leaves ovate to elliptic, 3-10 cm. long, singly or doubly sharply serrate, often slightly lobed, shining beneath, sticky when young; flowers opening with the leaves on the current years twigs; fruit with wings usually wider than the body.

Two subspecies occur in our area: *Alnus viridis* subsp. *fruticosa* (Rupr.) Nyman

(Siberian alder) and the more common *Alnus viridis* subsp. *sinuata* A. Love & D. Love (Sitka alder) [=*Alnus sinuata* (Regel) Rydb.]. They can be separated on the basis of the paper thin doubly serrate leaves of subsp. *sinuata* and the thicker, usually singly serrate ones of subsp. *fruticosa*.

Moist, often gravelly areas, along streams and lakes and in open woods.

Betula

Trees or rarely shrubs with smooth bark which often peels off in papery sheets; leaves simple, petioled, toothed or crenate, deciduous, with early deciduous stipules; staminate catkins pendulous, elongated, often clustered, flowers 3 to a scale, the scale consisting of a bract united at base with 2 bractlets; stamens 1-4; pistillate flowers usually 3 per scale; fruit a 2-winged samara.

1a Tall trees; leaves ovate with 5-9 pairs of lateral veins.............. *B. papyrifera*
1b Shrubs, less than 4 m. in height; leaves with 2-6 pairs of lateral veins
  2a Twigs warty with large resinous glands .... *B. glandulosa*
  2b Twigs not warty, glands small .... *B. pumila*

Betula glandulosa Michx. — Dwarf birch

Shrubs, often spreading, reaching 3 m. in height; bark dark brown; twigs warty with large resinous glands; leaf blades 0.5-3 cm. long, obovate to nearly orbicular, crenate-dentate, usually with resinous glands; wings of the fruit narrower than the body.

Subalpine to alpine summits.

Betula papyrifera Marsh. — Paper birch

Trees to 20 m. in height; bark of mature trees usually chalky white; twigs rarely glandular; leaf blades 4-12 cm. long, ovate, to elliptic-ovate, serrate to dentate, resinous-glandular; wings of the fruit as broad or broader than the body.

Usually in moist woods; in our range this species is known only from western Washington.

Betula pumila L. — Western low birch

Shrubs, 1-4 m. tall; bark reddish-brown; twigs with small resinous glands, especially near the nodes; leaf blades 2.5-7 cm. long, elliptic or broadly obovate to nearly orbicular, narrowed to rounded at base, crenate to dentate, usually with resinous glands; fruit with wings slightly narrower than the body.

Sphagnum bogs and swampy areas.

Corylus

Shrubs or trees with alternate simple deciduous leaves; inflorescences appearing before the leaves; staminate flowers 3 per scale in long slender catkins; pistillate flowers 3 per bract, clustered in a scaly bud; ovary inferior; fruit a nut, enclosed in a foliaceous involucre.

This genus is sometimes treated in a separate family, the **Corylaceae**.

Corylus cornuta Marsh. subsp. *californica* (DC.) E. Murray
Western hazel

Generally tall shrubs, 3-7 m., but sometimes shorter; twigs with glandular hairs; leaf blades 3.5-10 cm. long, nearly orbicular or obovate, rarely ovate, often more or less lobed, finely to coarsely doubly toothed, rough above, velvety beneath; nuts usually in clusters of 2-4, involucre enclosing nut forming a sheath extending beyond it, usually densely covered with short stiff hairs below, sometimes nearly smooth.

Common; usually in moist places along stream banks and in wooded areas.

FAGACEAE
Oak Family

Evergreen or deciduous trees or shrubs with alternate simple leaves; stamens and pistils borne in separate flowers; staminate flowers in catkins or catkin-like spikes; pistillate flowers 1-3 surrounded by a cup-like arrangement of scales, borne at the base of the staminate catkins or in very short separate catkin-like spikes; ovary inferior, 1-3-loculed, each locule with 2 ovules at first, only 1 ovule maturing; fruit a nut borne in cup of scales or in a spiny bur.

Deciduous trees; fruit a nut in a scaly cup (acorn) .... **Quercus**
Evergreen trees or shrubs; fruit of 1-3 nuts enclosed in a spiny bur.................
.... **Chrysolepis**

Chrysolepis

Evergreen trees or shrubs; inflorescence erect or spreading, sometimes branching, bearing staminate flowers in 3's, each with 6-15 stamens, calyx deeply 5-6-parted; pistillate flowers 1-3 in a cup of scales, these borne at the base of staminate flowers or on short separate spikes; fruit of 1-3 nuts borne in a spiny bur.

Chrysolepis chrysophylla (Dougl. ex Hook.) Hejlmq.
Giant golden chinkapin

Shrubs or trees, sometimes reaching a height of 45 m., round-topped; bark thick, furrowed; twigs densely covered with yellowish hairs; leaves oblong to lanceolate, the blades 5-15 cm. long, tapering at both ends, dark green above and golden-brown to pale yellow beneath, with a covering of minute scales; burs spiny, containing 1 or sometimes 2 nuts.

[=*Castanopsis chrysophylla* (Dougl. ex Hook.) DC.]

Foothills of the Coast Range and the Cascade Mountains.

Quercus

Evergreen or deciduous (ours) trees or shrubs with hard wood; flowers inconspicuous, greenish; staminate spikes catkin-like, slender, pendant, stamens 2-12; pistillate flowers each surrounded by a scaly cup-like structure, styles 3-6; fruit an acorn.

Lobes of leaf obtuse, not bristle-tipped .... *Q. garryana*
Lobes of leaf acute to acuminate, bristle-tipped .... *Q. kelloggii*

Quercus garryana Dougl. ex Hook.
Oregon white oak or Garry oak

Trees 8-20 m. tall, with thin, whitish to light gray deeply checked bark; leaves 5-13 cm. long (occasionally longer), more or less obovate, moderately to deeply pinnately 5-14-lobed, leathery, dark green and usually shiny above, pale or rusty beneath with fine pubescence; acorn cup shallow, of overlapping finely pubescent scales, nut smooth and shining, ovoid to rounded.

Common; hillsides and open areas, also in mixed forests.

Quercus kelloggii Newb. California black oak

Trees to 25 m. tall, with dark brown to black bark, checked on old trunks; leaves sharply cut into about 7-11 lobes, these again generally lobed, each lobe ending in a bristle, upper surface shining green, lower paler, both surfaces sometimes whitened at first by minute hairs; cup of acorn deep, nut 2-3.5 cm. long, minutely pubescent at first.

Foothills and mixed forests in the mountains; Lane County, Oregon and south.

URTICACEAE
Nettle Family

Annual or perennial herbs (ours) with simple alternate or opposite leaves; flowers small, greenish, bisexual or more often unisexual, in axillary spikes, panicles or cymes; perianth segments 2-5, nearly separate, or united; stamens 4-5; ovary superior, 1-loculed; style and stigma 1; fruit an achene, free or enclosed by the calyx.

Urtica

Annual or perennial herbs with stinging hairs, and opposite petioled leaves; stamens and pistils in separate flowers on the same plant or on separate plants; staminate flowers with 4 perianth segments attached under a rudiment of a pistil; pistillate flowers with 4 perianth segments in 2 pairs, the inner larger than outer and enclosing the achene in fruit.

Annuals .... *U. urens*
Rhizomatous perennials .... *U. dioica*

Urtica dioica L. Stinging nettle

Monoecious or dioecious perennials from creeping rhizomes; stems 5-30 dm. tall; foliage from nearly glabrous to profusely grayish-hairy and with stinging hairs; leaves 6-20 cm. long, 2-12 cm. broad, broadly to narrowly ovate, heart-shaped at base, coarsely toothed; inflorescence slender and clustered in the leaf axils.

Two subspecies occur in our area: subsp. *dioica* is dioecious and subsp. *gracilis* (Ait.) Seland. which is much more common and is monoecious.

Mostly in shady moist places.

Urtica urens L. Dog nettle

Monoecious annuals from a slender taproot; stems 1-6 dm. tall; leaf blades elliptic to ovate, 1-4 cm. long, coarsely and deeply toothed, pubescent with stinging hairs; flower clusters shorter than the leaves.

Weedy, but only occasionally introduced in our area.

Urtica dioica

SANTALACEAE
Sandalwood Family

Herbs, shrubs or trees parasitic on the roots of other plants; flowers generally greenish, sepals 4-5-lobed, petals lacking; ovary superior to inferior, 1-loculed; fruit nut-like or drupaceous.

Comandra

Perennials from creeping rhizomes; leaves alternate, entire; flowers small, in cymes, terminal or in the upper leaf axils; calyx 5-lobed; ovary inferior; fruit drupaceous, 1-seeded, tipped by the persistent calyx lobes.

Comandra umbellata (L.) Nutt. Bastard toad-flax

Stems clustered, woody at base, from extensive slender rhizomes, 1-4 dm. tall, erect, or somewhat trailing, with erect branches; leaves simple, entire, mucronate, oblong to lanceolate or elliptic, short-petioled or nearly sessile, green or more often glaucous; flowers 3-6 mm. long, in terminal compound cymes, green with whitish to purplish tips; fruit globose, capped with the persistent calyx lobes.

A partial parasite, the roots attaching themselves to those of other plants. Dry, often rocky, ground.

VISCACEAE
Mistletoe Family

Plants evergreen; hemiparasites on trees or shrubs; leaves opposite, conspicuous or reduced to scales; stamens and pistils in separate flowers on separate plants (ours), flowers small and greenish, perianth segments 2-7; ovary inferior, 1-loculed; fruit a berry with a mucilaginous substance enclosing the seed(s). Formerly included in the **Loranthaceae**.

Leaves scale-like; berry stalked .... **Arceuthobium**
Leaves not scale-like (ours); berry sessile .... **Phoradendron**

Arceuthobium

Parasitic plants, yellow or brownish, with fragile stems and leaves reduced to paired scales; flowers solitary or clustered; staminate and pistillate flowers on separate plants; perianth of staminate flowers generally 3-lobed, 1 sessile anther on each lobe; perianth of pistillate flowers 2-toothed, persistent; berry fleshy, eventually breaking away at the base.

Since most species of **Arceuthobium** in our region appear generally to be confined to a single genus of Gymnosperms as hosts, they are listed below in this relationship, without detailed description:

Arceuthobium abietinum (Engelm.) Hawksw. & Wiens (Fir dwarf mistletoe)—on true firs (*Abies*)

Arceuthobium americanum Nutt. ex Engelm.(Lodgepole pine dwarf mistletoe)—parasitic on Lodgepole pine (*Pinus contorta*); and sometimes on Knobcone pine (*Pinus attenuata*).

Arceuthobium campylopodum Engelm. (Western dwarf mistletoe)—parasitic on Western yellow pine (*Pinus ponderosa*) and sometimes on Knobcone pine (*Pinus attenuata*).

Arceuthobium douglasii Engelm. (Douglas-fir dwarf mistletoe)—on Douglas-fir (*Pseudotsuga menziesii*).

Arceuthobium tsugensis (Rosend.) G. Jones (Hemlock dwarf mistletoe)—on Hemlock (*Tsuga*).

The two species, ***A. americanum*** and ***A. campylopodum***, which sometimes occur on ***Pinus attenuata***, may be distinguished by the time of flowering: ***A. americanum*** flowers in the spring and ***A. campylopodum*** in late summer.

Phoradendron

Stems freely branched, woody at least at the base; leaves yellowish-green, leathery, the blade expanded or small and scale-like; flowers borne in short spikes in the axils of the leaves; perianth of the staminate flowers 3-lobed, anthers borne at base of lobes; perianth of the pistillate flowers 3-toothed, persistent at the summit of the ovary; berry semi-transparent.

Phoradendron villosum

Phoradendron villosum (Nutt.) Nutt.
Common or Oak mistletoe

Plants woody at the base, sometimes becoming 1 m. or more in height and in diameter, borne on the branches of the host; bark corrugated; young stems and leaves densely covered with short hairs forming a greenish velvety layer, but becoming more or less glabrate; leaves and stems at first yellowish green, becoming darker; leaves oblong to obovate, thick, 3-5 veins apparent on under surface, obtuse at apex, tapering to a short petiole; flowers in short spikes in leaf axils; berries pearl-like, slightly longer than broad, tipped by the 3 perianth lobes, gelatinous within.

The gelatinous substance in the berry causes the seeds to adhere to branches of trees where they have been carried by birds which eat the outer parts, thus insuring their planting on new hosts.

[=*Phoradendron flavescens* (Pursh) Nutt. var. *villosum* (Nutt.) Engelm.]

Common on *Quercus* (Oak) in our area.

ARISTOLOCHIACEAE
Dutchman's-Pipe Family

Herbs or twining shrubs with simple alternate leaves; sepals usually 3 (1-6), petal-like, partly united; petals reduced to scales or absent; ovary inferior or superior, 1-6-loculed; fruit a capsule.

Asarum

Perennial herbs with fragrant scaly rhizomes and long-petioled leaves, these basal or alternate but in an apparent pair at tip, the solitary flower borne between them on or near the surface of the ground; sepals 3; stamens 12; ovary inferior, 6-loculed; fruit a fleshy capsule.

Asarum caudatum Lindl. Wild ginger

Rhizomes horizontal and with 1 to several alternate broad clasping scale-like leaves, and near the tip 2 foliage leaves of nearly equal length, petioles 3.5-18 cm. long, pubescent, leaf blades broadly heart-shaped, palmately veined, deep green, somewhat thick, puberulent, particularly on the veins; flowers borne singly between the 2 leaves, apparently in their axils; the 3 calyx lobes spreading, greenish to very dark purple, extended into long tails at the apex, the tube 12 mm. or more wide, yellowish and purple striped within; stamens 12, more or less united around the pistil.

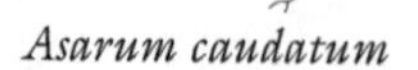

Asarum caudatum

The fragrant rhizomes creep readily through the leaf mold, sending up the foliage leaves at intervals. The flowers are often hidden by the debris of the forest floor and by the leaves.

In shady moist woods.

POLYGONACEAE
Buckwheat Family

Herbs or shrubs with simple leaves, stems often swollen at the nodes and usually with scarious sheathing stipules; flowers small, bearing both stamens and pistils or only stamens or pistils; perianth undifferentiated, 3-9 lobed; stamens 2-9; ovary superior, 1-loculed, 1-seeded; styles or stigmas 2 or 3; fruit an achene, 3-angled or lens-shaped.

1a Stipules absent; flowers borne in involucres............................ **Eriogonum**
1b Stipules scarious and sheathing the stem; flowers not in involucres
  2a Stigmas 2; leaves broadly reniform; perianth segments 4, nearly separate .... **Oxyria**
  2b Stigmas usually 3, sometimes 2 or 4, if 2, then the leaves not broadly reniform
    3a Perianth segments usually 6, 1 or more often bearing a grain-like callosity; stamens 6... **Rumex**
    3b Perianth segments usually 5, without callosities; stamens 3-8 ....**Polygonum**

Eriogonum

Annual or perennial herbs, or sometimes shrubs; leaves basal, alternate, or whorled; flowers small, borne with bractlets in toothed or lobed involucres, these often arranged in large showy inflorescences; perianth usually petal-like, of 6 divisions; stamens 9; styles 3; fruit an achene, usually 3-angled.

Other species than those listed below, ascend the Cascades from the east side, and may on occasion be found in our limits.

1a Perianth not stalk-like at the base
  2a Not mat-forming; perianth segments similar ....*E. nudum*
  2b Mat-forming; perianth segments usually slightly dissimilar ....*E. ovalifolium*
1b Perianth narrowed into a stalk-like extension at base (this short in *E. pyrolaefolium*)
  3a Involucres with short erect teeth.............................. *E. pyrolaefolium*
  3b Involucres with reflexed lobes
    4a Leaf blades 3-25 cm. long, truncate to cordate to cordate-sagittate at the base.. *E. compositum*
    4b Leaf blades 0.5-4 cm. long, tapered at the base........ *E. umbellatum*

Eriogonum compositum **Dougl.** — Northern buckwheat

Stems stout, 1.5-7 dm. tall, glabrous or nearly so; leaves all basal, long-petioled, the blades 3-25 cm. long, ovate, obtuse or acute at apex, truncate to cordate or cordate-sagittate at the base, the under surface felt-like, upper surface green or somewhat woolly; bracts conspicuous, slender or oblanceolate; umbel 5-10-rayed, the rays bearing a head-like umbel of several involucres, or of secondary rays each bearing a head; involucres woolly; perianth long stalk-like at base, white or yellow; achenes with some pubescence at the tip.

In gravelly soil; uncommon in our limits.

Eriogonum nudum Dougl. Barestem buckwheat

Perennial herbs 1-20 dm. tall; stems stout, hollow, often 3-branched at each node into a large spreading cymose inflorescence; leaves mostly in a basal rosette, sometimes a few borne on the stem, oblong or broadly ovate, obtuse or broadly acute at apex, obtuse or cordate at base, wavy-margined, woolly beneath, usually becoming glabrate above, often purplish, long-petioled; involucre tubular, 5-veined, fringed, the bractlets thread-like, very woolly; flowers white or yellowish or slightly rose-tinged, borne in groups with in clustered or solitary involucres; perianth spreading, not stalk-like at base, of 6 divisions, the outer sometimes smaller, the inner occasionally slightly pubescent on the outside, smooth or pubescent within; anthers often pinkish, filaments arising from a tuft of hairs; styles long and slender; achenes glabrous.

This is a very variable species, a number of more or less distinct varieties being recognized.

Sandy or gravelly soil from near the coast to the high Cascades.

Eriogonum ovalifolium Nutt. Cushion buckwheat

Mat-forming perennials, 1.5-3.5 dm. in height and up to 0.5-4 dm. in diameter; leaves basal, petioles long and slender, leaf blades elliptic to nearly orbicular, densely white tomentose, at least on the under surface; flowering stems leafless; involucres in a head-like cluster; perianth white to yellow or purple, the segments nearly free, not stalk-like at the base; achenes glabrous.

This is a variable species with several varieties which appear to intergrade.

Dry areas, often at high elevations in the mountains.

Eriogonum pyrolaefolium Hook. var. *coryphaeum* Torr. & Gray

Alpine buckwheat

Flowering stems several, 5-20 cm. tall, from a tufted base; leaves basal, the petioles as long as or shorter than the blades, the blades broadly to narrowly ovate, 1-4 cm. long, more or less felt-like beneath (ours) to nearly glabrous, green and slightly hairy above, cuneate, rounded, or nearly cordate at base; bracts 2, linear or oblanceolate; umbels simple, loose with few rays or densely head-like; involucres pubescent; perianth with a short stalk-like base, pubescent on the veins; flowers cream- to rose-colored; achenes pubescent at the tip.

In the mountains at 1,600 m. and above.

Eriogonum umbellatum Torr. Sulfur flower

Perennials from a woody taproot with nearly prostrate mat-forming branches and erect flowering stems 0.5-4 dm. tall; leaves clustered on the lower stem, variable is size and shape, the blades mostly 0.5-4 cm. long, long-petioled, lower surface usually gray-tomentose, upper surface nearly glabrous or woolly; inflorescence simple to compound umbels, the primary divisions subtended by a whorl of leaf-like bracts; involucres woolly to crisp-pubescent; perianth (in ours) deep yellow, fading to orange, long stalk-like at the base; achenes glabrous.

Extremely variable species with numerous named varieties.

In dry, open, often rocky areas, more common east of the Cascade crest.

Oxyria

A monotypic genus.

Oxyria digyna (L.) Hill — Mountain sorrel

Alpine perennial herbs; stems several, more or less tufted, 1-5 dm. tall; leaves mostly basal, leaf blades 1-5 cm. wide, broadly reniform, somewhat fleshy, petioles 2-10 cm. long, stipules fused and loosely sheathing the stems; flowers small, greenish, borne in a panicle; perianth segments 4, nearly separate, the outer 2 spreading, the inner 2 erect; stamens 6; stigmas 2; fruit a thin compressed and conspicuously winged achene.

On rocky slopes in the mountains.

Polygonum

Annual or perennial herbs to vines or shrubs with entire alternate leaves and fused scarious stipules forming sheaths around the stem above the node; pedicels jointed; flowers red, white or greenish; perianth usually 5-cleft or parted; stamens 3-8; fruit a flattened or 3-4-angled achene partially to completely enclosed by the persistent perianth.

1a Stem twining, vine-like; annuals; leaves cordate or sagittate..*P. convolvulus*
1b Stem not twining; leaves usually not cordate or sagittate, but if so, then plants perennial
  2a Stems stout, erect, much-branched, 1 m. or more in height
    3a Annuals .... *P. lapathifolium*
    3b Perennials
      4a Plants from a thick caudex; native in the mountains.. *P. phytolaccaefolium*
      4b Plants from extensive creeping rhizomes; escaped from cultivation
        5a The 3 outer perianth segments not keeled or winged; leaf blades 1-2 dm. long, long-acuminate.......................... *P. polystachyum*
        5b The 3 outer perianth segments keeled or winged
          6a Leaf blades 1-1.5 dm. long, abruptly pointed at the tip; stems 1-3 m. tall ....*P. cuspidatum*
          6b Leaf blades 1-3.5 dm. long, acuminate; stems 2-4 m. tall...*P. sachalinense*
  2b Stems, if erect, less than 1 m. in height
    7a Plants growing in water or very wet places
      8a Leaves mostly basal, long-petiolate, cauline leaves much reduced, sessile or nearly so
        9a Inflorescence usually at least 3 times as long as broad; lower flowers often replaced by bulblets............................*P. viviparum*
        9b Inflorescence less than 3 times as long as broad; flowers not replaced by bulblets .... *P. bistortoides*

8b Basal and cauline leaves similar, short-petiolate or sessile
10a Perianth glandular-punctate
11a Achenes dark, shiny.. *P. punctatum*
11b Achenes light brown, dull.................................. *P. hydropiper*
10b Perianth not glandular-punctate, although sometimes glandular-pubescent
12a Perennials
13a Flowers pink to rose-colored; marginal cilia of the stipules 1 mm. long or less .... *P. amphibium*
13b Flowers greenish-white; marginal cilia of the stipules 1.5 mm. or more in length...................................... *P. hydropiperoides*
12b Annuals
14a Stipules 2-lobed or lacerate; leaves about 1 mm. broad .... *P. kelloggii*
14b Stipules not 2-lobed or lacerate; leaves much broader
15a Stipules strigose and bristly margined; leaf blades usually with a dark central spot .... *P. persicaria*
15b Stipules glabrous or nearly so, except for the ciliate margins; leaf blades without a dark central spot .... *P. lapathifolium*
7b Plants growing in dry, or at least not conspicuously wet, places
16a Stems more or less prostrate
17a Plants woody at the base, found only on sandy beaches .... *P. paronychia*
17b Not woody at the base nor confined to sandy beaches
18a Petiole jointed; leaves without a dark central spot.... *P. aviculare*
18b Petiole not jointed; leaves usually with a dark central spot .... *P. persicaria*
16b Stems erect or nearly so
19a Perennials; alpine or subalpine
20a Flowers in conspicuous panicles............... *P. phytolaccaefolium*
20b Flowers in small axillary clusters........................... *P. newberryi*
19b Annuals
21a Pedicels becoming sharply recurved, the flowers (at least the older ones) reflexed
22a Perianth segments 2-2.5 mm. long.................. *P. cascadense*
22b Perianth segments 2.5-4 mm. long..................... *P. douglasii*
21b Pedicels erect
23a Stipules not lacerate on the margins, although sometimes bristly; nodes usually conspicuously swollen
24a Stipules strigose and bristly margined; leaf blades usually with a dark central spot .... *P. persicaria*
24b Stipules glabrous or nearly so, except for the ciliate margins; leaf blades without a dark central spot .... *P. lapathifolium*
23b Stipules lacerate on the margins; nodes not conspicuously swollen
25a Perianth segments 3-4.5 mm. long *P. spergulariaeforme*
25b Perianth less than 3 mm. long.

26a Fertile stamens only 3; leaves linear, about 1 mm. wide........... *P. kelloggii*
26b Fertile stamens 5-8; leaves broader
 27a Perianth segments free almost to the base and with a pink or green midvein.. *P. cascadense*
 27b Perianth segments fused for at least 1/3 their length .... *P. minimum*

Polygonum amphibium L. Water smartweed

Aquatic to amphibious rhizomatous perennials, rooting at the nodes, glabrous, or pubescent when young, the stems arising from the mud of ponds, the leafy ends floating on the surface of the water; leaves elliptic to lanceolate, the petioles from short to long, blades thick, shining, obtuse or nearly acute at both base and apex, often asymmetrical at the base; sheathing stipules cylindrical, often with a few bristles at upper edge; inflorescence of 1-2 terminal spike-like panicles; flowers pink or deep rose-colored; style about as long as the ovary, 2-cleft; achenes lenticular, 2.5-3 mm. long, dark brown, smooth and shining or granular and dull.

An extremely variable species with named varieties.

Ditches, swampy areas, stream banks and lake shores.

Polygonum aviculare L. Prostrate knotweed

Annuals with long branching, wiry, mostly prostrate stems; leaves green, oblong, narrow, acute at both ends, 6-12.5 mm. long; stipules lacerate; flowers small, white or pinkish with green veins, borne in the leaf axils almost along the full length of the stem; stamens 8; styles 3, short.

There are several species often segregated from this taxon, but for our purposes it seems best to treat it as one extremely variable species.

A common weed in gardens but more often found in hard beaten paths and yards.

Polygonum aviculare

Polygonum bistortoides Pursh American or Western bistort

Perennials from a short thick rhizome; stems 1-7.5 dm. tall, bearing dense terminal spike-like racemes; basal leaves 1.5-3 dm. long, 1/2 this length being petiole, leaf blades oblong to lanceolate to oblanceolate, acute at apex, abruptly narrowed at base, 1-3.5 cm. broad, smooth above, more or less puberulent beneath, cauline leaves reduced in size and sessile; stipules brown, 2-6 cm. long; inflorescence 1-5 cm. long with numerous brown papery bracts; flowers pure white to pinkish with the anthers and base of perianth greenish; stamens 8, exserted; styles 3; achene 3-sided, 3-4.5 mm. long, light brown and shiny.

Common in wet meadows, swampy areas, marshes, stream banks and alpine slopes, but also occasionally found in marshes in the Willamette Valley.

Polygonum convolvulus

Polygonum cascadense W. H. Baker Cascades Polygonum

Annuals; stems wiry, angled, 5-15 cm. tall; leaves oblanceolate to obovate, 0.5-2 cm. long, narrowed to a jointed petiole-like base, margins often revolute; stipules lacerate; flowers borne in the leaf axils, the pedicels sometimes reflexed; perianth segments 2-2.5 mm. long, nearly distinct, each segment with a pink to green midvein and a white-margin; stamens 8; achenes 3-angled, black and shiny.

Dry rocky areas in the Cascades.

Polygonum convolvulus L. Wild buckwheat or Black bindweed

Plant twining or prostrate, sometimes covering an area of 0.5-3 m. in diameter; stems wiry, smooth or puberulent; leaf blades broadly arrow-shaped, with rounded or pointed lobes at the base, acute or acuminate at apex, 2-6.5 cm. long, the petioles as long as or shorter than the blades; stipules short and inconspicuous; flowers in long slender terminal and axillary racemes; perianth 5-parted to about the middle, green to pinkish; achene 3-angled, somewhat depressed between the angles, black, minutely roughened.

An introduced weed often found in gardens.

Polygonum cuspidatum Sieb. & Zucc. Japanese knotweed

Dioecious rhizomatous perennials; stems erect, much-branched, 1-3 m. in height; stipules short, entire, often early deciduous; leaves petioled, the blades 10-15 cm. long, broadly ovate, tip abruptly acuminate, base cuneate to truncate or subcordate; inflorescence an open panicle; perianth segments greenish-white to white, the three outer segments keeled; achenes 3-angled, dark, smooth and shiny.

Escaped from cultivation and established in waste areas.

Polygonum cuspidatum

Polygonum douglasii Greene Douglas' knotweed

Annuals, 2-5 dm. tall, somewhat branching, the branches erect, slender; leaves linear or narrowly lanceolate, 2.5-5 cm. long; sheaths papery, becoming shreddy at apex; flowers reddish or white, borne in axils of scattered reduced leaves along slender branches, the pedicels soon turned downward, the perianth segments with green midveins; achenes black.

Dry or rocky ground.

Polygonum hydropiper L.

Usually annuals, but sometimes persisting for a second year; stems bright green, or somewhat reddish, glabrous or nearly so, simple or branched; leaves ovate, oblong or lanceolate, 2-10 cm. long, acute or acuminate at each end, minutely ciliate-margined; stipules slightly pubescent, fringed; inflorescence of 1-several terminal, often drooping, spike-like racemes; flowers greenish to greenish-white or pink margined in bud, on short erect pedicels; perianth segments 2.5-4.5 mm. long, gland-dotted; stamens 4 or 6; style short, 3-parted; achenes about 3 mm. long, lens-shaped or somewhat 3-angled, finely pitted, dull brown.

A weedy species of wet places, often in ditches and swampy areas.

Polygonum hydropiperoides Michx.
Waterpepper

Perennials from slender rhizomes and often rooting at the nodes; leaves slightly reduced upwards, the lower short-petiolate and the upper nearly sessile, leaf blades lanceolate, 5-14 cm. long, the apex acute or acuminate, the base acute; stipules 1-2 cm. long, strigose and the margins bristly; inflorescence of 2-several spike-like panicles or racemes; perianth segments 5, 2.5-3 mm. long, greenish to white or rarely pale pink; stamens 8, achenes 3-angled, dark brown to black, smooth, shiny.

Polygonum hydropiperoides

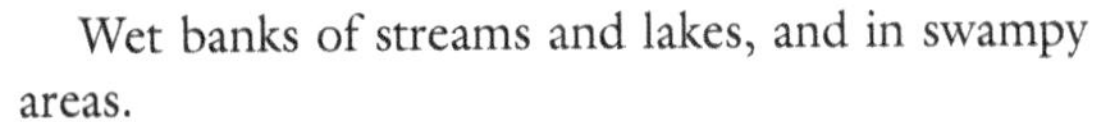

Wet banks of streams and lakes, and in swampy areas.

Polygonum kelloggii Greene — Kellogg's knotweed

Annuals with stems 2.5-20 cm. tall, often much-branched; leaves linear, about 0.5-2.5 cm. long, about 1 mm. broad; stipules papery, 2-lobed or often shreddy at apex; flowers nearly sessile in somewhat dense thick clusters at the ends of the branches, also some flowers usually borne in the lower leaf axils; floral bracts linear or short lanceolate, 1-veined or sometimes appearing 3-veined by the rolling back of the margins in some bracts, the margins often whitish; perianth segments white to pinkish with a green midvein, 1.5-2.5 mm. long; anther bearing stamens 3, the filament bases dilated, often with 2-5 sterile filaments; style short, 3-cleft; achene 3-angled, broad at base, yellowish to dark brown, smooth to striate.

This species is sometimes treated as *Polygonum polygaloides* Meiss. subsp. *kelloggii* (Greene) J. Hickman.

Found in the Willamette Valley, also occurring in the mountains, but more common east of our area.

Polygonum lapathifolium L. — Dock-leaved smartweed or Willowweed

Annuals, more or less erect, 3-15 dm. tall, often freely branched, the nodes conspicuously swollen; leaves pale green, lanceolate, acuminate, nearly glabrous to densely pubescent; stipules fragile, brown, 1-2 cm. long, the upper margins ciliate and usually glandular; inflorescence branches spike-like, usually nodding; perianth segments 4 or 5, the outer prominently 3-veined, pink to white (rarely greenish); stamens 6, not exserted; styles 2; achenes dark, flattened, with at least one side concave.

This is a highly variable species.

A weed of moist waste places, especially in sandy or gravelly ground; sometimes introduced into gardens in river loam. Native of Europe.

Polygonum minimum Wats. Broadleaf knotweed

Slender leafy annuals; stems wiry, 4-30 cm. tall, erect or spreading, simple or somewhat branched; leaves elliptic to nearly orbicular, 6-12.5 mm. long, generally obtuse and apiculate; flowers axillary, inconspicuous; achene 3-angled, smooth, shining.

Moist places in the mountains.

Polygonum newberryi Small Newberry's fleeceflower

Perennials; stems fleshy, stout, glabrous or pubescent, simple or somewhat branched; leaf blades 1-4.5 cm. long, dull green, ovate or oblong, often broadly so, obtuse or acute at apex, broad at the base but lower leaves generally decurrent down the petiole, glabrous or the under surface minutely velvety; stipules funnel-shaped, mostly entire, brown, brittle; flowers greenish, in axillary racemes or spikes; stamens 8, included; style 3-parted; achenes 3-angled, brown, shining.

In the mountains in rocky ground.

Polygonum paronychia C. & S. Black knotweed

Stems nearly prostrate, more or less woody, generally much-branched at the base, 3-9 dm. long, from woody rhizomes, leafy above, roughened below by the remains of old stipules; leaves stiff, glabrous, somewhat shining, elliptic, 2.5 cm. or less long, the margins rolled back; stipules papery, brown below, silvery at tips, at first lanceolate, becoming lacerate; flowers white or pink, borne in dense leafy spikes at ends of branches, the perianth segments with green branching midveins; achenes black and shiny.

On sandy ocean beaches.

Polygonum persicaria L. Smartweed or Lady's thumb

Annuals with stems nearly erect to more or less prostrate at base and often rooting at the nodes; leaves long, narrow, widest near middle, long-pointed at apex, tapering at base to very short petioles, blades usually with a central dark-purplish spot; sheathing stipules strongly veined, strigose and bristly margined; spikes short, erect; flowers pink to red or rarely white; stamens generally 6, shorter than the perianth; achenes flat to 3-angled, 2-2.5 mm. long, brown to black, shiny.

Weedy in moist places, often in marshy areas or along streams.

Polygonum phytolaccaefolium Meisn. ex Small Alpine knotweed

Perennials; stems 6-20 dm. tall, branching; leaves lanceolate to ovate, 3-20 cm. long, 1-4 cm. wide, acuminate, glabrous, or somewhat pubescent on margins and veins; flowers small, greenish to white, generally in conspicuous terminal panicles; stamens 8, some often sterile; achenes sharply 3-angled, brown, smooth, shiny.

In moist meadows or drier sites at 1,600 m. and above in the Cascades.

Polygonum polystachyum Meisn. Himalayan knotweed

Rhizomatous perennials; stems stout, erect, 1.5-2 m. in height; lower leaves petioled, becoming subsessile upwards, leaf blades 1-2 dm. long, usually densely soft-pubescent at least on the lower surface, ovate-lanceolate, long acuminate, the base truncate to cordate or with 2 small lobes; inflorescence an open many-flowered panicle; perianth segments white.

Escaped from cultivation and established in moist areas.

Polygonum punctatum Elliott Dotted water smartweed

Annuals or perennials; stems decumbent to erect, sometimes rooting at the nodes, 3-10 dm. tall, glabrous or nearly so; leaf blades lanceolate, cuneate at the base and acuminate at the tip, short-petiolate, 3-15 cm. long and 5-20 mm. wide, glandular-punctate; stipules 9-17 mm. long, glabrous to strigose, with marginal bristles; perianth 2.5-3.5 mm. long, the segments 5, strongly glandular-punctate, greenish, usually white- or rarely pink-margined; stamens 6-8; styles 2-3; achene 2-3 mm. long, flat or 3-angled, dark brown to black, shiny.

An extremely variable species.

Moist areas, ditches, marshes and ponds; not common in our area.

Polygonum sachalinense Maxim. Giant knotweed

Rhizomatous perennials, often forming dense clumps; stems stout, erect, 2-4 m. in height; leaves petioled, leaf blades 1-3.5 dm. long, ovate, cordate at the base, acuminate at the tip; inflorescence in open panicles borne in the axils of the upper leaves; flowers greenish-white to white, outer perianth segments keeled to winged; achenes 3-angled, about 3 mm. long, black, smooth and shiny.

Escaped from cultivation and established in waste areas.

Polygonum spergulariaeforme Meisn. Fall or Spurry knotweed

Annuals; stems more or less branched, erect or spreading, 1.5-3 dm. tall; leaves narrowly oblanceolate or linear, 1-4 cm. long, the margins generally rolled back; sheaths papery, shreddy at apex; flowers pinkish or white, clustered in the upper axils, often forming somewhat dense racemes or spikes, the perianth segments with green midveins; stamens 8, 4 of the filaments cylindrical, the alternating 4 dilated at the base; achenes 3-angled, 3-4 mm. long, black, shiny.

In sandy to gravelly soil; from the coast to lower elevations in the mountains.

Polygonum viviparum L. Alpine bistort

Perennials from a bulb-like caudex; stems solitary or sometimes several, erect, 1-4 dm. tall; basal leaves long-petioled, the blades oblong to lanceolate, 2.5-10 cm. long, 8-25 mm. broad, acute at apex, more or less cordate at base or sometimes cuneate, cauline leaves somewhat smaller, the lowermost short-petioled, the upper sessile, margins mostly rolled back; sheathing stipules cylindrical, 1.2-6.5 cm. long, somewhat lacerate at the broader top; flowers pinkish or white, in a terminal spike-like raceme, the lower flowers usually replaced by conical bulblets; stamens 8, exserted; style deeply 3-parted, exserted; achenes 3-4 mm. long, dark brown.

Wet meadows, stream banks and alpine slopes.

Rumex

Annuals, biennials or perennials, some slightly woody at base, with basal and alternate cauline leaves; stems grooved, often reddish-brown-tinged; stipules sheathing but often early deciduous; perianth segments 6 or rarely 4, united only at base, the 3 outer spreading, the 3 inner larger and later enclosing the achene, often bearing callosities on the back; stamens 6; styles 3; achenes 3-angled.

1a Staminate and pistillate flowers on separate plants; leaves arrowhead-shaped .... *R. acetosella*
1b Stamens and pistils in the same flower; leaves not arrow-head shaped
  2a Annuals or biennials; leaves linear to narrowly lanceolate; mostly in wet, saline areas .... *R. maritimus*
  2b Perennials
    3a Inner perianth segments with teeth or bristles over 1 mm. long
      4a Basal leaf blades narrowly lanceolate; panicle branches spreading .... *R. pulcher*
      4b Basal leaf blades broadly ovate to ovate-oblong; panicle branches ascending to erect .... *R. obtusifolius*
    3b Inner perianth segments entire or nearly so
      5a Leaves mostly cauline; stems branching from most nodes, often prostate to ascending .... *R. salicifolius*
      5b Leaves basal as well as cauline; stems essentially unbranched below the inflorescence, erect
        6a Inner perianth segments without conspicuous callosities .... *R. occidentalis*
        6b At least 1 of the inner perianth segments with a conspicuous callosity
          7a Margins of all the leaves strongly crisped; inflorescence of nearly contiguous whorls of flowers .... *R. crispus*
          7b Margins of the leaves not all crisped, none strongly so; flowers in distinct whorls .... *R. conglomeratus*

Rumex acetosella L. — Red sorrel or Sour dock

Stems from creeping rhizomes; lower leaves somewhat arrowhead-shaped with 2 lobes at the base, upper leaves smaller, narrower, without lobes; stamens and pistils borne on different plants; staminate panicle broad, yellowish, bearing an abundance of pollen; pistillate panicle narrower, red, the pistillate flowers somewhat smaller than the staminate; achenes red, becoming dark brown at maturity, 3-angled, granular.

Introduced from Europe; common. Prefers acid soil. The leaves are very acid, like rhubarb, which also belongs to this family.

Rumex conglomeratus Murr. Clustered or Green dock

Perennials from a stout taproot; stems ridged, leafy, slender, often much-branched, 5-15 dm. tall; leaves oblong to lance-ovate, tapered or truncate to cordate at the base, up to 3 dm. long and up to 1 dm. wide, margins of the lower leaves more or less crisp or wavy, the upper leaves reduced and cuneate at the base; flowering branches very slender, generally leafy at least below; flowers in distinct whorls, short-pediceled, the pedicels jointed near the middle and mostly bent; inner perianth segments 2-3 mm. long, margins usually entire, each generally bearing a conspicuous light-colored, smooth callosity.

Introduced from Europe; rather common in low moist areas.

Rumex crispus L. Curly dock

Perennials from a stout taproot; stems stout, 4.5-15 dm. tall; basal leaves 1-5 dm. long, up to 5 cm. broad, oblong-lanceolate, long-petiolate, the margins strongly-crisped, cuneate to rounded at the base, upper leaves reduced but also crisp-margined; inflorescence of nearly contiguous whorls of flowers; perianth segments enclosing the achene 4-5 mm. long, with entire or denticulate margins, usually each inner segment, or rarely, only 1 or 2, bearing an oblong callosity.

Common weed; often abundant about deserted buildings, yards and in disturbed sites.

Rumex maritimus L. Golden or Seaside dock

Annuals or biennials, 0.5-10 dm. tall; leaves mainly cauline, the blades linear to narrowly lanceolate, 4-15 cm. long, the margins more or less crisped, leaves reduced upwards; inflorescence leafy, flowers in dense whorls; inner perianth segments acuminate tipped and with 2-several slender teeth or bristles on the margins, each inner segment with a dorsal callosity; achenes 1-2 mm. long.

(=*Rumex fueginus* Phil.)

Mostly in wet, saline areas.

Rumex acetosella

Rumex conglomeratus

Rumex crispus

Rumex obtusifolius L. Broad-leaved or Bitter dock

Stout perennials, 6-15 dm. tall; stems scabrid, slender; lower leaves broadly ovate to ovate-oblong, sometimes slightly wavy-margined, heart-shaped or truncate at base, reaching 7 dm. in length, leaves gradually reduced upward on the stem, those near the top tapering at both ends; whorls of flowers not crowded; 3 inner perianth segments each bearing 4-10 teeth near base, large callosity 1, sometimes with 2 small ones.

Weedy, rather common in moist areas.

Rumex occidentalis Wats. Western dock

Stout perennials, little branched, 5-20 dm. tall; leaf blades thick, long-ovate, mostly heart-shaped at base, 1-3.5 dm. long, the petioles of the basal leaves longer than the blades; flowers in a narrow dense panicle, becoming pinkish in fruit; inner perianth segments strongly-veined and without teeth, and usually without callosities.

Wet meadows and swampy areas and other moist places.

Rumex pulcher L. Fiddle dock

Perennials; stems usually borne singly, simple or branched, 4-9 dm. tall; leaves lanceolate, often narrowed below the middle, the basal leaves cordate to cuneate at the base, up to 12 cm. long, cauline leaves few and much-reduced; flowers borne in distinct whorls on stout recurved pedicels; perianth segments greenish-brown, the inner segments sharply 2-5-toothed, 1 or more bearing a warty callosity; achenes smooth.

Usually in disturbed areas; introduced from the Mediterranean region.

Rumex salicifolius Weinm. Willow or Narrow-leaved Dock

Prostrate to ascending perennials from a stout taproot; stems branching from most nodes, 2-10 dm. long; leaves mainly cauline and not greatly reduced upwards, but more crowded below, the lower short-petiolate, the upper subsessile, leaf blades 3-15 cm. long, linear to oblong-elliptic, margins flat to undulate, but not crisped; inflorescence various, with the whorls of flowers nearly contiguous to distinct; inner perianth segments 2-4.5 mm. long with 1 to all 3 bearing callosities.

An extremely variable species with numerous named subspecies and varieties which appear to intergrade.

Moist areas from coastal sand dunes to montane meadows.

CHENOPODIACEAE
Goosefoot Family

Herbs or shrubs, often fleshy or white-mealy, with alternate (rarely opposite) leaves, these sometimes much-reduced; flowers very small, generally inconspicuous, with either stamens or pistils or both; sepals separate or slightly united at base or absent in some pistillate flowers; petals absent; ovary superior, 1-loculed, 1-seeded; styles 1-3; fruit usually a utricle.

1a Stems jointed and succulent; the leaves reduced and scale-like; salt marsh plants .... **Salicornia**
1b Stems not jointed and succulent
  2a Leaves cylindrical or awl-shaped, inflorescence bracts spine-tipped.......... **Salsola**
  2b Leaves not cylindrical or awl-shaped; inflorescence bracts not spine-tipped
    3a Flowers imperfect; pistillate flowers without a calyx, but with 2 bracts .... **Atriplex**
    3b Flowers perfect, the calyx usually 5-parted .... **Chenopodium**

Atriplex

Herbs or shrubs, generally covered with mealy scales; leaves alternate or opposite; stamens and pistils in separate flowers, on the same or separate plants; staminate flowers with a 3-5-parted calyx with 3-5 stamens; pistillate flowers usually merely a naked pistil between 2 leaf-like bracts, these bracts free or united, much enlarged in fruit; styles 2.

Atriplex patula L. — Beach salt-bush

Annuals; stems erect or decumbent, 2-10 dm. tall; leaves fleshy, opposite below and alternate above, lanceolate or hastate, slightly scurfy; flowers in terminal spikes or panicles; staminate flowers with a 5-parted calyx; fruiting bracts lanceolate to triangular, usually toothed and crested on the back.

There are several named varieties, but they tend to intergrade.

In saline soil, especially in salt marshes along the coast.

Chenopodium

Annual or perennial herbs with alternate petioled leaves, often fleshy, mealy or scaly or glandular-pubescent; flowers inconspicuous, greenish, bearing both stamens and pistils, clustered in simple or compound axillary or terminal spikes; sepals generally 5, united or not, usually persistent and enclosing the fruit; stamens usually 5; styles generally 2; fruit a utricle.

1a Herbage glandular and strongly aromatic
  2a Flowers sessile, bearing sessile glands .... *C. ambrosioides*
  2b Flowers short-pedicelled, the glands stalked.......... *C. botrys*
1b Herbage not glandular nor aromatic
  3a Fruiting calyx fleshy, often red, usually 3-parted; on saline soil; not common in our range.......... *C. rubrum*
  3b Fruiting calyx dry, usually 5-parted
    4a Plants generally conspicuously white-mealy; common.......... *C. album*
    4b Plants generally dark green at least on the upper surface of the leaves; not common is our range .... *C. murale*

Chenopodium album L. — Lamb's quarters

Erect branching annuals, 2-12 dm. tall, more or less white mealy; leaves fleshy, ovate, wedge-shaped at base, acute at apex, with 3 conspicuous veins at base, wavy or toothed-margined to irregularly lobed, or upper entire; flowers small, in narrow spikes borne in panicles in the upper leaf axils and terminating the stem, the leaves gradually decreasing in size upward, becoming bracts of the inflorescence; calyx usually becoming keeled and enclosing the fruit.

A common weed of gardens and waste areas.

Chenopodium ambrosioides L. — Mexican tea

Glandular and strongly aromatic; stems erect 2-12 dm. tall; herbage covered with numerous sessile yellow glands, also more or less pubescent; lower leaf blades lanceolate to ovate, toothed to pinnately lobed, upper leaves often nearly entire; flowers sessile and more or less glomerate in bracteate axillary spikes, these sometimes branched; calyx enclosing the fruit.

In disturbed sites, often along stream banks, but also in drier sites.

Chenopodium botrys L. — Jerusalem oak

Strong-smelling annuals, 1.5-6 dm. tall, densely glandular-pubescent; leaf blades oblong to elliptic or ovate, shallowly to deeply lobed or pinnatifid, reduced upwards and sometimes becoming entire or merely wavy margined; flowers numerous in curved racemes or cymes; sepals fused only at the base.

Weedy species, in disturbed areas and along stream banks.

Chenopodium murale L. — Nettle-leaved goosefoot

Annuals; stems more or less erect, loosely branched, 2-7 dm. tall, leafy; leaves thin, somewhat rhombic-ovate, acute at apex, tapered or truncate at base, coarsely and unevenly toothed, dark green on the upper surface, slightly white powdery to mealy beneath; inflorescence of terminal and axillary panicles or sometimes cymes; calyx lobes scarcely keeled, more or less enclosing the fruit.

Chenopodium album

An occasional weed in waste places; introduced from Europe.

Chenopodium rubrum L. — Red goosefoot

Decumbent to nearly erect, much-branched annuals, 1-8 dm. tall, usually glabrous except sometimes in the inflorescence; leaves only gradually reduced upwards, leaf blades deltoid to ovate, often somewhat hastate, cuneate at the base, the margins subentire to toothed or shallowly lobed, sometimes more or less powdery on the under surface; flowers sessile and glomerate in simple or compound spikes; calyx usually 3- or rarely 5-lobed or parted.

(=*Chenopodium humile* Hook.)

On moist saline soil.

Salicornia

Low succulent plants of salt marshes, with jointed stems and opposite branches, the leaves reduced and scale-like, sessile, the pair more or less fused at the base; flowers perfect or unisexual in spikes formed from the thickened upper joints, the flowers generally 3 per axil and nearly embedded; stamens 1-2; styles 2; fruit flattened.

Annuals; central flower borne well above the lateral ones .... *S. europea*
Rhizomatous perennials; flowers borne at more or less the same level .... *S. virginica*

Salicornia europea L. European glasswort
Spreading to ascending or erect annuals, 0.5-5 dm. tall, much-branched, at least above; flowers in fleshy spikes 1-7 cm. long; flowers usually borne in threes with the central flower borne well above the lateral ones.

Salt marshes and alkaline flats; not common.

Salicornia virginica L. Woody glasswort or Pickle-weed
Much-branched rhizomatous perennials; the prostrate jointed stems up to 1 m. in length, often rooting at the nodes, the flowering stems erect or merely ascending, 5-30 cm. tall, stems often tinged with purple or red; spikes 1-5 cm. long, slightly more slender than the stem; flowers borne at more or less the same level.

Salt marshes and alkaline flats along the coast.

Salsola

Herbs or sometimes shrubby, with somewhat fleshy cylindrical or awl-shaped leaves; flowers axillary, perfect, each flower subtended by a stiff spine-like bract and 2 bractlets; calyx 4-5-parted, horizontally winged on the back; stamens generally 5, styles 2; fruit enclosed by the persistent calyx.

Salsola tragus L. Russian thistle or Tumbleweed
Annuals, usually much-branched and rounded in outline; stems striate, purplish; early lower leaves linear, succulent, spine-tipped, the later upper leaves, which become the bracts of the inflorescence, stiff, awl-shaped, spine-like, each flanked by 2 spine-like bractlets; calyx forming a winged envelope over the mature fruit.

(*Salsola kali* L. var. *tenuifolia* Taush. misapplied)

Weedy; only occasionally found within our limits; much more common east of the Cascades.

AMARANTHACEAE
Amaranth Family

Coarse herbs or, in tropical areas, trees; leaves alternate or opposite, simple; flowers small, generally arranged in spikes, cymes or panicles, each flower subtended by 1-3 membranous or scarious bracts, these often sharp-pointed; flowers perfect or imperfect, lacking petals, sepals 0-5, often scarious; stamens 1-5; ovary superior, styles 1-5; fruit usually a circumscissile capsule, but sometimes drupaceous or a utricle.

Amaranthus

Monoecious or dioecious annuals with alternate entire leaves and small greenish or purplish flowers in spikes; sepals 3-5 or rarely 1, papery; stamens

2-5; ovary superior, 1-loculed; stigmas 2-3; fruit 1-seeded, circumscissile or splitting irregularly at maturity.

1a Flowers in small axillary clusters only; sepals 3 .... *A. albus*
1b Flowers in dense terminal and axillary spikes; sepals 4-5
  2a Plants nearly glabrous to puberulent.................................... *A. powellii*
  2b Plants villous, at least above, with multicellular hairs....... *A. retroflexus*

Amaranthus albus L. — Tumbleweed

Monoecious; stems erect or ascending, much-branched from the base and bushy, up to 7 dm. in height; leaf blades 1-6 cm. long, obovate to elliptic, abruptly narrowed to a slender petiole; flowers in axillary clusters with spiny scarious-margined bracts; staminate sepals 3, stamens 2-3; pistillate sepals 3.

In disturbed sites.

Amaranthus powellii Wats. — Powell's Amaranth

Monoecious; stems erect, up to 2 m. in height; leaves deltoid-elliptic to lanceolate; flowers in a terminal panicle of spikes and axillary clusters, bracts spine-tipped; staminate flowers with 4-5 sepals and 3-5 stamens; pistillate flowers with 5 sepals and 3 styles.

A common weed in our area.

Amaranthus retroflexus L. — Red-root pigweed

Monoecious or dioecious; stems erect, more or less branched from the base, pubescent, somewhat ridged, both stems and roots often reddish; leaf blades lanceolate to rhombic, 2-10 cm. long, long-petiolate, pubescent on the under surface, at least on the veins; staminate flowers with 4-5 sepals, 2-4 mm. long and 4-5 stamens; pistillate flowers with 5 sepals, 2.5-4 mm. long, these become reflexed, styles 3.

A common garden weed, also in wet fields, road side ditches and other disturbed sites.

AIZOACEAE
Carpet Weed Family

Annual or perennial herbs or subshrubs; herbage often succulent; leaves simple, alternate, opposite or whorled; flowers solitary or cymose, usually perfect; sepals 3-9; petals lacking or many in several whorls; stamens 1-many, the outer often petaloid; ovary superior or inferior; fruit a capsule, berry or nut.

Mollugo

Annuals or perennials; stems prostrate to ascending; leaves opposite, alternate or whorled; flowers small; petals lacking; sepals 5, often petaloid; ovary superior, styles 3; fruit a loculicidal capsule with numerous seeds.

Some authors put this genus in a separate family, the **Molluginaceae.**

Mollugo verticillata L. Carpet weed

Prostrate mat-forming annuals, much-branched from the base; leaves in whorls of 3-6, oblanceolate to spatulate, 5-30 mm. long, 1-6 mm. wide, entire, short-petiolate; flowers small, borne in the leaf axils; sepals 1.5-3 mm. long, white on the inner surface, prominently 3-veined; stamens 3-5; stigmas 3; capsule 3-4 mm. long.

Moist, often disturbed areas and river banks.

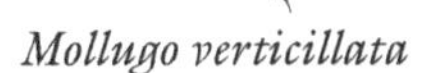

Mollugo verticillata

NYCTAGINACEAE
Four O'clock Family

Somewhat succulent herbs (ours) to shrubs or trees with simple, opposite, entire leaves and somewhat swollen nodes; flowers regular, perfect, in axillary or terminal clusters, these often subtended by bracts which form an involucre; calyx corolla-like, 4-5-lobed, the upper part falling after flowering, leaving the persistent base enclosing the ovary; petals none; stamens various in number; ovary superior, 1-loculed, 1-ovuled; fruit an achene closely enclosed in the generally hardened calyx base (ours).

Abronia

Sticky-glandular annual or perennial herbs; leaves opposite, thick, petioled, one leaf of each pair generally larger than the other; flowers showy in peduncled axillary or terminal heads or umbels, these subtended by an involucre of 4 to 6 bracts; calyx 4-5-lobed, tubular, spreading saucer-like above, usually petaloid; stamens shorter than the calyx tube, attached unequally within; style shorter than the calyx; persistent base of calyx usually 2-5-winged, enclosing the achene.

Flowers yellow .... *A. latifolia*
Flowers pink .... *A. umbellata*

Abronia latifolia Esch.
Yellow sand-verbena

Abronia latifolia

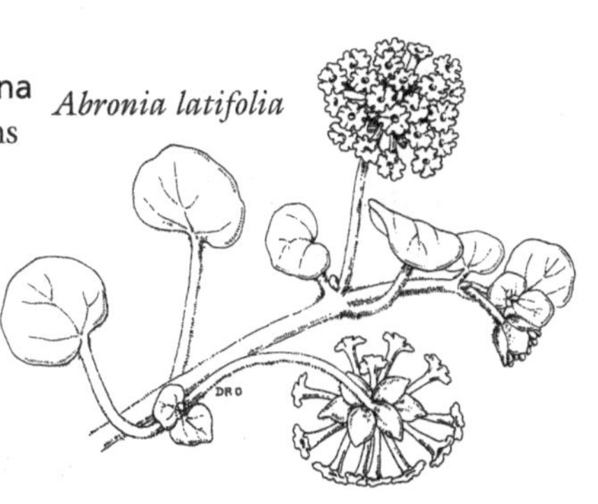

Perennials from a thick deep root; stems 3-10 dm. long, prostrate, sticky glandular-pubescent; leaves thick, broadly ovate to nearly reniform, abruptly narrowed to the petiole, the blades 1.5-4 cm. long; peduncles longer than the leaves; flowers yellow, fragrant, in showy heads subtended by 5 broadly ovate bracts; fruit generally with five thickened wings which are not as broad as the body. The sticky stem and foliage are generally encrusted with sand.

Along ocean beaches.

Abronia umbellata Lam. subsp. *breviflora* (Standl.) Munz
Pink sand-verbena

Habit of plant like *A. latifolia*; leaves oblong or ovate, thin, but fleshy; involucral bracts lanceolate; flowers pink; wings of fruit thin, at least as broad as the body.

Along ocean beaches; not abundant in our limits.

PORTULACACEAE
Purslane Family

Annuals or perennials with simple succulent leaves; sepals usually 2 (ours), generally separate (rarely as many as 8 in *Lewisia*); petals commonly 5, separate or slightly united at base, soon withering; stamens few to many, opposite the petals when of the same number; ovary usually superior (inferior in *Portulaca*), 1-loculed, with free-central placentation; fruit a capsule, opening by valves or circumscissile.

1a Flowers borne in dense curved clusters crowded on a short flowering stem; petals 4 .... **Calyptridium**
1b Flowers not so arranged; petals usually 5 or more
  2a Capsules circumscissile
    3a Ovary superior; circumscissile dehiscence near the base of the capsule; sepals free; plants montane .... **Lewisia**
    3b Ovary inferior; the circumscissile dehiscence near the middle of the capsule; sepals fused at the base .... **Portulaca**
  2b Capsules longitudinally dehiscent
    4a Inflorescence leafy.......... **Calandrinia**
    4b Inflorescence ebracteate or with small bracts
      5a Cauline leaves 1 pair, these often fused around the stem .... **Claytonia**
      5b Cauline leaves more than 2, not fused.......... **Montia**

Calandrinia

Small somewhat succulent annuals or perennials with alternate leaves and flowers in a leafy or bracteate raceme or panicle; sepals 2; petals usually 5; stamens 3-15; style 3-branched; capsule 3-valved; seeds numerous.

Calandrinia ciliata

Calandrinia ciliata (R. & P.) DC. — Red maids

Annuals; stems several to many from the base, erect or decumbent, 0.5-3.5 dm. long, glabrous or ciliate on the angles; leaves fleshy, linear to oblanceolate, the basal long-petioled, the upper cauline leaves sessile, often ciliate, sepals broad at the base, tapering to an acute apex, sometimes ciliate; petals red to magenta; capsule enclosed by the persistent sepals.

Occasional in cultivated fields and orchards.

Calyptridium

Fleshy herbs with the leaves tufted at the crown of a fleshy root or at the ends of short branches; flowers borne in dense curved clusters crowded on a short flowering stem; sepals 2; petals 2-4, somewhat unequal; stamens 1-3; capsule 2-valved, 1-many-seeded, the seeds black, shiny.

Calyptridium umbellatum (Torr.) Greene — Pussypaws

Low perennials, the crown generally branched, each branch short, bearing a cluster of leaves; leaves thick, spatulate, 1.5-7 cm. long; peduncles generally spreading, 1-10 cm. long; flowers white to rose color, in dense clusters; sepals 3-8 mm. long, papery; petals 4, about equal to the sepals; stamens 3.

(=*Spraguea umbellata* Torr.)

Dry soil in the Cascades.

Claytonia

Succulent annual or perennial herbs; basal leaves 1-several (or rarely absent); cauline leaves generally 2, opposite or nearly so, free or fused into a disk surrounding the stem; flowers often showy, white to deep rose, borne in 1-sided racemes; sepals 2; petals usually 5; stamens 5, opposite and basally fused to the petals; style 1, stigmas 3; capsule 3-valved; seeds few.

1a Perennials
- 2a Base corm-like .... *C. lanceolata*
- 2b Rhizomatous
  - 3a Rhizomes and stolons short; petals pale pink with darker pink veins .... *C. sibirica*
  - 3b Rhizomes long, often much-branched; petals white .... *C. cordifolia*

1b Annuals
- 4a Cauline and basal leaves linear .... *C. exigua*
- 4b Cauline and basal leaves not both linear
  - 5a Cauline leaves not fused, cauline leaves lanceolate to ovate .... *C. sibirica*
  - 5b Cauline leaves fused at least along one margin, often on both sides
    - 6a Basal leaf blades elliptic to deltoid, less than 2 cm. long, smaller towards the center of the rosette, cauline leaves often unequally fused .... *C. rubra*
    - 6b Basal leaf blades rhombic-obovate to linear-spatulate, nearly equal throughout the rosette, cauline leaves usually equally fused .... *C. perfoliata*

Claytonia cordifolia Wats.

Rhizomatous perennials; stems 1-4 dm. tall; basal leaves several, long-petioled, the blades 1-9 cm. long, broadly ovate and subcordate, cauline leaves 1 pair, 1-4 cm. long, sessile, broadly ovate; raceme loose, several-flowered, without bracts; flowers white, the petals nearly 12 mm. long, somewhat notched; capsule 4-5 mm. long.

[=*Montia cordifolia* (Wats.) Pax and Hoffm.]

Along wooded streams, wet meadows and swampy areas in the mountains.

Claytonia perfoliata

Claytonia exigua Torr. & Gray subsp. *exigua*

Fleshy glaucous annuals; stems 2.5-15 cm. long, erect or spreading; both basal and cauline leaves linear, the cauline pair united by bases around the stem on one side or nearly free, generally nearly as long as the raceme; sepals about 1/2 the length of the white or pink petals.

[=*Montia spathulata* (Dougl.) Howell]

In a variety of habitats but not common.

Claytonia lanceolata Pursh Narrow-leaved spring beauty

Perennials from a large (1-2 cm.) corm-like base; stems one to several, 3-8 cm. tall; basal leaves succulent, oblanceolate or obovate, narrowed to long petioles, cauline leaves 2, lanceolate or narrowly ovate; flowers often in one-sided racemes; petals (in ours) white or pale pink with deeper rose markings.

Dry slopes in the mountains, often closely following the melting of winter snow.

Claytonia perfoliata Donn ex Willd. Miner's lettuce

Annuals from a slender taproot; stems 1-40 cm. tall; herbage sometimes reddish; basal leaves from linear-spatulate to rhombic-obovate, cauline leaves 2, opposite, fused at least on one side, more often on both sides, forming a disk; sepals 1.5-5.5 mm. long; petals 5, 2-6 mm. long, white or pale pink; capsule 1.5-4 mm. long.

This is an extremely variable species with numerous named varieties and subspecies which tend to intergrade. When *C. perfoliata* hybridizes with *C. sibirica* it may produce a fertile offspring called *Claytonia washingtoniana* (Suksd.) Suksd.

[=*Montia perfoliata* (Donn) Howell]

Moist, shady woods, also sometimes more or less weedy.

Claytonia rubra (Howell) Tidestrom

Annuals; stems 1-15 cm. tall; leaves 1-8 cm. long, the blades elliptic to deltoid, less than 2 cm. long, cauline leaves 2, fused at least on one side, more or less round with two squarish corners; sepals 1.5-3 mm. long; petals 2-3.5 mm. long, white to pale pink; capsule 2-3 mm. long.

In moist areas, often in sandy soil or in disturbed habitats.

Claytonia sibirica L. Candy flower

Fleshy annuals, or more often, perennial from short rhizomes and/or stolons; stems 1-6 dm. tall; basal leaf blades lanceolate to deltoid, 1-8 cm. long, 3-40 mm. broad, long-petiolate, cauline leaves 2, opposite, sessile, 1-8 cm. long; racemes many-flowered, bracteate; sepals 2.5-6 mm. long; petals 5, 6-12 mm. long, usually pale pink with darker pink veins; stamens 3-5; capsule 2.5-3.5 mm. long.

Claytonia sibirica

When *C. sibirica* hybridizes with *C. perfoliata* it may produce a fertile offspring called *Claytonia washingtoniana* (Suksd.) Suksd.

[=*Montia sibirica* (L.) Howell]

Common in moist woods, especially along stream banks.

Lewisia

Low succulent perennials; leaves usually in a basal rosette (cauline in *L. triphylla*); flowering stems 1 to several; flowers showy; sepals 2-9; petals 4-18; stamens 3 to many; stigmas 2-8; capsule circumscissile near the base; seeds 2 to many, shining.

1a Cauline leaves 2-5, basal leaves lacking; plants from a deep-seated corm .... *L. triphylla*

1b Leaves mostly basal; plants not from a corm

2a Flowering stems with a many-flowered panicle .... *L. columbiana*

2b Flowers solitary or rarely 2 per stem .... *L. pygmaea*

Lewisia columbiana (Howell ex Gray) Robins. var. *columbiana*
Columbia Lewisia

Roots fleshy, from the base of a thickened caudex, the clustered leaves growing from its apex; leaves narrow, oblanceolate, somewhat fleshy, 2.5-5 cm. long; flowering stems erect, stout, 1-3 dm. tall; bracts small, glandular-margined; flowers rose-colored, in spreading panicles; sepals 2, very small, glandular-dentate; petals 7-11; stamens 5-6.

In the Cascade and Olympic Mountains.

Lewisia columbiana (Howell ex Gray) Robins. var. *rupicola* (Eng.) C. L. Hitchc.

Succulent perennials; basal leaves many, linear-lanceolate to spatulate, the apex obtuse; scapes short, sometimes many-flowered, the bracts glandular-toothed; sepals 2; petals usually pink, veined with red; stamens 5-6.

From Mt. Rainier to the Olympics and Saddle Mountain, Clatsop County, Oregon; now popular in cultivation.

Lewisia pygmaea (Gray) Robin. Dwarf Lewisia

Perennials from a turnip-shaped root; basal leaves linear to oblanceolate, 2-15 cm. long, 1-6 mm. wide; flowering stems usually several, 1-10 cm. long and with a pair of connate bracts at or below the middle, flowers usually solitary; sepals 2; petals 5-9, greenish white to pink or red, sometimes striped; stamens 4-12.

Open, often gravelly or rocky areas and along streams in the mountains.

Lewisia triphylla (Wats.) Robins. Three-leaved Lewisia

Dwarf perennials from a subterranean corm; cauline leaves 2-5, 1-6 cm. long, narrowly linear; flowers few to many; sepals 2; petals 5-9, white or pink; stamens 3-5.

High mountains, often following the melting of snow.

Montia

Glabrous succulent annuals or perennials; cauline leaves more than 2, alternate or opposite, entire; inflorescence a 1-sided raceme; petals 3-5; stamens 2-5; ovary 1-loculed with basal placentation, style 1, stigmas 3; capsules 3-valved; seeds 1-3.

1a Perennials
  2a Leaves alternate .... *M. parvifolia*
  2b Leaves opposite .... *M. chamissoi*
1b Annuals
  3a Leaves opposite .... *M. fontana*
  3b Leaves alternate
    4a Leaves broadly ovate, petioled .... *M. diffusa*
    4b Leaves more narrow
      5a Matted and usually rooting at the nodes .... *M. howellii*
      5b Not matted and if rooting at the lower nodes, the stems nearly erect
        6a Sepals 1.5-2.5 mm. long; lower leaves 5-40 mm. long .... *M. dichotoma*
        6b Sepals 2.5-7 mm. long; lower leaves 1-10 cm. long .... *M. linearis*

Montia chamissoi (Sprengel) Greene — Water Montia

Rhizomatous perennials with stolons and usually bearing bulblets; stems 2-30 cm. long, often floating or erect to prostate; leaves opposite, oblanceolate to obovate, 0.5-5 cm. long, 3-15 mm. wide; flowers sometimes replaced by bulbils; sepals 1.5-3 mm. long; petals 5-9 mm. long, white to pink; stamens 5; capsule 1-1.5 mm. long.

Wet areas.

Montia dichotoma (Nutt.) Greene — Dwarf Montia

Annuals, 2-8 cm. tall; leaves alternate, linear, 5-40 mm. long, about 0.5 mm. wide; flowers sometimes cleistogamous; sepals 1.5-2.5 mm. long; petals usually 5, about equalling the sepals or slightly longer, white; stamens 3; capsule 1.5-2 mm. long.

Moist areas.

Montia diffusa (Nutt.) Greene — Branching Montia

Annuals; stems much-branched, spreading, 5-15 cm. long; basal leaves long-petioled, the blades broadly ovate, abruptly narrowed to the petiole, cauline leaves several, similar in shape, the petioles becoming shorter above; racemes terminal and lateral, somewhat corymbose; petals white or pink, notched; capsule 2-3 mm. long.

Shady woods; not common.

Montia fontana L. — Water chickweed

Annuals, often rooting at the nodes; stems 2-30 cm. long, often floating or erect to prostate; leaves opposite, linear to oblanceolate, 3-20 mm. long; flowers sometimes cleistogamous; sepals 1-1.5 mm. long; corolla 1-2 mm.

long, the petals united below; stamens 3-5; capsule 1-2 mm. long; seeds minute, black, spiny (as seen under a lens).

A highly variable species with several varieties or subspecies distinguishable only by seed characters.

Ditches, ponds, streams, vernal pools and other wet areas.

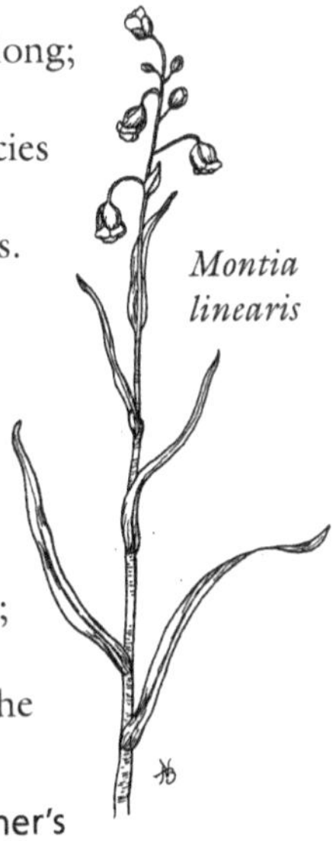

Montia linearis

Montia howellii Wats. Howell's Montia

Mat-forming annuals, usually rooting at the nodes; stems 2-9 cm. long, spreading, leaves alternate, linear-spatulate or narrowly oblanceolate, 5-20 mm. long, 0.5-2 mm. wide; flowers in clusters, opposite the leaves in the axil of a scarious bract, often cleistogamous; sepals 1-1.5 mm. long, petals sometimes lacking, if present, 2-5, white, 1-1.5 mm. long; stamens 2-3; capsule approximately 1 mm. long, 3-valved.

Vernal pools and other moist, often gravelly areas, in the early spring.

Montia linearis (Dougl.) Greene Narrow-leaved miner's lettuce

Simple or branched erect annuals; stems 2.5-20 cm. tall; leaves alternate, linear, clasping at the base, 1-10 cm. long and 1-3 mm. wide; racemes usually 1-sided, terminal or axillary; the flowers mostly nodding in bud and again in fruit; sepals 2.5-7 mm. long; petals 5, 4-7 mm. long, white to pale pink; stamens 3; capsule 3-4 mm. long; seeds lens-shaped, black, shining, smooth or very minutely roughened.

Wet or moist ground, often in vernally wet fields.

Montia parvifolia (DC.) Greene Small-leaved Montia

Succulent perennials usually producing stolons, often forming large patches; stems 5-40 cm. tall, usually bearing axillary bulblets; leaves alternate, the basal leaves 1-6 cm. long, the blades obovate to nearly orbicular, 2-20 mm. wide, cauline leaves and leaves of the stolons similar but much-reduced; sepals 2-3.5 mm. long; petals 5, 7-15 mm. long, pale pink or white with pink veins; stamens 5; capsule 2-3 mm. long, 3-valved; seeds black, smooth or minutely warty.

The var. *flagellaris* (Bong.) C.L. Hitchc. occurs along the coast and can be distinguished from var. *parvifolia* by it broader basal leaves (over 5 mm. wide) and larger flowers (petals 10-15 mm. long).

Moist banks and mossy rocks along streams.

Portulaca

Succulent annuals with alternate leaves; flowers opening only in sunlight, axillary or in terminal clusters; ovary inferior, its upper half deciduous at maturity with the upper portion of the floral tube; sepals 2; petals 4-6, generally 5, yellow (in ours); stamens 4-20; capsule circumscissile; seeds many.

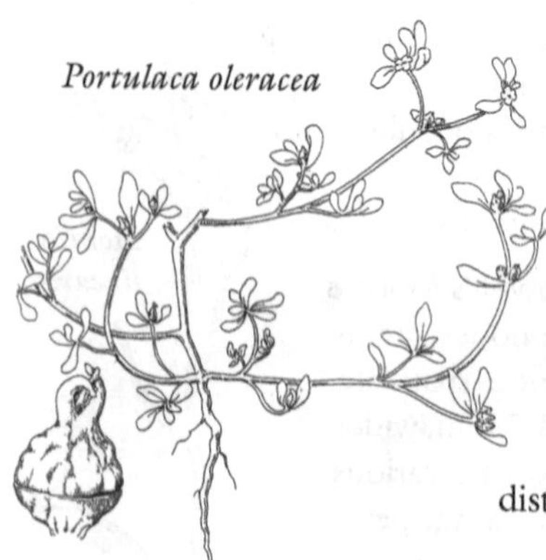

Portulaca oleracea L. Common purslane
Glabrous succulent annuals; stems prostrate to spreading, several from the base, branching, 0.5-3 dm. long; leaves oblanceolate to obovate, alternate, but by internodal shortening of the branch tips, appearing more or less whorled beneath the terminal flower clusters; sepals 3-4 mm. long, slightly longer than the yellow petals; stamens 6-10.

A common garden weed, also found in other disturbed areas; introduced from Europe.

CARYOPHYLLACEAE
Pink Family

Herbs with opposite leaves at swollen nodes; flowers regular; sepals and petals usually 5, sometimes 4, or the petals rarely absent; stamens 1-10, alternate with petals when of the same number; ovary superior, typically 1-loculed, at least above; styles 1-5; placentation usually free-central or basal; fruit a capsule, opening at the apex by teeth or valves, or rarely an indehiscent utricle.

1a Succulent seashore perennial herbs; flowers with a conspicuous gland-disk.. .... **Honkenya**
1b Not as above in all points
  2a Fruit a 1-seeded utricle; calyx persistent in fruit enclosing the utricle
    3a Sepals united only at the base, spine-tipped; stipules present........ **Cardionema**
    3b Sepals united about 1/2 their length, not spine-tipped; stipules absent .... **Scleranthus**
  2b Fruit a several-seeded capsule
    4a Sepals united into a conspicuous tube
      5a Styles 2; teeth or valves of the capsule typically 4
        6a Petals with 2 appendages between the claw and the blade .... **Saponaria**
        6b Petals without appendages .... **Dianthus**
      5b Styles 3-5; teeth or valves of the capsule 3-10
        7a Calyx lobes longer than the tube; petals without appendages ... . **Agrostemma**
        7b Calyx lobes shorter than the tube; petals usually with appendages at the junction of the claw and the blade
          8a Styles 5; plants densely gray-silky; petals red; flowers perfect (ours) .... **Lychnis**
          8b Styles usually 3, rarely 4-5, if 5, then the species dioecious.. .... **Silene**
    4b Sepals free or nearly so
      9a Leaves with conspicuous scarious stipules
        10a Styles 5; capsule dehiscent by 5 valves........ **Spergula**

10b Styles 3; capsule dehiscent by 3 valves. .... **Spergularia**
9b Leaves without stipules
11a Petals deeply 2-parted, often nearly to the base
12a Capsule cylindrical, opening by 10 teeth; styles 5 .... **Cerastium**
12b Capsule ovoid, with 6, 8 or 10 teeth or valves; styles usually 3 .... **Stellaria**
11b Petals entire to merely notched, but not deeply 2-parted (sometimes lacking)
13a Capsule opening by 4 or 5 valves .... **Sagina**
13b Capsule opening by 3, 6 or 8 valves or teeth
14a Petals 4; capsule teeth 8 .... **Moenchia**
14b Petals 5 (sometimes lacking); capsule valves and/or teeth 3 or 6
15a Capsule valves and teeth 3 .... **Minuartia**
15b Capsule valves and/or teeth 6
16a Plants rhizomatous .... **Moehringia**
16b Plants not rhizomatous .... **Arenaria**

Agrostemma

Annuals or biennials; flowers large; calyx tube strongly 10-ribbed, 5-lobed, the lobes much elongated, exceeding the corolla; petals 5, long-clawed; styles 5; capsule 5-valved.

Agrostemma githago L. Corn cockle

Stems erect, 1 m. or less tall, usually with a few ascending branches, pubescence silky; leaves 5-14 cm. long, linear-lanceolate, sessile, connected at base around stem by a thin membrane, silky pubescent, especially on margins; flowers terminal, calyx-tube 10-ribbed, lobes slender, 2-4 cm. long; petals broad, purplish-red, showy; stamens exserted.

Occasional weed in fields and disturbed areas; introduced from Europe.

Agrostemma githago

Arenaria

Annuals or perennials; flowers white, in cymes or axillary; sepals 5, nearly free; petals 5 entire or notched, or lacking; stamens usually 10; styles 3; capsule 6-toothed.

Perennials; leaves linear .... *A. capillaris*
Annuals; leaves ovate to lanceolate .... *A. serpyllifolia*

Arenaria capillaris Poir. var. *americana* (Maguire) R. J. Davis
Mountain sandwort

Mat-forming perennials, usually glandular, the slender erect flowering stems arising 7.5-30 cm. high from branching decumbent bases; leaves linear, 1-6 cm. long, tufted near the base, several pairs borne on the erect stems, these reduced upward to bracts; flowers white, pediceled, compact or in spreading cymes; sepals scarious-margined; petals longer than sepals; capsule equal to or slightly longer than calyx.

(=*Arenaria formosa* Fisch.)
Mountains.

Arenaria serpyllifolia L. Thyme-leaved sandwort

Pubescent annuals; stems much-branched, spreading, 5-30 cm. long; leaves small, 3-9 mm. long, ovate to lanceolate, pointed at apex, nearly sessile, scarcely reduced upward; flowers axillary or in leafy cymes, pediceled; sepals acute, pubescent, silvery-margined, about 4 mm. long, several- veined; capsule about equaling calyx.

Naturalized from Europe; often more or less weedy.

Cardionema

Low perennials with tufted stems and short narrow spine-tipped leaves and papery stipules; flowers small, clustered in the leaf axils; sepals 5, spiny-tipped; petals very small or absent; ovary 1-loculed, 1-seeded, becoming a utricle enclosed within the persistent calyx.

Cardionema ramosissimum (Weinm.) Nels. & McBr. Sandmat

Plants forming dense mats, the stems 5-30 cm. long; leaves numerous, linear with a needle-like tip, the papery stipules giving the plant a silvery appearance.

(=*Pentacaena ramosissima* Hook. and Arn.)
Dry sand along the coast.

Cerastium

Pubescent herbs; flowers white, borne in cymes; sepals 5, free; petals 5, notched to deeply 2-lobed or lacking; stamens 5 or 10; styles 5; capsule cylindrical, sometimes curved, opening at tip by 10 teeth to allow escape of seeds.

1a Annuals; stems all erect...*C. glomeratum*
1b Perennials; some stems decumbent or matted, others erect and flowering
  2a Petals about equalling or slightly longer than the sepals ....*C. fontanum*
  2b Petals nearly twice as long as the sepals..................................*C. arvense*

Cerastium arvense L. Field chickweed

Stems erect or matted, decumbent from slender rhizomes; leaves linear to lanceolate, or the lower obovate to oblanceolate, all sessile or nearly so, usually glandular-pubescent or sometimes nearly glabrous; cyme terminal, often compound; sepals narrow, glandular-pubescent, thin and membranous on the margins; petals white, showy, 1-1.5 cm. long, deeply notched, about twice as long as sepals.

Common on open slopes, rocky cliffs and meadows; from the coast to high in the mountains.

Cerastium fontanum Baumg. subsp. *vulgare* (Hartm.) Greuter & Burdet
Mouse-eared chickweed

Perennials but sometimes flowering the first year; stems several to many from base, some stems prostrate, often rooting at the nodes, densely pubescent, usually glandular above, the flowering stems more or less erect; leaves oblanceolate to spatulate, those of the erect stems larger; petals about equalling or slightly longer than the sepals; capsule cylindrical, about twice as long as sepals, somewhat curved, the 10 teeth narrow.

(*Cerastium vulgatum* L. misapplied)

Common weed.

Cerastium glomeratum Thuill.

Annual mouse-eared chickweed

Annuals; stems erect, more or less branching, pubescent, often sticky-glandular, particularly near the top; leaves broadly or narrowly obovate to elliptic, sessile, slightly connected at bases around stem, pubescent; flowers in congested cymes; petals shorter than or nearly equal to the sepals; capsule nearly twice as long as sepals, somewhat curved and narrowed at tip.

(*Cerastium viscosum* L. misapplied)

Common weed.

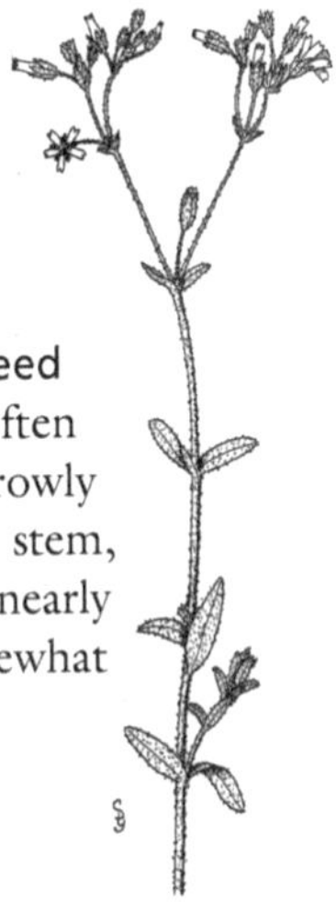

Cerastium fontanum

Dianthus

Erect annual or perennial herbs, generally with narrow leaves and with purple, pink, white, or variously variegated flowers borne in cymes or solitary; calyx tubular, ridged, 5-toothed, bracteate at the base; petals 5, long-clawed, variously toothed; stamens 10; styles 2; capsule elongated, stalked, splitting from 4 apical teeth.

Dianthus armeria L. Deptford pink

Annuals; stems stiffly erect, pubescent, 1.5-6 dm. tall, dark green; basal leaves lanceolate, cauline leaves linear, 2.5-10 cm. long; flowers in simple or compound terminal cymes; petal claws 1.5 cm. long, the expanded blades 4-6 mm. long, pink with small white dots, the apex toothed.

Naturalized from Europe; occasionally found in fields and along roadsides.

Honkenya

A monotypic genus.

Honkenya peploides (L.) Ehrb. Seabeach sandwort

Fleshy seashore perennial herbs with fleshy ovate or obovate clasping leaves and small axillary flowers; flower parts attached to a conspicuous 10-lobed gland-disk; stamens 8-10; styles 3-6; capsule somewhat fleshy.

[=*Ammodenia peploides* (L.) Rupr.]

Along the coast, usually on beaches.

Lychnis

Biennial or perennial herbs; stems erect; leaves basal and cauline; inflorescence a terminal few-flowered cyme; sepals 5, fused, 10-ribbed; petals 5, long-clawed, the blade notched, 2 appendages borne at the junction of the blade and the claw; styles 4-5; capsule opening by 4-5 teeth.

Lychnis coronaria (L.) Desr. Mullein pink

Densely gray-silky perennials; stems branched above; leaves oblanceolate to narrowly elliptic, 5-15 cm. long; flowers showy; petals deep red to reddish-purple, the 2 appendages linear and acute; styles 5.

Native of Europe; widely cultivated and escaped and naturalized in a few areas.

Minuartia

Erect to mat-forming annuals or perennials; leaves linear, awl-shaped or oblong, 1-3-veined; sepals 5, nearly free; petals 5 or lacking; styles 3; capsule 3-toothed.

1a Delicate annuals .... *M. tenella*

1b Perennials

2a Margins of the sepals incurved and hood-like, the tip obtuse........ *M. obtusiloba*

2b Margins of the sepals not incurved, the tip acuminate........ *M. nuttallii*

Minuartia nuttallii (Pax) Briq. Nuttall's sandwort

Perennials; stems leafy, much-branched and often mat-forming, densely glandular-pubescent, producing erect flowering stems 2-20 cm. long; leaves stiff, narrow, 4-15 mm. long; flowers numerous in somewhat spreading cymes; sepals acuminate; petals present; styles 3; capsule shorter than the sepals.

(=*Arenaria nuttallii* Pax)

High mountains.

Minuartia obtusiloba (Rydb.) House Arctic or Siberian sandwort

Low tufted perennials; stems decumbent, sending up erect, glandular-pubescent branches, 2-20 cm. tall from dense rosette-like clusters of leaves; leaves 1.5-9 mm. long, stiff, narrow, glabrous or somewhat pubescent; flowers terminal or axillary, often solitary, about 9 mm. high; sepals obtuse, the margins incurved, glandular-pubescent; petals white, equal to or longer than the sepals; capsule about twice as long as the sepals.

[=*Arenaria obtusiloba* (Rydb.) Fern.]

Alpine and subalpine meadows and talus slopes.

Minuartia tenella (Nutt.) Mattf. Slender sandwort

Annuals; stems slender, nearly simple to dichotomously branched, minutely glandular-pubescent above; leaves 6-12.5 mm. long, the flattened bases connate; flowers long-pediceled, white; sepals broad, obtuse, glandular, silvery-margined, 3-veined, shorter than the petals.

(=*Arenaria tenella* Nutt.)

Dry open ground.

Moehringia

Rhizomatous perennials; leaves lanceolate to elliptic; flowers solitary or 2-5, terminal or axillary; sepals 5, nearly free, petals 5, entire; styles 3; capsule with 6 more or less recurved teeth.

Sepals obtuse, 2.5-3 mm. long .... *M. lateriflora*
Sepals acute or acuminate, averaging over 3 mm. long........... *M. macrophylla*

Moehringia lateriflora (L.) Fenzel
Bluntleaf sandwort
Rhizomatous perennials; some of the stems decumbent, usually also with erect flowering stems 5-20 dm. in height; leaves oblong to lanceolate or oblanceolate, 1-4 cm. long, sessile or nearly so; flowers in terminal and lateral few-flowered cymes; sepals obtuse, white-margined, about one-half as long as the petals; petals white.

(=*Arenaria lateriflora* L.)

Moist to dry areas in the mountains.

Moehringia macrophylla

Moehringia macrophylla (Hook.) Fenzl
Bigleaf sandwort
Stems puberulent, erect, often weakly so, from slender rhizomes; leaves 1-5 cm. long, elliptic to lanceolate, tapering at each end, puberulent to glabrous; flowers mostly in few-flowered cymes or solitary; sepals scarious-margined and/or ciliate, acute or acuminate; petals nearly orbicular.

(=*Arenaria macrophylla* Hook.)

Mostly in coniferous woods.

Moenchia

Taprooted annuals; leaves linear to lanceolate; flowers solitary or in few-flowered terminal cymes; sepals usually 4, free; petals 4, entire; stamens 4 or 8; styles usually 4; capsule 8-toothed.

Moenchia erecta (L.) Gaertner, Meyer & Scherb.
Annuals; herbage glabrous and glaucous; basal leaves petioled, oblanceolate, cauline leaves sessile, linear to linear-lanceolate; petals white, lanceolate.

Moist disturbed areas.

Sagina

Low, usually matted annuals or perennials with short slender leaves and pediceled terminal or axillary flowers; sepals 4-5; petals white, usually shorter than the sepals, or absent; stamens 4-10; ovary 1-loculed; capsule splitting to the base by 4-5 valves, many-seeded.

1a Annuals, without a basal rosette of leaves; glandular-pubescent at least on the sepals
  2a Leaf margins ciliate at the base; sepals usually 4, petals absent or minute .... *S. apetala*
  2b Leaf margins not ciliate; sepals usually 5, petals commonly present, about equal to the sepals .... *S. decumbens*
1b Biennials or perennials, usually with a basal rosette of leaves; plants glabrous
  3a Leaves succulent; seashore plants .... *S. maxima*
  3b Leaves not succulent .... *S. procumbens*

Sagina apetala Ard. Dwarf pearlwort

Delicate annuals; stems 2.5-8 cm. tall, minutely glandular; leaves 2-9 mm. long, ciliate at the base; sepals 4 or rarely 5, glandular-pubescent; petals commonly absent, rarely 4, minute.

Moist ground often in disturbed areas.

Sagina decumbens (Elliott) Torr. subsp. *occidentalis* (Wats.) G. Crow Western pearlwort

Annuals; stems slender, 5-15 cm. tall; leaves 4-20 mm. long, or the lower leaves longer, linear, glabrous; sepals usually 5, finely glandular-pubescent; petals commonly present, about equal to the sepals.

(=*Sagina occidentalis* Wats.)

Vernal pools and other moist areas.

Sagina maxima Gray subsp. *crassicaulis* (Wats.) G. Crow Seacoast pearlwort

Glabrous succulent biennials or perennials with a basal rosette of leaves; stems prostrate to ascending; cauline leaves 5-20 mm. long, basal leaves somewhat longer, broadly linear; sepals 5; petals 5; stamens 10.

(=*Sagina crassicaulis* Wats.)

Sandy or rocky areas along the coast.

Sagina procumbens L. Arctic pearlwort

Glabrous biennials or perennials, usually with a basal rosette of leaves and prostrate to ascending stems to 15 cm. in length; leaves linear, 3-20 mm. long; sepals 4-5, obtuse; petals, if present, shorter than the sepals.

More or less weedy.

Saponaria

Erect rhizomatous perennials with opposite entire leaves; flowers numerous, usually showy, in clustered axillary and terminal cymes; sepals 5, fused into a tubular calyx; petals 5, entire or notched, long-clawed, with 2 appendages between the claw and the blade; stamens 10; ovary 1- or incompletely 2- or 4-loculed; styles 2 or rarely 3.

Saponaria officinalis L. Bouncing Bet

Stems stout, branching, often purplish, smooth, 3-9 dm. tall; leaves glabrous, ovate, long-elliptic or lanceolate, the lower 7.5-10 cm. long, with 3 conspicuous longitudinal veins, the upper shorter; flowers pink or whitish; claws of petals longer than the blades, the blades obovate, notched, 1-1.5 cm. long, the 2 linear appendages 1-2 mm. long; stamens conspicuously longer than the calyx tube; styles 2.

An escape from gardens, native of Europe; well established and particularly abundant in the Willamette Valley.

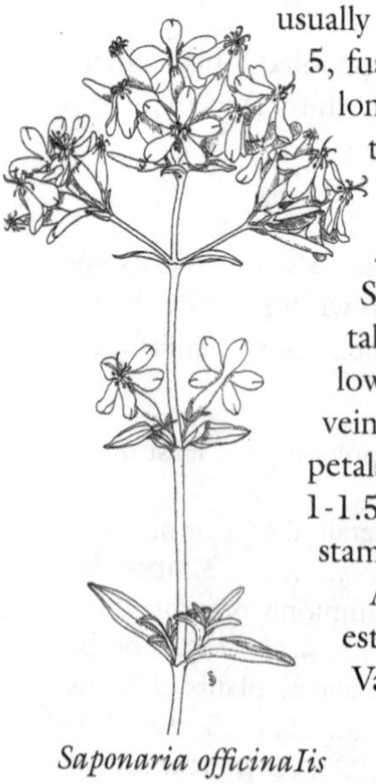

Saponaria officinalis

Scleranthus

Low annuals or perennials; stems dichotomously branched; leaves opposite and the bases more or less fused; flowers small, greenish; calyx 4-5-lobed the tube becoming thickened and enclosing the fruit; petals lacking; stamens 1-10; styles 2-branched; fruit a utricle.

Scleranthus annuus L. Knawel

Annuals; stems prostrate to ascending, up to 15 cm. long; leaves linear, 4-18 mm. long, the tip sharp-pointed, scarious and ciliate at the fused bases; flowers 3-4.5 mm. long; calyx tube 10-veined, abruptly expanded above, becoming thickened and hardened in fruit, the lobes lanceolate to narrowly triangular.

An occasional weed of waste places; introduced from Europe.

Silene

Annual or perennial herbs, often sticky-glandular; sepals united into a 5-toothed cylindrical or bell-shaped calyx; petals 5, long-clawed, usually with appendages at the junction of the blade and the claw; stamens 10, styles usually 3, rarely 4-5; capsule opening by 3, 6 or 10 valves or teeth; seeds numerous.

1a Styles 5; dioecious species
  2a Flowers red or purplish-red, leaves elliptic to ovate.................. *S. dioica*
  2b Flowers white; leaves oblanceolate to lanceolate.................... *S. latifolia*
1b Styles 3-4; flowers often perfect
  3a Annuals
    4a Flowers borne in a 1-sided raceme with leafy bracts........... *S. gallica*
    4b Flowers in cymes
      5a Calyx pubescent, conspicuously veined; flowers few in a simple or compound cyme.. *S. noctiflora*
      5b Calyx glabrous, inconspicuously veined; flowers in panicled cymes .... *S. antirrhina*
  3b Perennials
    6a Calyx glabrous, except sometimes for ciliate lobes
      7a Petals reddish to pink; tufted alpine species less than 7 cm. tall .... *S. acaulis*
      7b Petals white; weedy species, over 15 cm. tall................ *S. vulgaris*
    6b Calyx pubescent
      8a Flowers very showy, pink, 3-4 cm. in diameter; petals usually 4-lobed .... *S. hookeri*
      8b Flowers not as above
        9a Calyx more or less cylindrical in flower
          10a Plants sticky glandular-pubescent above and on the calyx .. .. *S. scouleri*
          10b Plants not sticky glandular............................ *S. douglasii*
        9b Calyx bell-shaped in flower
          11a Calyx conspicuously veined, low tufted alpine species........ .... *S. suksdorfii*
          11b Calyx not conspicuously veined
            12a Flowers nodding; blades of the petals 4-8-lobed........... .... *S. campanulata*
            12b Flowers erect or ascending; blades of the petals 2-4-lobed... *S. menziesii*

Silene acaulis L. — Moss campion

Tufted mat-like perennials 2.5-6 cm. tall or rarely more; leaves 4-15 mm. long, linear, sessile, the bases broader and overlapping; flowers often imperfect, borne singly at ends of short branches; calyx bell-shaped, 10-veined; petals reddish to pink, emarginate or entire; styles 3; capsule 3-loculed.

In the high Cascades.

Silene antirrhina L. — Sleepy catch-fly

Annuals; stems erect, 1.5-7.5 dm. tall, simple or branched, glabrous or slightly pubescent, glandular, at least between the upper nodes; lower leaves oblanceolate, nearly glabrous or with minute somewhat stiff pubescence, the upper leaves narrower; flowers in panicled cymes; calyx tubular, 4-10 mm. long, inconspicuously 10-veined, the ribs green, calyx teeth sometimes purplish, the calyx becoming much enlarged laterally by the development of the capsule; petals pink to white, broad at apex, 2-lobed, usually slightly longer than calyx; styles 3; seeds purplish-black, minutely spiny. The stems which are more or less sticky, are often covered with the fallen seeds.

A variable weedy species.

Silene campanulata Wats. subsp. *glandulosa* C. L. Hitchc. & Maguire

Much-branched perennials; stems 1.5-4 dm. tall, glabrous to puberulent, glandular or not; leaves lanceolate or the upper ovate; flowers axillary and terminal, usually nodding; calyx puberulent, glandular or not, inconspicuously 10-veined; petals 4-8-lobed, white to pink; styles 3.

Wooded areas.

Silene dioica (L.) Clairv. — Red campion

Dioecious biennials or short-lived perennials; stems 0.3-1 m. tall, pubescent; leaves broadly elliptic to ovate, 4-12 cm. long and up to 4 cm. broad, pubescent; calyx often purplish, pubescent; petals red to purplish-red; styles 5; capsule globose.

(= *Lychnis dioica* L.)

Introduced from Europe; established in waste areas and road sides.

Silene douglasii Hook.

Tufted perennials; stems decumbent, sending up erect branches 1-5 dm. or rarely taller, the plant generally pubescent throughout and sometimes glandular; lower leaves oblanceolate or spatulate, the upper narrower, becoming narrowly lanceolate to linear, 1.2-7.5 cm. long, sometimes somewhat curved; flowers few, in a cyme; calyx in flower 15 mm. or less long, somewhat cylindrical, sometimes inflated, often purplish, conspicuously 10-veined; petals white or pink to purplish-tinged, 2-lobed or rarely 4-lobed, noticeably longer than the calyx; styles 3-4.

There are several named varieties, which seem best separated on a geographical basis. Variety ***oraria*** (Peck) C. L. Hitchc. & Maguire. A fleshy leaved variety with the upper cauline leaves reduced only slightly, if at all; found only along the coast. Variety ***rupinae*** Kephart & Sturgeon has narrow cauline leaves and is found in rocky areas in the Columbia River Gorge. Variety ***douglasii*** is found in the mountains.

Silene gallica L. English catch-fly

Annuals, 3-5 dm. tall, erect, more or less branched, the branches slender, pubescent with short white hairs, glandular above; lower leaves oblanceolate or spatulate, mucronate, narrowed at base into a petiole, the upper leaves narrower, 1.5-4 cm. long; flowers sessile or short-pediceled axillary and terminal in a 1-sided raceme-like cyme with leafy bracts; calyx slender in flower, 0.5-1.6 cm. long, 10-veined, veins purple, calyx becoming inflated as the capsule develops; petals white or pinkish, toothed or entire, extending slightly beyond the calyx; styles 3.

Introduced from Europe, especially common along coast.

Silene hookeri Nutt. Indian pink

Slender rhizomes arising from the crown of the perennial root, giving rise to 1 or more weakly erect or decumbent stems 0.5-2.5 dm. long; leaves mostly obovate or oblanceolate, long-tapering at base, acute or acuminate at apex, each pair united around the stem by a thin membrane, gray-woolly to nearly smooth; flowers large, in terminal and axillary few-flowered cymes or solitary bright pink to pale pink, the petals deeply 4-cleft, tl appendages in the throat conspicuous; styles 3.

Common in open woods and on brushy hillsides.

Silene hookeri

Silene latifolia Poir. subsp. *alba* (Mill.) Greuter & Burdet White campion

Dioecious biennials or perennials; stems to 1 m. tall, pubescent and usually glandular above; leaves to 10 cm. long and 2.5 cm. broad, oblanceolate to lanceolate, gradually reduced upwards, the lower petioled, the upper sessile; staminate calyx 10-veined; pistillate calyx about 20-veined and becoming much inflated in fruit; petals white, 2-lobed; styles 5.

(=*Lynchis alba* Mill.)

Weedy in fields and along road sides; introduced from Europe.

Silene menziesii Hook.

Low mat-forming perennials; stems 0.5-3 dm. tall, rarely taller, weakly ascending, more or less branching, glandular at least above, pubescent; leaves variable, elliptic to lanceolate or oblanceolate, 2.5-6 cm. long, 3-20 mm. wide, sessile or nearly so; flowers in leaf axils, or the ends of branches forming leafy cymes; calyx bell-shaped, about 6 mm. long, pubescent, inconspicuously 10-veined; petals white, 2-4-lobed, longer than calyx; styles 3-4.

In wooded areas; not common in our range.

Silene noctiflora L. Night-flowering catchfly

Stout erect annuals; stems 2-8 dm. tall, simple or branching, sticky from the glandular-pubescence, at least above; lower leaves 5-12 cm. long, oblanceolate or obovate, narrowed to winged petioles, upper leaves elliptic or lanceolate, nearly or quite sessile, 2-8 cm. long; flowers long-pediceled, few in a compound or simple cyme; calyx glandular-pubescent, cylindrical in flower, becoming much enlarged in fruit, ovoid, to 2 cm. long (including teeth), conspicuously 10-veined, the veins branching above the middle, green, the rest of calyx whitish, teeth slender; petals white, 2-cleft; styles 3.

Weedy species, especially in fields; introduced from Europe.

Silene scouleri Hook.

Perennials; stems erect, simple, pubescent, sticky glandular above, 1.5-8 dm. tall; lower leaves lanceolate, oblanceolate, or spatulate, long-petioled; cauline leaves gradually reduced, becoming sessile, sheathing above the somewhat swollen nodes, lanceolate or linear, 2-9 cm. long; inflorescence of terminal and axillary cymes, or flowers sometimes solitary in the axils; calyx cylindrical, 1-2 cm. long, glandular-pubescent, conspicuously 10-veined; petals white to pink, 2-4-lobed; styles 3-4.

Variety *pacifica* (Eastw.) C. L. Hitchc. is strictly coastal, while var. *scouleri* does not usually occur on the coast.

Usually in dry, often rocky, areas.

Silene suksdorfii Robins.

Tufted perennials; stems more or less prostrate at base, sending up erect branches 5-12 cm. tall; herbage finely pubescent and sometimes glandular; lower leaves linear to oblanceolate, obtuse or barely acute, the upper leaves linear-lanceolate, about 2 pairs; inflorescence glandular, solitary at the ends of branches, or consisting of few-flowered terminal cymes; calyx bell-shaped, conspicuously 10-veined, the veins often purplish; petals white to greenish- or purplish-tinged, 2-lobed; styles 3-4.

In the high Cascades.

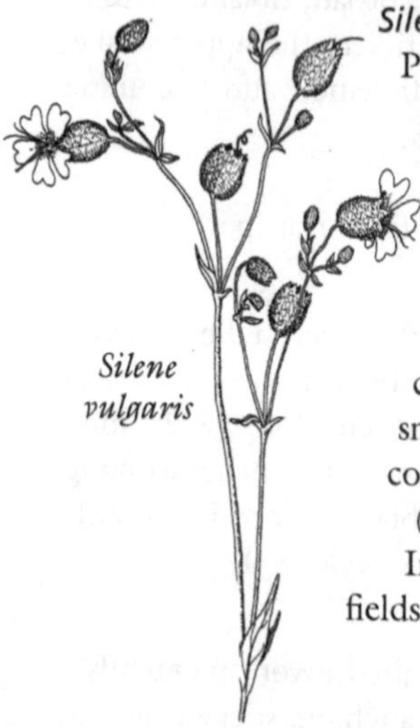

Silene vulgaris (Moench) Garcke Bladder campion

Perennials; stems more or less branched from the base, glaucous, glabrous or somewhat pubescent, 1.5-10 dm. tall; lower leaf pairs sheathing, leaf blades ovate to obovate, acute, reduced to bracts above; flowers in loose simple or compound cymes, erect or drooping; calyx 5-toothed, at first somewhat tubular bell-shaped, becoming much inflated, reaching 2 cm. in length, the main veins about 15, connected by small cross veins, these often purplish and more or less conspicuous; petals white, 2-cleft; styles 3.

(=*Silene cucubalus* Wibel)

Introduced, sometimes found as a weed in cultivated fields.

Silene vulgaris

Spergula

Spergula arvensis
a. seed
b. habit

Annual herbs with opposite leaves bearing clusters of similar-sized leaves in their axils, thus appearing whorled; stipules papery; flowers in terminal panicled cymes; sepals 5; petals 5, white; stamens 5-10; ovary 1-loculed; capsule 5-valved.

Spergula arvensis L. Starwort or Sandspurry
Stems much-branched, weakly ascending, 1-6 dm. tall; herbage glandular-pubescent; leaves 1-5 cm. long, linear, appearing as widely spaced whorls; flowers nodding in fruit; sepals scarious-margined; petals white, about equal to or shorter than the sepals; stamens usually 10.

A weed of gardens and cultivated fields.

Spergularia

Small annual or perennial herbs with mostly linear leaves and papery stipules; flowers in leafy terminal cymes and sometimes axillary; sepals 5; petals 5, entire; stamens 2-10; styles 3; capsule 3-valved; seeds often wing-margined.

1a Leaves not succulent; weedy species............*S. rubra*
1b Leaves succulent; not weedy
  2a Annuals.; stipules 1-3 mm. long..*S. marina*
  2b Perennials, from a thick root; stipules 4-11 mm. long.... *S. macrotheca*

Spergularia macrotheca (Hornem.) Heynh. Beach sandspurry
Perennials with large roots; herbage glandular-pubescent; stems ascending or prostrate; leaves succulent; stipules 4-11 mm. long, narrowly triangular, long acuminate; petals white or pink, shorter than the sepals.

Coastal, often in salt marshes.

Spergularia marina (L.) Griseb. Saltmarsh sandspurry
Annuals; stems generally ascending, sometimes prostrate, more or less glandular; leaves linear, succulent; stipules broadly triangular, 1-3 mm. long; flowers in branching cymes; petals white to pink, nearly as long as sepals.

Coastal and inland in saline and alkaline areas.

Spergularia rubra (L.) J. & C. Presl Red sandspurry
Annuals or sometimes over-wintering; herbage more or less glandular-pubescent; stems prostrate, forming mats 1-4 dm. in diameter; leaves 5-25 mm. long, linear, with conspicuous silvery stipules, these 3.5-5 mm. long; inflorescence about 1/2 as long as the stems; flowers pediceled, pink or red, rarely white; sepals pubescent, with papery margins; stamens 6-10.

A common weedy species; introduced from Europe.

Stellaria

Annuals or perennials; flowers perfect; sepals free, 4 or more often 5; petals 5 white, usually deeply 2-parted, or sometimes lacking; stamens 3-10; styles usually 3; capsule 1-loculed, short, ovoid or broadly oblong, opening by 6, 8 or 10 valves.

1a Bracts of cymes not leaf-like; reduced, scarious
  2a Basal leaves petioled; delicate annuals..*S. nitens*
  2b Leaves all sessile; perennials
    3a Leaves narrowed somewhat gradually at base; seeds smooth, shiny ....*S. longifolia*
    3b Leaves broadest at or just above base; seeds rough, dull ....*S. graminea*
1b Flowers axillary or in leafy cymes
  4a Stem with a single line of hairs; garden weeds; annuals............ *S. media*
  4b Stems usually glabrous, but never with a singe line of hairs; perennials
    5a Prostrate seaside species with thick succulent leaves....... *S. humifusa*
    5b Not prostrate seaside species; leaves not succulent
      6a Lower leaves petioled, margins crisped; sepals distinctly 3-veined; flowers solitary in the leaf axils .... *S. crispa*
      6b Leaves all sessile, not crisped; sepals not distinctly 3-veined; flowers in axillary or terminal cymes
        7a Leaves ovate to elliptic .... *S. calycantha*
        7b Leaves lanceolate to oblong-lanceolate ....*S. borealis*

Stellaria borealis Bigel.
Weak-stemmed much-branched perennials 1.5-4.5 dm., or sometimes longer; leaves lanceolate to oblong-lanceolate, 1-4.5 cm. long, acute or acuminate; flowers long-pediceled in leafy terminal or axillary cymes; sepals 5, acute, silvery-margined; petals usually 5 or sometimes lacking.
  Swampy areas, stream banks, meadows and damp woods.

Stellaria calycantha (Ledeb.) Bong. — Northern starwort
Erect to prostrate rhizomatous perennials; stems 5-25 cm. long; leaves ovate to elliptic, 3-25 mm. long, the margins sometimes ciliate; inflorescence leafy-bracteate; sepals 5, 1.5-3 mm. long, scarious-margined; petals, if present, 5, up to half as long as the sepals.
  Wet meadows, boggy areas, wet banks and damp woods.

Stellaria crispa Cham. & Schlecht. — Crisped starwort
Glabrous perennials; stems weak, 1-5 dm. long, more or less simple; the lowest leaves short-petioled, the middle and upper leaves sessile, lanceolate to ovate, 5-25 mm. long, acute or acuminate, generally crisped on the margins; bracts leaf-like; sepals 5, 3-veined, silvery-margined; petals often lacking; capsule longer than the sepals.
  Moist shady places.

Stellaria graminea L. — Lesser starwort
Perennials; stems weak, slender, 4-angled, 3-6 dm. long; leaves sessile, lanceolate, the upper about 2.5 cm. long, broadest just above the base; flowers pediceled, in loose cymes, spreading; bracts papery; sepals 3-veined, silvery-margined, about 3 mm. long; petals usually longer than the sepals; capsule

about equalling or somewhat longer than sepals; seeds rough-surfaced.

Weedy in moist places; introduced from Europe.

Stellaria humifusa Rottb. Low chickweed

Perennials; stems creeping and rooting at the nodes, up to 4 dm. long; leaves succulent, narrowly lanceolate or oblong, acute or obtuse at the apex, 6-18 mm. long; flowers axillary; sepals 3.5-5 mm. long; petals usually slightly longer than the sepals; capsule nearly of the same length as the sepals.

Along the coast, usually in salt marshes.

Stellaria longifolia Muhl. Long-leaved starwort

Rhizomatous perennials; stems trailing to ascending, 1.5-6 dm. long, 4-angled; leaves narrowly lanceolate to linear or the lower sometimes somewhat spatulate, 1.5-5 cm. long, to 4 mm. wide, sessile, the margins minutely roughened; flowers pediceled, many in a loose cyme, the pedicels eventually turning downward; sepals 5, acute, 3-veined, scarious-margined; petals cleft nearly to the base, as long as or somewhat longer than the sepals; capsule about twice as long as the sepals; seeds smooth.

Moist meadows and stream banks.

Stellaria media (L.) Cyr. Common chickweed

Annuals; stems weak, decumbent, rooting at the lower nodes, often matted, with a single line of hairs; leaves broadly ovate, acute at apex, long-petioled below, sessile above, petioles pubescent; flowers white, the pedicels turned backward in fruit; petals deeply 2-cleft, shorter than sepals; anthers reddish; capsule ovoid, slightly longer than sepals.

A common garden weed. This is one of the earliest plants to blossom in the spring, and a field of chickweed is nearly as fragrant as a field of white clover, and as popular with the bees.

Stellaria media

Stellaria nitens Nutt. Shining chickweed

Annuals; stems very slender, shining, more or less erect, 5-24 cm. tall, glabrous, at least above; leaves borne on lower half of the stem; basal leaves ovate, petioled, cauline leaves linear, sessile; at least some of the flowers long-pediceled in loose cymes, the bracts small, papery; sepals 5, 3-4 mm. long, conspicuously 3-veined, acute to acuminate, silvery-margined; petals 1/3 as long as sepals or absent; seeds finely roughened.

Stream banks to grassy open areas.

NYMPHAEACEAE
Water-lily Family

Aquatic perennials from large horizontal rhizomes or tubers; leaves alternate, long-petiolate, leaf blades usually floating; flowers axillary, solitary, usually perfect; sepals 3-many, sometimes petaloid; petals 0-many; stamens many, spirally arranged, often transitional to the petals; pistils 1-many, ovary superior or inferior, sometimes sunken in the receptacle; fruit indehiscent or tardily and irregularly dehiscent, variously interpreted as berry-like, follicular or capsular.

Flowers yellow; leaves oblong to ovate, 1-4.5 dm. long and 0.75-3.5 dm. wide .... **Nuphar**
Flowers white or pink; leaves nearly orbicular, 0.5-2.5 dm. broad and equally as long .... **Nymphaea**

Nuphar

Perennials from large, usually branched rhizomes; leaves erect or floating, long-petiolate, deeply cordate at the base; flowers long-pedunculate; sepals 5-14; petals 10-20, smaller than the sepals; stamens many; pistil one, compound, ovary superior, expanded at the top into a disk-like stigmatic surface; fruit leathery and many-seeded.

Nuphar luteum (L.) Sibth. & Sm. subsp. *polysepalum* (Engelm.) Beal
Western yellow pond-lily

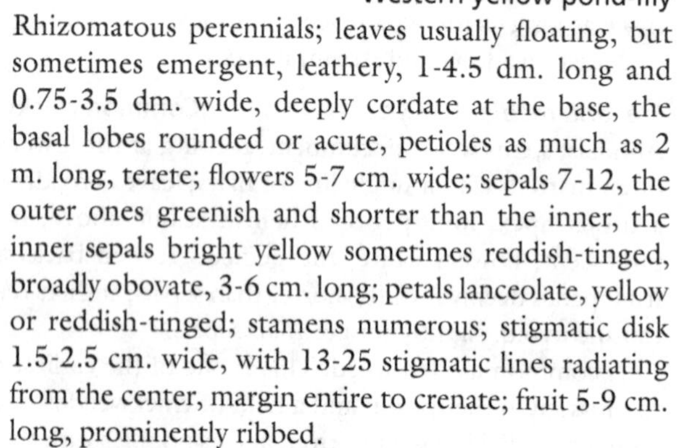

Nuphar luteum

Rhizomatous perennials; leaves usually floating, but sometimes emergent, leathery, 1-4.5 dm. long and 0.75-3.5 dm. wide, deeply cordate at the base, the basal lobes rounded or acute, petioles as much as 2 m. long, terete; flowers 5-7 cm. wide; sepals 7-12, the outer ones greenish and shorter than the inner, the inner sepals bright yellow sometimes reddish-tinged, broadly obovate, 3-6 cm. long; petals lanceolate, yellow or reddish-tinged; stamens numerous; stigmatic disk 1.5-2.5 cm. wide, with 13-25 stigmatic lines radiating from the center, margin entire to crenate; fruit 5-9 cm. long, prominently ribbed.

(=*Nuphar polysepalum* Engelm.)

Ponds, lakes, sloughs and slow-moving streams.

Nymphaea

Rhizomatous or tuberous perennials; leaves floating, deeply cordate to peltate; flowers solitary, long-pedunculate; sepals usually 4, green; petals 12-many, white to brightly colored; stamens numerous and the outer transitional to the petals; pistil one, compound; ovary inferior; stigmas 10-30, spreading out from a central depression; fruit irregularly dehiscent.

Nymphaea odorata Ait. Fragrant water-lily

Rhizomatous perennials; leaves floating, nearly orbicular, 5-25 cm. broad; flowers fragrant, floating or emergent; sepals 4, lanceolate to ovate, 3-6.5 cm. long; petals 20-30, white or pinkish, similar in shape to the sepals and equalling them or slightly longer; stamens 50-100; fruit 2.5-3 cm. long.

Quiet water, in scattered localities where it has apparently escaped from cultivation; introduced from the eastern United States.

CABOMBACEAE
Water-shield Family

Rhizomatous perennials; leaves floating and/or submerged, alternate, opposite or whorled, entire to dissected; flowers perfect, solitary; sepals and petals 2-4; stamens 3-36; fruit often a leathery follicle or indehiscent.

Plants with opposite or whorled dissected submerged leaves and small peltate alternate floating leaves .... **Cabomba**
Leaves all alternate, entire, peltate and floating .... **Brasenia**

Brasenia

A monotypic genus.

Brasenia schreberi J.F. Gmelin
Water-shield or Water-target

Rhizomatous perennials; entire plant, except for the upper surface of the leaves covered with a thick gelatinous coating; stems 3-20 dm. long, slender, much-branched; leaves floating, long-petiolate, peltate, nearly orbicular to broadly-elliptic, entire, green above, purplish beneath; flowers solitary in the leaf axils, perfect; sepals and petals 3, linear-oblong, 10-18 mm. long, purplish; stamens 12-36; pistils 4-20; fruit 6-8 mm. long, leathery, indehiscent.

Still and slow moving water.

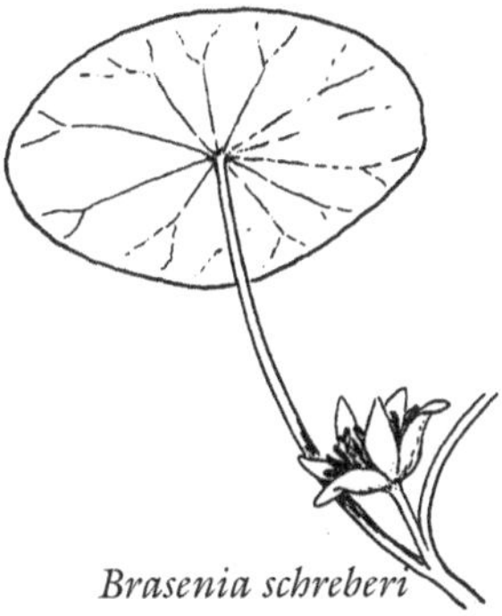

Brasenia schreberi

Cabomba

Rhizomatous perennials; stems slender; leaves mainly submerged, opposite or whorled and dissected, but a few smaller, upper alternate leaves peltate and floating; flowers solitary, axillary; sepals and petals 3; stamens 3-6; pistils 2-4; fruit indehiscent.

Cabomba caroliniana Gray — Fanwort

Stems up to 2 m. long; plants with a thin gelatinous coating; submerged leaves opposite or whorled, palmately dissected into numerous linear segments; floating leaves alternate, entire, peltate, elliptic and often basally notched; flowers solitary on long, axillary peduncles, sepals and petals 3, white to cream-colored with yellow basal spots; stamens 6; pistils 2-4; fruit indehiscent.

Introduced into a few lakes and ponds in our area; native of the central, eastern and southern United States.

CERATOPHYLLACEAE
Hornwort Family

Submerged monoecious annuals or perennials; stems slender, much-branched; leaves whorled, sessile, repeatedly forked into narrow divisions; flowers solitary in the leaf axils subtended by an 8-15-segmented structure variously interpreted as a calyx or an involucre; stamens 10-20, the anthers with 2-3 sharp, terminal points; pistillate flowers consisting of a simple, sessile, 1-loculed, 1-ovuled ovary; fruit an achene, the style forming a beak.

Ceratophyllum

Characters those of the family.

Leaves forked once or twice, the divisions conspicuously serrate on one side.... .... *C. demersum*

Leaves forked 2 to 4 times, the divisions only obscurely serrulate..................... .... *C. echinatum*

Ceratophyllum demersum

Ceratophyllum demersum L. Common Hornwort or Coontail
Forming large mats under the surface of the water; stems much-branched, up to 4 m. in length; leaves 5-12 per whorl, 1-2.5 cm. long, finely dichotomously dissected into linear or flattened segments, these conspicuously serrate on one side; body of the achene 4-6 mm. long with a long persistent style and 2 basal spines.

Ponds, lakes and slow-moving streams.

Ceratophyllum echinatum Gray
Similar to *C. demersum*, except the leaves 2-4-forked, the divisions only obscurely serrulate and the surface of the achene tuberculate.

Ponds, lakes and slow-moving streams; apparently introduced in lakes and ponds along the coast.

RANUNCULACEAE
Buttercup Family

Mostly herbs with alternate or basal leaves (or stems woody below, with opposite leaves), generally with acrid juice; flower parts all distinct; sepals 2-5, sometimes petal-like; petals 4 or more, or absent; stamens generally numerous; pistils several (rarely 1), superior, 1-loculed; fruit usually dry, rarely fleshy.

1a Climbing woody vines (ours) with opposite leaves; achenes long-tailed, plumose .... **Clematis**
1b Herbs
  2a Small plants; sepals 5, spurred; receptacle becoming elongated as achenes mature; leaves all basal.. **Myosurus**
  2b Without the above combination of characters

3a Dioecious (ours); petals none; sepals not petal-like (in ours) ..... **Thalictrum**
3b Flowers perfect
 4a Flowers irregular
  5a Upper sepal spurred at base .... **Delphinium**
  5b Upper sepal helmet-shaped, not spurred .... **Aconitum**
 4b Flowers regular
  6a Petals 5, spurred at base; sepals petal-like .... **Aquilegia**
  6b Petals not spurred
   7a Fruits fleshy, berry-like; flowers small, white, in a terminal raceme .... **Actaea**
   7b Fruits dry, achenes or follicles
    8a Petals present
     9a Flowers many, small, white, in paniculate-racemes .... **Cimicifuga**
     9b Flowers not as above
      10a Petals narrow, hooded; fruits follicles .... **Coptis**
      10b Petals not hooded; fruits achenes............ **Ranunculus**
    8b Petals absent; sepals often petal-like
     11a Cauline leaves in 1 whorl of 3 .... **Anemone**
     11b Cauline leaves not in 1 whorl of 3
      12a Flowers small, in paniculate-racemes .... **Cimicifuga**
      12b Flowers not in racemes
       13a Sepals 4; fruits of inflated achenes ..... **Trautvetteria**
       13b Sepals 5-10; fruits of follicles
        14a Leaves simple, cordate, obcordate or reniform .... **Caltha**
        14b Leaves decompound .... **Isopyrum**

Aconitum

Perennial herbs with palmately lobed or divided leaves, and showy irregular flowers; sepals 5, the upper helmet-shaped; petals 2-5, the 2 upper covered by the helmet, the 3 lower, if present, much reduced; stamens 25-50; pistils 3-5, becoming many-seeded follicles.

Aconitum columbianum Nutt. Monkshood or Aconite

Tuberous-rooted perennials; stems erect to trailing, 3-30 dm. tall; cauline leaf blades with 3-7 deep divisions, segment margins variously toothed or cleft; flowers racemose or paniculate, usually blue, but sometimes white or bluish-tinged; sepals 6-16 mm. long, the hood 10-30 mm. high; follicles 3-5, 1-2 cm. long. A subspecies bearing bulbils in the leaf axils and/or in the inflorescence is called: *A. columbianum* subsp. *viviparium* (Greene) Brink.

[*Aconitum howellii* A. Nels. = *Aconitum columbianum* subsp. *viviparum* (Greene) Brink]

Usually along streams or in bogs and wet meadows or other moist areas in the mountains.

Aconitum columbianum

Actaea

Rhizomatous perennials; stems with a few large decompound leaves; inflorescence a terminal raceme of small white flowers; sepals usually 3-5, petal-like; petals indefinite in number, usually 4-10 (ours) or none; stamens 15-50, white, showy; pistil 1, becoming a berry.

Actaea rubra (Ait.) Willd. Baneberry

Stems 1 to several, 4-12 dm. tall; leaves all cauline, 2-7 dm. long; leaflets mostly broad, ovate, appearing palmately veined, toothed; occasionally with short lateral racemes in axils of upper leaves in addition to longer terminal raceme; sepals creamy white or the tips pink-tinged; petals variable in number (4-10), narrow, white; berries red or white, straight on one side, curved on the other.

Moist shady places. The berries are poisonous.

Actaea rubra

Anemone

Perennial herbs with basal leaves arising from a caudex or a rhizome, cauline leaves forming a single whorl of 3; sepals 5 or more, petal-like; petals absent; stamens many; pistils many, developing into achenes.

1a Styles long, plumose .... *A. occidentalis*
1b Styles glabrous or pubescent, but not plumose
  2a Cauline leaves simple, toothed, sessile .... *A. deltoidea*
  2b Cauline leaves compound
    3a Plants generally tufted, not rhizomatous; achenes woolly
      4a Flowers white with a bluish tinge; styles about as long as achenes .... *A. drummondii*
      4b Flowers yellowish or reddish; styles half as long as achenes .... *A. multifida*
    3b Plants rhizomatous, not at all tufted; achenes not woolly, although sometimes pubescent
      5a Sepals 3.5-10 mm. long; stamens usually 10-35, in 1 series .... *A. lyallii*
      5b Sepals usually 10-20 mm. long; stamens 30-75, in 1 or 2 series
        6a Plants not of bogs; leaflets thin; sepals blue to purple; stamens 30-60 .... *A. oregana* var. *oregana*
        6b Plants of coastal bogs; leaflets thick; sepals white to pinkish; stamens mostly 60-75 .... *A. oregana* var. *felix*

Anemone deltoidea

Anemone deltoidea Hook. White windflower

Stems 1-3 dm. tall from long slender creeping rhizomes; basal leaves long-petioled, divided into 3 leaflets, each short-stalked and coarsely toothed, cauline leaves simple, coarsely toothed, broadly ovate, 3.5-8 cm. long; flowers solitary; sepals generally 5, creamy white, 1.2-2.5 cm. long; achenes somewhat inflated, pubescent, with short straight styles.

Shady moist woods.

Anemone drummondii Wats. Drummond's Anemone

Rhizomes short, covered with stringy remains of old roots; plants more or less pubescent at first, becoming glabrous or nearly so; stems 1-3 dm. tall; leaves decompound, the segments mostly linear, villous; flowers 2.5-4.5 cm. broad, white with a bluish-tinge; achenes densely woolly.

High mountains.

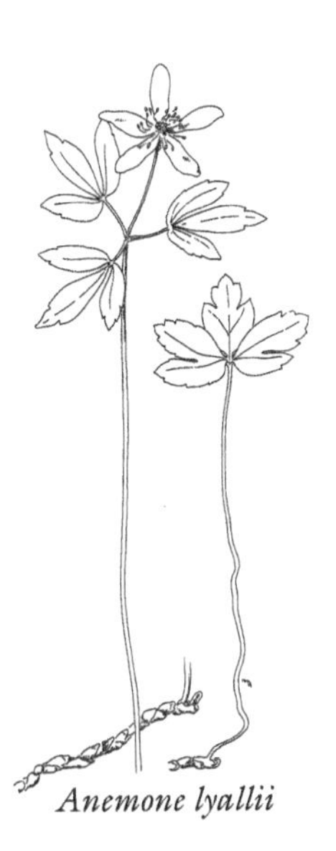
Anemone lyallii

Anemone lyallii Brit. Small windflower

Stems slender, 0.5-3.5 dm. tall from slender white horizontal scaly rhizomes with a single basal leaf with 3 leaflets, long-petioled, rarely on the plant at flowering time, cauline leaves short-petioled, the leaflets coarsely toothed especially near the tip, the lateral leaflets more or less cleft; flowers white, bluish, or pinkish, 1.2-3.5 cm. broad; stamens 10-35; achenes pubescent, short-beaked.

Anemone lyallii and *Anemone oregana* may intergrade.

Moist woods.

Anemone multifida Poir. Pacific or Cliff Anemone

Plants more or less tufted, silky pubescent, 1.5-7 dm. tall; leaves decompound, the segments linear; flowers 2-2.5 cm. in diameter, yellowish or tinged with red, blue or purple; sepals pubescent on the lower surface; achenes woolly, borne in a globose cluster.

(=*Anemone globosa* Nutt.)

From the foothills to subalpine areas.

Anemone occidentalis Wats. Pasque flower or Old man of the mountains

Plants more or less silky-pubescent, especially at first; stems 0.5-6 dm. tall; leaves generally long-silky beneath and unopened at flowering, the leaves spreading and the stem elongating as the fruits mature; lower leaves longer petioled than upper, all decompound, the ultimate segments narrow; flowers solitary; sepals 2-3 cm. long, white, often bluish or purplish at base; achenes numerous, with long plumose tails, usually in a large nearly globose head.

Alpine.

Anemone oregana Gray var. *felix* (Peck) C. L. Hitchc. Bog Anemone
Stems 0.5-3 dm. tall; rhizomes usually dark; leaves thick, generally broadly ovate to orbicular, the leaflets often deeply lobed or cleft; sepals white or with rose-purple markings on the margins or the back; stamens mostly 60-75; achenes densely pubescent.

(=*Anemone felix* Peck)

Coastal bogs from Lincoln Co., Oregon to Grays Harbor Co., Washington.

Anemone oregana Gray var. *oregana* Oregon Anemone
Stems slender, 0.5-4 dm. tall, from a slender horizontal brittle whitish, easily fragmenting rhizome; terminal leaflets of cauline leaves 3-lobed, lateral leaflets generally 2-lobed, all toothed or more or less cleft; sepals blue, purple or lavender; stamens 30-60; pistils 15 to 25, pubescent.

Anemone oregana and *Anemone lyallii* may intergrade.

Moist open woods in the mountains.

Aquilegia

Perennial herbs with mostly basal leaves divided in 3's and solitary flowers in the leaf axils; sepals petal-like; petals spurred at base; stamens many; pistils 5-10, becoming follicles.

Aquilegia formosa Fisch. ex DC. Columbine
Stems 3-10 dm. tall, generally from a dense tuft of basal leaves; sepals and spurs of petals deep scarlet (rarely crimson), blade of petals yellow; flowers mostly nodding.

Common in slightly shaded somewhat moist places.

Aquilegia formosa

Caltha

Perennial herbs with thick vertical or horizontal rhizomes with papery scales; leaves simple; flowers solitary or in cymes; sepals 5-12, white, bluish, or yellow; petals absent; stamens many; pistils 5 to many; fruits of beaked follicles.

Stems leafy; flowers yellow ....*C. palustris*
Leaves basal or with a single cauline leaf; flowers usually white .... *C. leptosepala*

Caltha leptosepala DC. White marsh marigold
Stems 0.5-2.5 dm. tall; leaves all basal or with a single cauline leaf, leaves thick, reniform or cordate to nearly orbicular, margins entire to toothed; flowers 1.5-4 cm. in diameter, showy; sepals usually white, petal-like; follicles 4-15.

(Includes *Caltha biflora* DC.)

In marshes and shallow ponds in the mountains.

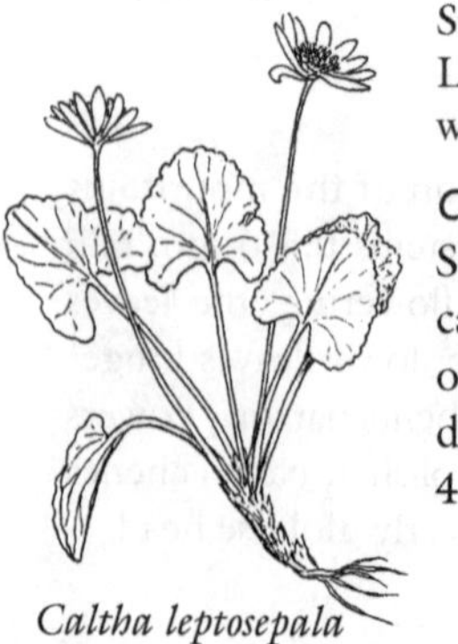

Caltha leptosepala

Caltha palustris L. Marsh marigold or Cowslip

Stems erect or ascending, sometimes rooting at the lower nodes; both basal and cauline leaves reniform or rounded-cordate; sepals yellow, 6-25 mm. long.

(=*Caltha asarifolia* DC.)

Coastal bogs.

Cimicifuga

Erect rhizomatous perennials with alternate decompound leaves and small white flowers in panicles with raceme-like branches; sepals 2-6, petal-like, early deciduous; petals small, 1-8, 2-toothed, or absent; stamens numerous; pistils 1-8, becoming follicles.

Petals absent; leaflets not deeply toothed; follicles sessile or nearly so.. *C. elata*
Petals 1-5; leaflets deeply toothed; follicles stipitate .... *C. laciniata*

Cimicifuga elata Nutt. Tall bugbane

Stems 7.5-18 dm. tall; leaves large, the leaflets several-lobed, cordate, irregularly toothed; sepals 5, white or pink; petals absent; follicles 1-3, about 8-12 mm. long, sessile or nearly so.

Wooded areas and shaded stream banks.

Cimicifuga laciniata Wats. Cut-leaved bugbane

Stems 10-13 dm. tall; leaflets deeply and sharply lobed and toothed; sepals 4-5, white; petals 1-5, white to yellowish; follicles 7-13 mm. long, about 5 mm. broad, stipitate.

Moist woods, wet meadows and stream banks in low altitudes in the mountains.

Clematis

Perennial herbs or somewhat woody vines, climbing by tendril-like petioles or leaf rachises, or plants erect; leaves opposite, simple or compound; flowers perfect or imperfect; sepals generally 4, petal-like; petals generally absent or minute; stamens many; pistils many; achenes in a conspicuous cluster, tipped by the long, generally plumose styles.

Flowers perfect .... *C. vitalba*
Flowers unisexual .... *C. ligusticifolia*

Clematis ligusticifolia Nutt. Wild Clematis

Dioecious climbing vines, woody below, stems reaching 10 m. or more in length; leaves glabrous, long-petioled, of 5-7 leaflets, the leaflets long- stalked and often 3-lobed or divided, the leaves frequently deformed by the curving of the petioles in climbing; flowers unisexual, in axillary inflorescences; sepals of both staminate and pistillate flowers 6-10 mm. long, white to cream-colored, densely but minutely pubescent; flowers of pistillate plants followed by panicles of long-tailed plumose achenes.

Moist woods and along stream banks.

Clematis vitalba L. Travelers-joy

Stems climbing to 12 m. high; leaves with 5 leaflets; flowers perfect, sepals spreading, white to cream-colored, about 1 cm. long.

Naturalized in a few localities, mostly along road sides.

Coptis

Small herbs with basal leaves and flower stems from slender rhizomes or stolons, rhizomes and roots a deep golden yellow beneath the bark; leaves compound; flowers white, 1-8 in an umbel-like cluster; sepals petal-like, 5-8; petals 5-7, narrow below, broadened above, hooded; stamens 10 to many; pistils short-stalked, 4-15; fruit of stalked follicles.

Coptis laciniata

Coptis laciniata Gray Goldthread

Peduncles generally 2-4-flowered, sometimes 7-8-flowered; leaves divided into 3 leaflets, leaflets generally 3-5-lobed, cleft, or divided and sharply toothed; sepals thread-like to linear-lanceolate, 6-12 mm. long to 1 mm. broad; petals shorter, narrow; follicles inflated, 6-12 mm. long, stalked.

Moist woods, wet cliffs and stream banks; Cascades and west.

Delphinium

Herbs with palmately lobed or divided leaves and irregular flowers in terminal racemes; sepals 5, generally petal-like, the upper one spurred at the base; petals 4, in 2 dissimilar pairs, the upper pair spurred, the nectar-bearing spurs enclosed by the spur of upper sepal; stamens 25-40; pistils usually 3, many-seeded, developing into follicles.

1a Sepals white or cream-colored
- 2a Lateral sepals 12-18 mm. long, spurs 14-20 mm. long. *D. pavonaceum*
- 2b Lateral sepals 9-12 mm. long, spurs 9-14 mm. long... *D. leucophaeum*

1b Sepals blue or bluish-purple
- 3a Lateral sepals 18-25 mm. long; plants 5-15 dm. tall from coarse fibrous roots .... *D. trolliifolium*
- 3b Lateral sepals averaging less than 18 mm. long; roots fibrous or tuber-like
  - 4a Plants 0.75-2.5 m. tall; stems hollow; herbage glabrous, except for the under surface of the leaves, glaucous........ *D. glaucum*
  - 4b Plants averaging less than 0.75 m. tall; not glaucous
    - 5a Lower petals with a notch 3-4 mm. deep; roots fibrous; alpine or subalpine plants........ *D. glareosum*
    - 5b Lower petals not deeply notched; roots tuber-like; plants of lower elevations
      - 6a Lateral sepals 6-12 mm. long, spurs mostly 7-12 mm. long; pedicels not much longer than the flowers........ *D. nuttallii*
      - 6b Lateral sepals 9-19 mm. long, spurs 9-17 mm. long; pedicels, at least the lower, much longer than the flowers

7a Pubescence not of fine incurved hairs, often glandular in the inflorescence.. ....*D. menziesii*
7b Pubescence of fine incuved hairs, not glandular ....*D. oreganum*

Delphinium glareosum Greene Olympic Mountain or Rockslide larkspur
Stems 1-4 dm. tall, glabrous to glandular-pubescent, from fleshy-fibrous roots; leaves deeply 3-5-cleft, again divided into oblong segments; racemes 6-20-flowered; sepals deep blue, the lateral pair 12-18 mm. long, spur about the same length as the sepals or slightly longer; upper petals white or blue, lower petals blue and deeply notched; follicles spreading at maturity.
Talus slopes at high elevations in the Cascades.

Delphinium glaucum Wats. Mountain larkspur
Stems erect, stout, hollow, 0.75-2.5 m. tall; plants glaucous and glabrous except on under surface of leaves and in the inflorescence; lower leaves large, orbicular, 5-9-parted, the segments narrowly and acutely lobed and toothed, upper leaves principally narrowly 3-cleft, both upper and lower leaves green above, pale and usually minutely white-pubescent beneath; flowers numerous in a dense terminal raceme, purplish to pale blue; lateral sepals 6-14 (18) mm. long, spur straight, 10-19 mm. long; follicles erect or spreading at maturity.
Meadows, bogs, open woods and stream banks in the mountains.

Delphinium leucophaeum Greene White rock larkspur
Perennials from tuber-like roots; leaves usually with 5 main lobes; sepals white to cream-colored, the lateral pair 9-12 mm. long, spur 9-14 mm. long; lower petals white, upper petals bluish-tinged, not glandular; follicles ascending or appressed at maturity.
Dry bluffs, cliffs and rocky areas; Willamette Valley; uncommon.

Delphinium menziesii DC. Field larkspur
Stems erect, 1-6 dm. tall, arising from a cluster of more or less globose or elongated tuber-like roots, stems nearly smooth or more often minutely pubescent, the hairs sometimes glandular on the peduncles; basal leaves generally pubescent, nearly orbicular in outline, 1.5-5 (8) cm. in diameter, palmately cleft nearly to the base, the segments again cut into narrow lobes, the lobes callous-tipped, segments of upper leaves narrower and fewer; racemes mostly few-flowered; sepals blue or purplish-blue, the lateral pair 12-19 mm. long, the spur straight, 11-17 mm. long; lower petals purplish, upper generally white or veined with purple; follicles usually pubescent, spreading at maturity.
Fields and hillsides; common.

Delphinium menziesii

Delphinium nuttallii **Gray** **Nuttall's larkspur**

Stems from tuber-like roots, 2-6 dm. tall, finely pubescent, at least above, stems leafy, leaves 3-6-cleft and again divided into linear segments; racemes short, usually densely flowered; sepals dark blue or bluish-purple, the lateral pair 6-12 mm. long, spur straight and about the same length as the sepals; upper petals white to light blue, lower petals deep blue; follicles ascending or appressed at maturity.

Moist to rocky areas; western Washington and the Columbia River Gorge.

Delphinium oreganum **Howell** **Oregon larkspur**

Stems from a tuber-like base, erect, simple or sparingly branched, 4-7.5 dm. tall, puberulent; leaf blades 3-5-parted, 3-10 cm. broad, long-petioled, pubescent; racemes loosely flowered; sepals dark blue to purple, rarely pinkish, the lateral pair 9-14 mm. long, spur straight or slightly curved, about as long as the sepals; lateral petals blue, rarely pinkish; follicles pubescent, spreading at maturity.

Meadows and open woods; Willamette Valley, and foothills of the Coast Range and Cascade Mountains.

Delphinium pavonaceum **Ewan** **Peacock larkspur**

Similar to *D. leucophaeum*, but larger; the lateral sepals 12-18 mm. long, spur 14-20 mm. long; at least the lower petals glandular-pubescent; follicles spreading at maturity.

Low moist areas and roadsides; Willamette Valley; uncommon.

Delphinium trolliifolium **Gray** **Wood, Tall or Poison larkspur**

Stems coarse, hollow, 5-15 dm. tall, smooth or minutely pubescent above, arising from elongated roots; basal leaves cut three-fourths of the distance to base into 5-9 broad segments, these coarsely toothed, upper leaves cut into fewer, narrower, and generally deeper segments, mostly smooth or slightly pubescent on the veins, 8-15 cm. in diameter; racemes generally many-flowered; sepals deep clear blue, the lateral pair 18-25 mm. long, the spur 16-23 mm. long, straight or more often curved downward, usually somewhat wrinkled transversely; upper petals whitish; follicles erect to spreading at maturity.

Delphinium trolliifolium

In moist open woods and along stream margins.

Isopyrum

Perennial herbs with decompound leaves and white solitary or clustered flowers; sepals 5 or 6, petal-like; petals 5 or none (ours); stamens many; pistils 2-20, becoming follicles.

Stems 3-10 dm. tall; flowers 3-10 in a simple or compound umbel-like inflorescence .... *I. hallii*

Stems 0.3-1.5 dm. tall; flowers solitary or rarely two .... *I. stipitatum*

Isopyrum hallii Gray

Stems erect, few-leaved, 3-10 dm. tall, both basal and lower cauline leaves 2 or 3 times ternate, the leaflets cuneate, coarsely lobed and toothed, reduced upwards; flowers showy, creamy white to pinkish, long-peduncled in an umbel-like inflorescence; stamens with white thread-like filaments, these slightly swollen beneath the anthers.

[=*Enemion hallii* (Gray) J.R. Drummond & Hutchinson]

Moist shady places, especially in the mountains.

Isopyrum stipitatum Gray

Stems 3-15 cm. tall, arising from a cluster of fusiform tubers; leaves at least twice ternate, nearly equaling the stem, the leaflets narrow; peduncle generally solitary, usually single-flowered; sepals white, narrow; stamens usually 10; follicles 6-12, stipitate.

[=*Enemion stipitatum* (Gray) J.R. Drummond & Hutchinson]

Moist shady places; uncommon in our area.

Myosurus

Small annuals with all basal leaves; flowering stems each bearing a single flower; sepals usually 5, spurred; petals 5 or rarely none, whitish or yellowish, with a long nectary-bearing claw; achenes many, borne on a receptacle which elongates as they mature, becoming of such size and shape as to give the plant the name of "mouse tail."

Myosurus minimus L. — Mouse-tail

Leaves linear, 2.5-11 cm. long; flowering stems several, 3-16 cm. tall, the fruiting receptacles spike-like, about 1.5-5 cm. long; achenes with flattened somewhat 4-sided keeled back, generally ending in a short beak.

Moist soil; often in vernal pools.

Ranunculus

Annual or perennial herbs; leaves simple or compound, the cauline leaves alternate; flowers in cymes or solitary, generally yellow (rarely white, green or pink); sepals 3-6; petals 0-22 with a basal nectar pit usually covered by a scale; stamens usually many; pistils 4-250, becoming achenes or rarely utricles, the seed attached at the base.

For several of the species described below, more or less well-defined varieties have been distinguished by specialists in this genus. Within the limits of our text, however, these divisions seem unnecessary, and with a few exceptions have not been included.

1a Mostly submerged plants, generally with both floating and submerged leaves, the submerged leaves filiformly dissected
- 2a Achenes 2-7; receptacle glabrous; floating leaves, when present, narrowly 3-lobed .... *R. lobbii*
- 2b Achenes more than 10; receptacle usually hispid; floating leaves, when present, broadly 3-lobed .... *R. aquatilis*

1b Plants of dry or wet places, but not submerged with finely dissected leaves
3a Leaves entire or somewhat toothed, not deeply-lobed or compound
4a Roots tuberous; sepals 3 .... *R. ficaria*
4b Roots not tuberous
5a Stems rooting at some or all nodes or stoloniferous
6a Leaf blades linear to narrowly lanceolate ....*R. flammula*
6b Leaf blades cordate to ovate .... *R. cymbalaria*
5b Stems erect, not rooting at the nodes
7a Cauline leaves long-lanceolate to oblanceolate ..*R. alismifolius*
7b Cauline leaves ovate .... *R. populago*
3b Leaves lobed at least 1/2 the way to the midrib or compound
8a Achenes spiny or rough-hispid; annuals
9a Achenes merely rough-hispid ....*R. parviflorus*
9b Achenes spiny
10a Achenes with a broad smooth margin, the face spiny ....
.... *R. muricatus*
10b Achene not conspicuously margined, but margin and face spiny-tubercled .... *R. arvensis*
8b Achenes not spiny or rough-hispid
11a Petals averaging less than 7 mm. in length, usually not much longer than the sepals
12a Leaves glabrous; annuals of wet areas; head of achenes cylindrical .... *R. sceleratus*
12b Leaves usually pubescent, but if glabrous, then plants perennial
13a Basal leaves not compound; flowers usually with fewer than 5 petals .... *R. uncinatus*
13b Basal leaves, or some of them compound
14a Head of achenes globose or ovoid; petals about as wide as long .... *R. macounii*
14b Head of achenes cylindrical; petals about half as wide as long ....*R. pensylvanicus*
11b Petals averaging over 7 mm. long, usually much longer than the sepals
15a Basal leaves pinnately compound or divided. ....
....*R. orthorhynchus*
15b Basal leaves palmately compound or lobed or leaflets only 3
16a Stems creeping, rooting at the lower nodes; basal leaves compound .... *R. repens*
16b Stems erect, not rooting at the lower nodes
17a Base of plant bulbous and corm-like..........*R. bulbosus*
17a Base of plant not bulbous and corm-like
18a Glabrous high mountain species; head of achenes cylindrical or ovoid ....*R. eschscholtzii*
18b Lowland plants; usually pubescent; head of achenes globose or hemispheric
19a Lower leaves 5-7-parted, the segments again cleft into narrow lobes; sepals spreading .... *R. acris*
19b Lower leaves mostly 3-lobed, these again lobed or toothed; tips of the sepals reflexed..... *R. occidentalis*

Ranunculus acris L. — Blister buttercup

Stems pubescent, 3-9 dm. tall; lower leaves petioled, deeply 3-7-parted, the segments mostly deeply, narrowly and acutely lobed, upper leaves short-petioled, 3-to 5-lobed; flowers bright yellow, about 2.5 cm. in diameter; petals longer than the sepals; achene-bearing receptacle globose; achenes somewhat flattened, the beaks very short, straight or slightly curved backward.

Introduced, uncommon.

Ranunculus alismifolius Geyer ex Benth. — Plantain-leaved buttercup

Stems stout, more or less hollow, erect, somewhat branching, 2-6 dm. tall, mostly smooth; leaves 5 -15 cm. long, slender, long-lanceolate to oblanceolate, the basal long-petioled, cauline leaves short-petioled or sessile, margins entire or somewhat toothed; flowers deeply saucer-shaped, petals 5-14 mm. long, deep yellow, rounded; stamens and pistils forming a compact ball in the center; achenes short-beaked, the beaks straight or weakly curved.

This is a variable species with many named varieties.

(Also spelled *alismaefolius*)

Marshes, wet meadows and stream banks.

Ranunculus aquatilis L. — Water buttercup

Perennials, growing entirely under water except flowers and stem tips in var. ***diffusus*** Wither., or with floating leaves in var. ***aquatilis***; floating leaves broadly 3-lobed, the lobes toothed, submerged leaves many times divided into thread-like segments; flowers white, 6-17 mm. broad, the petals thin and translucent; sepals and generally also the styles falling away; achenes 10-25, or sometimes more, in a compact globose head, wrinkled transversely, usually pubescent with short hairs.

Ponds, slow streams, ditches and lake margins.

Ranunculus aquatilis
Floating leaves of a. *R. aquatilis*, b. *R. lobbii*

Ranunculus arvensis
a. achene
b. lower leaf

Ranunculus arvensis L. Hunger-weed

Annuals; stems erect, 2-6 dm. tall; lower leaves long-petioled, sometimes entire, or few-lobed, cauline leaves petioled to nearly sessile, mostly several times divided into long narrow segments; petals yellow, 5-8 mm. long; achene-bearing receptacles not elongated; achenes conspicuously spiny-tubercled, long-beaked.

A weedy species; introduced from Europe.

Ranunculus bulbosus L. Bulbous buttercup

Pubescent perennials; stems 1.5-6 dm. tall, erect from a thickened bulbous corm-like base; basal leaves long-petioled, ternately compound, the leaflets deeply 3-lobed, the margins lobed or toothed, cauline leaves reduced upwards; petals yellow, 7-13 mm. long; achenes with a stout flattened beak.

A weedy species; introduced from Europe.

Ranunculus cymbalaria Pursh Seaside buttercup

Perennials, creeping by stolons, the flowering stems 0.5-4 dm. tall, erect or ascending; leaves mostly basal, cordate, ovate, reniform, or oblong, coarsely few-toothed or shallowly lobed, petioles longer than the blades; flowers 6-15 mm. in diameter, the petals about the length of the sepals; achene-bearing receptacle elongated; achenes small, broad, short-beaked.

In moist, often saline, places.

Ranunculus eschscholtzii Schlecht. Subalpine buttercup

Nearly glabrous perennials; stems 5-25 cm. tall; lower leaves somewhat long-petioled, the blades reniform to broadly cordate, deeply 3-5-parted, the segments generally again lobed, the lobes rounded or obtuse except in var. ***suksdorfii*** (Gray) Benson where they are acute, cauline leaves short-petioled to nearly sessile, 3-5-cleft or parted, the segments generally oblong, entire; flowers pale yellow; petals 6-16 mm. long, often nearly as broad; achene bearing receptacle elongated, achenes rounded, beaked, the beaks straight or curved outward.

Mountain meadows and talus slopes.

Ranunculus ficaria L. Pilewort

Tuberous-rooted perennials; stems 1-3 dm. long, erect or decumbent, often producing bulbils in the leaf axils; basal leaves long-petiolate, cordate to broadly deltoid, 2-4 cm. long, the margins entire to crenate; sepals 3, 4-9 mm. long; petals 8-12, yellow but fading to white, 1-1.5 cm. long; achenes beakless.

This species is extremely variable.

Naturalized in moist disturbed areas and along stream banks. Introduced from Europe as a garden plant.

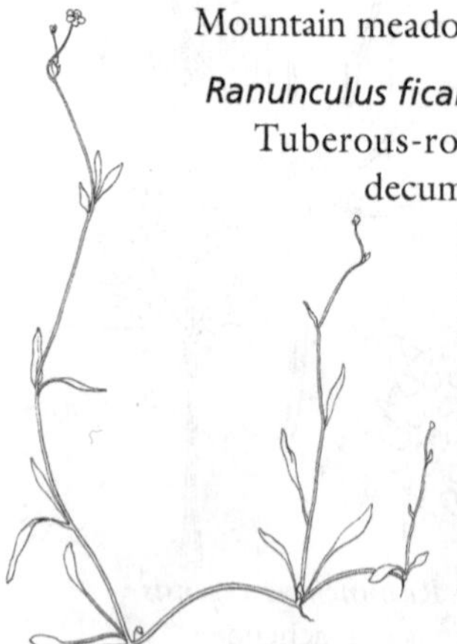

Ranunculus flammula

Ranunculus flammula L. Creeping spearwort

Stems very slender, rooting at all or most of the nodes; leaves entire, long, narrow, broadest near the apex, tapering at base to long petioles to nearly linear and not much wider than the petioles; petals 2.5-7 mm. long, pale yellow; achenes with a straight or curved beak.

Margins of lakes and streams.

Ranunculus lobbii Gray Leafy water buttercup

Annuals with mostly floating leaves cut into 3 spreading oblong lobes, notched or entire, submerged leaves, filiformly dissected; petals 4-6 mm. long, white; style long, thread-like; achenes 2-7, wrinkled.

Slow streams and vernal pools.

Ranunculus macounii Britt. Macoun's buttercup

Perennials; stems stout, 3-6 dm. tall, nearly glabrous or with spreading pubescence; leaves palmately compound with 3 leaflets, these generally lobed or irregularly toothed, petiolulate; flowers yellow, up to 1.2 cm. in diameter; achene bearing receptacle globose or slightly elongated; achenes with a distinct margin, sides smooth, short-beaked, the beaks stout and straight or very slightly curved backward.

R. macounii Britt. var. ***oreganus*** (Gray) K.C. Davis has glabrous herbage.

Moist shady places along the Columbia River and northward; more common east of the Cascades.

Ranunculus muricatus L. Spiny-fruited buttercup

Annuals; stems somewhat succulent. 0.5-6 dm. tall, erect or spreading; leaves long-petioled, mostly 3-5-lobed, the lobes again lobed or toothed or those of upper leaves entire; petals pale yellow, 4-8 mm. long; achene bearing receptacle not elongated; achenes smooth margined, the faces densely spiny, the beak stout, curved slightly backward.

A weedy species introduced from Europe.

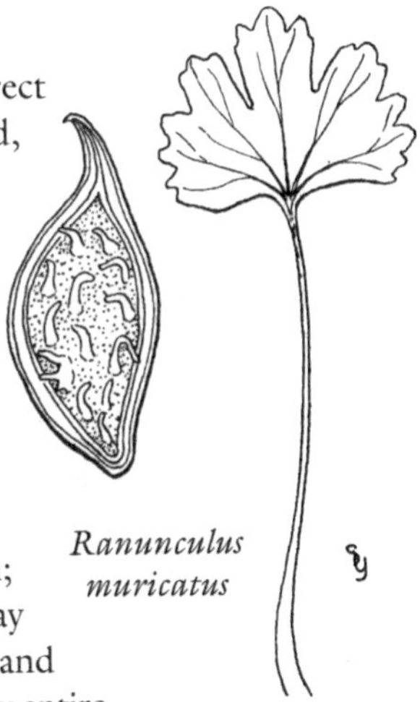

Ranunculus muricatus

Ranunculus occidentalis Nutt. Field buttercup

Perennials; stems generally branched, erect, 1-4.5 dm. tall, with spreading pubescence, or nearly smooth; basal and lower cauline leaves pubescent, cut half-way to base or deeper into broad segments, these lobed and toothed, upper cauline leaves entire or cut into narrow entire segments; flowers 1-2.5 cm. in diameter, deep yellow; achenes with short beaks curved slightly backward at tips.

Common everywhere in open fields.

Ranunculus orthorhynchus Hook. Bird-foot buttercup

Perennials; stems generally several arising from a cluster of long-petioled basal leaves, erect, pubescent or nearly glabrous; leaves pinnately compound, the lateral leaflets opposite or often irregularly placed, leaflets of upper

Ranunculus occidentalis *Ranunculus orthorhynchus*

leaves mostly narrow, more or less 3-lobed or irregularly toothed, generally densely pubescent on the petioles and lower surface of blades, sometimes nearly glabrous; flowers large, bright yellow; petals 1-2 cm. long, sometimes purplish-brown or red-tinged beneath; achenes thick, the nearly straight beak longer than the body.

Common in marshy places.

Ranunculus parviflorus L.

Annuals; stems 1-4 dm. tall, erect; herbage soft pubescent; basal and lower cauline leaves long-petioled, the uppermost cauline ones sessile, leaf blades deeply 3-lobed to coarsely toothed or the uppermost entire; sepals reflexed; petals 1-2 mm. long, less than 1 mm. broad; achenes in a globose head, beak recurved at the tip.

Road sides, fields and waste places; introduced from Europe.

Ranunculus pensylvanicus L. f. Bristly buttercup

Annual or short-lived perennials; stems erect, hollow, 3-6 dm. tall or more, densely pubescent; basal leaves ternately compound, the leaflets deeply lobed, the margins toothed, cauline leaves reduced upwards; petals yellow, 2-4 mm. long; achenes glabrous, with a stout nearly straight beak.

Wet areas, only in western Washington in our range.

Ranunculus populago Greene Mountain buttercup

Perennials; stems solitary, erect or ascending, 1-2.5 dm. tall, usually glabrous, somewhat succulent; lower leaves round-cordate to ovate, obtuse or acute, with long petioles which are slender above and sheathing below, or some leaves reduced to petioles with blades minute or none, cauline leaves long-ovate, the uppermost nearly sessile, those below narrowed to winged petioles, both basal and cauline leaves with several basal veins; petals 4-9 mm. long, deep yellow; achenes plump, short-beaked.

Moist mountain meadows and stream banks.

Ranunculus repens L. Creeping buttercup

Perennials; stems prostrate at base, creeping, rooting at nodes; leaves with more or less stiff pubescence, the blades somewhat triangular in outline, the lower generally divided into lobed and toothed leaflets, on long petioles, upper leaves lobed; petals 6-18 mm. long, 5-12 mm. wide, spreading saucer-like, deep golden-yellow; achenes with a stout curved beak.

In marshy areas, meadows, lawns and ditches, particularly near the coast; introduced from Europe.

Ranunculus repens

Ranunculus sceleratus L. Biting buttercup

Annuals; stems branching, 1.5-6 dm. tall; lower leaves thick, more or less cordate, palmately 3-5-lobed, the lobes coarsely and irregularly again lobed, long-petioled, the cauline leaves shorter petioled to nearly sessile, deeply 3-lobed or divided, the divisions several toothed or entire; sepals yellowish; petals pale yellow, flowers 6-12 mm. broad; achene bearing receptacle elongated, achenes many, thick-margined, almost beakless.

Found on sandy banks of the Columbia, and elsewhere in wet habitats in our limits, but not common.

Ranunculus uncinatus

Ranunculus uncinatus D. Don Woods buttercup

Often rank growing, 2-6 dm. or more tall with coarse hollow stems and few leaves; leaves 3-parted, the segments commonly lobed or coarsely toothed, or upper leaves narrow and entire; flowers small, inconspicuous; petals often less than 5, 2-4.5 mm. long, pale yellow to nearly white; achenes smooth or with a few hairs, the beak strongly curved backward into a hook.

Moist woods or meadows, particularly along streams.

Thalictrum

Erect perennial herbs; leaves 1 to 4 times compound; flowers very small in panicles, corymbs, umbels, racemes or rarely solitary, perfect or imperfect, and often with stamens and pistils on different plants (ours); sepals 4-5, sometimes petal-like; petals none; stamens numerous, with slender filaments; achenes heavily veined to grooved or ribbed, sometimes inflated.

Achenes elliptic-fusiform, 5-10 mm. long, less than half as broad, prominently 6-veined.. *T. occidentale*
Achenes obovate, 4-6 mm. long, well over half as broad, irregularly reticulate . ... *T. polycarpum*

Thalictrum occidentale Gray Western meadow-rue
Dioecious species 3-10 dm. tall, glabrous to glandular pubescent; leaves ternately compound, leaflets 1.2-4 cm. long, deeply 3-lobed at apex, lobes often coarsely toothed; staminate flowers tassel-like, filaments filiform, usually purplish; styles deep red; achenes 6-9, slightly oblique, 5-10 mm. long, prominently 6-veined, spreading or turning backward in fruit.
Moist shady places.

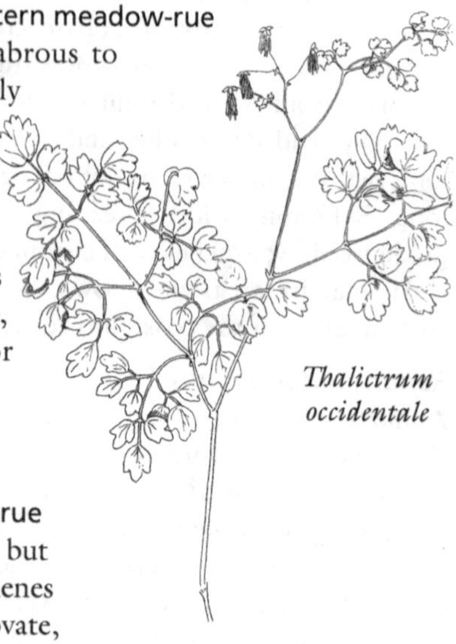

Thalictrum occidentale

Thalictrum polycarpum (Torr.) Wats.
Tall western meadow-rue
Somewhat resembling T. occidentale, but plants stouter, (4-24 dm. tall), the achenes 10-15, 4-6 mm. long, flattened, obovate, irregularly reticulate.
(=*Thalictrum alpinum* L. var. *polycarpum* Torr.)
Lowland, along streams.

Trautvetteria

Perennial herbs with palmately lobed leaves and corymbs of white flowers; sepals generally 4, petal-like, early deciduous; petals none; stamens numerous, white, the filaments linear below, enlarged beneath the anther; pistils numerous, becoming inflated achenes or utricles.

Trautvetteria caroliniensis (Walt.) Vail False Bugbane
Rhizomatous perennials 3-9 dm. tall; basal leaves often reaching 1.5-2 dm. in width, 5-7-cleft, sharply and irregularly serrate, long-petioled, the cauline leaves similar or fewer lobed, short-petioled to nearly sessile; sepals 3-6 mm. long, concave; stamens white; fruit papery, prominently veined.
(=*Trautvetteria grandis* Nutt.)
Moist places.

BERBERIDACEAE
Barberry Family

Shrubs or herbs with alternate, usually compound leaves; flowers with both stamens and pistils; sepals usually 6 or absent; petals 6, in 2 whorls of 3 or absent; stamens 6, opposite the petals, or more numerous in *Achlys*; anthers opening by uplifting valves; pistil one, superior, 1-loculed; fruit a berry or dry and dehiscent or indehiscent.

1a Shrubs; leaves evergreen .... **Berberis**
1b Herbs; leaves deciduous
  2a Leaflets 3, broad; flowers borne in a spike; sepals and petals absent .... **Achlys**
  2b Leaves several times compound; flowers in a loose raceme or panicle; sepals and petals present .... **Vancouveria**

Achlys

Perennial herbs with leaves and flower stalks arising from a creeping rhizome; leaves long-petioled, palmately divided into 3 leaflets; flowers very small, crowded in a terminal spike; sepals and petals absent; stamens and pistils present; stamens indefinite in number; fruit dry, 1-seeded.

Achlys triphylla (Sm.) DC. Vanillaleaf
Rhizomes scaly; flowering stalks 2-4.5 dm. tall; leaves long-petioled, leaflets 3, narrow at base, broad at apex, wavy-margined, 5 to 20 cm. broad; fruits crescent-shaped.
The foliage has a sweet vanilla-like odor when dried.
Shady moist woods.

Achlys triphylla

Berberis

Evergreen (ours) or deciduous shrubs, or low plants shrubby at base; leaves alternate and spiny, simple or pinnately compound (ours); flowers usually in racemes, yellow, each subtended by 3 small petal-like bractlets; sepals, petals and stamens 6, sepals petal-like; stigma disk-like; fruit a berry.

Leaflets 5 to 7, rarely 9, with a prominent midrib .... *B. aquifolium*
Leaflets 9 to 21, with several main veins from the base .... *B. nervosa*

Berberis aquifolium Pursh Tall Oregon-grape
Erect bushy shrubs 0.5-4.5 m. tall; leaves stiff, leathery, bright green and shining above, paler beneath, generally with 5-7 leaflets (rarely 9), leaflets long-ovate, 3-8 cm. long, the margins with sharp slender spines; racemes clustered in the leaf axils and at apex of stem, making

Berberis aquifolium

a dense showy inflorescence 3-10 cm. long; berries deep blue, grape-like.

This is the state flower of Oregon.

[=*Mahonia aquifolium* (Pursh) Nutt.]

Common, usually in open woods.

Berberis nervosa Pursh — Mountain Oregon-grape

Low, slow-growing shrubs 0.5-6 dm. tall, stems showing scars of preceding years' growth; leaves long, slender, with 9 to 21 leaflets, leaflets ovate-lanceolate, thick, leathery with slender spines on the margins, and with 5-7 prominent veins arising from base; racemes slender, 7-20 cm. long, several arising from one bud; bud scales prominent, persistent for several years, long-ovate, tapering to a long point; yellow flowers sometimes tinged with rose or purplish; berries blue with a white waxy coating.

Berberis nervosa

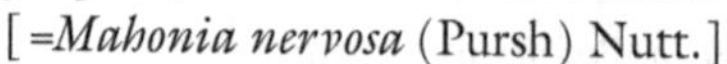

[=*Mahonia nervosa* (Pursh) Nutt.]

Common in the Coast Range and Cascades, generally occurring at higher elevations than *B. aquifolium.*

Vancouveria

Low perennial herbs (or somewhat woody at base) from slender rootstocks; leaves once or twice divided into 3's, mostly or all from base of the plant; flowers nodding, in a panicle; sepals petal-like, white or yellow, turned backward, in 2 series of 3, with 6-9 small bractlets at base; petals 6, shorter than the sepals, also reflexed, each with a hood-like nectar gland near apex; stamens 6, erect, surrounding the pistil, the filaments extending beyond the separate anthers; style and stigma 1; fruit a follicle.

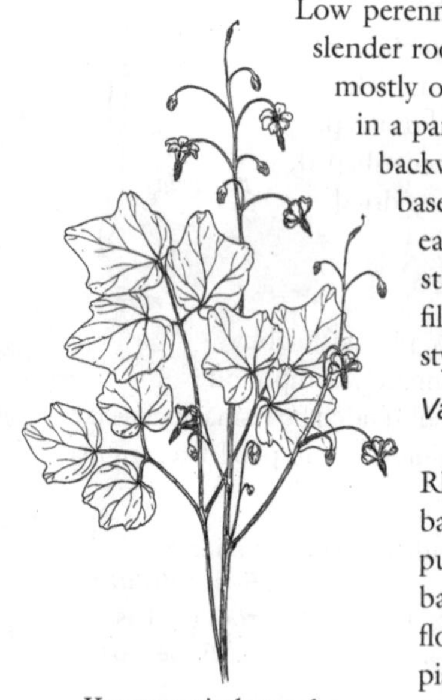

Vancouveria hexandra

Vancouveria hexandra (Hook.) Morr. & Dec. — Inside-out-flower

Rhizomatous perennials 2-5 dm. tall; leaves basal, shorter than the flowering stems, sparsely pubescent, leaflets somewhat heart-shaped at base, broadest above middle, mostly 3-lobed; flowers cream-colored; bractlets, stamens, and pistils covered with small stalked glands; follicle somewhat crescent-shaped, with a persistent style.

In shady woods or under thick shrubbery.

PAPAVERACEAE
Poppy Family

Herbs (ours) or shrubs with watery clear or colored juice; sepals 2-3, sometimes united into an early deciduous cap; petals twice as many as sepals or absent; pistil 1, compound, or rarely several simple pistils; ovary superior, 1-several-loculed.

Leaves alternate or basal, decompound; sepals united .... **Eschscholzia**
Leaves opposite or whorled; sepals separate .... **Meconella**

Eschscholzia

Annuals or perennials, with basal or alternate leaves several times divided into 3's; receptacle cup-like, surrounding the base of the pistil; sepals 2, united into a cap which is pushed off as petals unfold; petals 4 or rarely more; stamens 12-many; ovary slender, cylindrical, ribbed; capsule 2 valved, 1-loculed, many-seeded.

Eschscholzia californica Cham. California poppy

Stems erect or sometimes spreading, generally leafy, arising from tuft of basal leaves; leaves usually bluish-green, several times divided into very narrow segments; receptacle with an outer spreading and an inner erect rim; petals attached inside spreading rim, yellow or deep orange, or rarely white, broadest at apex, satiny in texture; capsule slender, 2.5-10 cm. long with a persistent rim at the base.

A highly variable species.

Common in open areas and waste places.

Eschscholzia californica

Meconella

Small annuals with mostly basal and opposite or whorled leaves and long-pediceled yellow or white flowers, nodding in bud; sepals 2-3; petals 4 or 6; stamens few or numerous; capsule 1-loculed, 3-valved.

Meconella oregana Nutt. White Meconella

Annuals to 8 cm. or more tall with several very slender branches; leaves thin, glabrous, opposite or some whorled, the lower 1.2-2.5 cm. long including the somewhat elongated petiole, the upper oblanceolate, sessile; flowers terminal or axillary on long thread-like pedicels; petals white; stamens 3-6 (12); capsule linear, 1 cm. or slightly more in length, generally twisted.

In sandy or gravelly places; uncommon.

FUMARIACEAE
Bleeding-heart Family

Glabrous herbs with alternate or basal usually decompound leaves; flowers perfect; sepals 2, very small; petals 4, in 2 series, one or both of the outer petals spurred or saccate at the base and spreading at the tips, the two inner petals often with crested and united tips; stamens 6, in 2 sets of 3's, opposite the outer petals, the filaments of each set partially or completely united; pistil compound; fruit a thin-walled, 2-valved, 1-loculed, several- to many-seeded capsule, or hard, indehiscent, and 1-seeded.

1a Leaves all basal (ours); outer petals saccate or spurred at the base. **Dicentra**
1b Leaves not all basal; only 1 outer petal spurred at base
  2a Flowers dark reddish-purple at least apically; fruit 1-seeded, indehiscent. .... **Fumaria**
  2b Flowers yellow or pink to lavender; fruit several-seeded, dehiscent .... **Corydalis**

Corydalis

Leafy-stemmed annual or perennial herbs; leaves usually large and decompound (ours); petals 4, in 2 pairs, only one of the outer petals spurred; fruit a capsule.

1a Flowers yellow; stems decumbent .... *C. aurea*
1b Flowers pink to lavender; stems erect
  2a Aquatic or subaquatic plants; spur about equaling the rest of the flower in length .... *C. aquae-gelidae*
  2b Not aquatic or subaquatic; spur averaging at least twice the length of the rest of the flower .... *C. scouleri*

Corydalis aquae-gelidae Peck and Wilson — Clackamas Corydalis

Perennial aquatic or subaquatic herbs; stems simple or branched, slender above, hollow below; leaves both basal and cauline, the lower equalling the stem, the blades divided four or more times, somewhat compact, the ultimate segments numerous, narrow, glaucous beneath; inflorescence simple or compound, many-flowered; corolla pinkish-lavender, deeper colored at the tip, spur of the upper petal about the length of its blade.

On the swampy margins of cold mountain lakes and along stream banks; uncommon.

Corydalis aurea Willd. — Golden Corydalis

Annuals or biennials; stems more or less decumbent, sometimes reaching 7 dm. in length; leaves glaucous, dissected into many short linear to elliptic segments; flowers in short racemes, yellow, the spur less than half as long as the rest of the flower; capsules narrow, deflexed, curving.

Within our limits, found chiefly in woods and thickets in the western foothills of the Cascades.

Corydalis scouleri

Corydalis scouleri Hook. Scouler's Corydalis
Rhizomatous perennials; stems 4.5-12 dm. tall, hollow; leaves glaucous, decompound, loosely spreading; inflorescence terminal; corollas almost uniformly pink, the spur about twice as long as the rest of the flower.

In shady and somewhat moist areas in the Coast Range and Cascade Mountains.

Dicentra

Annuals or perennials; leaves all basal (ours); herbage usually more or less glaucous; leaves ternately divided, then usually further compounded, either ternately or pinnately; flowers irregular; corolla somewhat flattened, saccate at the base or 2-spurred; fruit a capsule.

1a Scapes single-flowered, or rarely 2-flowered, 3-8 cm. tall .... *D. uniflora*
1b Scapes several-many-flowered; plants at least 10 cm. tall
  2a Corolla pink to lavender, cordate at the base with very short incurved spurs .... *D. formosa*
  2b Corolla generally white or slightly pinkish, yellow-tipped, spurs elongated, divergent or slightly incurved .... *D. cucullaria*

Dicentra cucullaria (L.) Bern. Dutchman's-breeches
Leaves and scapes erect from vertical short, thick, non-creeping rhizomes bearing numerous small tubers or bulblets; leaves long-petioled, the blades once or twice ternately divided, then further dissected, the ultimate segments linear; flowers nodding, 3 to 14 in a raceme borne on a scape 1-3 dm. tall; corolla generally white or slightly pinkish, yellow-tipped; capsules 0.7-2.5 cm. long.

In moist rocky or wooded slopes. Uncommon west of the Cascade Mountains except along the Columbia River and in Benton County, Oregon.

Dicentra formosa (Andr.) Walp. Bleeding-heart
Leaves and scapes erect from long creeping, fleshy, scaly rhizomes; leaves long-petioled, the blades ternately divided, then further dissected, the ultimate segments generally deeply-dissected; flowers pendent, usually 3-20 on a scape, scapes 1.5-5 dm. tall; corolla pink to lavender, 2-2.5 cm. long, spurs very short, incurved, forming a cordate base; capsules 2-3.5 cm. long.

Moist shady woods.

Dicentra formosa

Dicentra uniflora Kell. One-flowered Dicentra or Steer's-head

Dwarf perennials from a cluster of fleshy tapering roots or tubers; leaves few, petioles 2-8 cm. tall, blades ternately divided, then further lobed or dissected; scapes 1- or rarely 2-flowered; flowers white or pinkish, 1-2 cm. long, the upper half of each outer petal strongly divergent or recurved; capsule 1.5 cm. or less long, at maturity carried to the ground by the scape.

Growing on rocky or gravelly slopes in the Cascade Range.

Fumaria

Annuals; stems weak, often trailing or climbing; leaves decompound; flowers in a simple or compound inflorescence; one of the two outer petals spurred; ovary 1-loculed, 1-ovuled; fruit indehiscent.

Fumaria officinalis L. Fumitory

Glabrous and glaucous annuals; stems varying from erect to climbing, 1-7 dm. tall; corolla deep purplish-red at the tip, often paler below, 7-12 mm. long with a spur up to 4 mm. long; fruit nearly globose, 1.5-2.5 mm. in diameter, covered with small warty projections.

An introduction from Europe, occasionally found as a weed in gardens and fields.

BRASSICACEAE
Mustard Family

Herbs with alternate (rarely all apparently basal) leaves; flowers solitary or usually in racemes; flower-parts in 4's, separate; petals entire or cleft, spreading or turned backward at nearly right angles in the form of a cross; stamens generally 6 (rarely 4 or 2), commonly 4 long and 2 short, the long stamens opposite the petals, the short ones alternate with them; fruit a 2-loculed fruit which splits to allow escape of seeds, or 1-loculed and not splitting, or breaking transversely into 1-seeded sections; seeds in 1 or 2 rows in each locule.

The alternate name for this family is **Cruciferae**.

1a Fruits not splitting length-wise at maturity
  2a Fruits flattened, circular or short-oval, 1-loculed
    3a Fruits winged ....**Thysanocarpus**
    3b Fruits not winged...**Athysanus**
  2b Fruits not as above
    4a Fruits often constricted between seeds but not jointed ..... **Raphanus**
    4b Fruits of 2 distinct sections, jointed between them; plants of sea beaches
      5a Terminal section of the fruit globose, lower section undeveloped.. ....**Crambe**
      5b Terminal section of the fruit longer than wide, 4-angled ... **Cakile**
1b Fruits splitting length-wise at maturity
  6a Fruits less than 3 times as long as wide
    7a Small aquatic plants; leaves entire, all basal ....**Subularia**
    7b If aquatic, cauline leaves present
      8a Fleshy maritime species .... **Cochlearia**
      8b Not fleshy maritime species
        9a Fruits inflated, nearly round in cross-section .... **Rorippa**
        9b Fruits more or less flattened
          10a Fruits flattened parallel to the partition
            11a Flowers solitary.. **Idahoa**
            11b Flowers in racemes
              12a Fruits with a stipe-like base .... **Lunaria**
              12b Fruits sessile on the pedicel.......................... **Draba**
          10b Fruits flattened at right angles to the partition
            13a Seeds 1 in each locule .... **Lepidium**
            13b Seeds several in each locule
              14a Outer pair of petals longer than the inner pair .... **Teesdalia**
              14b Petals equal
                15a At least the lower leaves pinnately lobed or divided ....**Capsella**
                15b Leaves entire or merely toothed............**Thlaspi**
  6b Fruits more than 3 times as long as wide
    16a Flowers yellow, cream-colored, or orange
      17a Leaves entire or merely toothed
        18a Petals 1.5-3 cm. long .... **Erysimum**

18b Petals 0.5 cm. long or less .... **Draba**
17b At least lower leaves pinnately cleft or divided
19a Fruits distinctly beaked beyond the seeds
20a Valves of the fruit with 1 prominent midvein .... **Brassica**
20b Valves of the fruit 3-7-veined ....**Sinapis**
19b Fruits not conspicuously beaked, although the style sometimes appearing beak-like, but 3 mm. or less long
21a Seeds in 2 rows per locule .... **Rorippa**
21b Seeds in 1 row per locule
22a Fruits and stems terete .... **Sisymbrium**
22b Fruits and stems angled...**Barbarea**
16b Flowers white, pinkish or purplish
23a Seeds in 2 rows per locule
24a Aquatic or marsh herbs; at least lower leaves pinnately divided (in ours) .... **Rorippa**
24b Not aquatic; leaves not pinnately divided
25a Fruits 2-10 cm. long ....**Arabis**
25b Fruits less than 1.5 cm. long .... **Draba**
23b Seeds in 1 row per locule
26a Glabrous or pubescent with simple hairs
27a Valves of fruit not conspicuously veined their full length; plants often rhizomatous; leaves often divided...........................**Cardamine**
27b Valves of fruit distinctly 1-veined; leaves not divided ....**Arabis**
26b Pubescence, at least in part, of forked or branched hairs
28a Petals 1.8-2.5 cm. long .... **Hesperis**
28b Petals 1 cm. or less in length
29a Leaves pinnately divided..................................**Smelowskia**
29b Leaves not pinnately divided
30a Annuals, leaves nearly all basal; weedy species................... ....**Arabidopsis**
30b Usually biennials or perennials, if annual, not weedy......... ....**Arabis**

Arabidopsis

Annuals or perennials with a basal rosette of petioled leaves and alternate, sessile cauline leaves; flowers small, borne in axillary and terminal racemes; fruits dehiscent, linear, the valves 1-veined.

Arabidopsis thaliana (L.) Heynh. — Mouse-ear cress

Annuals 1-4 dm. tall; stems simple or freely branched; basal leaves 1-4 cm. long, oblanceolate, remotely serrulate, pubescence of both simple and forked hairs, cauline leaves few, lanceolate, nearly sessile, the upper usually glabrous; petals white, about 3 mm. long; fruits 1-1.5 cm. long, less than 1 mm. broad, glabrous; seeds in a single row in each locule.

Weedy in disturbed ground; introduced from Eurasia.

Arabis

Erect annual, biennial, or perennial herbs, or occasionally somewhat woody at the base; leaves commonly undivided, the basal petioled, the cauline generally sessile, often auriculate and clasping; flowers white, yellowish, or purple, sometimes showy; fruits dehiscent, slender, more or less flattened parallel to the partition, the valves generally 1-veined; seeds flattened, often winged, in 1 or 2 rows in each locule.

1a Plants of the valleys or foothills
  2a Glabrous and glaucous except at the base.............................. *A. glabra*
  2b Typically rough-pubescent throughout................................. *A. hirsuta*
1b Plants of the mountains or in the Columbia River Gorge
  3a Cauline leaves sagittate or auriculate at base
    4a Fruits erect.. *A. drummondii*
    4b Fruits reflexed to pendulous .... *A. holboellii*
  3b Cauline leaves not sagittate or auriculate at base
    5a Mature fruits 3-5 mm. wide, flattened; stems typically stellate-pubescent.. *A. platysperma*
    5b Mature fruits to 2 mm. wide; stems glabrous or nearly so .... *A. furcata*

Arabis drummondii Gray — Drummond's rockcress

Biennials or perennials; stems erect, mostly simple, 2-9 dm. tall, leafy; basal leaves petioled, 2-8 cm. long, the blades oblanceolate, glabrous or slightly pubescent with simple or branched hairs, at least the upper cauline leaves sagittate, glabrous; petals 7-10 mm. long, white or pinkish; fruits erect, slender, short-pediceled, 3.5-10 cm. long, 2-3 mm. wide.

In gravelly or rocky soils in the mountains.

Arabis furcata Wats. — Fork-haired rockcress

Caespitose perennials; stems slender, 0.5-4 dm. tall; basal leaves petioled, the blades 2-5 cm. long, obovate, or spatulate, entire or few-toothed, nearly glabrous or somewhat stellate-pubescent, the cauline leaves narrowly or broadly oblong, sessile; petals white, spreading above the calyx; fruits narrow, erect or nearly so, 2-4 cm. long, about 1.5-2 mm. wide.

Alpine and subalpine areas in the Cascades and the Olympics, also in the Columbia River Gorge.

Arabis glabra (L.) Bernh. — Tower mustard

Biennials; stems 3-15 dm. tall, slender, generally unbranched, from a long slender tapering root; whole plant somewhat bluish-green, and glabrous except at the base; lower leaves broadest above middle, narrowing to the base, usually coarsely toothed or lobed, veins and margins more or less covered with mostly 2-forked hairs, cauline leaves clasping, with auriculate lobes, lanceolate, somewhat toothed or with entire margins; flowers very small, greenish or yellowish-white; fruits narrow, slightly flattened, 6-10 cm. long, erect, sometimes appressed to stem; seeds in 2 somewhat irregular rows in each locule.

Dry fields, stream banks and in open woods.

Arabis hirsuta (L.) Scop. Hairy rockcress

Annuals, biennials or short-lived perennials; stems 1.5-7 dm. tall, erect, mostly unbranched, somewhat stiff-hairy at least below, bearing a raceme of white to pinkish flowers; basal leaves oblong to spatulate, narrowed at base to short winged petioles, cauline leaves long-ovate, the upper becoming narrower, nearly erect, auriculate, entire or coarsely dentate, at least the lower surface more or less short-hairy; petals 3-9 mm. long, white to pinkish; fruits 3-6 cm. long, glabrous, flattened, erect to ascending.

Found occasionally on sunny sandy bluffs along the coast or along rivers.

Arabis holboellii Hornem. Holboell's rockcress

Biennials or short-lived perennials; stems 1-9 dm. tall, rough at least below with branched hairs; basal leaves 1.5-5 cm. long, narrowly oblanceolate to spatulate, entire to toothed, densely pubescent with branched hairs, cauline leaves, or at least the uppermost, usually auriculate and clasping; petals pinkish-purple or sometimes nearly white, 5-10 mm. long; fruits 3-7 cm. long, 1-2.5 mm. broad, glabrous, straight or slightly curved, reflexed to pendulous.

Widespread variable species, but not common in our range.

Arabis hirsuta

Arabis platysperma Gray Flatseed rockcress

Caespitose perennials; stems more or less erect, 0.5-4 dm. tall from a somewhat woody branching caudex, the whole plant minutely pubescent or nearly glabrous; leaves entire, the basal oblanceolate or spatulate, narrowed to a petiole, the cauline leaves lanceolate, the upper ones sessile; petals white to pink or purplish; fruits flattened, erect or nearly so, 3-7 cm. long, 3-6 mm. wide.

High Cascades.

Athysanus

Small pubescent annuals; stems branching near the base; flowers very small in slender racemes; petals white or absent; fruits on recurved pedicels, small, flattened, orbicular or short-oval, 1-seeded, indehiscent.

Athysanus pusillus (Hook.) Greene Sandweed

Low, sometimes nearly prostrate annuals, generally with slender spreading branches; leaves mostly basal, entire or few-toothed, pubescent with simple and/or branched hairs; sepals purplish, about 1 mm. long; petals white, 1-2 mm. long, or sometimes lacking; fruits 2-2.5 mm. in diameter, pubescent.

Dry sandy or rocky ground.

Barbarea

Biennial or perennial herbs with angular stems; basal leaves pinnately lobed to compound, usually with a large terminal segment; flowers yellow; fruits dehiscent, slender, somewhat 4-angled, style short, beak-like in fruit; seeds in a single row in each locule.

Petals 3-5 mm. long; beak-like style stout in fruit, 0.5-1 mm. long .... B. *orthoceras*

Petals 6-8 mm. long; beak-like style slender in fruit, 1.5-3 mm. long .... B. *vulgaris*

Barbarea orthoceras Ledeb. Winter cress

Biennials; stems erect, 1-7 dm. tall, mostly smooth, dark green or often dark purplish; leaves all pinnately lobed with a large terminal segment and small unequal irregularly placed lateral segments, smooth, dark green, somewhat succulent, reduced upwards, often clasping; petals 3-5 mm. long, yellow; fruits 1.5-5 cm. long, linear, 4-angled, erect or somewhat spreading, narrowed to a short stout beak-like style.

Rather common along road sides and on waste ground, particularly in moist areas.

Barbarea orthoceras

Barbarea vulgaris R. Br. Common winter cress or Yellow rocket

Biennials; stems erect, 2-8 dm. tall, branched above; basal leaves with 1-4 pairs of lateral lobes, upper cauline leaves entire to dentate; petals 6-8 mm. long, yellow; fruits 1-3 cm. long, erect or ascending with a slender beak-like style.

Disturbed areas; introduced from Eurasia.

Brassica

Annuals or biennials, smooth or with scattered stiff hairs; lower leaves pinnately parted or divided, with large terminal segments, upper leaves becoming entire; flowers in racemes; petals yellow, rarely white, alternating at base with 4 minute dark green glands; stamens 6; fruits cylindrical, with a stout beak; seeds globose, in 1 row in each locule.

1a Cauline leaves clasping the stem .... B. *rapa*

1b Cauline leaves not clasping the stem.

  2a Fruits closely appressed to the stem .... B. *nigra*

  2b Fruits ascending, not appressed to the stem .... B. *juncea*

Brassica juncea (L.) Czern. Indian mustard

Annuals, mostly glabrous, pale green to glaucous; stems erect, 4-10 dm. tall; lower leaves pinnately cleft, the terminal segment larger, coarsely dentate, leaves becoming much smaller and nearly to completely entire above, the uppermost nearly linear at bases of lower flowers, all leaves tapered to the base, not clasping; flowers pale yellow, 8-12 mm. long in terminal and sometimes lateral racemes; fruits ascending but not at all appressed, the beak seedless, making up about one-fourth the length of the fruit.

Weedy in disturbed areas; introduced from Eurasia.

Brassica nigra (L.) Koch Black mustard

Annuals; stems 3-20 dm. tall, sometimes taller, dark green, mostly glabrous; leaves all petioled, the lower pinnately parted or divided, with a very large lobed and sharply toothed terminal segment and small lateral segments or leaflets, leaves less lobed upward, sometimes becoming entire; petals narrow-clawed, 7-11 mm. long; fruits 1-2 cm. long, erect, appressed closely to stem, irregularly cylindrical, slightly 4-angled.

Fields and road sides; introduced from Europe; not as abundant as *B. rapa*.

Brassica rapa L.

Field mustard, Wild turnip or Yellow mustard

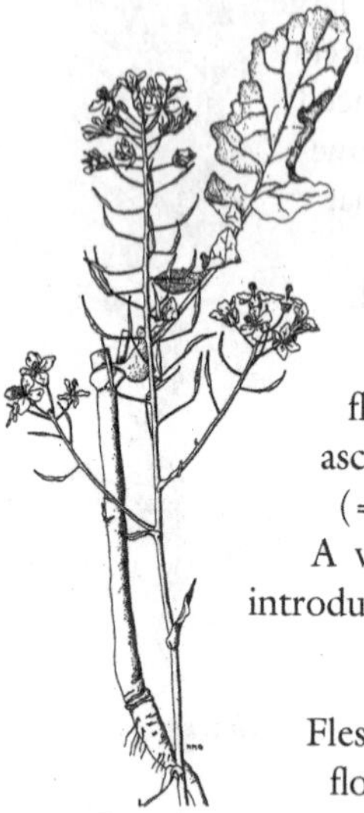

Taprooted annuals; stems erect 2-10 dm. tall, simple to freely branched; lower leaves pinnately parted, with large terminal and small lateral segments, irregularly toothed, broadly clasping at base, usually with scattered stiff hairs from cushion-like bases on upper surface, leaves becoming less divided upward, until entire with broad clasping bases, tapering toward apex, generally bluish-green and glabrous; flowers 6-12 mm. long, golden yellow, very fragrant; fruits ascending to spreading, stout, cylindrical, 3-7 cm. long. (=*Brassica campestris* L.)

A very common weed especially of fields and gardens; introduced from Europe.

Brassica rapa

Cakile

Fleshy seaside annuals or perennials with purplish or white flowers; fruits fleshy at first, becoming corky, jointed at about the middle, each segment indehiscent, one-seeded, or the lower sometimes seedless; upper segment separating at maturity.

Leaves deeply pinnately lobed; fruits with 2 lateral projections at the joint between the segments .... *C. maritima*
Leaves merely irregularly toothed or shallowly lobed; fruits without projections at the joint .... *C. edentula*

Cakile edentula (Bigel.) Hook. Sea rocket

Annuals; stems freely branching, ascending or erect, up to 8 dm. tall; leaves fleshy, obovate, or oblanceolate, the margins wavy to irregularly toothed or shallowly lobed; petals white to pale lavender; fruit 1-3 cm. long, upper segment slightly longer than the lower, 4-angled, narrowed into a flattened beak, the lower segment cylindrical.

Sea beaches.

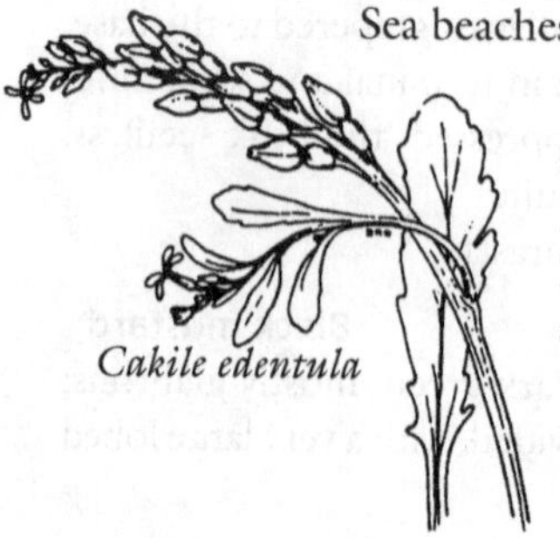

Cakile maritima Scop. European sea rocket

Similar to *Cakile edentula*; leaves deeply pinnately lobed; upper segment of fruit much longer than the lower, obtuse to acute, with 2 lateral projections at the joint between the segments.

Naturalized on sea beaches; native of Europe.

Cakile edentula

Capsella

Annuals or biennials with pinnately-parted leaves and small white or pink flowers; fruits several-seeded, more or less flattened, elliptical or broadly heart-shaped at apex.

Capsella bursa-pastoris (L.) Medic. Shepherd's-purse

Taprooted annuals; stems erect, 0.5-5 dm. tall, somewhat pubescent at least below, arising from a rosette of basal leaves; lower leaves petioled, pinnately cleft or parted, with the terminal lobe the largest, or rarely subentire, upper leaves toothed or entire, broadly clasping at the base; fruits flattened perpendicular to the partition, broadly heart-shaped, 4-8 mm. long.

Capsella bursa-pastoris

A common weed of waste land and gardens.

Cardamine

Annuals, biennials or perennials; leaves entire or generally parted to compound, sometimes with a basal rosette; flowers white, pink, rose or lavender; fruits slender, usually flattened, often explosively dehiscent; seeds in a single row in each locule.

1a Leaves all simple and not deeply divided.
  2a Stems mostly 2-12 cm. tall; leaves pinnately veined .... *C. bellidifolia*
  2b Stems over 15 cm. tall; leaves palmately veined .... *C. cordifolia*
1b At least some of the leaves compound or deeply divided.
  3a Plants from a taproot or fibrous roots; petals 2-4 mm. long, flowers inconspicuous
    4a Leaflets not stalked, at least of the upper leaves, linear to lanceolate; basal leaves few .... *C. pensylvanica*
    4b Leaflets distinctly stalked, usually orbicular to broadly obovate
      5a Petals 2-4 mm. long; fruits 6-24-seeded; stems 1-few, erect .... *C. oligosperma*
      5b Petals 1.5-2 mm. long; fruits 22-36-seeded; stems several, the outer decumbent .... *C. hirsuta*
  3b Plants rhizomatous; flowers often showy
    6a Petals 3-7 mm. long
      7a Basal leaves mostly simple, entire to shallowly lobed, rarely with 2 small lateral leaflets .... *C. breweri*
      7b Basal leaves compound with 3-7 leaflets .... *C. occidentalis*
    6b Petals averaging over 7 mm. long
      8a Leaflets generally 7 to 15 .... *C. penduliflora*
      8b Leaflets typically 3 to 5 (rarely 7)
        9a Stems averaging less than 2 dm. tall; petals pink to purple .... *C. nuttallii*
        9b Stems averaging over 2 dm. tall; flowers white to pinkish
          10a Rhizomes mostly less than 1 cm. long, tuber-like .... *C. californica*
          10b Rhizomes elongate, over 2 cm. long, not tuber-like .... *C. angulata*

Cardamine angulata Hook. Wood bittercress

Stems 3-10 dm. tall, weakly erect, usually simple, from a creeping rhizome; leaves with mostly 3 (sometimes 5) segments or leaflets, these nearly equal in the basal leaves, smooth or pubescent, palmately veined, toothed or shallowly lobed, upper leaves with a large terminal segment; flowers 8-14 mm. long, white, showy.

Moist woods, often along stream banks.

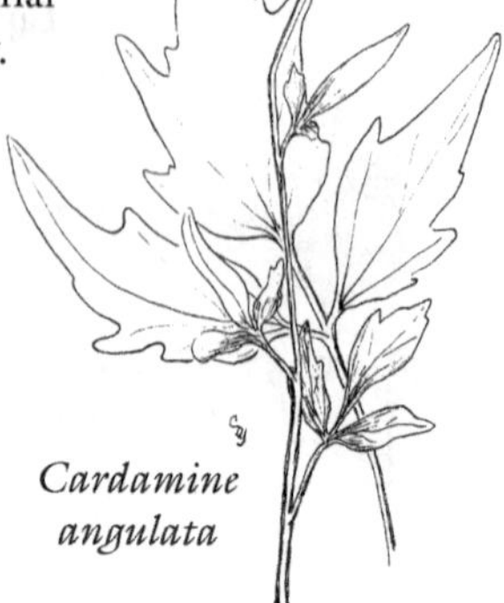

Cardamine angulata

Cardamine bellidifolia L.

Alpine or Mountain bittercress

Glabrous tufted perennials 2-12 cm. tall, from a branching base; leaves simple, ovate or oblong, generally obtuse, mostly entire, rarely 1-4-toothed, petioles slender, cauline leaves few, if any; flowers several in a short raceme; petals 3-5 mm. long, white; fruit 1.5-3.5 cm. long, erect.

High Cascades.

Cardamine breweri Wats. Brewer's bittercress

Perennials; stems 1.5-7 dm. tall, weakly erect, from creeping rhizomes; basal leaves simple, cordate, coarsely sinuate, or occasionally compound with 2 small lateral leaflets, cauline leaves mostly with 3 leaflets, rarely 5, leaflets of lower cauline leaves rounded, often sinuate, those of upper leaves longer, the terminal leaflet the largest; flowers white, 3-7 mm. long; fruits 1.5-3 cm. long, erect or nearly so.

Wet places in the mountains and more rarely along the coast.

Cardamine californica (Nutt. ex Torr. & Gray) Greene

Perennials from short thick rhizomes; stems 2-7 dm. tall; basal leaves typically with 3 leaflets (occasionally simple to 7-foliolate), leaflets ovate to cordate, cauline leaves alternate, the lower long-petioled, the upper becoming sessile or nearly so, with 3-5 leaflets, the margins variously toothed to entire; flowers white to pinkish, 8-14 mm. long; fruits 2-5 cm. long, erect.

A variable species with several varieties in our area.

Open woods and hillsides.

Cardamine cordifolia Gray Large mountain bittercress

Perennials; stems 1.5-6 dm. long, erect, from long slender creeping rhizomes; leaves simple, entire or shallowly sinuate to dentate, minutely mucronate at the vein tips, cordate, palmately veined, all petioled, the upper petioles shorter; flowers white, 6-12 mm. long, in a terminal raceme; fruits 2-3.5 cm. long, spreading to erect.

Stream banks or alpine meadows in the Cascades.

Cardamine hirsuta L.
Annuals with several decumbent to ascending stems up to 3 dm. tall from a basal rosette; leaves pinnately compound with 4-6 lateral leaflets and a larger terminal leaflet, cauline leaves much reduced, all the leaflets tapered to a short-stalked, often pubescent on the upper surface, petioles hirsute; petals white, 1.5-2 mm. long; fruits 22-36-seeded.

Introduced from Europe; weedy in lawns and along roadsides, especially in moist areas; not common, but often confused with *C. oligosperma*.

Cardamine nuttallii Greene — Spring beauty
Perennials from short rhizomes; stems usually less than 2 dm. tall; basal leaf single, simple to palmately compound with 3-5 leaflets, cauline leaves usually 2, 2-5-foliolate, the leaflets oblong, margins entire or conspicuously mucronate and callous-tipped; petals pink to purple, 1-1.5 cm. long; fruits 2-4 cm. long, erect.

A variable species with several varieties in our area.

(=*Dentaria tenella* Pursh and *Cardamine pulcherrima* Greene)

Moist shaded areas.

Cardamine nuttallii

Cardamine occidentalis (Wats. ex Robins.) T.J. Howell — Western bittercress
Perennials from short thick rhizomes; stems 2-4 dm. tall, usually branched, sometimes decumbent and rooting at the nodes; basal leaves pinnate with 3-7 leaflets, the terminal leaflet the largest, upper cauline leaves sessile or subsessile; petals 4-6 mm. long, white; fruits 2-3 cm. long, erect.

Wet, often muddy banks along lakes and streams; from the Coast Range to the coast.

Cardamine oligosperma Nutt. — Little western bittercress
Taprooted annuals or biennials; stems simple or branching, 0.5-4 dm. tall, smooth or sparsely pubescent, arising from a basal rosette of leaves; basal and cauline leaves similar, pinnately compound, leaflets 3-11, mostly suborbicular to oblanceolate, 3-30 mm. long (rarely more), usually with a single shallow lobe on each side near the apex, or the terminal leaflet sometimes 2 or more lobed on each side, all the leaflets short-stalked; petals 2-4 mm. long, white; fruits 1.5-3 cm. long, slender, ascending to erect, 6-24-seeded.

Common in moist places; often weedy.

Cardamine penduliflora Schulz — Meadow bittercress
Glabrous perennials from tuberous rhizomes; stems 1.5-4 dm. long, weakly erect, often rooting at the nodes, simple or sparingly branched; leaves pinnately compound, lateral leaflets in several pairs, ovate to narrowly oblong, terminal leaflets larger, oblong or broadly oblanceolate, entire or wavy; flowers white, rarely pinkish, in terminal racemes, appearing corymbose at first; petals 7-12 mm. long, several times as long as the sepals; fruits 2-4 cm. long, erect or nearly so.

Wet meadows, swampy ground and marshes.

Cardamine pensylvanica Muhl. ex Willd. Bittercress
Biennials or short-lived perennials; stems simple or branched, 1-7 dm. long; leaves pinnately compound; leaflets 5 to 13, the terminal leaflet the largest, leaflets of lower leaves roundish, upper linear to lanceolate, generally sessile; petals 2-4 mm. long, white; fruits slender 2-3 cm. long, 20-40-seeded.

Shady moist places, stream banks and lake margins.

Cochlearia

Annual or perennial fleshy seacoast herbs with small white, yellow or purple flowers; fruits dehiscent, more or less globose to oblong; seeds few, in 2 rows in each locule.

Cochlearia officinalis L. var. *arctica* (Schlecht. ex DC.) Gelert Scurvygrass
Fleshy, branching, glabrous herbs about 3 dm. tall or less; stems erect or more often decumbent; lower leaves deltate, long-petioled, entire or with strongly wavy margins, upper leaves obscurely toothed, ovate or oblong, sessile or nearly so; flowers white; fruit globose or ovoid, 3-7 mm. long, smooth or slightly wrinkled, 2-valved.

Along the coast, often on off shore rocks.

Crambe

Coarse annuals or perennials often with a branching crown; herbage generally fleshy; stems thick, sparingly branched; leaves very large, more or less lobed or toothed; the 2 outer stamens simple, the 4 inner often with a tooth on the inner side of the filament; fruit of 2 segments, each 1-loculed, 1-seeded, the terminal segment fertile, becoming globose, the basal segment remaining undeveloped.

Crambe maritima L. Sea-cabbage or Sea-kale
Annuals, the crown sometimes branched, bearing long-petioled, thick fleshy, glaucous, cabbage-like leaves, the blades reaching 4.5 dm. or more in length, and over 3 dm. in width, with large shallow lobes and irregular teeth, the margins wavy, the midrib and large veins very thick and fleshy, the base more or less heart-shaped, upper leaves becoming smaller, less wavy margined, and wedge-shaped at the base; flowering stems 2.5-7.5 dm. tall, bearing panicle-like flower clusters the branches of which elongate into slender racemes, with developing fruits below as the buds open successively above; buds yellow; flowers pediceled, not quite iso-diametric, 1-3.5 cm. in diameter, white with centers at first yellowish, changing to purplish; sepals greenish, broadly ovate or elliptic, 3 mm. long; petals several times as long as sepals, nearly round or broadly elliptic, abruptly narrowed to a slender claw, arranged in overlapping pairs; ovary 2-loculed, each locule 1-ovuled, the ovule of the lower not maturing; pedicels elongating in fruit; fruit of 2 segments, the lower thickened, stalk-like, containing a rudimentary seed; the upper globose or nearly so, appearing near maturity like a glaucous green grape, 1-seeded.

A plant of the Atlantic seacoast of Europe, sparingly introduced on the beaches of Oregon.

Draba

Small annuals or perennials with white or yellow flowers; fruits dehiscent, flattened parallel to the partition, oval to oblong or linear, with seeds in 2 rows per locule.

1a Petals white, mostly annuals
  2a Leaves all basal; petals cleft to the middle .... *D. verna*
  2b Cauline leaves present; petals somewhat notched
    3a Fruits linear, slightly curved .... *D. stenoloba*
    3b Fruits narrowly elliptical, not curved........................ *D. brachycarpa*
1b Petals yellow, mostly perennials
  4a Fruits ovate-oblong, densely stellate pubescent .... *D. aureola*
  4b Fruits linear to narrowly ovate, usually glabrous
    5a Stems pubescent with simple or once-forked hairs .... *D. albertina*
    5b Stems pubescent, at least below, of branched hairs .... *D. stenoloba*

Draba albertina Greene

Stems erect, 0.5-3 dm. tall, pubescent with simple hairs; basal leaves obovate, entire or nearly so, 0.5-3.5 cm. long, at least the lower surface pubescent with branched hairs, cauline leaves few, sessile, lanceolate; petals 2-3 mm. long, yellow; fruits 5-15 mm. long, narrowly ovate.

Meadows and open areas in the mountains.

Draba aureola Wats. — Golden Draba

Perennials; stems simple or sometimes branched, 0.5-1.5 dm. tall, whole plant densely stellate pubescent; leaves crowded below, oblanceolate to oblong; flowers in a crowded raceme; petals 3.5-5 mm. long, yellow; fruits ovate-oblong, 5-15 mm. long, stellate pubescent.

Known from several points in the high Cascades, including Mt. Rainier, Mt. Hood, and the Three Sisters.

Draba brachycarpa Nutt. ex Torr. & Gray — Shortpod whitlow-grass

Slender annuals; stems simple or branched, erect or spreading, 0.5-2 dm. tall; leaves 5-15 mm. long, lanceolate, elliptic, or ovate, entire or nearly so, stellate pubescent, nearly or quite sessile; petals white, 1-3 mm. long; fruits 2-6 mm. long, narrowly elliptical, acute, minutely tipped, generally glabrous.

Not common in our limits, but known from the Willamette Valley.

Draba stenoloba Ledeb. — Alaska whitlow-grass

Annuals or short-lived perennials; stems one or several, slender, 0.5-4 dm. tall; leaves mostly basal, 1-4 cm. long, oblanceolate to oblong or ovate, more or less pubescent, entire or toothed, basal leaves narrowed to petioles, cauline leaves sessile or nearly so; racemes mostly few-flowered; petals 2-4.5 mm. long, yellow at first, often fading to white; fruits linear, usually slightly curved, about 1 cm. in length, usually glabrous.

Mountains of Washington.

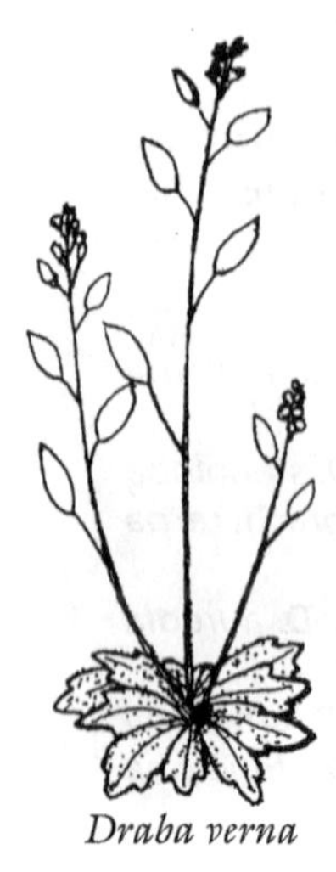
Draba verna

Draba verna L. — Spring whitlow-grass

Small annuals; stems 4-20 cm. tall, arising from a basal rosette of leaves; leaves oval to obovate, entire to distantly toothed, tapering into a petiole at base or sessile, 6-25 mm. long; petals 2-3 mm. long, white, cleft to the middle, twice as long as the sepals; fruits 3-10 mm. long, oval to obovate, the 2 valves falling away early in fruit, leaving the partition membrane.

A weedy species, common on hillsides in spring, making conspicuous white patches in moist places when in blossom, in spite of the diminutive size of the individual plant.

Erysimum

Annual, biennial or perennial herbs to subshrubs with simple leaves and cream-colored to yellow or orange to pink or purple generally showy flowers; fruits dehiscent, flattened or 4-angled, pubescent; seeds in 1 row in each locule (ours).

Fruits flattened, often twisted; montane species .... *E. arenicola*
Fruits 4-angled; in the foothills or the lowlands .... *E. capitatum*

Erysimum arenicola Wats. — Sand-dwelling wallflower

Pubescent perennials, 1-5 dm. tall; basal leaves linear to oblanceolate, cauline leaves similar; petals yellow, 1.5-2.5 cm. long; fruits 4-8 cm. long, flattened, often twisted, style beak-like.

High in the mountains.

Erysimum capitatum (Dougl.) Greene var. *capitatum* — Wallflower

Biennials or short-lived perennials; stems 2-10 dm. tall, stout or slender, leafy, whole plant more or less grayish-green with short, stiff, simple or branched hairs; basal leaves 4-15 cm. long, many, narrow or rarely spatulate, petioled below, sessile above, entire or dentate; flowers orange to yellow, 1.5-3 cm. long, in terminal and sometimes axillary racemes; fruits 4-angled, erect, ascending or spreading.

(*Erysimum asperum* misapplied)

A widespread highly variable species.

In a variety of habitats.

Idahoa

Dwarf annuals with leaves in a basal rosette, and several flowering stems bearing small solitary white flowers; fruits orbicular or nearly so, flattened, the 2 valves falling away leaving a papery partition attached to the peduncle; seeds several, winged.

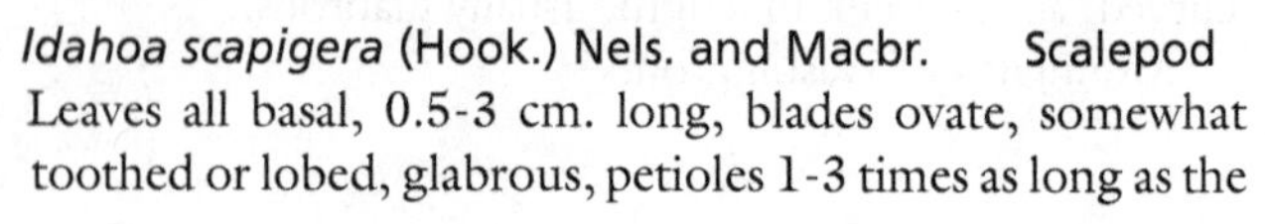

Idahoa scapigera (Hook.) Nels. and Macbr. — Scalepod

Leaves all basal, 0.5-3 cm. long, blades ovate, somewhat toothed or lobed, glabrous, petioles 1-3 times as long as the

Idahoa scapigera
a. flower
b. partition of fruit

blades; peduncles 2-13 cm. tall bearing a solitary flower; sepals purplish-red; petals white; fruits 6-12 mm. in diameter.

Springy places about rock outcroppings; not common west of the Cascades.

Hesperis

Biennials or perennials; flowers showy, white to purple; fruits dehiscent, linear to terete or 4-angled, the valves 1-3-veined, seeds in 1 row per locule.

Hesperis matronalis L. Dame's rocket

Stems 3-14 dm. tall, leafy, pubescent; leaves 5-20 cm. long, lanceolate to ovate-lanceolate, serrate-dentate, the lower long-petioled the upper often becoming sessile; petals 1.8-2.5 cm. long, white to purple; fruits 4-10 cm. long, terete, often twisted.

Escaped from cultivation and widely naturalized; introduced from Europe.

Lepidium

Annual to perennial herbs or shrubs; leaves toothed, pinnately cleft, or entire; flowers small white or rarely yellow or the petals sometime lacking; fruits dehiscent, round, ovate, or obovate, flattened perpendicular to the partition, often winged at the apex; seeds 1 per locule.

1a Cauline leaves not auriculate-clasping .... *L. virginicum*
1b Cauline leaves auriculate-clasping
  2a Anthers cream-colored; annuals .... *L. campestre*
  2b Anthers purple; perennials .... *L. heterophyllum*

Lepidium campestre (L.) R. Br. Field peppergrass

Annuals, usually grayish-green; stems leafy, 2-5 dm. tall, erect and branched above; basal leaves oblanceolate, petioled, cauline leaves broader, the upper sessile and auriculate-clasping; flowers numerous, small; petals 2-2.5 mm. long, white; fruits 5-6 mm. long, more or less ovate, covered with scale-like hairs, the margins and tips winged.

Weedy along road sides in fields and disturbed areas; introduced from Europe.

Lepidium heterophyllum Benth.

Perennials; stems 1.5-4.5 dm. tall, often decumbent, pubescent; basal leaves oblanceolate, petioled, cauline leaves auriculate-clasping, sessile, oblong to deltoid; petals 2.5-3 mm. long, white; anthers purple; fruits 4-7 mm. long, ovate-oblong, winged for 1/3 their length.

Weedy in disturbed areas; introduced from Europe.

Lepidium virginicum L. var. *menziesii* (DC.) C.L. Hitchc.
Tall peppergrass

Annuals or biennials; stems 1.5-7 dm. tall, often weakly erect, branched, minutely pubescent, at least above; lower leaves petioled, somewhat pubescent, pinnately cleft, the segments more or less toothed, upper leaves often nearly entire; petals 1-3 mm. long, white; fruits nearly orbicular, notched at apex, about 3 mm. broad, long-pediceled.

(=*Lepidium menziesii* DC.)
Northern coast.

Lunaria

Annuals, biennials or perennials; leaves cordate to rounded at the base and deeply dentate; flowers showy, purplish (rarely white); fruits flattened parallel to the partition; seeds in 2 rows in each locule.

Lunaria annua L. Money plant or Honesty
Pubescent annuals or biennials; stems much-branched, 5-10 dm. in height; leaf blades 3-10 cm. long, lower leaves petioled, the uppermost cauline leaves sessile; petals about 2 cm. long, bluish-purple, rarely white; fruits with a stipe-like base, suborbicular to oblong-ovate, 3.5-4.5 cm. long.

An escape from cultivation, often found along road sides and disturbed areas; introduced from Europe.

Raphanus

Annuals or biennials; lower leaves pinnately compound or parted, the terminal lobe or leaflet larger than the lateral, lower leaves petioled, the upper subsessile or short-petioled; fruit thick, spongy within, more or less constricted between seeds, not splitting longitudinally to discharge seeds, but sometimes breaking transversely.

Fruits deeply constricted between the seeds; petals usually yellow to white....... .... *R. raphanistrum*
Fruits only slightly, if at all, constricted between the seeds; flowers commonly purple, but varying to white .... *R. sativus*

Raphanus raphanistrum L. Jointed charlock
Plant somewhat similar to R. *sativus*; stems 3-8 dm. tall, pubescent below, glabrous above; petals yellow to white, often purplish veined; fruits long, narrow, deeply constricted between seeds like a string of beads, long-beaked, 4-12-seeded.

Weedy, but less common than R. *sativus.*

Raphanus sativus L. Wild radish
Taprooted annuals or biennials; stems 4-12 dm. tall, freely branching, smooth or sparsely pubescent; lower leaves pinnately parted, with several small lateral segments, the terminal much larger and round, segments wavy-margined or with low rounded teeth, upper leaves toothed, undivided or with a few small lateral segments; petals 15-25 mm. long, purple or rarely white, the veins conspicuous; fruits 2-6 cm. long, 1-3- (or rarely 5-seeded), only slightly constricted between seeds, usually inflated on 1 side, straight on the other, narrowed abruptly to a long beak.

Common in waste places. The garden radish "run wild."

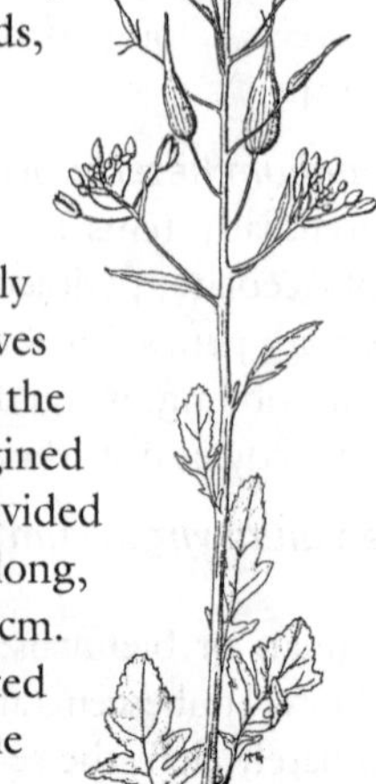

Raphanus sativus

Rorippa

Herbs growing mostly in the water or wet places; leaves entire to toothed, or pinnately lobed or divided; petals yellow or white or rarely absent; fruits dehiscent, cylindrical, to globose with minute seeds usually in 2 rows in each locule.

1a Flowers white .... *R. nasturtium-aquaticum*
1b Flowers yellow
  2a Perennials with creeping underground roots or slender rhizomes; petals 2.5-5.5 mm. long
    3a Leaves usually deeply lobed, the lobes entire or with a few teeth; native species, often in alkaline soil .... *R. sinuata*
    3b Leaves deeply lobed to pinnatifid; usually sharply toothed; weedy species .... *R. sylvestris*
  2b Annuals or biennials (or if rarely a short-lived perennial then taprooted); petals 1-3.5 mm. long
    4a Fruits linear, distinctly curved, pedicels about half as long as the fruits .... *R. curvisiliqua*
    4b Fruits oblong, straight or slightly curved, pedicels about as long as the fruits .... *R. palustris*

Rorippa curvisiliqua (Hook.) Bess. Western yellowcress
Annuals or biennials; stems erect or spreading, 1-4 dm. long, branched above, often purplish; leaves 2-8 cm. long, usually pinnately cleft or parted, the segments narrow or broader and lobed or toothed, green or purplish; flowers 1-2 mm. long, yellow, borne in racemes at the ends of branches; fruits 6-15 mm. long, terete, curved upward, abruptly narrowed to a very short style.

Common in vernal pools, marshes, stream banks and wet meadows.

Rorippa curvisiliqua

Rorippa nasturtium-aquaticum (L.) Hayek Water-cress
Glabrous perennials; stems 1-8 dm. long, leafy, rooting at the nodes, often partially submerged or floating; leaves 4-6 cm. long, mostly pinnately compound with 3-11 ovate to oblong or nearly orbicular leaflets, the terminal leaflet often cordate and usually much larger than the laterals; petals 3-4 mm. long, white or tinged with purple; fruits 1-2.5 cm. long, oblong, straight or slightly curved.

Shallow quiet water, often in road side ditches; introduced from Europe.

A similar looking species, *Rorippa microphylla* (Boenn. ex Reichenb.) Hylander ex Love & Love, has been reported to be introduced here. It differs from *R. nasturtium-aquaticum* in having mature siliques 1-1.5 mm. wide with seeds in one row per locule as opposed to the 2-3 mm. wide siliques of *R. nasturtium-aquaticum* which has seeds in two rows in each locule.

Rorippa palustris (L.) Bess. Marsh yellowcress

Annuals to short-lived perennials; stems erect, 1-10 dm. tall, erect to decumbent; leaves short-petiolate to sessile, 5-30 cm. long, more or less oblanceolate, dentate to pinnately lobed; petals 1-3.5 mm. long, yellow; fruits 3-14 mm. long, straight or slightly curved.

[=*Rorippa islandica* (Oed.) Barb.]

Wet areas.

Rorippa sinuata (Torr. & Gray) Hitchc. Spreading yellowcress

Perennials; stems prostrate to ascending, 1-5 dm. long; basal and lower cauline leaf blades 3-8 cm. long on short, winged petioles, oblanceolate, dentate to deeply pinnately lobed, upper cauline leaves becoming sessile and more or less clasping the stem; petals 2.5-5.5 mm. long, light yellow; fruits 5-15 mm. long, about 1.5 mm wide, often curved.

Lake shores and wet areas, often in alkaline soils.

Rorippa sylvestris (L.) Bess. Creeping yellowcress

Perennials from slender rhizomes; stems 1.5-8 dm. tall, erect to decumbent, usually much-branched; leaves ovate to oblong or oblanceolate in outline, deeply lobed or pinnately divided; petals yellow, 3-5.5 mm. long; fruits 1-2.5 cm. long, linear.

Occasional in waste ground.

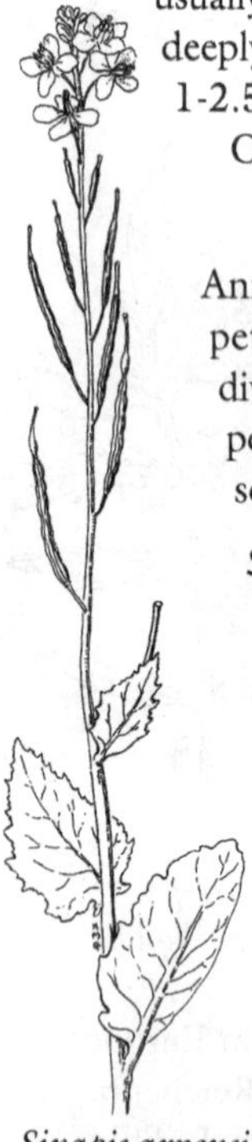

Sinapis

Annuals or rarely perennials with erect leafy stems; lower leaves petioled, the upper short-petioled to sessile, pinnately lobed or divided, the terminal lobes larger than the lateral; sepals yellowish; petals yellow; fruits dehiscent, linear to oblong, strongly beaked, seeds in a single row in each locule.

Sinapis arvensis L. Charlock

Annuals 2-6 dm. tall; lower leaves with a large terminal lobe, usually irregularly toothed, upper leaves simple, toothed, sessile; petals about 1 cm. long; fruits 2-3.5 cm. long, 5-12-seeded, slightly constricted between the seeds.

[=*Brassica kaber* (DC.) Wheeler]

Weedy in disturbed areas and fields; introduced from the Mediterranean region.

Sinapis arvensis

Sisymbrium

Erect annuals or biennials with pinnately parted or compound leaves and small yellow flowers; fruits dehiscent, linear to cylindrical, seeds in 1 row per locule (ours).

Fruits 5-10 cm. long, spreading............................ *S. altissimum*

Fruits less than 2 cm. long, closely appressed to the stem .... *S. officinale*

Sisymbrium altissimum *Sisymbrium officinale*

Sisymbrium altissimum L. Jim Hill mustard

Freely branched annuals; stems erect, 3-15 dm. tall; leaves lanceolate to oblanceolate in outline, the upper cauline leaves becoming pinnately compound with linear leaflets; petals pale yellow, 6-8 mm. long; fruits linear, nearly cylindrical, 5-10 cm. long, spreading on pedicels of nearly equal diameter.

Weedy in disturbed areas; introduced from Europe.

Sisymbrium officinale (L.) Scop. Hedge mustard

Stems stiffly erect, 3-10 dm. tall, with ungainly spreading branches above, somewhat stiff-hairy; leaves slightly rough-hairy, pinnately parted or divided, with a large terminal segment, lower leaves petioled, upper cauline leaves usually much reduced and sessile; petals 3-4 mm. long, pale yellow; fruits awl-shaped, 8-18 mm. long, very slender, tapering to tip, erect, closely appressed to the stem.

Common in waste places; introduced from Europe.

Smelowskia

Low tufted perennials usually with whitish stellate-pubescent stems and foliage; leaves once or twice pinnately cleft; flowers small whitish, yellowish or purplish, borne in racemes; fruit dehiscent, small, nearly terete to somewhat flattened at right angles to the partition, the valves keeled.

Fruits linear, 5-12 mm. long, 1.5-2.5 mm. wide .... *S. calycina*
Fruits ovate to oblong, 3-6 mm. long, about 4 mm. wide .... *S. ovalis*

Smelowskia calycina Mey.

Perennials; stem woody at base, branching, densely covered with old leaves; stems tufted, 5-20 cm. tall; herbage grayish from dense minute simple and stellate hairs; basal leaves petioled, blades pinnately lobed or compound, the segments narrow, ciliate, cauline leaves few and becoming sessile; flowers small, white or purplish-tinged; fruits 5-12 mm. long, linear, sometimes slightly curved.

High Cascades.

Smelowskia ovalis Jones

Resembling *S. calycina,* but the leaves not ciliate and the fruits shorter and ovate to narrowly oblong.

Usually in subalpine or alpine rocky areas in the high Cascades and the Olympics.

Subularia

Small glabrous annuals with linear to subulate leaves; inflorescence a raceme; flowers small; petals, if present, white; stamens 6; style absent; fruits dehiscent, obovate or ellipsoid, inflated, the valves 1-ribbed; seeds several, in 2 rows in each locule.

Subularia aquatica L. Awlwort

Glabrous annuals with subulate leaves 1-7 cm. long, all basal; scapes 2-15 cm. tall; flowers 2-8, racemose; sepals and petals about 1 mm. long; fruits 2-3 mm. long, elliptic to obovate.

Generally submerged or on muddy banks on the margins of ponds, lakes and streams in the mountains; not common.

Teesdalia

Annuals; leaves mostly all basal, usually pinnately dissected; flowers borne in racemes; petals small, white; fruits dehiscent, keeled and winged above, seeds 2 per locule.

Teesdalia nudicaulis (L.) R. Br. Shepherd's cress

Glabrous annuals; stems 0.5-2.5 dm. tall; leaves mostly in a basal rosette, the blades entire to pinnatifid, cauline leaves if present, few and much reduced; petals only about 1 mm. long, white; fruits 3-4 mm. long, oblong-obovate.

On beaches and in sandy ground along the coast and inland in the Willamette Valley.

Thlaspi

Glabrous annual or perennial herbs with entire or toothed leaves, the cauline leaves auriculate-clasping; flowers mostly white, small; fruits dehiscent, somewhat flattened at right angles to the partition, the valves keeled and often winged; seeds 2 or more in one row in each locule.

Annuals; fruits 10-17 mm. long .... *T. arvense*
Perennials; fruits 5-10 mm. long .... *T. montanum*

Thlaspi arvense L. Fanweed

Annuals; stems 1-5 dm. tall; basal leaves oblanceolate, short-petioled, cauline leaves gradually reduced upwards and becoming sessile and auriculate; petals 3-4 mm. long, white; fruits 10-17 mm. long, strongly compressed, oblong-obcordate, notched, winged-margined.

Weedy in fields and waste places; introduced from Europe.

Thalspi montanum L. Perennial pennycress

Perennials; stems several, slender, 1-4 dm. tall; basal leaves spatulate, petioled, 1.2-3.5 cm. long, cauline leaves oblong or ovate, sessile, auriculate-clasping, slightly extended at both sides of base; flowers small, white; fruits 5-10 mm. long, narrowly winged above, slightly notched.

(=*Thlaspi glaucum* Nels.)

In the mountains.

Thysanocarpus

Slender erect annuals with simple sessile leaves and very small white or purplish flowers; fruit indehiscent, flattened, 1-seeded, 1-loculed, surrounded by a flat nearly circular wing with conspicuous radiating veins, and/or with perforations.

Thysanocarpus curvipes Hook. Sand fringepod

Stems 1.5-8 dm. tall, slender, somewhat pubescent below, glabrous above; basal leaves 1.5-5 cm. long, in an apparent rosette, coarsely dentate, narrowed at base into a short petiole, cauline leaves mostly entire, small, auriculate, sessile; fruits pendulous, 5-8 mm. long, slightly longer than broad, the wing entire or wavy margined or perforated.

Dry ground or moist meadows to open woods; more common east of the Cascades.

SARRACENIACEAE
Pitcher-plant Family

Perennials; leaves all basal, pitcher-like or tubular, enlarged above and modified for trapping insects, partially filled with a liquid apparently possessing digestive properties; flowers solitary; sepals and petals generally 5; stamens many; pistil 1, compound; ovary superior; style 1, usually peltate; fruit a many-seeded loculicidal capsule.

Darlingtonia californica

Darlingtonia

A monotypic genus.

Darlingtonia californica Torr.
California Pitcher-plant or Cobra-plant

Rhizomatous perennials; leaves 1-9 dm. long, tubular, greenish-yellow with purple mottling and translucent spots on the hood and with two yellowish to purple-green appendages, the tube bearing stiff, downward pointing hairs on the inside; peduncle up to 1 m. tall with a solitary, more or less nodding, showy flower; sepals yellowish-green, veins purple, 4-6 cm. long; petals purple, 2-4 cm. long; stamens 12-15; style deeply 5-lobed, stigmas 5; capsules 2.5-4.5 cm. long.

Bogs and seepages areas; more common in southwestern Oregon.

DROSERACEAE
Sundew Family

Annuals, perennials or even subshrubs; leaves generally in a basal rosette, usually coiled in the bud, the blades with gland-tipped hairs or sensitive bristles for trapping insects; flowers one to several on a long peduncle; calyx 4- or more often 5-lobed; petals usually 5; stamens 4-20; pistil 1, compound, ovary superior, styles 3-5, often bifid; fruit a loculicidal capsule.

Drosera

Annual or perennial herbs, usually reddish in color; leaves long-petiolate, the upper surface of the blade bearing gland-tipped hairs which trap insects, the leaf blade slowly put perceptibly folding over the prey; flowers borne in a slender 1-sided raceme; sepals and petals generally 5; stamens 5; capsule 3-5-valved.

Leaf blades nearly orbicular, leaves lying close to the ground .. *D. rotundifolia*
Leaf blades at least twice as long as broad, leaves erect or ascending .... *D. anglica*

Drosera anglica Huds. Long- or Narrow-leaved sundew

Perennials; leaves erect or ascending, the blades twice as long as wide or longer, oblong-oblanceolate to spatulate, 1-3.5 cm. long, 2-7 mm. wide, petioles 1.5-8 cm. long; scapes up to 2 dm. tall, 2-7-flowered; calyx 4-6 mm. long; petals 8-12 mm. long, white.

In bogs and swampy areas.

Drosera rotundifolia L. Round-leaved sundew

Rosette of leaves lying close to the ground; leaf blades more or less orbicular, 5-12 mm. long and equally as wide, fringed with red glandular hairs; flowering stems 5-25 cm. tall; sepals nearly as long as petals.

In bogs and swampy areas, usually in *Sphagnum*, along the coast and in the mountains.

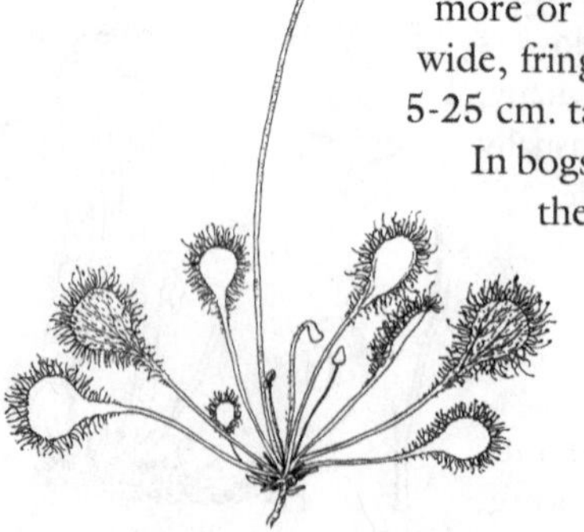

Drosera rotundifolia

CRASSULACEAE
Stone-crop Family

Fleshy annuals, perennials or subshrubs; leaves usually simple, alternate or opposite; inflorescence axillary or cymose; sepals 3-5; petals usually 3-5, free or fused at the base; stamens usually as many as the petals or twice as many; pistils usually 3-5; fruit of follicles, sometimes united at base.

Dwarf annuals; flowers axillary .... **Crassula**
Perennials; flowers in cymes .... **Sedum**

Crassula

Mostly small annuals, sometimes submerged or later stranded in drying pools; leaves opposite, fleshy, the bases fused and more or less sheathing; sepals 3-5, fused at the base; petals 3-5, free or slightly fused at the base; stamens of the same number as the petals; pistils 3-5; follicles 3-5.

Crassula aquatica (L.) Schonl. — Pigmy-weed
Stems decumbent, or if stranded, erect, 2-10 cm. long, often branched at the base and rooted at the lower nodes; leaves linear to oblanceolate, 2-6 mm. long; flowers axillary, usually on long pedicels; sepals 4, 0.5-1.5 mm. long; petals usually 4, whitish, 1-2 mm. long; follicles erect.
(=*Tillaea aquatica* L.)
Vernal pools, mud flats and salt marshes.

Sedum

Usually glabrous perennials (ours), generally low, spreading mat-like; flowers in cymes; petals 5, separate or slightly united; stamens 10; pistils 4-5; follicles several to a flower and separate or slightly united at base.

1a Leaves of the flowering stems opposite; mature pistils and follicles widely spreading .... *S. divergens*
1b Leaves of the flowering stems alternate
  2a Leaves widest at or below the middle, tapering to the apex
    3a Leaves strongly keeled; upper cauline leaves bearing propagules in the axils .... *S. stenopetalum*
    3b Leaves not strongly keeled .... *S. lanceolatum*
  2b Leaves widest above the middle, tapering to the base
    4a Petals separate or nearly so, acute, 7-10 mm. long .... *S. spathulifolium*
    4b Petals conspicuously united below
      5a Petals long acuminate, yellow; inflorescence usually 3-branched .... *S. oreganum*
      5b Petals obtuse or merely acute, buff-colored; inflorescence panicle-like .... *S. oregonense*

Sedum divergens Wats. — Cascade stonecrop
Stems leafy, prostrate at base, the tips ascending, 5-15 cm. long; leaves opposite, very thick and fleshy, roundish in outline, sessile, about 4-10 mm.

long; flowers yellow in a small compact simple or compound cyme; mature pistils and follicles widely spreading.

Alpine to subalpine in the Olympics and Cascades.

Sedum lanceolatum Torr. Lance-leaved stonecrop

Stems 5-20 cm. tall, erect from a cushion-like base of densely leafy branches from the spreading rhizomes; rosette leaves linear to broadly lanceolate to nearly cylindrical, fleshy, sessile, 5-30 mm. long, cauline leaves 3-10 mm. long; flowers yellow in somewhat compact compound cymes; pistils and follicles nearly erect to spreading.

Exceedingly variable species.

Rocky places from sea level to the mountains.

Sedum oreganum Nutt. Western stonecrop

Strongly rhizomatous glaucous perennials; stems horizontal at base, tips erect; flowering stems 3-20 cm. tall; leaves spatulate, flattened, but very fleshy; flowers yellow, often becoming pinkish; petals long acuminate, fused at base, longer than stamens; mature pistils and follicles strictly erect.

Rocky to gravelly areas; Coast Range and Cascades.

Sedum oregonense (Wats.) Peck Creamy stonecrop

Strongly rhizomatous perennials; rosette leaves 1-4 cm. long, the tips rounded or slightly notched, spatulate-ovate, cauline leaves 5-15 mm. long; petals buff-colored, fused at the base, approximate in length to stamens and mature pistils; pistils and follicles erect.

Rocky areas; Cascades.

Sedum spathulifolium Hook. Broad-leaved stonecrop

Plants fleshy, with stolons which send up rosettes at intervals and take root; leaves thick, fleshy, spatulate, glaucous, those on the stem shorter and thicker than in the rosette and sometimes reddish; decumbent stems arising from rosette, often reddish, bearing open bracteate cymes of flowers, bracts fleshy; sepals 5, glaucous, whitish, united at base; petals 5, yellow, narrow, twice as long as sepals or more; stamens 10, 5 opposite petals and attached to them, 5 opposite sepals and attached to calyx at base; pistils 5, free or perhaps slightly united at base, not widely spreading, narrowing to slender styles with capitate stigmas; follicles widely spreading at maturity.

In crevices of rocks; from the coast to the foothills of the Cascades.

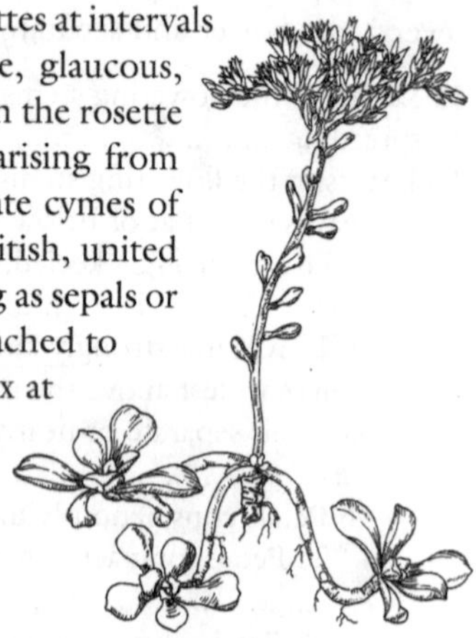

Sedum spathulifolium

Sedum stenopetalum Pursh — Worm-leaf stonecrop

Flowering stems erect from stout rhizomes, 0.5-3 dm. tall; leaves conspicuous in apparent rosettes at the ends of sterile branches, narrow, linear-lanceolate, acute at apex, 5-15 mm. long, at first dark green or often reddish in spring, and somewhat fleshy, though slightly flattened, later drying papery, the bases dilated and somewhat sheathing; leaf axils on flowering stems commonly bearing short leafy branches like small rosettes, which drop off and become means of propagation; cymes not conspicuously compound, branches generally less than 2.5 cm. long; petals deep yellow, narrow, lanceolate, longer than the stamens; pistils and follicles erect to widely spreading.

Rock crevices. Not common in our limits.

SAXIFRAGACEAE
Saxifrage Family

Annual or perennial herbs or subshrubs; leaves usually simple; flowers with a superior ovary and a floral tube or with an inferior ovary; calyx sometimes petaloid, calyx lobes usually 5, rarely 4 or 6; petals 4-6, rarely lacking; stamens 3-10; pistils, if simple, usually 2, or 1 and compound; fruit of 2 follicles or a capsule.

1a Fertile stamens 8 or 10
  2a Stamens 8 (ours); petals none .... **Chrysosplenium**
  2b Stamens 10; petals generally present
    3a Leaves peltate, nearly orbicular, up to 4 dm. wide .... **Darmera**
    3b Leaves not peltate
      4a Pistil 2-7-loculed; placentation axile; or pistils more than 1
        5a Matted rhizomatous plants of the high mountains; leaves elliptic, leathery, crenate-serrate except at the base; carpels distinct almost to the base .... **Leptarrhena**
        5b Not as above in all points
          6a Ovary 5-7-loculed; fruit a 5-7-horned capsule.................... .... **Penthorum**
          6b Ovary 2- (rarely 3-) loculed............................... **Saxifraga**
      4b Ovary 1-loculed; placentation parietal; pistil 1
        7a Petals entire .... **Tiarella**
        7b Petals pinnately or palmately cleft or fringed
          8a Petals pinnately cleft .... **Tellima**
          8b Petals palmately cleft (ours)........................ **Lithophragma**
1b Fertile stamens 5 or fewer
  9a Stamens 3 or 2; petals 4 or 5, filiform, attached to calyx tube.... **Tolmiea**
  9b Stamens 5 (ours); petals not as above
    10a Sterile stamens clustered at base of petals; flower solitary, terminal... .... **Parnassia**
    10b Sterile stamens none; flowers several to many
      11a Ovary superior .... **Bolandra**
      11b Ovary inferior

12a Bulblet-bearing at base and lower leaf axils .... **Suksdorfia**
12b Not bulblet-bearing at base
13a Petals pinnatifid or 3-5-cleft into linear lobes; inflorescence a raceme or a spike.. **Mitella**
13b Petals entire or rarely absent; inflorescence paniculate or cymose
14a Placentation parietal.. **Heuchera**
14b Placentae axile
15a Plants with stolons and rhizomes; petals 4-7 mm. long, early deciduous.. **Boykinia**
15b Plants with rhizomes, but lacking stolons; petals less than 3 mm. long, persistent... **Sullivantia**

Bolandra

Perennial herbs with bulblet-bearing creeping rhizomes; stems leafy; leaves alternate, palmately veined, stipulate; inflorescence open, panicle-like; calyx 5-lobed; petals 5, small; pistil 1, ovary superior, compound; fruit a loculicidal capsule.

Bolandra oregana Wats.

Stems leafy, glandular-pubescent, 1.5-6 dm. tall, bulblet-bearing at the base; leaf blades 2-7 cm. broad, reniform or cordate, irregularly lobed and toothed, glabrous or nearly so, grading into the bracts of the inflorescence which are more of less clasping, 1-3 cm. long and crenate-dentate; flowers in a loose panicled cyme; calyx tube about 6 mm. long, conspicuously veined, becoming broad, the narrow teeth about as long as the tube; petals narrow, purple; ovary superior.

Wet cliffs along the Willamette and Columbia Rivers and eastward.

Boykinia

Rhizomatous perennials; leaves alternate, 3-many-lobed, sharply toothed and with foliaceous or bristle-like stipules; flowers in axillary and terminal panicled cymes; petals early deciduous; stamens 5; ovary inferior, 2-loculed; fruit a capsule.

Stipules, at least those of the upper leaves, 8 mm. or more long, leaf-like......... .... *B. major*
Stipules 2-4 mm. long, all reduced and scarious or consisting of several bristles .... *B. occidentalis*

Boykinia major Gray

Mountain or Greater Boykinia

Perennials from scaly rhizomes; stems pubescent and glandular, 2-10 dm. tall; leaves less than 20 cm. wide, the lower long-petiolate, the upper nearly sessile, unequally 3-7-lobed half way to the base, the lobes again toothed or shallowly lobed, stipules leaf-like; inflorescence dense; calyx 2-6 mm. long; petals 4-7 mm. long.

Stream banks and wet meadows.

Boykinia major

Boykinia occidentalis Torr. & Gray — Slender or Coast Boykinia

Rhizomatous perennials; stems 1.5-6 dm. tall, more or less reddish-pubescent, especially at base, slender, with scattered leaves, stipules reduced to bristles; basal leaves cordate to reniform, 5-7-lobed, coarsely toothed, 2-8 cm. wide, long-petioled, cauline leaves less lobed, the upper cuneate, becoming bracts; inflorescence open; petals white, 5-6 mm. long, ephemeral.

This is an extremely variable species in the amount and color of the pubescence.

[=*Boykinia elata* (Nutt.) Greene]

Stream banks and moist forests.

Chrysosplenium

Prostrate perennials; stems succulent, rooting at the nodes; leaves opposite or alternate; flowers inconspicuous, solitary in the leaf axils or in leafy cymes; calyx lobes 4; petals lacking; stamens 4 or 8; ovary inferior, compound, 1-loculed; fruit a capsule.

Chrysosplenium glechomaefolium Nutt. — Golden saxifrage

Glabrous perennials; stems up to 2 dm. long, rooting at the nodes, usually mat-forming; leaves, all but the uppermost, opposite, leaf blades broadly ovate, crenate-serrate, 3-15 mm. wide and up to 2 cm. long on petioles up to 1 cm. long; flowers solitary in the upper leaf axils; calyx lobes 1.5-2 mm. long; stamens 8; capsule 2-beaked.

Shady moist habitats.

Darmera

A monotypic genus.

Darmera peltata (Torr.) Voss — Indian rhubarb or Umbrella plant

Perennials from scaly rhizomes; leaves all basal on petioles up to 1.5 m. long, blades peltate, nearly orbicular, to 4 dm. wide, deeply 7-15-lobed and irregularly toothed; inflorescence appearing before the leaves, 3-20 dm. long, more or less flat-topped; calyx lobes reflexed, 2.5-4 mm. long; petals 4.5-7 mm. long, white to pink; stamens 10; pistils 2; fruit of 2 follicles, 6-12 mm. long, purplish.

[=*Peltiplyllum peltatum* (Torr.) Engl.]

Usually on rocks in streams.

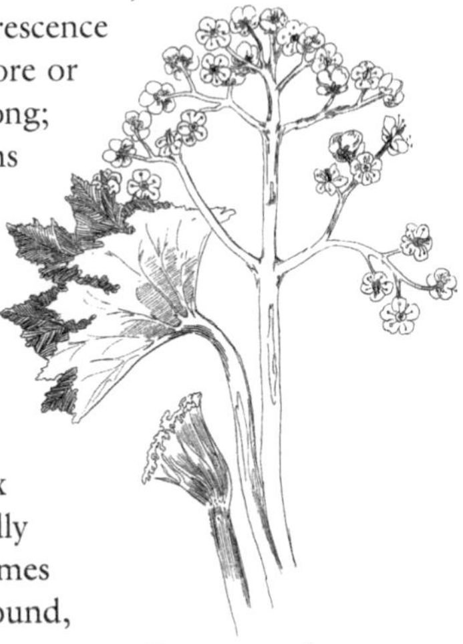

Heuchera

Rhizomatous perennials; leaves basal, palmately lobed, stipulate; flowers usually numerous; calyx lobes usually 5; petals usually 5, but occasionally fewer or lacking; stamens generally 5, but sometimes one or more rudimentary; ovary inferior, compound, 1-loculed; fruit a many-seeded capsule.

Darmera peltata

1a Inflorescence a dense spike-like panicle
  2a Petioles and lower stems villous, but not glandular.......... *H. chlorantha*
  2b Petioles and stems glabrous or glandular-pubescent.......... *H. cylindrica*
1b Inflorescence open, not spike-like
  3a Petioles usually villous; lobes of the leaf blades rounded or, if acute, the leaf blades longer than broad.. *H. micrantha*
  3b Petioles usually glabrous; leaf blades acutely lobed, blades broader than long.. *H. glabra*

Heuchera chlorantha Piper — Tall alum-root

Leaves all basal, the blades orbicular to reniform, heart-shaped at the base, with 7-9 broadly rounded and toothed lobes, pubescent, at least on the veins beneath, petioles brown-hairy; flowering stems 4-10 dm. tall, pubescent; flowers in a dense spike-like panicle; calyx top-shaped below, broader above, greenish; petals very small or absent.

Moist places or shady hillsides.

Heuchera cylindrica Dougl. ex Benth. — Glandular alum-root

Perennials from short stout rhizomes; herbage more or less minutely glandular; leaves many, nearly all basal, long-petioled, the blades 1-3.5 cm. or more long, mostly longer than wide, thick, 3-5-lobed, the lobes rounded and round-toothed; flowers in dense spike-like panicles; calyx top-shaped at base, spreading above, densely glandular, often purplish at base, greenish above, and subtended by purplish bracts; petals usually lacking.

(=*Heuchera ovalifolia* Nutt.)

In rocky places.

Heuchera glabra Willd. — Smooth alum-root

Leaves long-petioled, mostly basal; leaf blades thin, orbicular in outline, with 5-7 sharply toothed lobes, these again lobed and toothed, green and shining above, paler and glandular-pubescent beneath, cauline leaves 1-4, similar to basal leaves, but reduced in size; flowers small in an open panicle with cyme-like clusters; calyx top-shaped, 5-toothed, glandular-pubescent at least below; petals white, longer than calyx, slender, broader above.

Rocky cliffs and stream banks.

Heuchera micrantha Dougl. — Small-flowered alum-root

Heuchera micrantha

Leaves mostly basal, long-petioled, the petioles and lower stems bearing long white hairs that turn brown upon drying, leaf blades more or less pubescent, reniform to broadly oblong or ovate, 5-11-lobed, the lobes rounded, mucronate; flowering stems loosely branched, the branches bearing open cymes, the pedicels thread-like; calyx bell-shaped, with short lobes, sparsely pubescent; petals white to pinkish, long spatulate.

In moist rock outcroppings or gravelly stream banks.

Leptarrhena

A monotypic genus.

Leptarrhena pyrolifolia (D. Don) R. Br. Leather-leaf saxifrage

Perennials from stout rhizomes covered below with old leaf bases, giving rise to closely arranged basal leaves and almost leafless flowering stems; leaves 2.5-15 cm. long, the blades leathery, elliptic, crenate-serrate except at base, nearly glabrous above, under surface pale, blades narrowed to short petioles; flowering stems 1-5 dm. tall, bearing 1 or 2 reduced sessile leaves and a terminal dense glandular panicle of small flowers; petals somewhat unequal; pistils 2; follicles 5-6 mm. long.

Stream banks, moist meadows and wet subalpine slopes, from the coast to the Olympic and Cascade mountains.

Lithophragma

Rhizomes slender, bearing numerous rice-grain sized bulblets; leaves generally mostly basal; flowers in a raceme; petals (ours) deeply cut into narrow segments; stamens 10; fruit a 1-loculed capsule.

Lithophragma parviflorum (Hook.) Nutt. ex Torr. & Gray
Ragged starflower or Woodland star

Stems 0.5-5 dm. tall, glandular-pubescent; both cauline and basal leaves palmately parted, then again cleft or lobed, the ultimate segments toothed or entire, the basal leaves large and long-petioled, the cauline generally smaller and with shorter petioles; flowers in racemes; calyx top-shaped; petals pink or white, long-clawed, 3-5-lobed at the apex; ovary inferior.

Moist areas.

Mitella

Perennials from scaly rhizomes and sometimes stoloniferous; leaves mostly basal (cauline leaves present in *M. caulescens*); inflorescence usually a simple, often one-sided raceme or spike; calyx lobes 5; petals 5, often dissected, stamens 5 or 10; ovary inferior, compound, 1-loculed; fruit a capsule, appearing circumscissile.

1a Petals 3-5-lobed at apex or rarely entire, white to pink
  2a Leaves with triangular lobes, the margins entire or nearly so.................. .... *M. diversifolia*
  2b Leaves with shallow rounded lobes, crenately-toothed........... *M. trifida*
1b Petals pinnately divided into filiform segments, green to greenish-yellow
  3a Cauline leaves present.. *M. caulescens*
  3b Leaves all basal
    4a Stamens opposite the petals.. *M. pentandra*
    4b Stamens alternate the petals

5a Leaves round-cordate, the lobes crenate .... *M. breweri*
5b Leaves ovate to oblong the lobes sharply-toothed......................... *M. ovalis*

Mitella breweri Gray — Brewer's mitrewort

Stems 1-4 dm. tall; leaves all basal, the petioles more or less pubescent, the blades glabrous or nearly so, shallowly 7-11-lobed, crenate-serrate; petals 1.5-2 mm. long, yellowish-green, divided into 5-9 filiform segments; stamens opposite the sepals.

Moist shaded areas and in drier sites in the mountains.

Mitella caulescens Nutt. — Leafy-stemmed bishop's cap

Stems 1.5-4.5 dm. tall; basal leaves thin, long-petioled, nearly orbicular, to cordate in outline, 3-7-lobed, stiff-hairy, 2-7 cm. broad, cauline leaves similar, smaller, with shorter petioles; flowers very small in a slender raceme; petals yellowish-green, often purple-tinged at the base, dissected into 4-8 filiform segments.

A delicate little plant growing in swampy areas, wet meadows and wet woods.

Mitella caulescens

Mitella diversifolia Greene — Angle-leaved mitrewort

Glandular-pubescent perennials 1.5-5 dm. tall; leaves mostly basal, the blades 3-6 cm. wide, ovate to cordate, deeply 5-7-lobed, the margins nearly entire to slightly crenate; petals 3-5-lobed at the tip, white to pink.

Stream banks and moist wooded areas.

Mitella ovalis Greene — Small bishop's cap

Stems 1.5-3.5 dm. tall, hirsute below and becoming glandular-pubescent above; leaves all basal, the blades 1.5-5 cm. wide, cordate-ovate to oblong, 5-9-lobed, sharply serrate, conspicuously veined on the under surface; racemes 20-60-flowered; petals about 1.5 mm. long, cut into 4 to 7 filiform segments, yellowish-green.

Bogs, stream banks and other wet shaded areas.

Mitella ovalis

Mitella pentandra Hook. — Alpine mitrewort

Stems 1-4 dm. tall; leaves all basal, subglabrous to hirsute, the blades ovate-cordate, shallowly 5-9-lobed and somewhat finely toothed, 2-5 cm. long; petals green, pinnately divided into 4-10 filiform segments; stamens opposite the petals.

Stream banks, wet meadows and moist woods in the mountains.

Mitella trifida Graham Three-tooth mitrewort

Stems pilose or hirsute below and becoming glandular-pubescent above; 1.5-4.5 dm. tall; leaves usually all basal, more or less pubescent, especially the petioles, blades nearly orbicular to cordate-ovate, 2-8 cm. wide, very shallowly 5-7-lobed, crenately-toothed; petals 1-3.5 mm. long, white, often purplish-tinged, mostly 3-lobed at apex or rarely entire, narrowed at base.

Wet, shaded areas to subalpine slopes.

Parnassia

Glabrous perennials with tufted basal leaves; stems usually bearing one leaf or bract and one terminal flower; petals white, with greenish or yellowish veins, sometimes fringed on the lower half and bearing at base a cluster of 5 sterile stamens that are fringed or divided and tipped with glandular knobs; fertile stamens 5; ovary compound, 1-loculed; fruit a capsule.

This genus is occasionally put in a separate family, the **Parnassiaceae**.

Leaf blades reniform to cordate at the base; calyx lobes toothed ...*P. fimbriata*
Leaf blades truncate or tapered to the base; calyx lobes entire........... *P. cirrata*

Parnassia cirrata Piper

Stems from short rhizomes and with a single clasping bract borne above the middle; leaves 3-18 cm. long, the blades 1-6 cm. long, nearly orbicular to broadly ovate, truncate or tapered to the base; petals 8-15 mm. long, fringed below; sterile stamens 3-6 mm. long, lobed, the tips spheric.

Wet areas in the mountains.

Parnassia fimbriata Konig. Grass-of-Parnassus

Stems 1.5-5 dm. tall from stout rhizomes and bearing a single cordate-clasping bract above the middle; leaves all basal, leaf blades 1.5-5 cm. long, on petioles 1-14 cm. long, nearly orbicular to reniform; petals 7-12 mm. long, white, fringed on the lower half and become entire to erose above; sterile stamens scale-like, flared and fringed above the middle.

Bogs, stream banks, wet rocky crevices and wet meadows in the mountains.

Penthorum

Perennials; leaves alternate; flowers in scorpioid cymes; calyx 5-7-lobed; petals usually lacking; stamens 10; ovary 5-7-loculed; fruit a 5-7-angled, 5-7-horned capsule. Sometimes segregated into a monogeneric family the **Penthoraceae**.

Penthorum sedoides L.

Stoloniferous perennials; stems decumbent at the base, 1.5-10 dm. tall, simple or branched; leaves alternate, up to 15 cm. long and 4 cm. wide, broadly lanceolate, acute to acuminate, serrate, tapered at the base to a short petiole; flowers borne in scorpioid cymes; calyx 5-7-lobed; petals usually lacking; stamens 10; ovary 5-7-loculed, forming a 5-7-angled, 5-7-horned, capsule.

Wet areas, including fresh water tidal flats; apparently introduced along the Columbia River in Washington and Oregon.

Saxifraga

Perennials (ours); leaves simple, basal or alternate, rarely opposite; calyx lobes usually 5; petals usually 5; stamens 10; pistils 1 or 2, ovary superior or inferior; fruit of 2 follicles or a capsule.

1a Low more or less tufted plants, the perennial branches leaf bearing
  2a Leaves 3- or more-lobed...*S. caespitosa*
  2b Leaves entire
    3a Leaves somewhat leathery, the margins ciliate; petals orange- to purple-dotted...*S. bronchialis*
    3b Leaves fleshy, cilate only at the base; flowers not orange- to purple-dotted...*S. tolmiei*
1b Plants not with tufted leaf bearing branches
  4a Cauline leaves as large as the basal ones...............................*S. nuttallii*
  4b Cauline leaves much reduced or leaves all or nearly all basal
    5a Leaves nearly orbicular to reniform
      6a Leaves regularly and sharply dentate; petals with 2 basal spots..*S. arguta*
      6b Leaves irregularly crenate-dentate; petals without spots..*S. mertensiana*
    5b Leaves longer than broad
      7a Leaf blades gradually tapered to an expanded, almost indistinct petiole, blades usually more than twice as long as wide; plants of wet habitats
        8a Flowers sometimes modified into leafy bulblets; normal flowers with petals 3-8 mm. long, unequal, the upper 3 petals with one or more basal spots ....*S. ferruginea*
        8b Flowers not modified; petals similar, 1-5 mm. long, nearly equal in size and shape
          9a Leaves nearly entire to denticulate; common.....*S. oregana*
          9b Leaves strongly toothed or dentate uncommon and restricted to the Coast Range ....*S. hitchcockiana*
      7b Leaf blades subcordate to abruptly tapering to a usually well-defined petiole
        10a Leaves serrate or dentate; filaments more or less club-shaped, widest near the top or the middle
          11a Leaves rounded to subcordate at the base; petals with a pair of basal spots...*S. marshallii*
          11b Leaves tapered to the base; petals not spotted...*S. rufidula*
        10b Leaves entire or nearly so; filaments widest near the base
          12a Often bulblet bearing at the base; sepals ascending to erect ....*S. integrifolia*
          12b Never bulblet bearing; sepals strongly reflexed to spreading ....*S. nidifica*

Saxifraga arguta D. Don — Brook saxifrage

Perennials from long horizontal rhizomes; flowering stems 1.5-6 dm. tall; leaves all basal, leaf blades nearly orbicular, coarsely dentate, the teeth gland-tipped, 1.5-8 cm. wide, 1-6 cm. long, on petioles usually several times longer than the blades, glabrous or nearly so; inflorescence of large, open panicled cymes; petals white, sometimes unequal, with 2 yellow spots near middle; scapes, pedicels, and capsules generally purplish.

(It is possible that the correct name for our species is *Saxifraga odontoloma* Piper)

Moist places at low to high altitudes in the mountains.

Saxifraga bronchialis L. var. *vespertina* (Small) Rosend.
Matted saxifrage

Stems more or less ascending, leafy, the branches short; leaf blades oblong to oblanceolate, ciliate on the margins, 3-18 mm. long, crowded; flowering stems 5-15 cm. tall, slender, bearing one or a few flowers; petals white, orange- to purple-dotted, slender pediceled; capsule usually purplish, the beaks divergent.

Wet banks to dry cliffs in the Coast Range and the Cascades.

Saxifraga caespitosa L.
Tufted saxifrage

Stem producing short perennial branches densely covered with leaves, the dead brown leaves remaining on the previous year's growth, leaves sessile, generally 3-lobed, the narrow lobes often again divided, cauline leaves few, becoming smaller and entire upward, leaves and stem glandular-pubescent; petals twice as long as calyx, thick, white, broad, slightly notched at apex; stamens 10, those opposite the petals longer; ovary inferior; capsule 2-beaked.

Saxifraga caespitosa
a, b, c leaf forms

Moist places along the coast and on the Columbia River, also alpine.

Saxifraga ferruginea Graham
Rusty saxifrage

Flowering stems 1-6 dm. tall; leaves all basal, 2-14 cm. long, 3-30 mm. wide, oblanceolate, 5-17-toothed, only gradually tapered to ill-defined winged petioles, the blades pilose to merely ciliate; inflorescence an open panicle of cymes, glandular-pubescent, at least above, bracts of the inflorescence almost leaf-like to reduced and linear; flowers sometimes modified into leafy bulblets; normal flowers with petals 3-8 mm. long, white or rarely purplish, usually unequal, the upper 3 petals with one or more basal spots.

Wet stream banks in the mountains.

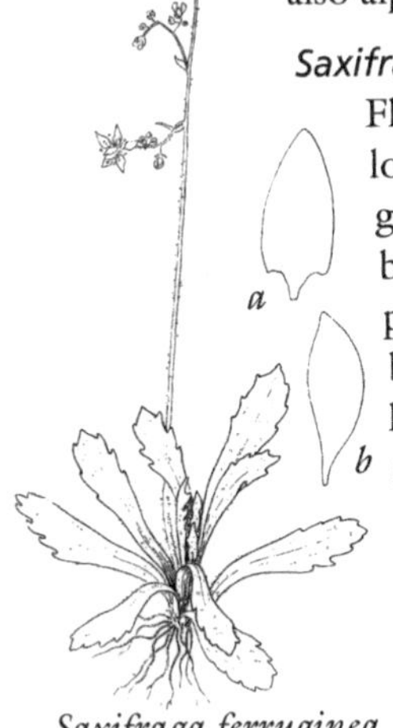

Saxifraga ferruginea
a, b petals

Saxifraga hitchcockiana Elvander — Hitchcock's saxifrage

Rhizomatous perennials; scapes 1-2 dm. tall, glandular-pubescent; leaf blades 4-12 cm. long, obovate to elliptic, the margins toothed and ciliate, densely pubescent on both surfaces, gradually tapered to the indistinct petiole; inflorescence large, open; sepals reflexed; petals white, 2-5 mm. long.

(=*Saxifraga occidentalis* Wats. var. *latipetiolata* C.L. Hitchc.)

In wet rocky areas in the mountains of Clatsop and Tillamook counties, Oregon.

Saxifraga integrifolia Hook. — Broad-leaved saxifrage

Rhizomatous perennials, often bulblet bearing; flowering stems 1.5-3 dm. tall, glandular-pubescent with reddish to purplish glands, at least above; leaves all basal, leaf blades 1.5-4.5 cm. long, 0.5-6 cm. wide, ovate to lanceolate or even deltoid, usually entire, but occasionally remotely serrulate, usually ciliate on the margins and often rusty-tomentose on the under surface, tapered to petioles 0.5-4 cm. long; inflorescence cymose-paniculate, often more or less compact; sepals 1.2-2.5 mm. long, ascending to erect; petals 1.5-4 mm. long, white to greenish or yellowish and sometimes purple-tinged.

Wet meadows, and other, often vernally wet, habitats.

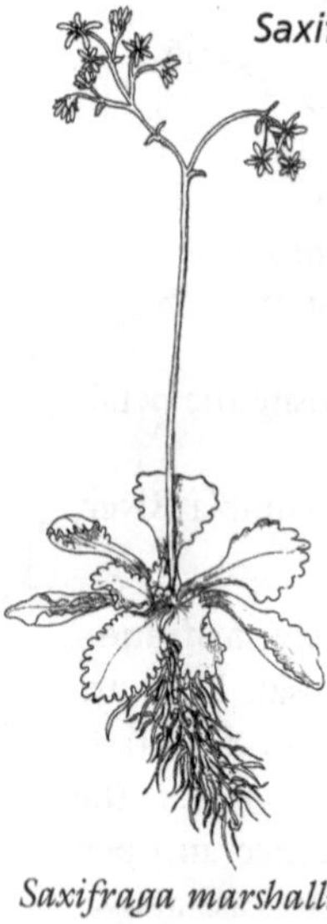

Saxifraga marshallii Greene — Marshall's saxifrage

Flowering stems 1-4 dm. tall, glandular-pubescent with reddish hairs, at least above; leaves all basal, blades 2-5 cm. long, 0.5-3 cm. wide, ovate to elliptic-oblong, serrate, at least the smaller leaves rusty-tomentose on the under surface, petioles often longer than the blades; inflorescence open, the lower bracts somewhat leaf-like; calyx lobes reflexed; petals 2-3.5 mm. long, white with a pair of basal yellow spots.

Wet banks and wet rocky crevices.

Saxifraga marshallii

Saxifraga mertensiana Bong. — Wood saxifrage

Stems 1-5 dm. tall, with a few cauline leaves, but most of the leaves basal, pubescent, glandular, long-petioled, blades orbicular to reniform, many-lobed, the lobes often 3-toothed; inflorescence a loose panicled cyme, the branches bearing a flower at the tips and slender, clustered bulblets below; sepals reflexed; petals white, 3-5 mm. long; filaments club-shaped; fruit a capsule.

Rock crevices in the mountains.

Saxifraga nidifica Greene var. *claytoniifolia* (Canby ex Small) Elvander

Rhizomatous perennials; scapes glandular-pubescent, 2-3.5 dm. tall; inflorescence open; leaves 7-15 cm. long, the blades 4-10 cm. long, 1.5-3 cm. wide, glabrous or sparsely pubescent, broadly ovate to deltoid, the margins entire or nearly so, petioles 4-10 cm. long; sepals spreading to reflexed; petals 2-3 mm. long, white to greenish-white.

In moist habitats, often along streams.

Saxifraga nuttallii Small — Nuttall's saxifrage

Perennials from slender rhizomes and stolons; stems slender, 0.5-4 dm. long, weakly erect to trailing; leaves alternate, succulent, the lower nearly entire, the blades obovate, 1.5-4 mm. long, the upper larger, up to 2 cm. long, and 1 cm. wide, usually with 3 shallow lobes across the top of the blade; inflorescence of bracteate cymes or lower flowers solitary; petals 3-6 mm. long, white, pink-veined; capsules 3.5-5 mm. long.

Wet mossy banks.

Saxifraga oregana Howell — Marsh saxifrage

Rhizomatous perennials; scapes 2.5-10 dm. in height, glandular-pubescent, at least above; leaves 7-25 cm. long, on winged and almost indistinct petioles, leaf blades oblanceolate to obovate, nearly entire to denticulate, ciliate, sparsely pubescent to nearly glabrous; inflorescence elongated, compact to open; petals white to greenish, sometimes unequal, 1-4 mm. long.

This species is extremely variable in the size of the inflorescence and the length of the petals.

Stream banks, wet meadows, seepage areas, marshes and bogs.

Saxifraga rufidula (Small) Macoun

Flowering stems 3-15 cm. tall, arising from a rosette of leaves; leaves with broad petioles as long as or shorter than the blades, blades oblong or ovate 2-4 cm. long, coarsely and regularly dentate, thick, green above, more or less conspicuously clothed beneath with rusty hairs, especially when young; flowering stems pubescent at least at base, bearing a generally spreading compound cyme; flowers white, about 6 mm. wide; filaments of stamens broadest near the top.

[=*Saxifraga occidentalis* Wats. subsp. *rufidula* (Small) Baci.]

Rock crevices in the mountains, also along the Willamette River and on the coast.

Saxifraga tolmiei Torr. & Gray — Alpine saxifrage

Perennials from a more or less woody caudex; leafy stems trailing and mat-forming; leaves entire, glabrous except usually long-ciliate at the base, succulent and nearly terete, 3-12 mm. long, spatulate to oblanceolate; flowering stems erect, 3-8 cm. tall, flowers solitary or up to 4 in a cyme; calyx saucer-shaped, the lobes 2-3 mm. long; petals white, 3-6 mm. long; capsules often purple mottled.

In alpine meadows, near streams or on moist talus slopes.

Suksdorfia

Perennials with short, bulbiferous rhizomes; middle and upper cauline leaves stipulate, differing in shape from the lower cauline and basal leaves; calyx lobes 5; petals 5; stamens 5; ovary inferior; fruit a capsule.

Suksdorfia violacea Gray
Stems 1-2.5 dm. tall, bulblet bearing at base and in the axils of lower leaves, glandular-pubescent, few-leaved; blades of the basal leaves cordate or reniform, 3-5-parted, the segments coarsely lobed, minutely glandular-pubescent to nearly glabrous, long-petioled, the upper leaves sessile, clasping, 3-lobed, stipulate, the uppermost becoming entire bracts; flowers white or violet-tinged, in a terminal cyme; ovary inferior.

Wet rocky banks in the Columbia Gorge and eastward.

Sullivantia

Slender perennial herbs with leaves mostly basal; flowers in cymes; calyx lobes triangular; petals white; stamens 5; ovary inferior; fruit a capsule.

Sullivantia oregana Wats.
Plants with rhizomes and stolons; stems 1-2 dm. tall, glandular-pubescent above, bearing 1-3 leaves; basal and lower cauline leaves reniform in outline, several lobed, the lobes coarsely and sharply toothed, thin, bright green, glabrous or nearly so, long-petioled, upper cauline leaves few-lobed, shorter-petioled, the uppermost becoming small sessile bracts; flowers few in a loose cyme; petals a little longer than the calyx lobes.

Wet cliffs in the Columbia Gorge and along the lower Willamette River.

Tellima

A monotypic genus.

Tellima grandiflora (Pursh) Dougl. ex Lindl. Fringe-cups
Perennials from scaly rhizomes; stems hirsute and glandular, 4-10 dm. tall; basal leaves long-petiolate, the blades from nearly reniform to cordate-ovate, 3-10 cm. broad and 2-6 cm. long, shallowly 3-7-lobed, crenate-dentate, cauline leaves 1-3, reduced; inflorescence a loosely-flowered, more or less one-sided raceme; calyx 5-11 mm. long, the lobes 2-3 mm. long; petals 3-7 mm. long, mostly 5-7-lobed, greenish-white to red; stamens 10; ovary inferior, compound, 1-loculed; fruit a beaked capsule.

(includes *Tellima odorata* Howell)

Stream banks and other moist habitats.

Tiarella

Perennials from scaly rhizomes; leaves palmately-lobed or compound; inflorescence a panicle or a raceme; petals 5, white, filiform; stamens 10; ovary superior, compound, 1-loculed below, but divided over half way into 2 unequal, sterile projections; fruit a capsule, dehiscent along the unequal sterile segments.

Tiarella trifoliata L. Coolwort
Perennials; stems 1.5-6 dm. tall; leaves basal and cauline, reduced upwards, leaf blades 3-12 cm. wide, simple and 3-5-lobed to trifoliolate, sharply toothed; calyx lobes 1.5-2.5 mm. long; petals 3-4 mm. long. Two varieties of this species occur in our area: var. *trifoliata* with compound leaves and var. *unifoliata* (Hook.) Kuntz with simple leaves.

Stream banks and moist woods.

Tolmiea

A monotypic genus.

Tolmiea menziesii (Pursh) Torr. & Gray
Youth-on-age or Pig-a-back plant

Perennials from scaly rhizomes; stems 1-10 dm. tall, hirsute and glandular, at least in the inflorescence; leaves both basal and cauline, blades 3-10 cm. wide and about as long, shallowly 5-7-lobed, more or less ovate, sharply toothed; stipules well developed, leaf-like on the basal leaves, but reduced upwards, reproducing vegetatively by the development of plantlets at the base of the leaf blades; racemes many-flowered; calyx cleft between the lower 2 lobes, upper 3 lobes larger; petals 4, 8-12 mm. long, filiform, brown; stamens 3, unequal; pistil 1, ovary superior, compound, 1-loculed; fruit a 2-beaked capsule.

Tolmiea menziesii

Stream banks, seepage areas and moist woods.

GROSSULARIACEAE
Gooseberry or Currant Family

Shrubs; leaves alternate, simple, mostly palmately 3-5-lobed, usually without stipules; flowers saucer-shaped to tubular; calyx lobes, petals and stamens usually 5; ovary inferior; fruit a berry. Members of this family were formerly included in the **Saxifragaceae.**

Ribes

Shrubs, unarmed or with nodal spines; leaves alternate, palmately veined and lobed; flowers greenish-white, white, yellow, red or purple; petals 5, usually smaller than the calyx lobes; ovary 1-loculed, placentation parietal; fruit a globose berry.

1a Stems without spines or prickles
- 2a Flowers bell-shaped or tubular
  - 3a Plants not sticky-glandular; flowers pink to red (rarely white); tip of anthers without a gland .... *R. sanguineum*
  - 3b Plants sticky-glandular; flowers greenish to whitish and tinged with red; tip of anthers with a cup-like glandular depression .... *R. viscosissimum*
- 2b Flowers saucer-shaped
  - 4a Racemes drooping .... *R. howellii*
  - 4b Racemes erect or ascending
    - 5a Ovary and under side of the leaves with sessile glands; foliage strong-scented .... *R. bracteosum*

5b Ovary and under side of the leaves with stalked glands; foliage not strongly scented .... *R. laxiflorum*

1b Stems with spines and/or prickles

6a Spines weak; racemes 5-18-flowered; fruit breaking stemless from the cluster .... *R. lacustre*

6b Spines generally stout; inflorescence less than 5-flowered; fruit breaking with a distinct stalk from the cluster

7a Flowers inconspicuous; berries black, glabrous .... *R. divaricatum*

7b Flowers showy; berries pubescent or prickly

8a Leaves glabrous or nearly so; berries spiny .... *R. roezlii*

8b Leaves pubescent and glandular, at least beneath

9a Young stems bristly; anthers white .... *R. menziesii*

9b Young stems not bristly; anthers purple .... *R. lobbii*

Ribes bracteosum Dougl. ex Hook. — Stinking black currant

Unarmed shrubs 1.5-3 m. tall, erect or straggling; leaves 4-12 cm. or more in diameter, 5-7-lobed, irregularly toothed, heart-shaped at base, bearing sessile yellow glands on the under surface; racemes narrow, mostly erect, 10-25 cm. long; flowers greenish, inconspicuous, saucer-shaped; berries black, glandular, with whitish waxy coating.

Shady moist places, often along streams.

Ribes divaricatum Dougl. — Coast black or Straggly gooseberry

Shrubs 1-3 m. tall, with long straggling branches, spines at nodes stout and long, bristles between nodes rarely present; leaves 2-6 cm. broad, sometimes broader than long, pubescent above and on the margins, sparsely pubescent on veins beneath, 3-5-lobed, the lobes rather regularly toothed; racemes 2-4-flowered, with slender pedicels; flowers greenish to reddish or purplish, inconspicuous, bell-shaped; berry black, smooth.

Open woods and along streams from the west slope of the Cascades to the coast.

Ribes howellii Greene

Unarmed shrubs, erect or spreading, 6-15 dm. tall; herbage finely puberulent and usually glandular; leaf blades 2.5-8 cm. broad, cordate, 3-5-lobed, the lobes irregularly toothed; racemes drooping, 5-15-flowered; flowers saucer-shaped, inconspicuous, the small petals red to purplish; fruits black with a white waxy bloom, glandular pubescent.

Forming thickets at and below timberline in the mountains.

Ribes lacustre (Pers.) Poir. — Swamp or Prickly currant

Erect or prostrate shrubs, 1-2 m. tall, the stems bearing weak spines at the nodes, and sometimes prickly between; leaves 1-7 cm. broad, smooth, deeply divided into 3-7 lobes, these again lobed or toothed; racemes erect at first, later drooping, 5-18-flowered; calyx saucer-shaped, green to purplish; petals minute, fan-shaped, pink to purple; berry black, bearing stalked glands.

Common, generally growing in moist places.

Ribes laxiflorum Pursh Coast trailing currant
Unarmed shrubs, usually trailing, 1-7 m. long; leaf blades 5-7-lobed, 4-10 cm. in diameter, smooth and shining dark green above, paler beneath, pubescent at least on veins below and often also glandular; racemes 6- 18-flowered; flowers greenish, usually reddish-tinged, saucer-shaped, inconspicuous; berries black with a waxy coating and stalked glands.

Moist forested areas in the mountains and burned-over areas near the coast.

Ribes lobbii Gray Gummy gooseberry
Shrubs 5-15 dm. tall, erect, spreading; spines 3-forked, stout, borne only at the nodes; stems and leaves sparingly clothed with glandular hairs; leaf blades 1.5-6 cm. broad, the under surface and petioles somewhat sticky from the glands; flowers solitary or in 2's, large, showy; calyx purplish red, with stamens extending beyond calyx lobes; petals white to pinkish; berry densely covered with glandular hairs.

Moist open slopes and burned-over areas in the Cascades to the coast.

Ribes lobbii

Ribes menziesii Pursh Coast prickly-fruited gooseberry
Shrubs 0.5-3 m. tall, erect, spreading; spines at nodes usually 3, straight, long, stout, sharp, also with numerous bristles on young shoots; leaves truncate or heart-shaped at base, with 3-5 lobes, these toothed or cut, 1.5-4 cm. broad, pubescent on the veins below and covered between veins with stalked glands, sometimes pubescent above; flowers solitary or up to 3 in a cluster, drooping, showy, purple, the pedicel and young ovary glandular; berry reddish-purple, densely covered with weak bristles, some of which are gland-tipped.

Lincoln County, Oregon, and southward along the coast.

Ribes roezlii Regel var. *cruentum* (Greene) Rehder
Shiny-leaved gooseberry
Shrubs 0.5-1.5 m. tall, erect, freely branched, bearing 1-3 spines at the nodes; leaves thick, 1-2.5 cm. broad, smooth and shiny, shallowly 3-5-lobed; flowers solitary or in 2's, showy; calyx-lobes purplish-red, shorter than styles but slightly longer than the stamens; berries reddish-purple, densely spiny, also with stalked glands.

(=*Ribes cruentum* Greene)

Foothills to high elevations in the Cascades from Lane County, Oregon southward.

Ribes sanguineum Pursh Red-flowering currant
Unarmed shrubs 1-4 m. tall; bark thin, brown; leaves 2-6 cm. broad, more or less orbicular in outline, 3-5-lobed, irregularly toothed, conspicuously veined, deep green above, paler and often pubescent beneath, sometimes glandular at first; flowers deep red to

Ribes sanguineum

nearly white; calyx glandular-pubescent below; berry bluish-black with whitish waxy coating and stalked glands.

Common; from the coast to the Cascades.

Ribes viscosissimum Pursh — Sticky currant

Unarmed shrubs, 1-2 m. tall, sticky-glandular and usually also pubescent; leaf blades 2-10 cm. broad, 3-5-lobed; calyx bell-shaped, greenish or whitish and often reddish-tinged; fruits bluish-black.

Moist to dry slopes; more common east of the Cascades.

HYDRANGEACEAE
Hydrangea Family

Trees, shrubs, but also subshrubs that are trailing and vine-like; leaves opposite, simple, without stipules; calyx lobes 4-10; petals 4-6, rarely more; stamens 5-many; ovary inferior, typically 3-5-loculed; fruit a capsule. Members of this family are sometimes included in the **Saxifragaceae** or put in the **Philadelphaceae.**

Subshrubs with prostrate main stems; stamens 8-12 .... **Whipplea**
Erect shrubs over 1.5 m. tall; stamens 25-50 .... **Philadelphus**

Philadelphus

Shrubs with opposite leaves; flowers white, showy, terminating the branches in simple or compound racemes, cymes or panicles; calyx lobes and petals 4-5; stamens numerous; ovary inferior, 4-loculed (in ours); fruit a more or less woody capsule.

Philadelphus lewisii Pursh — Mock orange or Syringa

Shrubs 1.5-5 m. tall; leaves opposite, ovate, smooth to somewhat pubescent, with 3-5 conspicuous veins from near the base, leaf margins entire to slightly toothed; flowers very fragrant; calyx lobes markedly mucronate; petals creamy white, 8-25 mm. long; stamens 25-50.

Philadelphus lewisii

This is an extremely variable species.

Usually along stream banks, but also in drier sites.

Whipplea

A monotypic genus.

Whipplea modesta Torr. — Whipplevine

Stems slender, woody, trailing, up to 1 m. long, giving rise to short ascending branches; leaves opposite, 1-4 cm. long, ovate to elliptic, with 3 main veins from the base, pubescent, somewhat toothed, short-petioled to sessile; inflorescence dense, the flowers small, short-pediceled; petals 4-6, 3-6 mm.

long, white at first then usually greenish; stamens 8-12; capsule small, depressed-globose.

Open woods.

ROSACEAE
Rose Family

Trees, shrubs, or annual to perennial herbs; leaves usually alternate, stipules generally present, at least at first; flowers regular, usually perfect; sepals partly united forming a 4-5-lobed calyx; petals 5 (rarely to 10) or absent; stamens 1-many, commonly more than 10, attached with the petals to the floral cup; pistils simple, 1 to many, free, or compound and united with the floral cup; fruit dry or fleshy.

1a Annual to perennial herbs
  2a Petals none
    3a Leaves palmately lobed .... **Aphanes**
    3b Leaves pinnately compound....................................... **Sanguisorba**
  2b Petals present
    4a Dioecious; flowers minute, in a large plume-like panicle; fruit a follicle .... **Aruncus**
    4b Flowers generally perfect, not borne in a plume-like panicle; fruit not a follicle
      5a Bractlets not present on the calyx; fruit an aggregate of druplets... .... **Rubus** (in part)
      5b Bractlets alternating with the calyx lobes; fruit of achenes
        6a Receptacle fleshy in fruit forming a strawberry........... **Fragaria**
        6b Receptacle dry in fruit
          7a Stamens 10; petals cream-colored to pinkish; leaves pinnately compound .... **Horkelia**
          7b Stamens either more than 10 or petals yellow or reddish
            8a Styles deciduous from the achenes ...**Potentilla** (in part)
            8b Style persistent upon the achene, often hooked or plumose .... **Geum**
1b Shrubs, subshrubs (sometimes very low and matted) or trees
  9a Fruit dry
    10a Low creeping plant; leaves once or twice dissected into linear lobes; follicles leathery.. **Luetkea**
    10b Not as above in all points
      11a Leaves compound .... **Potentilla** (in part)
      11b Leaves simple, sometimes lobed
        12a Leaves palmately lobed.................................... **Physocarpus**
        12b Leaves not palmately lobed
          13a Stamens well exserted; fruits several-seeded.......... **Spiraea**
          13b Stamens scarcely exserted; fruits 1-seeded.......**Holodiscus**
  9b Fruit fleshy
    14a Floral tube in fruit globose or urn-shaped, completely enclosing the achenes; leaves pinnately compound .... **Rosa**

14b Floral tube not as above in fruit; leaves rarely pinnately compound
15a Ovary superior
16a Pistil 1 .... **Prunus**
16b Pistils more than 1 per flower
17a Dioecious species; fruit a drupe .... **Oemleria**
17b Flowers generally perfect; fruit an aggregate, generally of many drupelets .... **Rubus** (in part)
15b Ovary inferior; fruit a pome, or pome-like
18a Leaves pinnately compound .... **Sorbus**
18b Leaves simple, although sometimes deeply lobed
19a Carpels becoming bony in fruit, each 1-seeded; stems with stout thorns .... **Crataegus**
19b Carpels becoming papery in fruit; stems usually unarmed
20a Flowers in racemes; locules of pome each partly divided by a secondary partition .... **Amelanchier**
20b Flowers in corymbs or cymes; locules without secondary partitions
21a Fruit pear-shaped, the flesh with grit cells .... **Pyrus**
21b Fruit apple-shaped, flesh without grit cells .... **Malus**

Amelanchier

Shrubs or trees with simple deciduous leaves; flowers in racemes, white; calyx 5-lobed, the lobes narrow; petals more or less erect; stamens many; ovary inferior, 4-5-loculed, the locules again divided by secondary partitions; style branches 4-5; fruit a small fleshy almost berry-like pome.

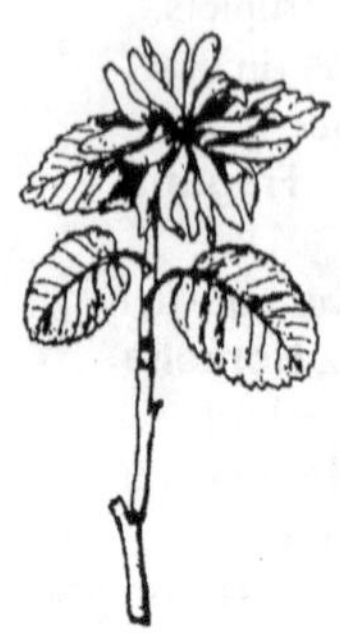

Amelanchier alnifolia

Amelanchier alnifolia Nutt. Service berry

Shrubs 1-10 m. tall; leaf blades 1.5-5 cm. long, oval to oblong, usually coarsely toothed especially above the middle, obtuse at apex, glabrous to sparsely pubescent above, paler beneath and sometimes densely pubescent; racemes short; petals white, long and narrow; fruit purplish to nearly black, nearly globose, 7-14 mm. in diameter. Variety ***semiintegrifolia*** (Hook.) C.L. Hitchc. accounts for most of our material, however, var. ***humptulipensis*** (G.N. Jones) C.L. Hitchc. is known from western Washington and differs in having nearly entire leaves and some flowers with only 4 styles.

This is an extremely variable species and has been variously divided into numerous taxa most of which seem to intergrade. Common in open woods.

Aphanes

Annuals with palmately lobed leaves and sheathing stipules; flowers small, clustered in the leaf axils and more or less hidden by the stipules; sepals 4, often alternating with minute bractlets; petals absent; stamens 1-4; pistils 1-4, ovary superior; achenes enclosed by the calyx tube.

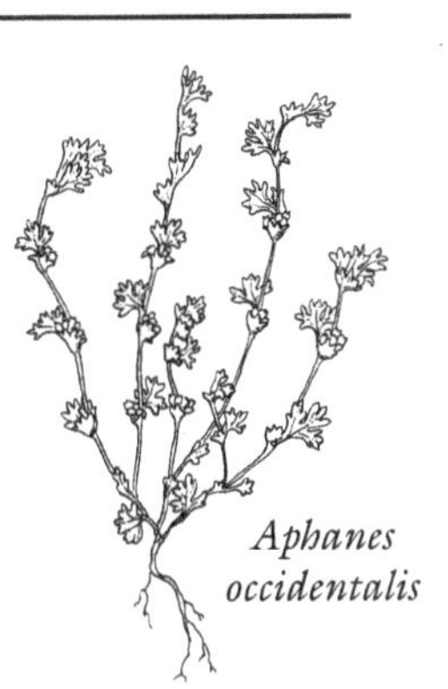
Aphanes occidentalis

Aphanes occidentalis (Nutt.) Rydb.
Western lady's-mantle

Stems 2-10 cm. long, rarely more, pubescent; leaves 2-12 mm. long, widest near apex, deeply 3-lobed, the lobes deeply-toothed; flowers borne at nearly every node; achenes glabrous.

(=*Alchemilla occidentalis* Nutt.)

Fields, waste ground and open woods.

Aruncus

A monotypic genus.

Aruncus dioicus (Walter) Fern. var. *pubescens* (Rydb.) Fern.
Goat's-beard

Rhizomatous perennials; stems stout, 1-2.5 m. tall; leaves compound, up to 6 dm. long, 2-3-pinnate, long-petiolate, leaflets to 15 cm. long and to 8 cm. wide, ovate to lanceolate, acuminate, sharply doubly-serrate, usually glabrous at least on the upper surface; flowers in plume-like panicles, unisexual and plants mostly with either staminate or pistillate flowers; sepals 5, 1-2 mm. long; petals 5, white to yellowish, 1-2 mm. long; pistillate flowers often with rudimentary stamens and 3-5 pistils; staminate flowers with 15-30 stamens; fruit of follicles, 3-5 mm. long.

(=*Aruncus sylvester* Kostel.)

In shady moist woods and logged-off lands and along streams. Often seen growing on new road grades, or on the upturned roots of trees unearthed by storms. A valuable addition to the wild garden.

Crataegus

Shrubs or small trees usually bearing thorns (spinose branches); leaves deciduous, simple, toothed to deeply lobed; flowers in corymbs; floral tube cup-shaped; petals 5; stamens 5-25; ovary inferior, 2-5-loculed; fruit fleshy, carpels each with a bony covering, each 1-seeded.

1a Leaves deeply 3-7-lobed; fruit scarlet, seeds one .... *C. monogyna*
1b Leaves nearly entire to variously toothed or shallowly-lobed; fruit black, seeds 2-several
  2a Stamens 10 or less; leaves typically shallowly-lobed at the apex .... *C. douglasii*
  2b Stamens more than 10; leaves toothed .... *C. suksdorfii*

Crataegus douglasii Lindl. Western hawthorn or Western black haw

Shrubs or small tree reaching 12 m. in height, bearing a few stout thorns, typically 1-2 cm. long; young twigs reddish; leaves mostly obovate, finely serrate to the middle, coarsely double serrate or lobed above the middle, the tips usually truncate, narrowing at base to the petiole, sparsely pubescent or glabrous; flowers 10-16 mm. or less broad; stamens 5-10; ovary pubescent, styles 3-5; fruit black; seeds usually 5.

In a variety of habitats in the Columbia River Gorge and the San Juan Islands; uncommon in the Willamette Valley.

Crataegus monogyna Jacq.

Shrubs or small trees to 10 m. in height; leaves deeply 3-7-lobes, the lobes narrow, usually entire or with only a few teeth near the apex; styles typically 1; fruit elliptic, scarlet, with a single seed.

Hybridization between this species and *C. suksdorfii* produces plants with black fruits and a variety of leaf shapes as well as a variable number of stamens and styles.

Escaped from cultivated and naturalized.

Crataegus suksdorfii (Sarg.) Kruschke

Similar to *Crataegus douglasii* , except thorns often shorter, 8-12 mm. long, and leaf blades typically not lobed, the tips acute or obtuse; stamens typically 19-20; ovary usually glabrous, styles 5; fruit black, seeds 5.

Hybridization between this species and *C. monogyna* produces plants with black fruits and a variety of leaf shapes as well as a variable number of stamens and styles.

Mesic habitats.

Fragaria

Perennial herbs with stolons and scaly rhizomes; leaves with 3 leaflets; flowers white, rarely pinkish, borne in bracteate cymes on leafless stems; calyx remaining with fruit, 5-lobed, with 5 alternate bractlets; petals broad; stamens 20 or more; pistils many, borne on an enlarged receptacle, style attached at side of ovary; fruit formed from the fleshy receptacle, with seed-like achenes in numerous pits on its surface.

1a Leaves leathery, deep green and shining above; petals 10-18 mm. long; strictly coastal.. *F. chiloensis*

1b Leaves not leathery, usually light green or bluish-green; petals less than 13 mm. long

2a Leaflets mostly bluish-green, glabrous and glaucous on the upper surface, terminal tooth of the leaflets usually surpassed by the adjacent lateral teeth . ...*F. virginiana*

2b Leaflets sparingly pubescent, bright yellow-green and not glaucous on the upper surface, terminal tooth of the leaflets usually surpassing the adjacent lateral teeth .... *F. vesca*

Fragaria chiloensis (L.) Duch. — Coast strawberry

Stems 3-20 cm. long; stolons usually reddish-tinged and pubescent; leaves dark green and shining above, pubescent beneath and on the petioles; flowering stems usually shorter than the leaves; flowers large, showy; fruit 1-1.5 cm. in diameter, achenes sometimes embedded in receptacle, but often near the surface.

Sand dunes and bluffs along the coast.

Fragaria vesca L. — Wood strawberry

Plants producing numerous stolons; leaflets thin to slightly thickened, bright green above, paler beneath and more densely pubescent; cymes 3-15-flowered;

calyx pilose; petals white or pale pink, 8-12 mm. long; fruit about 1 cm. broad, succulent.

Our material is referable to two varieties: var. ***bracteata*** (Heller) Davis in which the peduncles in fruit exceed the leaves and the achenes appear largely superficial and var. *crinita* (Rydb.) C.L. Hitchc. in which the peduncles in fruit are shorter than the leaves, and the achenes are largely embedded.

Moist woods, meadows and stream banks.

Fragaria virginiana Duch. var. *platypetala* (Rydb.) Hall
Mountain strawberry

Plants stoloniferous; leaflets bluish-green and usually glabrous and glaucous on the upper surfaces, pubescent beneath; calyx villous; petals white or pinkish, 6-13 mm. long; fruit about 1 cm. broad; achenes partially embedded in the receptacle.

Woods, meadows and stream banks; more common east of the Cascades.

Geum

Perennial herbs with pinnate basal leaves, the terminal lobe much larger than the several lateral lobes, cauline leaves few and reduced; flowers borne in cymes or solitary; calyx 5-lobed, alternating with bractlets; petals 5; stamens numerous; pistils many on an enlarged receptacle, becoming achenes, the styles persistent.

Flowers yellow; persistent portions of styles hooked in fruit .... *G. macrophyllum*

Flowers reddish; persistent portions of styles not hooked, plumose .... *G. triflorum*

Geum macrophyllum Willd. Bigleaf or Large avens

Stems stout, 3-10 dm. tall, pubescent; basal leaves 2-4 dm. long, with a very large somewhat round heart-shaped toothed terminal leaflet, the lateral leaflets more or less triangular in outline, toothed and sometimes lobed with smaller leaflets interspersed, pubescent; cymes few-flowered, pedicels long; sepals reflexed; petals deep yellow; achenes forming a dense ball, achenes tawny hairy on upper side, bearing a hooked and jointed persistent style.

Wet meadows, stream banks and moist wooded areas.

Geum macrophyllum

Geum triflorum Pursh var. *campanulatum* (Greene) C. L. Hitchc.
Red avens or Old man's whiskers

Stems 2-4 dm. tall; leaves mostly basal, clustered, grayish-pubescent, with many leaflets divided into narrow segments; flowers cymose; both calyx and corolla

Holodiscus discolor

reddish, petals longer than sepals; persistent styles of the achenes plumose.

Saddle Mt. of Clatsop Co., Oregon, also in the Olympics Mountains of Washington, and on the east side of the Cascades.

Holodiscus

Shrubs with alternate deciduous toothed or lobed leaves; flowers very small, white or cream-colored, in dense panicles; stamens 15-20; pistils 5; fruits dry, 1-seeded, pubescent.

Holodiscus discolor (Pursh) Maxim. Ocean spray
Shrubs 1-5 m. tall, the branches generally arching; leaf blades ovate, green above, whitish-pubescent beneath, 3-10 cm. long, very coarsely toothed above the base; panicle large, pyramidal, erect or somewhat drooping.

Common in open woods.

Horkelia

Usually pubescent and glandular perennials; leaves pinnately compound; inflorescence of open to dense cymes; flowers white to pinkish; floral tube more or less flat-bottomed; stamens 10; pistils 5-many; fruit of achenes.

Horkelia congesta Hook. Shaggy Horkelia
Stems 2-3.5 dm. tall, slender, from a basal tuft of leaves; whole plant more or less grayish-green, silky-pubescent and glandular; basal leaves pinnately compound with 5-9 leaflets, these oblong to oblanceolate, the apex generally cut into several linear lobes, upper leaves reduced, stipules cut into narrow segments; flowers white, about 1 cm. wide; calyx lobes often reflexed; petals round above, abruptly clawed.

[=*Potentilla congesta* (Hook.) Jepson]

Dry open hillsides; Willamette Valley and southward.

Luetkea

A monotypic genus.

Luetkea pectinata (Pursh) Kuntze Partridge foot
Woody stems creeping and mat-forming; basal leaves tufted; flowering stems leafy, 3-15 cm. tall; leaves fan-shaped at the top, once or twice ternately-parted into linear segments; flowers white, 6-8 mm. wide, in a slender raceme; sepals, petals, and pistils generally 5; stamens numerous; follicles leathery.

Mountain meadows and moist shady areas near timber line.

Malus

Trees or shrubs, sometimes thorny; leaves deciduous; flowers white or pink, in corymbs at the ends of short branches; stamens numerous; ovary inferior, 2-5-loculed, ovules 2 in each locule; fruit a pome.

Fruits 10-15 mm. long; at least some of the leaves lobed .... *M. fusca*
Fruits over 25 mm. long; leaves merely crenate-serrate .... *M. sylvestris*

Malus fusca (Raf.) C. Schneid.
Oregon or Wild crab apple

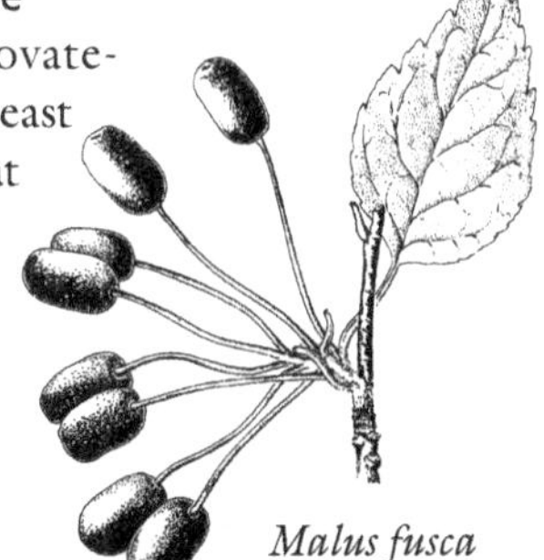
Malus fusca

Small trees or shrubs 3-12 m. tall; leaf blades ovate-lanceolate, pointed at apex, coarsely toothed and at least some of the leaves lobed, 4-10 cm. long, pubescent at least on the under surface; flowers 5-10 in a corymb; petals white or rarely pinkish; stamens about 20; fruits oblong-ovoid, 10-15 mm. long, yellowish with rosy tinting, becoming nearly black.

(=*Pyrus fusca* Raf. and *Malus diversifolia* Roem.)

Rather common in moist wooded areas from the coast to the foothills of the Cascades.

Malus sylvestris Miller — Apple

Round-headed trees to 15 m. in height; leaves elliptic to ovate, 4.5-10 cm. long, more or less acute at the apex, margins crenate-serrate, pubescent on both surfaces at first, but becoming glabrous above; flowers white, tinged with pink; fruit subglobose at least 2.5 cm. broad.

This is the cultivated apple that has escaped and become naturalized.

Oemleria

A monotypic genus.

Oemleria cerasiformis (Hook. & Arn.) Landon — Indian peach or Oso berry

Dioecious shrubs 1.5-5 m. tall; leaves simple, entire and with deciduous stipules, leaf blades thin, smooth, broadly oblong to obovate; staminate flowers more spreading and whiter than the pistillate; stamens 15, in 3 series, aborted pistils often present; pistillate flowers bearing aborted stamens, pistils 5, separate, often only 1 or 2 developing; drupes becoming peach-colored, changing to deep blue-black.

[=*Osmaronia cerasiformis* (Hook. & Arn.) Greene]

Roadsides, common.

The flowers are fragrant, and the broken stems have a characteristic pungent odor. This is generally our earliest shrub to blossom and develop foliage in the spring. The pistillate plants are often in full flower before the leaves appear.

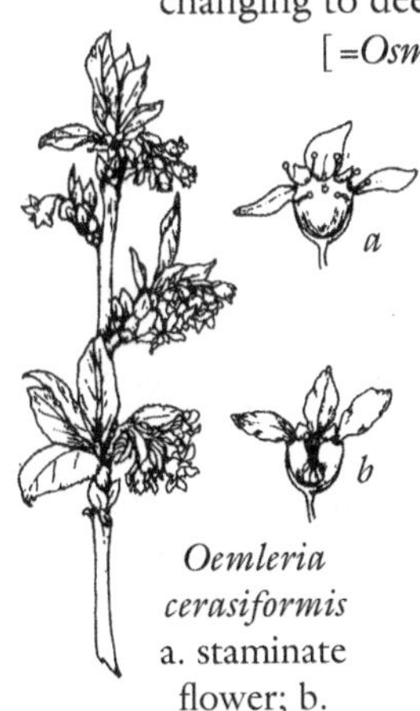

Oemleria cerasiformis a. staminate flower; b. pistillate flower

Physocarpus

Shrubs with deciduous leaves and thin reddish exfoliating bark; inflorescence a corymb; calyx lobes 5; petals 5, white; stamens 20-40; pistils 1-5; fruit of follicles.

Physocarpus capitatus

Physocarpus capitatus (Pursh) Ktze. Ninebark

Straggly shrubs 1.5-6 m. tall; branches angled, glabrous or minutely stellate-pubescent (seen with a hand-lens), the bark exfoliating in long, irregular strips; leaves petioled, the blades 3-10 cm. long, ovate to cordate, 3-5-palmately lobed, the margins doubly-toothed, glabrous or stellate-pubescent above, stellate-pubescent on the lower surface; calyx stellate-pubescent; petals 2.5-4 mm. long, white; stamens about 30; pistils 3-5; follicles inflated.

Common along stream banks, lake margins, swampy areas and in moist woods.

Potentilla

Annuals, perennials or shrubs; leaves mostly basal to alternate, pinnately or palmately compound; inflorescence solitary or more often a bracteate cyme; calyx lobes 5; petals generally 5; stamens many; pistils many, borne on an enlarged floral tube; fruit of achenes.

1a Petals dark red to purple; aquatic or semi-aquatic plants............ *P. palustris*
1b Petals yellow, or rarely white, not aquatic plants, though sometimes growing in wet places
  2a Shrubs; achenes pubescent.. *P. fruticosa*
  2b Plants herbaceous; achenes usually glabrous
    3a Low plants with long stolons; flowers solitary.................. *P. anserina*
    3b Plants without stolons, although often rhizomatous; flowers in cymes
      4a Usually annuals or biennials, if short-lived perennials then without rhizomes
        5a Petals 1.5-2 mm. long, about 1/2 as long as the sepals; stamens generally 10-15.. *P. rivalis*
        5b Petals 3-4 mm. long, from 3/4 as long to equalling the sepals; stamens 15-20 .... *P. norvegica*
      4b Rhizomatous perennials
        6a Leaflets 5-9
          7a Leaves palmately compound; styles attached near the tip of the ovary.. *P. gracilis*
          7b Leaves pinnately compound; styles attached below the middle of the ovary... *P. glandulosa*
        6b Leaflets 3; plants of mountain meadows
          8a Plants gray-villous.. *P. villosa*
          8b Plants nearly glabrous or crisped pubescent .... *P. flabellifolia*

Potentilla anserina L. Silver-weed

Nearly prostrate perennials spreading by stolons; leaves pinnately compound with 7-29 deeply toothed leaflets, main lateral leaflets alternating with much-reduced leaflets, white silky pubescent at least on the lower surface, 0.5-5 cm. long, petioles with sheathing stipules, those of the stolons dissected into linear segments; flowers solitary on peduncles 3-15 cm. long; petals yellow, 6-20 mm. long; stamens 20-25; achenes about 2 mm. long.

Most of the plants within our range belong to subsp. ***pacifica*** (How.) Rousi (=*Potentilla pacifica* How.), with subsp. ***anserina*** occurring mostly east of the Cascades and differing in having leaflets densely pubescent above, smaller leaflets and smaller petals.

Wet, often brackish areas of meadows, stream banks and salt marshes.

Potentilla anserina

Potentilla flabellifolia Hook. Fan-leaf cinquefoil

Perennials, stems 1-2.5 dm. tall from scaly rhizomes, plant nearly glabrous to crisped-pubescent; leaves long-petioled, compound with 3 leaflets, the leaflets 1-3.5 cm. long, fan-shaped or broadly obovate, upper half or two-thirds of blade coarsely and deeply toothed; flowers yellow, showy; bractlets ovate or oval, nearly as long as the sepals; petals to 1 cm. long; stamens about 20.

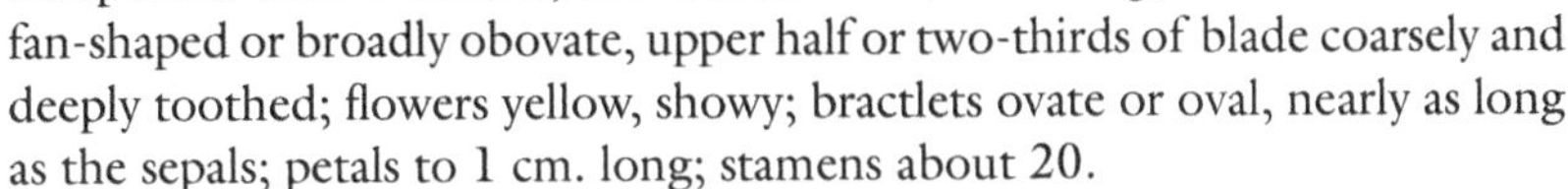

Wet alpine and subalpine meadows.

Potentilla fruticosa L. Shrubby cinquefoil

Woody stems freely branched, 1-15 dm. tall, very leafy; leaves pinnately compound, generally with 5-7 narrow leaflets, these 1-2.5 cm. long or less, lanceolate or oblong, more or less silky-pubescent, the margins entire and somewhat involute; flowers golden-yellow, reaching 2.5 cm. wide, on short terminal and axillary peduncles.

Foothills to subalpine meadows and rocky slopes.

Potentilla glandulosa L. Sticky cinquefoil

Rhizomatous perennials; stems erect, 1.5-7 dm. tall, glandular and somewhat villous; basal leaves pinnately compound, with 5-9 broadly obovate or nearly orbicular simply or doubly serrate leaflets, cauline leaves few and reduced; flowers in somewhat open cymes; calyx pubescent and glandular, the lobes lanceolate, usually longer than the narrower bractlets; petals 3-10 mm. long, broadly ovate, yellow to almost white.

This an extremely variable species with many ill-defined subspecies or varieties.

Common in moist to dry habitats, often in the open, but also in shaded places.

Potentilla gracilis Dougl. Five-finger or Cinquefoil

Stems brown-scaly at base, erect from a woody crown, 2-10 dm. tall, pubescent; leaves mostly palmately divided into 5-9 leaflets, basal leaves long-petioled, leaflets 2-12 cm. long, narrow, margins deeply and narrowly toothed, under surface usually densely white silky-pubescent, upper surface green with some hairs, leaflets of uppermost leaves

Potentilla gracilis

subtending flowering stems reduced in number and leaves short-petioled; flowers golden yellow, saucer-shaped, in leafy-bracteate open to dense cymes; petals 4-10 mm. long; peduncle, pedicel, and calyx usually silky-pubescent; stamens and achenes many.

This is a variable species with many ill-defined varieties.

Common in meadows, open woods from the coast to subalpine areas in the mountains and eastward.

Potentilla norvegica L. Norwegian cinquefoil

Annuals to biennials or perhaps short-lived perennials; stems stout, leafy throughout, 1-8 dm. tall, pubescent, often purplish; leaves mainly cauline, palmately divided into 3 leaflets, (rarely 5, palmately or pinnately, below), lower leaves long-petioled, upper sessile, all pubescent, leaflets obovate or oblong, coarsely toothed, stipules broad, generally coarsely toothed; receptacle, sepals, and bractlets pubescent; the sepals and bractlets nearly equal in length; petals 3-4 mm. long, yellow; style spindle-shaped, glandular near base; achenes roughened, often wrinkled at maturity.

(=*Potentilla monspeliensis* L.)

Occasional throughout our area, usually in moist places or disturbed sites.

Potentilla palustris (L.) Scop. Marsh cinquefoil

Rhizomatous perennials; stems reddish, prostrate to ascending, often floating, rooting at the nodes, up to 1 m. in length, usually glabrous below, but becoming glandular-pubescent above; leaves 5-20 cm. long, pinnately compound of 3-7 leaflets, leaflets 2-10 cm. long, obovate to oblong, serrate, at least the lower surface with long-silky pubescence; flowers in open cymes; calyx lobes 7-20 mm. long, greenish-purple, glandular and pilose; petals 2-6 mm. long, dark red to purple; stamens 20-25; achenes about 1.5 mm. long.

Bogs, marshes, wet meadows, lake margins and ditches.

Potentilla rivalis Nutt. River cinquefoil

Taprooted annuals or biennials; stems 2-5 dm. tall, pubescent, freely branched and bearing flowers for most of its length; basal and lower cauline leaves usually with 5 leaflets, terminal leaflet longer than the laterals, upper cauline leaves with only 3 leaflets, leaflets 1-5 cm. long, coarsely toothed; calyx lobes surpassing the petals; petals yellow, 1.5-2 mm. long; stamens 10-15; achenes about 8 mm. long.

(=*Potentilla leucocarpa* Rydb.)

Swampy areas to wet banks or moist disturbed habitats.

Potentilla villosa Pall. Villous cinquefoil

Somewhat tufted perennials 3-20 cm. tall; basal leaves long-petioled, compound with 3 leaflets, the leaflets 2-4 cm. long, silvery-silky above, white-woolly beneath, with prominent veins, nearly wedge-shaped, the lateral leaflets oblique, cauline leaves few, reduced; bractlets about the length of the sepals; petals 5-8 mm. long, yellow; stamens about 20.

In the mountains of Washington.

Prunus

Shrubs or small trees with simple alternate leaves; flowers solitary or in racemes, corymbs or umbels; calyx 5-toothed, deciduous; petals 5, white to pink or red; stamens 15 or more; pistil 1, ovary superior; fruit a drupe, generally 1-seeded.

1a Flowers in elongate racemes .... *P. virginiana*
1b Flowers in corymbs or umbels
  2a Flowers in corymbs; native species .... *P. emarginata*
  2b Flowers in umbels; escaped from cultivation
    3a Leaves pubescent on the lower surface, at least on the veins; fruit sweet... *P. avium*
    3b Leaves glabrous or nearly so; fruit sour .... *P. cerasus*

Prunus avium L. Sweet cherry

Pyramidal trees to 24 m. in height; leaves oblong-ovate, 6-15 cm. long, unequally toothed, acuminate, sparingly pubescent on the lower surface, at least on the veins; flowers white, borne in umbels; petals obovate; fruit sweet, subglobose, about 2.5 cm. in diameter.

Commonly cultivated; escaped and occasionally naturalized.

Prunus cerasus L. Sour cherry

Round-headed trees to 10 m. in height; leaves ovate to elliptic, 5-8 cm. long, finely toothed, glabrous or nearly so; petals suborbicular; fruit sour, depressed globose.

Commonly cultivated; escaped and occasionally naturalized.

Prunus emarginata (Dougl.) Walp. Wild cherry

Shrubs to trees, 2-15 m. tall; branches and trunk reddish brown with conspicuous lenticels on the twigs; leaf blades ovate or elliptic, generally obtuse, smooth or minutely pubescent beneath, finely toothed on the margins, 2-6 cm. long, generally with 1 or 2 glands on base of blade just above the junction with the petiole; flowers in small corymbs; fruits bright red, 8-10 mm. long, very bitter.

Margins of woods, or forming thickets along streams.

Prunus emarginata

Prunus virginiana L. var. *demissa* (Nutt.) Sarg. Western chokecherry

Shrubs or small trees 2-8 m. tall; leaf blades long-obovate, acute or short acuminate at apex, obtuse or subcordate at base, finely toothed on the margins, 4-9 cm. long, 1 or 2 glands commonly found on petiole just below the junction with the blade; flowers in long slender racemes; fruit dark red or purplish black.

Common in a variety of habitats.

Prunus virginiana

Pyrus

Trees or rarely shrubs, sometimes with thorns; leaves entire to toothed; flowers in umbel-like racemes; petals white to pinkish; stamens 20-30; ovary inferior, styles 2-5, free; fruit pear-shaped with a gritty texture in the flesh.

Pyrus communis L. — Common Pear

Trees to 20 m. in height, often spiny; leaves orbicular-ovate to elliptic, acute to short-acuminate, 2-8 cm. long, finely toothed, often pubescent when young, but becoming glabrous; fruit pear-shaped.

Commonly cultivated and escaped and naturalized.

Rosa

Prickly shrubs or vines with pinnately compound leaves; stipules united with the leaf base and persistent; flowers showy, solitary or more or less cymose; receptacle and floral tube globose or urn-shaped, enclosing the pistils and becoming fleshy; petals generally 5 (or numerous in cultivated forms), broad, borne with the numerous stamens on the floral tube; pistils several to many, separate and distinct, enclosed in the receptacle but not united with it; fruit called a "hip" consisting of the colored fleshy receptacle enclosing several hard achenes.

1a Prickles stout, recurved or hooked; calyx lobes reflexed, some with lateral lobes; introduced species
  2a Under surface of leaflets glabrous or nearly so .... *R. canina*
  2b Under surface of leaflets with stalked glands; herbage sweet-scented .... *R. eglanteria*
1b Prickles, if stout, seldom recurved or hooked; calyx lobes without lateral lobes; native species
  3a Calyx lobes deciduous from the small fruit .... *R. gymnocarpa*
  3b Calyx lobes generally persistent on the fruits
    4a Flowers usually borne singly; leaflets glandular; prickles stout, straight .... *R. nutkana*
    4b Flowers generally several in corymbose cymes (rarely 1); leaflets not glandular; prickles slender .... *R. pisocarpa*

Rosa canina L. — Dog rose

Stems 1-4 m. tall bearing stout flattened curved or hooked prickles; leaflets 5-7, broadly elliptic to nearly orbicular, acute to acuminate, serrate, often doubly so and the teeth often gland-tipped, usually glabrous; flowers solitary or in few-flowered cymes; sepals 1-2 cm. long, usually with lateral lobes, reflexed and soon deciduous; petals white to pink, 1.5-3 cm. long; fruit bright red, 1.5-2 cm. long, glabrous.

Introduced from Eurasia; naturalized especially along roadsides.

Rosa eglanteria L. — Sweetbrier

Stems 1-2 m. tall bearing stout flattened recurved or hooked prickles; leaflets 5-7, ovate or elliptic to suborbicular, coarsely toothed, the teeth gland-tipped, densely glandular beneath with glands stalked, sweet-scented; calyx-lobes with lateral lobes, glandular; petals 1-2 cm. long, light clear pink, usually with white bases; fruits 1-1.5 cm. long, urn-shaped, contracted into a neck, the calyx-lobes finally falling away.

Introduced from Europe; now abundant west of the Cascades.

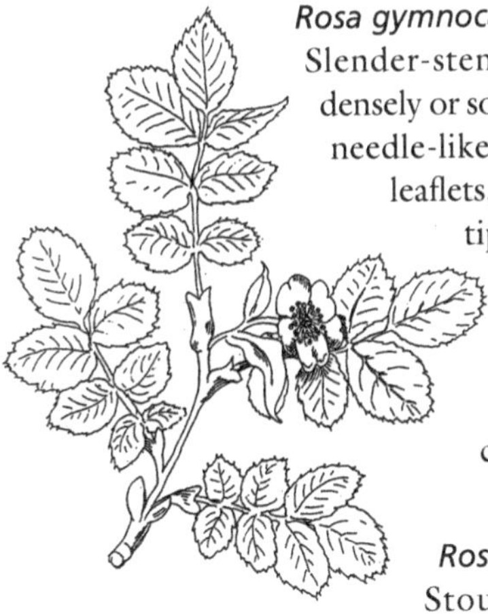

Rosa gymnocarpa Nutt. Wood rose

Slender-stemmed shrubs 3-15 dm. tall, generally densely or sometimes sparingly covered with numerous needle-like prickles; leaves divided into 5-9 small leaflets, these round-ovate to elliptic with gland-tipped teeth; flowers often borne singly; petals 1-1.5 cm. long, pink, usually with white bases; calyx-lobes sometimes with leafy tips, deciduous in fruit; fruits generally pear-shaped, rarely globose, 1 cm. or less long, bright red.

Woods or hillsides with other shrubs.

Rosa gymnocarpa

Rosa nutkana Presl Common wild rose

Stout erect shrubs, 0.5-3.5 m. tall, with broad-based stiff straight prickles and usually with smaller prickles intermixed; leaflets 5-9, ovate or elliptic, coarsely glandular-toothed with glandular hairs beneath or confined to the rachis; flowers typically borne singly; calyx lobes often leaf-like at the tips; petals 2.5-4 cm. long, deep pink with a slight magenta tinge; fruits globose to pear-shaped, 1-2 cm. broad, without a neck, calyx-lobes remaining attached.

Our largest and most widely distributed wild rose. Found associated with other shrubs in a variety of habitats.

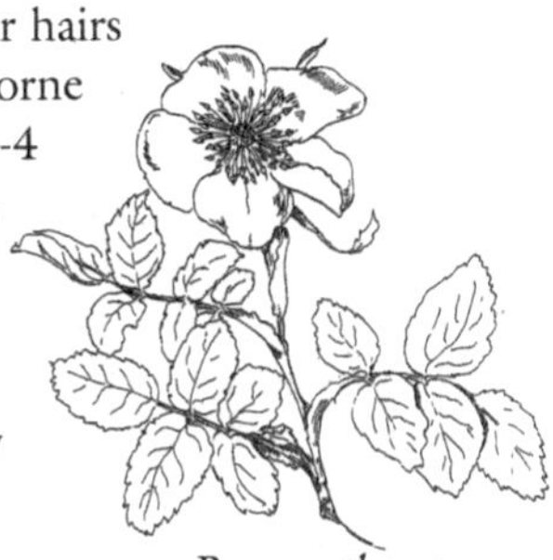

Rosa nutkana

Rosa pisocarpa Gray Cluster rose

Slender erect shrubs, 1-3 m. tall, bearing a few slender generally straight prickles; leaflets deep green above, paler beneath, with minute hairs, but not glandular; flowers in corymbose cymes; calyx-lobes elongated, bearing stalked glands on the under surface; petals 1-2 cm. long, deep pink with slight magenta tinge; fruits globose or slightly elongated, 6-12 mm. broad, with a short neck, the calyx lobes remaining attached.

Found associated with *R. nutkana* in the valleys, or alone at altitudes higher than generally inhabited by the latter.

Rubus

Semi-herbaceous or more often shrubs; stems sometimes trailing or climbing, prickly or smooth; leaves simple and lobed or compound, deciduous or evergreen; calyx 5-parted; petals 5 or more, white to red; stamens many; pistils many, united into a thimble-like covering over the elongated receptacle, forming an aggregate fruit of individual drupelets.

1a Stems not armed with prickles
  2a Stems erect, usually 1 m. or more tall
    3a Leaves compound with 3, or rarely 5, leaflets; flowers red to rose-colored.......... *R. spectabilis*
    3b Leaves simple, palmately lobed; flowers white .... *R. parviflorus*
  2b Stems trailing, semi-herbaceous
    4a Leaves compound, with 3-5 leaflets.......... *R. pedatus*
    4b Leaves deeply 3-5-lobed, but usually not compound.. *R. lasiococcus*
1b Stems armed with slender or stout prickles
  5a Leaves mostly or all simple; stems prostrate; fruit red .... *R. nivalis*
  5b Leaves typically compound
    6a Fruit separating, thimble-like, from receptacle at maturity (like raspberries)
      7a Flowers red to rose-colored; fruit yellow or orange-red .... *R. spectabilis*
      7b Flowers white; fruit black.......... *R. leucodermis*
    6b Fruit persistent upon the receptacle (blackberries)
      8a Leaflets deeply and sharply cleft.......... *R. laciniatus*
      8b Leaflets not deeply cleft
        9a Leaflets green beneath, typically 3; stems not angled *R. ursinus*
        9b Leaflets densely white-tomentose beneath, typically 5; stems angled .... *R. armeniacus*

Rubus armeniacus L. Himalaya blackberry

Stems thick, up to 10 m. long, 5-angled, arching, with flat curved prickles; leaflets palmately arranged, generally 5, green above, white-tomentose beneath, sharply toothed; buds densely white-woolly; petals 1-1.5 cm. long, white to pinkish; fruit black, glabrous.

(=*Rubus procerus* Muell. and *Rubus discolor* Weihe & Nees.)

Introduced species now well established in disturbed habitats.

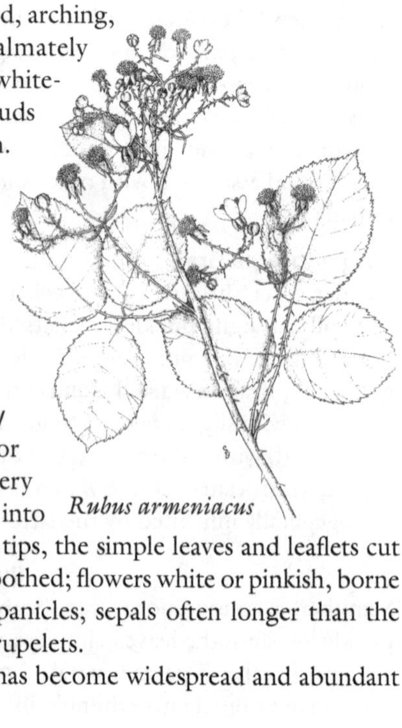

Rubus armeniacus

Rubus laciniatus Willd. Evergreen blackberry

Evergreen shrubs with long trailing or arching stems to 10 m. long, with very sharp prickles; leaves primarily divided into 3-5 leaflets, or simple near the branch tips, the simple leaves and leaflets cut into long slender lobes, these sharply toothed; flowers white or pinkish, borne singly in the leaf axils or in terminal panicles; sepals often longer than the petals; fruit shining black with large drupelets.

An escape from cultivation, which has become widespread and abundant in certain localities.

Rubus lasiococcus Gray — Dwarf bramble

Stems unarmed, trailing, rooting at the nodes; leaves deeply 3-5-lobed, but usually not compound; flowers solitary or 2 per stem; petals white, 5-8 mm. long; fruit red, puberulent.

Common in open woods in the mountains.

Rubus leucodermis Dougl. — Blackcap

Trailing or arching shrubs straggling at time of flowering; old stems strikingly light bluish due to a waxy coating over the red bark; leaves divided into 5-7 leaflets the first year, the second year's leaves with 3 leaflets; leaflets broadly ovate, coarsely toothed, mostly green above, paler beneath and covered with soft down; stems, petioles, pedicels, and sometimes the leaf veins, bearing stiff, broad-based, recurved spines; flowers solitary or in several flowered corymbs in the leaf axils or at ends of lateral branches; petals white; sepals longer than petals; fruit purplish-black with a whitish waxy coating, puberulent, sweet and mild-flavored.

Rather common, often associated with *R. ursinus.*

Rubus nivalis Dougl. — Snow bramble

Stems prostrate, with slender recurved prickles and rooting at the nodes; leaves mostly simple, the blades irregularly lobed and toothed; flowers white to pink to purple; fruits red, puberulent.

In moist shady areas in the mountains.

Rubus parviflorus Nutt. — Thimbleberry

Unarmed erect shrubs 1-3 m. tall; leaves palmately 5-lobed, or rarely 3-7-lobed, deeply cordate at base, raggedly toothed, pubescent, the veins, petioles, and stems glandular at least when young, leaf blades 0.5-2.5 dm. long; flowers 4-8 in a loose terminal corymb, white, rarely pinkish, 3-5 cm. broad; calyx-lobes broad at base but abruptly narrowed to a long appendage; petals 1-3 cm. long, white or rarely pinkish-tinged, obtuse; fruit scarlet, puberulent, sweet.

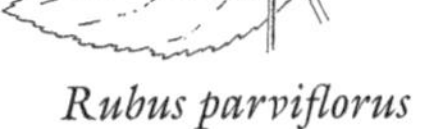

Rubus parviflorus

Common along streams and in open woods. One of our most attractive shrubs in foliage, blossom, and fruit. Has a pleasing fragrance in the rain.

Rubus pedatus Smith — Trailing raspberry

Unarmed low trailing plants with cord-like stems, rooting at the nodes; leaves palmately 3-5-foliolate, on slender petioles, leaflets cuneate at the base, coarsely double-serrate above, 1-3 cm. long; flowers white, solitary, long-peduncled; fruit red, glabrous.

Common in open moist woods.

Rubus spectabilis Pursh — Salmonberry

Erect shrubs with long sparingly branched stems 1-5 m. tall, with thin reddish shreddy bark and few prickles, or young stems very prickly; leaves divided into 3 (rarely 5) leaflets, these more or less lobed and irregularly toothed; flowers red or rose-colored, borne singly or in clusters of 2-4 in the leaf axils and at the ends of branches, or on short axillary branchlets, flowers sometimes appearing before the leaves unfold; fruit yellow at first, usually becoming dark red.

Along streams; common.

Rubus ursinus Cham.& Schlecht — Wild blackberry

Usually dioecious shrubs with trailing or climbing prickly stems 1-8 m. long; leaves with 3-5 leaflets or some simple and deeply lobed; leaflets ovate, coarsely toothed, pubescent or nearly glabrous; flowers 2-3 cm. in diameter, white, pistillate flowers with abortive stamens, the buds covered with short white hairs, and sparsely glandular; fruits shining black.

Common; particularly abundant in burned and logged-off areas. This is our only native blackberry. The fruit, though small, has an excellent flavor.

Sanguisorba

Annual or perennial herbs with alternate odd-pinnate stipulate leaves; flowers small, perfect or imperfect in long-peduncled head-like spikes; calyx lobes 4, deciduous; petals none; stamens 2-12 or more; pistil generally one, ovary superior, becoming an achene enclosed in the 4-angled floral tube.

1a Annuals or biennials; leaflets deeply cut .... *S. occidentalis*
1b Perennials; leaflets merely toothed
  2a Weedy species; stamens about 12 .... *S. minor*
  2b Plants of bogs or stream banks; stamens 2-4
    3a Spikes purplish.. *S. officinalis*
    3b Spikes white or greenish .... *S. sitchensis*

Sanguisorba minor Scop. — Garden burnet

Perennials 2-6 dm. tall, leafy, glabrous to sparsely pubescent; leaflets 7 to 21, ovate to nearly orbicular, coarsely toothed, the stipules crescent-shaped and toothed; flowers in short greenish-purplish spikes, often with imperfect flowers, the pistillate flowers above the staminate ones; stamens usually about 12; the floral tube 4-angled.

Introduced at one time as a forage plant in the Willamette Valley; now occasionally naturalized and weedy.

Sanguisorba occidentalis Nutt. — Western burnet

Taprooted annuals or biennials; stems branching, erect, 1-12 dm. tall; leaflets 9-15, cut into narrow lobes; spikes greenish, oblong, 0.5-3.5 cm. long; flowers perfect; stamens 2 or 4.

Common.

Sanguisorba occidentalis

Sanguisorba officinalis L. — Great burnet

Perennials from thick creeping rhizomes; stems 1-10 dm. tall; leaves mostly basal, 1-3 dm. long, leaflets 7-15, ovate or oblong, cordate at base, 1.5-5 cm. long, somewhat leathery, often somewhat folded and turned downward, coarsely toothed, the teeth mucronate; spikes 1 or several, 1-3 cm. long, oblong, purplish; stamens 2 or 4.

Bogs in the mountains and along the coast.

Sanguisorba sitchensis C.A. Mey. — Sitka burnet

Perennials from stout rhizomes; stems 2.5-10 dm. in height, leaflets 9-17, ovate to oblong, coarsely toothed; spikes 2-8 cm. long; flowers white or greenish; stamens 4.

Bogs and stream banks in the mountains.

Sorbus

Shrubs or small trees; leaves pinnately compound; flowers white in corymbs or cymes; fruits of small red pomes.

Sorbus sitchensis Roem. var. *grayi* (Wenzig) C.L. Hitchc. — Mountain ash

Shrubs, 1-5 m. tall; leaflets 7-15, lanceolate to ovate or oblong, sharply serrate usually only near the apex; flowers in cymes; petals 3-4 mm. long; fruits scarlet, about 1 cm. in diameter, in open cymes.

In the mountains.

Spiraea

Deciduous shrubs with simple, short-petioled leaves without stipules; flowers small, in compound corymbs or panicles; calyx lobes and petals 5; stamens numerous, borne with the petals on the floral tube; pistils usually 5, becoming follicles.

1a Inflorescence a panicle, longer than wide, not flat-topped .... *S. douglasii*
1b Inflorescence of flat-topped corymbs
  2a Flowers white or pinkish-tinged .... *S. betulifolia*
  2b Flowers rose- to lilac-colored .... *S. densiflora*

Spiraea betulifolia Pall. var. *lucida* (Dougl. ex Greene) C.L. Hitchc. — Shiny-leaf Spiraea

Low shrubs with woody stems often basally prostrate, sending up erect leafy stems 2-6 dm. tall; leaves ovate, elliptic or obovate, the lower fourth entire and rounded to the base, the upper three-fourths coarsely toothed, glabrous or nearly so; flowers small, borne in dense flat-topped corymbs; petals white, sometimes pinkish-tinged.

In the mountains.

Spiraea densiflora Nutt. Subalpine Spiraea

Stems erect to spreading, 2-10 dm. tall, branching above, the older stems with reddish bark; leaves 1.5-4 cm. long, pale green, glabrous or finely puberulent, broadly obovate or elliptic, short-petioled, obtuse, toothed above the middle; flowers very small, rose- to lilac-colored, in short dense corymbs.

In the mountains near timber line.

Spiraea douglasii Hook. Douglas' Spiraea or Hardhack

Erect shrubs 1-2 m. tall, bushy, with reddish bark; leaves 2.5-10 cm. long, oblong to obovate, toothed above the middle, green above but generally sparsely pubescent, whitish beneath and usually minutely woolly; panicles 5-30 cm. long, pyramidal to oblong, appearing powdery from the minute projecting stamens; flowers rose-colored to purplish, pleasantly fragrant.

Common in the borders of marshes, stream banks and wet meadows and occasionally found in drier situations.

FABACEAE
Pea Family

Herbs, shrubs, or trees, with alternate mostly compound leaves and entire leaflets, generally with stipules; sepals partly united, calyx often 2-lipped, generally 5-toothed; petals generally 5, rarely 1, separate, or certain petals sometimes partly united, regular or irregular (ours), usually with 1 banner, 2 wings, and a keel of 2 petals; stamens usually 10, separate, or more commonly united into a sheath about the pistil, or the upper stamen free; pistil 1, simple, ovary superior, 1-loculed, with 1 row of seeds, the fruit a legume, generally opening by splitting on both sides or rarely indehiscent.

The alternate name for this family is **Leguminosae**.

1a Trees or shrubs
  2a Leaves pinnately compound with at least 7 leaflets
    3a Petal 1; fruits indehiscent, 1-seeded .... **Amorpha**
    3b Petals 5; fruits dehiscent, 3-10-seeded .... **Robinia**
  2b Leaves palmately compound or leaflets 3 or reduced to spines
    4a Densely spiny; leaves reduced to spines and scales...................... **Ulex**
    4b Not spiny
      5a Leaflets 5-12; leaves palmately compound .... **Lupinus**
      5b Leaflets 3
        6a Upper lip of the calyx barely 2-lobed; style curved for the entire length or abruptly curved near the middle........................**Cytisus**
        6b Upper lip of the calyx strongly 2-lobed; style abruptly bent at the tip...**Genista**
1b Herbs (ours), sometimes slightly woody at the base
  7a Leaves of 2 or more than 3 leaflets; tendrils sometimes present
    8a Leaves palmately compound; leaflets 5 to many (ours)......... **Lupinus**
    8b Leaves pinnately compound
      9a Tendrils none; leaves typically odd-pinnate
        10a Flowers in umbels or axillary; not alpine .... **Lotus**
        10b Flowers in short spike-like racemes, alpine (ours) .... **Oxytropis**
      9b Tendrils present (sometimes reduced to a bristle); leaves even-pinnate; leaflets sometimes only 2
        11a Style filiform, with a tuft of hairs just below the apex .....**Vicia**
        11b Style flattened; hairy on upper side only .... **Lathyrus**
  7b Leaves with 3 leaflets (ours); tendrils absent
    12a Leaflets gland-dotted
      13a Calyx 6-8 mm. long, black-hairy .... **Rupertia**
      13b Calyx 2-2.5 mm. long, not black-hairy...................**Psoralidium**
    12b Leaflets not gland-dotted
      14a Stamens all free; flowers racemose, yellow............... **Thermopsis**
      14b Stamens in 2 sets, 9 united by the filaments and 1 free
        15a Flowers in numerous slender elongated usually 1-sided racemes; corollas small, deflexed, white or yellow ....**Melilotus**
        15b Not as above in all points

16a Flowers in umbels or axillary and solitary; leaflets entire.................. **Lotus**
16b Flowers typically not in umbels, but if so, the fruit spirally coiled and prickly; leaflets usually toothed
17a Fruits spirally coiled, usually prickly or merely kidney-shaped, well exserted from the calyx... **Medicago**
17b Fruits not curved or coiled nor prickly, mostly enclosed in the calyx..... ....**Trifolium**

Amorpha

Shrubs; leaves odd pinnate, stipulate, strongly resinous-dotted, the midrib of the leaflets often ending in a short bristle, leaflets with minute stipule-like appendages; flowers with only 1 petal (the banner) present; calyx 5-toothed; stamens 10; fruit indehiscent, 1-2-seeded, resinous-dotted.

Amorpha fruiticosa L. False Indigo

Aromatic shrubs, 1-4 m. in height, branches finely pubescent, at least when young; leaves long petiolate, with 9-31 leaflets, 1-6 cm. long, up to 2 cm. wide; banner 5-6 mm. long, blue to red-purple; fruit incurved, 5-8 mm. long.

Along stream banks, but also in drier wooded sites; mainly along the Columbia River in our area; introduced from the eastern United States.

Cytisus

Shrubs with green branches; leaves divided into 3 leaflets; flowers generally yellow or white; calyx 2-lipped, the lips barely 2-lobed; style gently curved for the entire length or abruptly curved near the middle; stamens 10, united by the filaments.

Flowers yellow, 15-25 mm. long .... *C. scoparius*
Flowers white, 9-11 mm. long .... *C. multiflorus*

Cytisus multiflorus (L'Her.) Sweet White Spanish broom

Freely branching shrub 1-4 m. tall; leaves with 3 leaflets at least below, but unifoliolate above, leaflets less than 1 cm. long, silky pubescent; flowers solitary or 2-3 in axillary clusters; corolla white, 9-11 mm. long; legumes 1.5-2.5 cm. long, pubescent.

Escaped from cultivation and established along roadsides.

Cytisus scoparius (L.) Link Scotch broom

Shrubs 1.5-3 m. tall, branches green, angled; leaflets 3, (or the uppermost leaves unifoliolate), obovate or oblanceolate, more or less pubescent at least when young; flowers showy, borne in the leaf axils; corollas 15-25 mm. long, deep yellow; legumes 2.5-4 cm. long, glabrous except for long hairs on the margins.

A European shrub originally introduced as an ornamental, now widely naturalized.

Var. *andreanus* Dipp. A form having dark crimson wings is an occasional escape from cultivation.

Genista

Shrubs or small trees with green branches and angled stems; leaves with 3 leaflets; flowers yellow; upper lip of the calyx strongly 2-lobed; style abruptly bent at the tip; stamens united by the filaments.

Genista monspessulana (L.) L. Johnson French broom

Shrubs up to 3 m. in height; branches leafy, silky pubescent; leaflets 3, cuneate-obovate, glabrous above, pubescent on the lower surface; flowers 3-10 in short racemes; corollas 1-1.5 cm. long, light yellow; legumes densely silky pubescent.

(=*Cytisus monspessulanus* L.)

Escaped from cultivation and now becoming widespread, especially along the coast.

Lathyrus

Annual or perennial herbs; stems angled to winged; leaves pinnately compound or rarely reduced to tendrils, main axis ending in a tendril or rarely a short bristle; flowers in 1-sided racemes or axillary; stamens with 9 filaments fused and 1 free; styles flattened and hairy only on the upper side; legumes usually flattened.

1a Leaf blades reduced to tendrils; stipules leaf-like, hastate at the base; flowers yellow, typically solitary in the leaf axils.. *L. aphaca*
1b Leaf blades not reduced to tendrils
  2a Leaflets 2
    3a Corollas 14 mm. long or less
      4a Flowers orange-red; peduncles 1-2 cm. long............*L. sphaericus*
      4b Flowers purple to lavender; peduncles 2-6 cm. long .... *L. angulatus*
    3b Corollas averaging over 15 mm. long; stems winged
      5a Annuals; peduncles 1-3-flowered; flowers showy, 2.5-3 cm. long, crimson or the banner purplish .... *L. tingitanus*
      5b Rhizomatous perennials; peduncles typically 4-15-flowered
        6a Stipules linear-lanceolate; flowers 14-18 mm. long; legumes 4-7 cm. long .... *L. sylvestris*
        6b Stipules usually ovate; flowers 15-25 mm. long; legumes 6-10 cm. long .... *L. latifolius*
  2b Leaflets more than 2
    7a Corollas creamy-yellowish, turning deep yellow-brown, 1-2 cm. long .... *L. holochlorus*
    7b Corollas purple to violet or rarely white
      8a Sea-beach or coastal marsh species
        9a Plants gray silky pubescent; leaflets 1.5-2 cm. long.................. .... *L. littoralis*
        9b Plants green, glabrous or if pubescent, not gray; leaflets 2-6 cm. long
          10a Stipules about equal to the leaflets; leaflets 6-12; flowers purple .... *L. japonicus*

10b Stipules less than 1/2 as long as the leaflets; leaflets typically 6, (4-8); flowers pinkish-purple or rarely white .... *L. palustris*
8b Not sea-beach or coastal marsh plants
11a Peduncles 1-2-flowered; leaflets 8-16, 1-2 cm. long; legumes 1.5-2.5 cm. long .... *L. torreyi*
11b Peduncles more than 2-flowered or leaflets or legumes longer
12a Stipules 1 cm. wide or more; leaflets 10-16 .... *L. polyphyllus*
12b Stipules less than 1 cm. wide; leaflets 12 or less
13a Corollas white.. *L. vestitus*
13b Corollas purple to lavender or pink...................... *L. nevadensis*

Lathyrus angulatus L. Grass pea

Erect to spreading annuals; stems angled; leaflets 2, narrowly lanceolate to linear, tendrils well developed; flowers solitary; corollas 10-12 mm. long, purple to lavender; legumes 3-4.5 cm. long, glabrous.

Road sides and other disturbed sites; introduced from Europe.

Lathyrus aphaca L. Yellow pea

Glabrous annuals; stems slender, 1-8 dm. tall; leaves reduced to tendrils and to the 2 leaf-like stipules which are 1-5 cm. long and hastate at the base; flowers usually solitary, rarely 2, borne in the leaf axils, yellow; corolla about 12 mm. long; legumes 2-4 cm. long, glabrous.

A European species naturalized west of the Cascades.

Lathyrus holochlorus (Piper) C.L. Hitchc. Thin-leaved pea

Rhizomatous perennials; stems 3-10 dm. tall, rather stout, glabrous, more or less winged; leaflets 6-10, broadly ovate to elliptic, obtuse and mucronate at apex, 2-5 cm. long, tendrils well developed; racemes 5-many-flowered, the flowers turned more or less downward from the apex of the pedicel; corolla creamy-yellowish, turning deep yellow-brown, 1-2 cm. long; legumes 3-5 cm. long, glabrous.

In the margins of woods and along fence rows.

Lathyrus japonicus Willd. Purple beach pea

Rhizomatous perennials; stems 1-15 dm. long, erect at first, later decumbent, pale green, usually glabrous, angled to winged; leaflets 6-12, elliptic or broadly ovate, 2-6 cm. long, generally obtuse with a minute point; stipules broadly ovate, as large as the leaflets, acute, somewhat clasping, hastate at base, the margins often toothed; peduncles 2-10-flowered, the 2 upper calyx teeth short, broadly triangular, the 3 lower teeth long and narrow; corolla purple, about 2 cm. long; legumes 3-7 cm. long, puberulent at least at first.

Sea-beaches and salt marshes.

Lathyrus latifolius L. Perennial sweet pea

Rhizomatous perennials; stems to 2 m. long, trailing or climbing, broadly winged; leaflets 2, lanceolate, up to 15 cm. long and 5 cm. wide, tendrils well developed; stipules usually entire, ovate; flowers white to pink or red, 1.5-2.5 cm. long; legumes to 1 dm. long.

Introduced from Europe; widely escaped from cultivation and naturalized.

Lathyrus littoralis (Nutt.) Endl. Silvery beach pea

Plants gray silky-pubescent throughout, spreading from creeping rhizomes; stems many, often prostrate, 1-6 dm. long; leaflets 1.5-2 cm. long, oblanceolate, 2-5 pairs, with a broad terminal flattened bristle; stipules larger than the leaflets, ovate and hastate; peduncles in the leaf axils, longer than the leaves, generally 3-10-flowered; flowers showy, 1-2 cm. long; the 2 upper calyx-lobes somewhat shorter than the lower, all shorter than the tube; banner petal purple or blue, the wings and keel nearly white; legumes broadly oblong, about 1.5-3.5 cm. long, villous.

Along ocean beaches.

Lathyrus nevadensis Wats.

Perennials; stems angled, slender, mostly erect, 1-9 dm. tall; leaflets 4-10, linear to ovate-oblong, mucronate, more or less pubescent beneath, tendrils often reduced and bristle-like; stipules narrow, usually sagittate; peduncles about as long as or shorter than the leaves, 1-7-flowered; flowers purple to lavender or pink, about 1-2.5 cm. long; calyx teeth all shorter than the tube; legumes 3-6 cm. long, glabrous.

This is a variable species with several varieties in our area.

Open woods.

Lathyrus palustris L. Marsh pea

Glabrous to pubescent perennials; stems slender, angled to narrowly winged, weakly erect, 3-10 dm. tall; leaflets 4-8, linear to ovate, mucronate, 2-6 cm. long, conspicuously veined, tendrils simple or more often branched; stipules usually sagittate, acuminate, narrow; peduncles longer than the leaves, 2-6-flowered; flowers pinkish-purple or rarely white, 1.5-2 cm. long; upper 2 calyx teeth shorter than the lower ones.

Marshes near the coast.

Lathyrus polyphyllus Nutt. Leafy pea

Glabrous perennials; stems stout, angled, weakly erect, 4-10 dm. tall; leaflets 10-16, ovate or oblong, obtuse, mucronate, thin, bright green above, paler beneath, 2.5-6 cm. long, tendrils slender, simple or branched; stipules broad, sagittate; flowers 5-13, bluish-purple to nearly white, 1.5-2 cm. long; legumes 4-6 cm. long.

In open woods and road sides.

Lathyrus sphaericus Retz. Vetchling

Glabrous annuals; stems slender, narrowly winged, freely branching, 1.5-6 dm. tall; leaflets 2, linear or narrowly lanceolate, 2-7 cm. long; stipules narrow, sagittate, somewhat curved; peduncles 1-flowered, the flowers 8-14 mm. long, orange-red; legumes narrow, 3-6 cm. long, striate.

Introduced from Europe; naturalized in disturbed habitats.

Lathyrus sphaericus

Lathyrus sylvestris L. — Narrow-leaved everlasting pea

Glabrous rhizomatous perennials 6-20 dm. long; stems trailing or climbing, broadly winged; leaflets 2, lanceolate, 4-15 cm. long, tendrils branched; stipules linear-lanceolate, 1-3 cm. long; flowers red to pinkish-purple, 14-18 mm. long; legumes 4-7 cm. long.

Introduced from Europe; widely established and weedy.

Lathyrus tingitanus L. — Tangier pea

Glabrous annuals; stems stout, narrowly winged, climbing, widely branching, 5-20 dm. long; leaflets 2, narrowly lanceolate to ovate, mucronate, 2-10 cm. long, conspicuously veined, tendrils compound; stipules narrow; peduncles 1-3-flowered; flowers showy, 2.5-3 cm. long, crimson or the banner purplish; legumes 6-10 cm. long, glabrous.

An escape from cultivation; widespread in our range, especially common in the Willamette Valley.

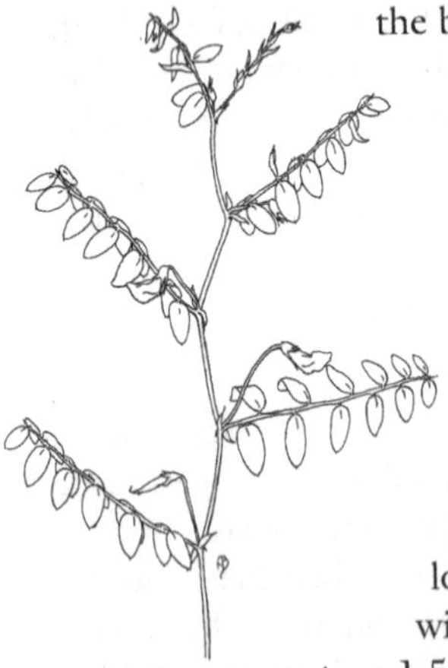

Lathyrus torreyi Gray — One-flowered pea

Rhizomatous perennials; stems slender, angled, leafy, erect, sometimes weakly so, 0.5-4 dm. tall; leaves bright green, not conspicuously pubescent; leaflets 8-16, 1-2 cm. long, ovate or oblong, mucronate; stipules lanceolate, sagittate; peduncles much shorter than the leaves, 1-2-flowered; flowers about 1 cm. long; calyx-teeth very slender, the 3 lower much longer than the tube; banner spreading, purplish, the wings and keel lilac-colored to nearly white; legumes 1.5-2.5 cm. long, puberulent.

Open woods, not common.

Lathyrus torreyi

Lathyrus vestitus Nutt. var. *ochropetalus* (Piper) Isley — Pacific pea

Perennials; stems erect to spreading, 2-20 dm. tall, typically glabrous; leaflets 8-12, 2-4.5 cm. long, ovate to elliptic, tendrils well developed; peduncles 6-12-flowered; corolla usually white, 14-16 mm. long; legumes 4-6 cm. long.

Moist areas in the Coast Range.

Lotus

Annuals to perennials; leaves 3-many-pinnate or rarely some simple; stipules gland-like or membranous to leaf-like; flowers axillary or in simple umbels; calyx bell-shaped or short-tubular; petals yellow or white, often pink- or red-tinged; stamens 10, 9 fused by their filaments and 1 free; fruit usually dehiscent.

1a Annuals; peduncles 1-2-flowered borne in the leaf axils
- 2a Flowers sessile or nearly so in the leaf axils; legumes not constricted between the seeds .... *L. denticulatus*
- 2b Flowers on peduncles nearly equalling or exceeding the leaves; legumes usually more or less constricted between the seeds
  - 3a Calyx-teeth shorter than the tube; leaflets 3-6 .... *L. micranthus*
  - 3b Calyx-teeth longer than the tube; leaflets 3 .... *L. purshianus*

1b Perennials; flowers in 3-many-flowered umbels
4a Leaflets 5, the lower pair basal and appearing as stipules, but stipules gland-like
5a Rhizomes lacking; stems solid; umbels 3-8-flowered.......................... *L. corniculatus*
5b Rhizomes present; stems hollow; umbels 8-12-flowered.................. *L. uliginosus*
4b Leaflets, if 5, the lower pair not basal and appearing as stipules
6a Stipules gland-like; petals yellow .... *L. nevadensis*
6b Stipules scarious and often early deciduous or leaflet-like; petals not all yellow
7a Bract of the peduncle borne well below the umbel
8a Flowers 6-12, white to pinkish .... *L. aboriginus*
8b Flowers 7-20, greenish-yellow often spotted with purple......... *L. crassifolius*
7b Bract of the peduncle, if present, at the base of the umbel
9a Wing petals pink to purple, becoming white; bracts usually 3-parted; plants coastal .... *L. formosissimus*
9b Wings white to cream-colored; bracts, if present, simple.......... *L. pinnatus*

Lotus aboriginus **Jeps.**

Perennials; stems 1-7 dm. tall, pubescent only when young; leaflets 9-15, 1.5-3 cm. long, elliptic, pale on the under surface; stipules triangular, inconspicuous and often early deciduous; flowers 6-12; corollas 1-1.5 cm. long, white to pinkish and streaked or tipped with red or purple; legumes 3-4 cm. long, glabrous.

[=*Lotus crassifolius* (Benth.) Greene var. *subglaber* (Ottley) C.L. Hitchc.]

From the Coast Range westward, often in logged areas.

Lotus corniculatus **L.** — **Bird's-foot trefoil**

Perennials; stems often trailing and rooting at the nodes, 0.5-4 dm. long; leaflets 5, the lower pair basal and appearing as stipules, but stipules gland-like, leaflets 5-20 mm. long, 2-7 mm. broad, obovate, usually ciliate and finely serrulate; flowers 3-8 in an umbel; calyx 4-8 mm. long; corolla 8-15 mm. long, yellow and often reddish-tinged; legumes 1.5-4 cm. long.

Introduced from Europe and well established in disturbed wet areas, but also in drier sites.

Lotus crassifolius **(Benth.) Greene var. *crassifolius***

Rhizomatous perennials; stems erect to spreading, up to 1 m. tall; leaflets 7-15, oblong to obovate; umbels 7-20-flowered; corolla 12-18 mm. long, greenish-yellow often spotted with purple; legumes 2.5-7 cm. long.

Woods, stream banks and disturbed sites; on both sides of the Cascades.

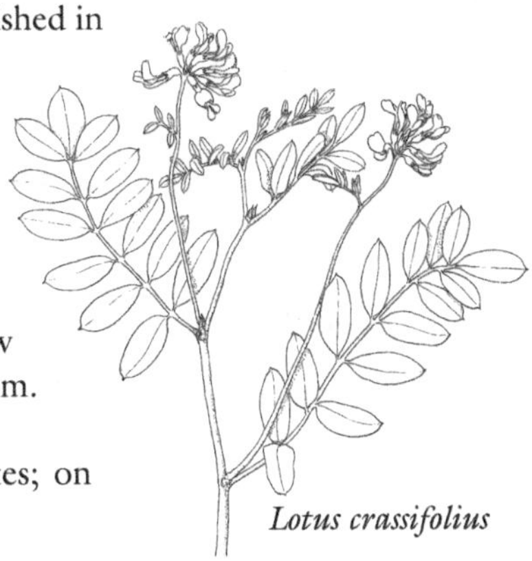

Lotus crassifolius

Lotus denticulatus (Drew) Greene
Annuals; stems generally branching from the base, erect, 1-5 dm. tall, pubescent to nearly glabrous, often glaucous; leaflets 2-4, mostly ovate, acute at apex, 1-2 cm. long, often with white hairs, especially on the lower side; flowers mostly solitary in the leaf axils, sessile or nearly so, without bracts; corollas cream-colored to yellow with a red to purplish banner; legumes oblong, straight, 1-2 cm. long, sparsely pubescent, not constricted between seeds.

Not common in our area, but occasionally found in open, often sandy areas.

Lotus formosissimus Greene — Seaside Lotus
Perennials from rhizomes and/or stolons; stems 1-5 dm. long, more or less trailing; leaflets 3-7, elliptic, 6-20 mm. long; stipules leaf-like; umbels 3-9-flowered; petals 1-1.5 cm. long, the banner yellow, the wings pink to purple, becoming white, the keel purple-tipped; legumes 2-3 cm. long.

Wet meadows, bogs, marshes, lake shores, stream banks and ditches, particularly along the coast.

Lotus micranthus Benth. — Slender trefoil
Annuals; stems generally several from the base, freely branching, wiry, and often reddish; leaflets 3-6, elliptic or obovate, obtuse, 4-12 mm. long, shining or somewhat pubescent; stipules gland-like; flowers subtended by a leafy bract, mostly solitary on somewhat elongated peduncles in the leaf axils; corollas pink or red and yellow; legumes very slender, shining, 3 cm. or less long, 4-9-seeded, constricted between the seeds.

Very common, usually in open disturbed areas.

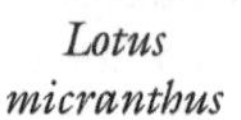

Lotus micranthus

Lotus nevadensis (Wats.) Greene — Nevada deervetch
Perennials; stems 1.5-6 dm. long, forming dense low clumps; foliage grayish-pubescent; leaflets 3-5, oblanceolate to obovate, acute, generally less than 1.5 cm. long; peduncles mostly shorter than the leaves; stipules gland-like; umbels 3-12-flowered, the bract with 1 to 3 leaflets; flowers yellow, the wings slender, conspicuously longer than the keel; fruit indehiscent, 1-3-seeded.

Dry ground, especially in sandy or rocky soils.

Lotus pinnatus Hook. — Butter-and-eggs
Perennials, 1-6 dm. tall; stems decumbent or erect; leaflets 5-9, oblong to obovate, 1-2.5 cm. long, mostly obovate, obtuse at apex, or the upper acute, mucronate; stipules scarious, 3-15 mm. long; umbels 4-12-flowered; corollas 10-13 mm. long, banner yellow, wings white to cream-colored, keel yellow or white; legumes 3-6 cm. long, glabrous, not constricted between the seeds.

Lotus pinnatus

Stream banks, lake margins, wet meadows, bogs and even ditches.

Lotus purshianus (Benth.) F. Clem. & E. Clem. Spanish-clover
Annuals; stems stiff, widely branched above, erect, loosely pubescent to nearly glabrous; leaflets generally 3, oblong or ovate-lanceolate, acute, 0.5-3 cm. long; stipules reduced to glands; peduncles about the length of the leaves, mostly 1-flowered; flowers 4-9 mm. long, cream-colored and red- to pink-tinged; mature legumes 1.5-4 cm. long, 4-8-seeded, slightly constricted between the seeds.

Common in dry ground.

Lotus uliginosus Schk.
Rhizomatous perennials; stems 3-9 dm. tall, more or less ascending; leaflets 5, the lower pair basal and appearing as stipules, but stipules gland-like, leaflets obovate to oblong, 1-2.5 cm. long; umbels 8-12-flowered with a 1-3-parted bract just below the umbel; corolla about 12 mm. long, yellow or the banner reddish; legumes 2-3 cm. long.

Usually in wet places; introduced from Europe.

Lupinus

Herbs or shrubs with palmately compound leaves; leaflets generally 5-17, entire; stipules fused with the petiole; flowers in racemes; calyx 2-lipped, the lips entire or toothed; lateral bractlets sometimes present; banner petal generally grooved through the center, with sides turned backward; stamens 10, united by the filaments, the filaments and anthers of alternate lengths; legumes usually flattened, 2-12-seeded.

1a Annuals.
  2a Banner as wide or wider than long; corollas 8-13 mm. long .... *L. affinis*
  2b Banner longer than wide; corollas 4-10 mm. long
    3a Banner scarcely reflexed from wings; keel blunt.......... *L. micranthus*
    3b Banner well reflexed; keel slender .... *L. bicolor*
1b Perennials
  4a Calyx distinctly spurred ....*L. arbustus*
  4b Calyx not spurred, though sometimes swollen
    5a Shrubs to 2 m. tall; flowers yellow; coastal .... *L. arboreus*
    5b Not yellow-flowered coastal shrubs
      6a At least some leaves with more than 10 leaflets; flowers 1-1.5 cm. long; stems erect, usually unbranched, often hollow... *L. polyphyllus*
      6b None of the leaves with more than 10 leaflets or otherwise not as above
        7a Keel not ciliate, but bent and early exposed for most of its length...*L. albicaulis*
        7b Keel ciliate on the upper margin for at least part of its length or nearly concealed
          8a Plants prostrate, densely pubescent, strictly coastal............. .... *L. littoralis*
          8b Plants either not prostrate or not coastal

9a Banner only slightly reflexed; corollas 8-12 mm. long .... *L. sulphureus*
9b Banner either well reflexed or the corollas longer
  10a Banner densely pubescent on the back .... *L. albifrons*
  10b Banner glabrous on the back
    11a Stems many from the base, typically unbranched; plants low, tufted, densely pubescent; inflorescence dense .... *L. lepidus*
    11b Not as above in all points
      12a Upper calyx lip much wider than the lower............. *L. latifolius*
      12b Upper calyx lip not much wider than the lower, although shorter .... *L. rivularis*

Lupinus affinis Agardh — Showy lupine

Annuals; stems erect or slightly decumbent, 1.5-5 dm. tall, more or less pubescent; leaflets 5-8, mostly broadly obovate, mucronate, pubescent on both surfaces, 2-5 cm. long; flowers 8-13 mm. long, deep blue, on pedicels 6 mm. or less long, base of banner white or yellow, changing to lavender, keel ciliate; legumes pubescent, broad, sometimes reaching 5 cm. in length, with 4-12 mottled seeds.

Limited in distribution, but sometimes occurring in large patches in open areas from the Coast Range westward. Very showy.

Lupinus albicaulis Dougl. ex Hook. — Silky-stemmed lupine

Perennials; stems stout, branched, several from a thick crown, 3-12 dm. tall, somewhat silky-pubescent, especially above; leaflets 5-10, 2-7 cm. long on petioles of nearly equal length, oblanceolate, acute or the larger sometimes obtuse, mucronate, somewhat silky-pubescent to glabrous; raceme of main stem generally more elongated than those of the branches, and usually flowering earlier; bracts about as long as pedicels, and these nearly equalling the corollas; pedicels and calyx silvery-pubescent; upper lip of calyx 2-toothed, lower lip longer, slender, entire or nearly so; corolla purplish, banner glabrous, keel bent at nearly right angles and early exposed for most of its length, not ciliate; legumes 2-5 cm. long, pubescent.

Common in dry open ground.

Lupinus albifrons Benth. — White-leaved lupine

Perennials; stems woody, generally freely branched and more or less prostrate at base, giving rise to ascending branches 2-5 dm. tall; leaflets 6-10, oblanceolate, mucronate, 1.5-3 cm. long, silky, the mature leaflets greenish, young leaflets often bronze-silky, petioles once or twice the length of the leaflets in the summer foliage or even longer on the lower leaves; racemes loose; flowers showy, 11-17 mm. long, wings and banner broad, purple or blue, banner center at first white or yellowish, changing to lavender, banner reflexed from wings for 1/2 or more of its length, keel ciliate; legumes 3-5 cm. long, pubescent; seeds 4-9, mottled.

Open, often sandy or rocky areas; Willamette Valley and southward.

Lupinus arboreus Sims — Yellow tree lupine

Shrubs to 2 m. tall with erect or decumbent branches; herbage silky-pubescent to nearly glabrous; leaflets 5-12, oblanceolate, 2-8 cm. long; flowers 13-18 mm. long, yellow (ours), keel ciliate; legumes 3-8 cm. long, pubescent; seeds 8-12, usually striped.

A native of the California coast, planted as a sand-binding species and naturalized along the coast in our area.

Lupinus arbustus Dougl. ex Lindl. — Spurred lupine

Slender-stemmed perennials 3-9 dm. tall, minutely silky but green; leaflets 6-10, 3-7 cm. long, oblanceolate, mostly acute, apiculate, pubescent on both surfaces, often less so above, petioles of the basal leaves 2 or 3 times as long as leaflets; flowers 8-14 mm. long, blue, white, or pink in long racemes; calyx silky, distinctly spurred above the pedicel; flowers blue, the banner lavender- or white-centered, not greatly reflexed from wings except at tip, the keel ciliate; legumes 2-3.5 cm. long.

(=*Lupinus laxiflorus* Dougl. ex Lindl.)

In the Columbia Gorge and the Cascades within our limits; more common east of the Cascades.

Lupinus bicolor Lindl. — Two-color lupine

Annuals 1-6 dm. tall; stems slender, simple or branching, leafy; leaflets generally 5-8, linear to spatulate, pubescent, petioles slender; flowers 5-10 mm. long, blue, the banner darker with the center at first white or yellow, changing to lavender, usually shorter than the wings and reflexed at nearly right angles from them, keel narrow, upturned at tip, ciliate; legumes oblong 1.5-3 cm. long, inconspicuously pubescent.

Open, often disturbed sites; not common in our area.

Lupinus latifolius Agardh — Broad-leaf lupine

Perennials; stems several, erect, 0.5-2 m. tall, with a thinly appressed pubescence; leaflets of lower leaves 6-10, elliptic to oblanceolate, generally distinctly apiculate, pubescent at least on the lower surface, 4-10 cm. long, the lower petioles two or three times as long as the leaflets; racemes long, interrupted, showy; flowers 10-18 mm. long, blue to purplish, the banner folded back well away from the wings, keel concealed by wings but the upper margin ciliate below the middle, dark blue or purple at the tip; legumes very silky when young, sparsely so at maturity, dark, 2-3.5 cm. long; seeds mottled.

This is an extremely variable species with numerous named varieties.

Common in open woods and along road sides.

Lupinus lepidus Dougl. ex Lindl. — Dwarf lupine

Low, decumbent, tufted perennials, many-stemmed, leafy, the whole plant more or less tawny or silvery with appressed-pubescence; leaflets 5-9, 0.5-4 cm. long, oblanceolate, short-apiculate, petioles slender; flowers 7-14 mm. long in dense spikes, dark-blue and purple; calyx densely silky-pubescent; banner slightly shorter than wings, eventually reflexed from them, keel slender, ciliate, dark purple, protruding at maturity; legumes 1-2 cm. long, 2-4-seeded.

A variable species with numerous named varieties.

In the mountains and at lower altitudes.

Lupinus littoralis Dougl. ex Lindl. Seaside lupine

Perennials from long thick taproots; stems mostly prostrate, reaching 0.5 m. or more in length, densely pubescent, leafy; leaflets 5-9, obovate, more or less silky on both sides or nearly glabrous above, 1-3 cm. long, generally shorter than the petioles; racemes numerous, short, pedicels 2-5 mm. long, slender, becoming longer and thicker in fruit; flowers 10-16 mm. long, blue to lavender and white; upper lip of calyx 2-toothed, lower lip nearly entire; banner petal glabrous, well reflexed from wings, keel ciliate from apex to below the middle; legumes to 4 cm. long, pubescent, 8-12 seeded.

On high sandy ledges along the coast.

Lupinus micranthus Dougl. ex Lindl. Small lupine

Annuals; stems branching from the base, erect, 1-5 dm. tall, covered with both short appressed and long spreading hairs; leaflets 5-8, oblanceolate to linear, pubescent mainly on the lower surface; flowers more or less in whorls, on short pedicels; calyx grayish-pubescent; corolla 5-8 mm. long, blue and white, the center of the banner turning violet, the sides scarcely turned backward, the keel ciliate; legumes 2-4 cm. long, pubescent at least at first; seeds 5-7, mottled.

Our most common lupine, found abundantly on hillsides, fields, road sides, and disturbed areas.

Lupinus polyphyllus Lindl. Many-leaved or Large-leaved lupine

Perennials; stem stout, erect, branching, 2-15 dm. tall; leaflets 5-17, but more commonly at least 9, 4-15 cm. long, elliptic to oblanceolate, dark green above, lighter beneath, pubescent; racemes 0.5-4 dm. long; flowers 1-1.5 cm. long, somewhat whorled; peduncle, pedicels, and calyx sparsely pubescent; corollas white, pink, lavender or bluish-purple, becoming brown with age; legumes sparsely pubescent, 3-9-seeded.

Very common in dry or wet meadows, stream banks, bogs, ditches and moist woods. Our largest and most conspicuous lupine. Very ornamental before the flowers begin to wither.

Lupinus polyphyllus

Lupinus rivularis Dougl. ex Lindl. Riverbank lupine

Perennials; stems 3-12 dm. tall, usually woody at the base, appressed silky-pubescent to nearly glabrous, weakly erect; leaflets 5-10, oblanceolate to oblong, apiculate, 2-4 cm. long, green but silky pubescent, becoming nearly glabrous above; flowers 10-16 mm. long, in loose racemes; corollas blue or partially so, banner well reflexed from the wings, keels ciliate, concealed by the wings except in plants from along the Columbia River; legumes mottled yellow and brown, 2-5 cm. in length, pubescent; seeds spotted.

Generally on streams banks or in open woods and on dunes.

Lupinus sulphureus Dougl. ex Lindl. var. *kincaidii* (C.P. Sm.) C.L. Hitchc. — Kincaid's lupine

Perennials with stout stems 3-10 dm. tall, the plants green but more or less appressed-pubescent, the hairs white or tawny; leaflets 6-11, oblanceolate, distinctly mucronate, 2-5 cm. long, pubescent at least on the lower surface; flowers more or less whorled in long slender racemes; corollas violet purple or paler, 8-12 mm., on pedicels nearly as long, banner shorter than the wings, not much reflexed, base of banner enclosing wing bases, keel sharply bent upward; legumes 2-3 cm. long, pubescent.

Var. *sulphureus* may occasionally enter our range; it typically has yellow flowers.

Willamette Valley.

Medicago

Annual or perennial herbs with compound leaves of 3 leaflets; flowers yellow, blue, or purple, in short spike-like racemes or heads; banner erect, longer than the wings; 9 stamens united into a sheath, the upper stamen free; fruit small, 1-several-seeded, curved or coiled, indehiscent.

1a Flowers blue or purple (rarely white); perennials .... *M. sativa*
1b Flowers yellow; annuals
  2a Fruit kidney-shaped, without prickles; 1-seeded .... *M. lupulina*
  2b Fruit several times coiled, usually with prickles; several-seeded
    3a Leaflets about as wide as long, with a large dark spot in the center.......... *M. arabica*
    3b Leaflets longer than wide, without a dark spot .... *M. polymorpha*

Medicago arabica (L.) Huds. — Spotted medic or Bur-clover

Annuals; stems 0.5-8 dm. long, prostrate or decumbent; leaflets obovate to obcordate, 1-3 cm. long, with a dark central spot, toothed on the upper half; flowers 4-6 mm. long, yellow; fruit spirally coiled with curved and hooked prickles.

Introduced and established, especially along the coast, but not common.

Medicago lupulina L. — Black medic

Pubescent annuals; stems weak, spreading, 1-4 dm. long; leaflets 0.5-2 cm. long, obovate, obtuse or slightly emarginate at apex, finely toothed near the tip; flowers 2-3 mm. long, deep yellow, in short dense racemes; fruits small, kidney-shaped, 1-seeded, without prickles, but often pubescent, becoming black at maturity.

Introduced from Europe; rather common in disturbed habitats.

Medicago polymorpha L. — Bur-clover

Annuals; stems prostrate to spreading, 0.5-4 dm. long, nearly smooth; leaflets 1-2 cm. long, obovate to cuneate, stipules with slender teeth; flowers 3.5-6 mm. long, yellow, in few-flowered clusters; fruits several times spirally coiled, flattened, wrinkled, the thickened margin usually bordered by a double row of hooked prickles.

(=*Medicago hispida* Gaertn.)

Introduced and well established in our area.

Medicago sativa L. Alfalfa

Perennials; stems erect and often bushy, 2-10 dm. tall; leaflets obovate, those of the upper leaves generally very narrow, the lower broader, 1-4 cm. long, finely toothed near the apex; flowers 8-11 mm. long, blue to purple, rarely white; fruits spirally coiled 2 or 3 times, unarmed.

A forage plant, escaped from cultivation; occasionally found along road sides.

Melilotus

Annual or biennial herbs with compound leaves of 3 leaflets, fragrant upon drying; flowers in narrow often 1-sided racemes, yellow or white, deflexed; stamens in 2 sets (9 and 1); fruits indehiscent, 1-4-seeded.

1a Flowers white .... *M. alba*
1b Flowers yellow
  2a Leaflets serrate nearly the entire length; corolla 4-7 mm. long .... *M. officinalis*
  2b Leaflets serrate only on the upper half; corolla 2-3 mm. long. *M. indica*

Melilotus alba Desr. White sweet-clover

Stems slender, unbranched below, 0.5-3 m. tall; leaflets lanceolate or oblanceolate or oblong, 1-3 cm. long with sharp low teeth; flowers white, 3.5-5 mm. long, in slender long-peduncled axillary racemes.

Introduced from Eurasia, weedy, often locally abundant.

Melilotus indica (L.) All. Small-flowered yellow sweet-clover

Plants up to 1 m. tall; leaflets oblanceolate, obovate or oblong, serrate only above the middle, apex truncate; flowers light yellow, 2-3 mm. long.

Occasional along the coast in disturbed areas; introduced from Eurasia.

Melilotus officinalis (L.) Lam. Yellow sweet-clover

Plants 0.5 to 3 m. tall; leaflets 1-2.5 cm. long, oblanceolate, obovate or oblong, finely but sharply serrate nearly the entire length; flowers pale yellow, 4-7 mm. long.

Disturbed areas and fields; more common east of the Cascades; introduced from Eurasia.

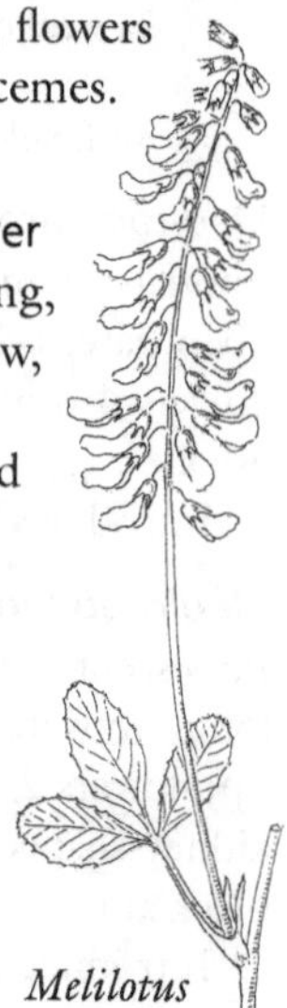

Melilotus officinalis

Oxytropis

Low perennial herbs with basal odd-pinnate leaves, generally with many leaflets; flowers purple, pink, white, or yellowish, in short spike-like racemes; stamens in 2 sets (9 and 1), keel extended at apex into a narrow beak; legumes often inflated, sometimes falsely nearly 2-loculed by the intrusion of the upper suture.

Oxytropis campestris (L.) DC. var. *gracilis* (A. Nels.) Barneby
Field crazyweed

Tufted low perennials with many pale green basal leaves covered with appressed white hairs; leaflets 18-29, the upper pairs opposite, the lower nearly so; peduncles 8-12-flowered, covered below with appressed white hairs, with appressed black hairs above; calyx bell-shaped, black-hairy; corolla yellowish, about 1.2 cm. long; legumes lanceolate, beaked, 1.5 cm. or less long, with mostly black but also some white hairs; seeds many.

In the mountains of western Washington, other varieties occurring eastward.

Psoralidium

Perennial herbs (ours); leaves divided into 3-5 leaflets with embedded translucent glands; inflorescence raceme-like; calyx 5-lobed; corolla blue, purple or bicolored; 9 stamens united by the filaments, 1 free; fruit usually glandular, 1-seeded, not splitting at maturity.

Psoralidium lanceolatum (Pursh) Rydb.
Lance-leaf scurf-pea

Rhizomatous perennials; stems erect, 2-6 dm. tall, much branched, pale green with dark glands; leaflets 3, lanceolate or oblanceolate to linear, usually greenish-yellow; inflorescence generally fewer than 10-flowered; flowers bluish-white, 4.5-8 mm. long; corolla much longer than calyx; fruit conspicuously glandular; seeds globose, brown, shining.

(=*Psoralea lanceolata* Pursh)

Usually in sandy soil; uncommon in our limits.

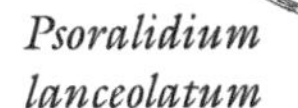

Psoralidium lanceolatum

Robinia

Shrubs or trees; leaves alternate, deciduous, odd-pinnate and with small stipules sometimes modified into spines; flowers borne in axillary racemes, these often drooping; stamens 10, 9 fused by the filaments and 1 free; fruit a legume.

Robinia pseudoacacia L. Black Locust

Trees to 25 m. in height; leaves pubescent when young, but usually becoming glabrate, leaflets 7-21, 2-5.5 cm. long, 1.5-3.5 cm. wide, ovate to oblanceolate, some of the leaves with stipular spines; flowers fragrant, 1.4-2 cm. long, white, the banner usually with a yellow spot at the base; legumes 6-10 cm. long.

Often along river banks, but also in drier habitats; introduced from the eastern United States, but well established throughout our area.

Rupertia

Perennials from a woody caudex; leaves compound, the leaflets 3; glandular-dotted; inflorescence raceme-like; flowers yellowish to cream-colored or white; stamens starting out all fused by the filaments, but one soon splitting off; fruit indehiscent, enclosed in an enlarged calyx.

Rupertia physodes (Dougl. ex Hook.) Grimes California-tea scurf-pea
Stems erect, 3-6 dm. tall, smooth or slightly pubescent; leaflets 3, triangular to ovate, generally pointed at the apex, pubescent to nearly glabrous; flowers borne in dense racemes in the leaf axils, bracts small; calyx at first narrow, becoming inflated as the fruit matures, dotted with embedded glands, and bearing conspicuous black hairs; corolla 10-14 mm. long, whitish or greenish, the keel sometimes tipped with purple; fruit obovate, flattened.

(=*Psoralea physodes* Dougl. ex Hook.)

Open woods.

Thermopsis

Perennial herbs somewhat resembling lupines; stems generally erect branching; leaves palmately divided into 3 entire leaflets; flowers yellow, borne in a raceme, each flower subtended by a permanent bract; calyx somewhat bell-shaped, 4-5-toothed, the 2 upper teeth nearly united; banner broad, shorter than the wings; stamens 10, separate; legumes flattened.

Thermopsis macrophylla Hook. & Arn. Golden-pea
Perennial herbs 3-20 dm. tall; leaves mostly long-petioled with large leaf-like stipules, leaflets 2.5-6 cm. long, generally broadest above the middle; flowers 1.5-2.5 cm. long, showy, in a terminal raceme, yellow; calyx 7-10 mm. long; stamens all separate; legumes 3-6 cm. long, pubescent at least at first.

(= *Thermopsis gracilis* Howell)

Open woods and burnt-over land, somewhat locally distributed.

Trifolium

Annual or perennial herbs with compound leaves of 3 leaflets (in a few species, more than 3); stipules conspicuous, united to the base of the petiole; flowers in heads, racemes or short spikes; calyx 5-toothed; 9 stamens united by the filaments into a sheath, the 10th free; fruits usually indehiscent, 1-6-seeded, remaining within the persistent calyx.

1a Heads subtended by an involucre
  2a Perennials, often rhizomatous; sometimes rooted at lower nodes .... *T. wormskjoldii*
  2b Annuals
    3a Involucre villous, usually deeply cup-shaped, at least half-enclosing the flowers
      4a Lobes of the involucre acute, nearly entire; calyx lobes as long as the tube .... *T. microcephalum*
      4b Lobes of the involucre toothed and bristle-tipped; calyx lobes less than 1/2 the length of the tube .... *T. microdon*
    3b Involucre not villous, nearly flat, or saucer-shaped, not enclosing the head
      5a Calyx lobes longer than the tube .... *T. variegatum*
      5b Calyx lobes about equal to or less than the length of the tube
        6a Flowers 8-17 mm. long; heads over 1 cm. broad .... *T. willdenovii*

6b Flowers 4-8 mm. long; heads averaging less than 1 cm. broad .... *T. oliganthum*

1b Involucre absent

7a Stems creeping, rooting at the nodes or stoloniferous

8a Inflorescence globose, many-flowered, all the flowers with petals....... .... *T.repens*

8b Inflorescence not globose, few-flowered, only the outer 2-6 flowers with petals... *T. subterraneum*

7b Stems not rooting at the nodes and not stoloniferous

9a Flowers deep or bright yellow

10a Banner broad; heads mostly more than 20-flowered .... *T. campestre*

10b Banner narrow; heads mostly less than 20-flowered... *T. dubium*

9b Flowers not deep yellow

11a Perennials

12a Flowers not early reflexed, if at all

13a Flowers pink or red .... *T. pratense*

13b Flowers creamy-white, rarely purple-tinged .... *T. longipes*

12b Flowers early reflexed

14a Calyx teeth woolly-villose .... *T. eriocephalum*

14b Calyx teeth not woolly-villose...................... *T. hybridum*

11b Annuals

15a Calyx strongly veined, becoming much inflated in fruit; flowers white to yellowish, becoming pink; inflorescence 4-6 cm. long, appearing prickly in fruit .... *T. vesiculosum*

15b Not as above in all points

16a Heads elongated, spike-like

17a Heads deep crimson, becoming 2.5-6 cm. long............ .... *T. incarnatum*

17b Heads dark purple, 1-2 cm. long ... *T. albopurpureum*

16b Heads not spike-like, though sometimes short-conical

18a Calyx lobes fimbriate, calyx 5-6 mm. long................... .... *T. ciliolatum*

18b Calyx lobes not fimbriate, calyx 3.5-5 mm. long .... *T. bifidum*

Trifolium albopurpureum Torr. & Gray

Pubescent annuals; stems several, erect or slightly decumbent, 0.5-6 dm. tall; leaflets 1-3 cm. long, narrowly or somewhat broadly obovate, notched at apex, sparsely covered with closely appressed hairs, bluntly toothed; heads spike-like; calyx lobes bristle-like, plumose; corollas 5-12 mm. long, dark purple, white tipped; fruits 1-seeded.

[=*Trifolium macraei* Hook. & Arn. var. *albopurpureum* (Torr. &Gray) Greene]

Fields and hillsides.

Trifolium bifidum Gray var. *decipiens* Greene — Pinole clover

Annuals 1-4 dm. tall; stems erect to decumbent, usually branched from the base, often pubescent; leaflets 1-2.5 cm. long, obovate, truncate to more often notched at the tip, toothed or entire; calyx teeth long, narrow; corolla 5-9 mm. long, pale yellow or pinkish-purple; fruits 1-2-seeded.

Open places.

Trifolium campestre Schreb. Hop clover

Stems generally several, decumbent to erect, pubescent; leaflets mostly broadly obovate, obtuse, mostly notched, pubescent on under side at least at first, 5-15 mm. long; heads globose or slightly elongated; flowers 3.5-6 mm. long, bright yellow, reflexed in age, the banner petal broad. (*Trifolium procumbens* L. misapplied)

Fields and road sides.

Trifolium campestre
a. flower

Trifolium ciliolatum Benth. Foothill clover

Annuals; stems 1-6 dm. long; leaflets obovate, minutely and sharply toothed, obtuse or truncate at apex, mucronate, 1-3.5 cm. long; heads broadly ovoid; flowers pink to purplish, becoming reflexed; calyx teeth lanceolate, acuminate, the margins white, scarious, fimbriate; fruits 1-seeded.

Grassy prairies and hillsides.

Trifolium dubium Sibth. Small hop clover

Annuals; stems 0.5-2 dm. tall, slender, 1 to many from base, erect or somewhat decumbent; leaflets 5-15 mm. long, rather narrowly obovate, obtuse, sometimes notched at apex, minutely toothed, more or less pubescent on under side on the midrib; heads slightly elongated; flowers 3-4 mm. long, yellow, reflexed in age, the banner folded; fruits 1-seeded.

Common weedy species; introduced from Europe.

Trifolium eriocephalum Nutt. Woolly clover

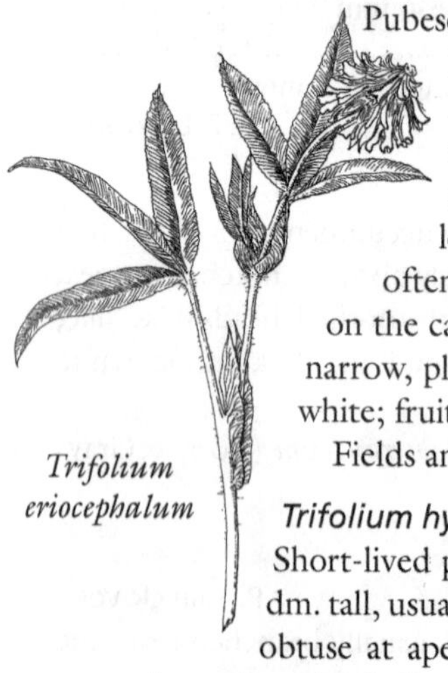

Pubescent perennials; stems erect or slightly decumbent, 1.5-6 dm. tall; leaflets narrowly elliptic or ovate, 1.5-7 cm. long, minutely toothed, mucronate, the veins close and conspicuous; stipules large, mostly entire, united with petiole for at least half their length, sometimes very pubescent; heads dense, often somewhat elongated, woolly from long hairs on the calyx lobes; flowers soon reflexed; calyx lobes very narrow, plume-like; corolla 8-17 mm. long, pale yellow to white; fruits pubescent, 1-4-seeded.

Fields and hillsides, rather common.

Trifolium eriocephalum

Trifolium hybridum L. Alsike clover

Short-lived perennials; stems several, ascending to erect, 1-8 dm. tall, usually more or less pubescent; leaflets ovate, obovate, obtuse at apex, 1-3 cm. long, somewhat pubescent; stipules conspicuously veined; heads 1.5-3 cm. broad, the flowers reflexed after maturity; corolla 5-11 mm. long, pink to reddish.

Introduced from Europe; cultivated and escaped and naturalized; common. Fragrant.

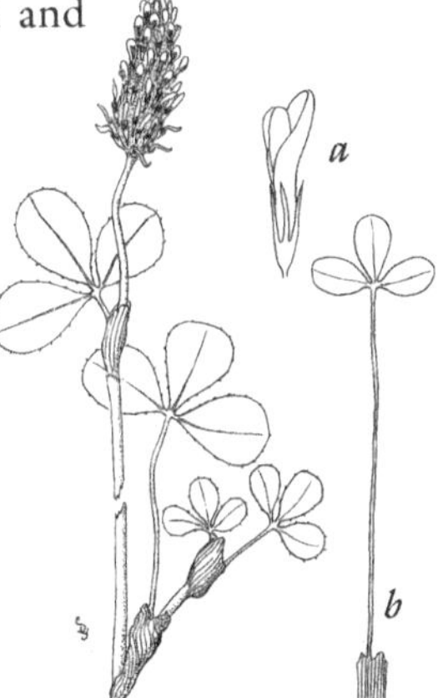

Trifolium incarnatum
a flower; b. basal leaf

Trifolium incarnatum L. Crimson clover
Annuals 1.5-8 dm. tall; stems erect, soft-pubescent; leaflets broadly obovate, obtuse, sometimes notched at apex, minutely toothed to nearly entire, somewhat pubescent, 1-3 cm. long; heads 3-6 cm. long, spike-like; calyx densely tawny-hairy, the lobes sharp and stiff; corolla 12-15 mm. long, deep crimson; fruits 1-seeded.

An escape from cultivation, sparingly naturalized. Showy.

Trifolium longipes Nutt. Long-stalked clover
Perennials, usually rhizomatous and/or stoloniferous, nearly glabrous to puberulent; stems stout, erect or ascending, 1-5 dm. tall; leaves long-petioled, leaflets of upper leaves lanceolate or oblong, mostly sharply toothed, conspicuously veined, leaflets of lower leaves short, oval or ovate to obovate, obtuse; stipules conspicuous, 1-3 cm. long, entire or somewhat toothed, mucronate; heads dense, subglobose; flowers white or purplish, 1-2 cm. long, long-peduncled; calyx-teeth slender, more or less villous.

Wet places in the mountains.

Trifolium microcephalum Pursh Small-headed clover
Annuals; stems erect, 1-7 dm. tall; leaflets 1-2.5 cm. long, soft-pubescent or nearly glabrous, narrowly obovate, mucronate, minutely toothed at widest part, margin entire at base; heads small; involucre villous, 6-10-lobed, the lobes tapering to long acute points, rarely toothed; flowers 4-7 mm. long, white or pink to lavender; fruits 1-2-seeded.

Moist meadows and stream banks to hillsides and fields.

Trifolium microdon Hook. & Arn. Thimble or cup clover
Pubescent annuals; stems several, mostly erect, or slightly decumbent, 1-5 dm. tall, branches slender; leaflets 5-15 mm. long, broadly or narrowly obcordate, pubescent at least beneath, minutely toothed; heads small; involucre deeply cup-shaped, at least half enclosing the head, becoming flattened in fruit, 10-12-lobed, the lobes toothed and bristle-tipped; flowers 4-6 mm. long, white or pale pink; fruits 1-2-seeded.

Hillsides and fields, common.

Trifolium oliganthum Steud. Few-flowered clover
Annuals 1-5 dm. tall; stems slender, more or less decumbent or weakly erect; leaves scattered, slender-petioled, upper leaflets linear, 5-30 mm. long, lower shorter, obovate or oblanceolate, all minutely serrate or nearly entire, acute, obtuse, or truncate at apex, mucronate; heads small, few-flowered; involucre sharply cleft at least half way to the base into linear teeth; flowers 4-8 mm. long, purple, white-tipped.

Common in moist places.

Trifolium pratense

Trifolium pratense L. Red clover

Taprooted perennials; stems several from the base, mostly erect, 2-10 dm. tall, more or less pubescent; lower leaves long-petioled, the petioles decreasing in length upward, absent from the leaves immediately beneath the head; leaflets 2-5 cm. long, broadly (or the upper narrowly) lanceolate to ovate or elliptic, minutely toothed or the margins entire, deep green, generally a whitish crescent near the center; stipules thin, conspicuously veined, the veins sometimes dark red or brown; heads generally slightly elongated; flowers 1-2 cm. long, deep magenta pink or reddish; true involucre none; fruits 1-2-seeded.

Escaped from cultivation and naturalized; common.

Trifolium repens L. White clover

Stoloniferous perennials; stems several from the base, creeping, glabrous or nearly so; leaves long-petioled, leaflets mostly broadly heart-shaped at apex, minutely toothed, 1-2.5 cm. long; heads mostly globose, 1-2 cm. or more broad, the flowers all reflexed after maturity; corolla 5-12 mm. long, white or slightly pinkish-tinged, becoming brownish with age; peduncles much longer than the leaves.

Commonly escaped from cultivation; introduced from Europe. Fragrant.

Trifolium subterraneum L. Subterranean clover

Pubescent annuals with prostrate creeping stems rooting at the nodes; leaflets 0.6-2 cm. long, obovate to obcordate; heads few-flowered; calyx tube red to purple, the teeth bristle-tipped; corolla 8-14 mm. long, white to cream-colored; flowers in part sterile, with a palmately parted calyx, developing above the fertile flowers as the fruits mature and becoming reflexed, turning the head into a bur-like structure which is buried below the surface of the ground by the downward growth of the peduncle.

Introduced from Europe, escaped from cultivation and naturalized.

Trifolium variegatum Nutt. White-tip clover

Annuals; stems prostate to erect, 1-7 dm. long; leaflets 5-30 mm. long, obovate to obcordate, with bristle-tipped marginal teeth; involucre several-lobed, the lobes laciniately cleft or toothed; corolla 3.5-20 mm. long, purple, generally white-tipped; fruit 1-2-seeded.

An extremely variable species.

Common in moist places, but also occurring in drier sites.

Trifolium vesiculosum Savi Arrowleaf clover

Stout annuals 2-6 dm. tall; leaflets obovate to lanceolate, 1.5-5 cm. long, finely toothed; inflorescence elongate, the flowers at first erect but becoming reflexed; calyx becoming inflated in fruit; corollas 1.5-2 cm. long, white to yellowish, becoming pink.

Native of Europe; cultivated and escaped and naturalized.

Trifolium willdenovii Sprengel   Three-toothed clover
Glabrous annuals; stems weak, more or less ascending, several from the base, 1-7 dm. long; stipules conspicuously veined, slender toothed; leaflets 1-5 cm. long, narrow, mostly minutely toothed, mucronate; heads 1-3 cm. broad, on long terminal and axillary peduncles; head subtended by a circular involucre mostly deeply cut into slender teeth; calyx lobes bristle-tipped, often 3-toothed; corolla 8-17 mm. long, purple, usually white- tipped.

(=*Trifolium tridentatum* Lindl.)

Common in fields and on hillsides, usually in moist areas.

Trifolium willdenovii
a. basal leaf; b. flower

Trifolium wormskjoldii Lehm.   Marsh clover
Glabrous perennials, often rhizomatous; stems generally weak, spreading, 0.5-8 dm. long, leaflets 1-3 cm. long, mostly obovate or oblanceolate, sometimes ovate, minutely and sharply toothed, obtuse or acute at apex, mucronate, bright green; heads large, purple, red or pink, lighter or white at corolla tips, subtended by a large saucer-shaped involucre, involucre deeply lobed or toothed; fruits 2-4-seeded.

(Also spelled *wormskioldii*)

Common in mountain meadows, along streams, or at margins of fresh or salt marshes, or often occurring on beaches along the coast.

Ulex europaeus

Ulex

Dark green shrubs with leaves mostly reduced to stiff spines; flowers showy, yellow, axillary; calyx 2-lipped, yellow; stamens united by filaments into one set; legumes explosively dehiscent.

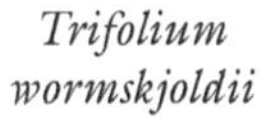

Trifolium wormskjoldii

Ulex europaeus L.   Gorse or Furze
Very spiny dark green shrubs 1-3 m. tall, the branches green, angled and spine-tipped; flowers 1.5-2 cm. long, yellow, showy, axillary, clustered at the stem tips; legumes 1-2 cm. long, densely pubescent.

Introduced from Europe as an ornamental; escaped from cultivation and well established, mostly along the coast.

Vicia

Annual or perennial herbs with angled, weak, sometimes climbing stems; leaves pinnately compound, ending in a tendril or a bristle; flowers in more or less 1-sided axillary racemes, or solitary in the leaf axils; calyx usually unequally 5-toothed; sides of the banner petal generally turned backward, wings united to keel at its middle; 9 stamens united, 1 free; style filiform bearing a tuft of hairs just below the apex; legumes flattened, 2-several-seeded.

1a Corollas 2.5-6 mm. long
  2a Legumes pubescent; peduncles 2-8-flowered .... *V. hirsuta*
  2b Legumes glabrous; flowers solitary or in 2's.................. *V. tetrasperma*
1b Corollas well over 6 mm. long
  3a Inflorescence a sessile or subsessile axillary cluster of 1-8 flowers
    4a Banner glabrous; corollas 1-3 cm. long, banner purple, wings reddish-purple .... *V. sativa*
    4b Banner pubescent on the back; corollas 1-2 cm. long, yellowish-white to purple .... *V. pannonica*
  3b Inflorescence borne on elongated axillary peduncles
    5a Annuals; racemes dense, many-flowered............................ *V. villosa*
    5b Perennials
      6a Flowers 1.5-2.5 cm. long, purple, lavender or blue; leaflets 8-16 .. .. *V. americana*
      6b Flowers 1.5 cm. long or less or leaflets more than 16
        7a Calyx 5-7 mm. long; flowers yellowish and tinged with purple, or rarely reddish-purple.. *V. gigantea*
        7b Calyx 2.5-4 mm. long; flowers dark blue or purplish *V. cracca*

Vicia americana Muhl. American vetch or Wild pea

Vicia americana

Rhizomatous perennials; stems trailing or climbing, 1.5-9 dm. long, strongly 4-angled, sometimes winged at the angles, glabrous or somewhat pubescent; leaflets 8-16, typically wedge-shaped, linear, oblong or ovate, nearly glabrous or densely pubescent, acute to square-cut at apex, mucronate to notched or toothed, 1-4 cm. long; peduncles 2-10-flowered, shorter than the leaves; corolla purple or lavender at first, changing to blue, 1.5-2.5 cm. long; lower calyx teeth longer than the upper; legumes 2-5 cm. long, dark-colored; seeds black.

A variable species.

Common in open areas, especially in disturbed habitats, but also found in moist wooded areas.

Vicia cracca L. Tufted vetch

Rhizomatous perennials; stems weak, 5-15 dm. long, more or less pubescent; leaflets 12-22, 1.5-3 cm. long, linear to oblong, mucronate, the tendrils well developed; flowers 1-1.5 cm. long, dark blue or purplish, many in a dense 1-sided raceme; legumes 1.5-3 cm. long, glabrous.

Introduced from Europe; naturalized in disturbed habitats; rather common in the Willamette Valley.

Vicia gigantea Hook. — Giant vetch

Perennials; stems stout, climbing, 6-20 dm. tall, glabrous or pubescent; leaflets 16-29, 1.5-4 cm. long, narrow, oblong or elliptic, apex mucronate; corolla yellowish and tinged with purple, or rarely reddish-purple, 12-18 mm. long; legumes dark, oblong, abruptly narrowed to a long tapering point, glabrous, 3-4-seeded, the seeds black.

Along streams, on coastal beaches and inland in open woods and along roadsides.

Vicia hirsuta (L.) S.F. Gray — Tiny vetch

Annuals, usually pubescent, stems slender, weak, branching, 3-7 dm. tall; leaflets 8-18, linear, truncate or retuse, mucronate, notched or toothed at apex, 0.5-2 cm. long; peduncles about as long as the leaves, 2-8-flowered; flowers 2.5-5 mm. long, whitish or bluish; legumes oblong 6-12 mm. long, pubescent, 2-3-seeded.

Introduced from Europe; established in disturbed habitats.

Vicia pannonica Crantz — Hungarian vetch

Villous annuals; stems 1.5-7 dm. long, prostrate to weakly ascending or climbing; leaflets 8-20, linear to oblong or ovate, tendrils short, usually simple; flowers 1-2 cm. long, yellowish-white to purple, in a 2-8-flowered short-peduncled or sessile raceme; legumes 2-3 cm. long, pubescent.

Introduced from Europe; escaped from cultivation and well established in our area.

Vicia sativa L. — Cultivated vetch or Tare

Stems 1-9 dm. tall, clinging by their tendrils to each other or to other neighboring plants for support; leaflets 6-14, glabrous or sometimes pubescent on the margins and under surface, linear to obovate, notched or square-cut at apex, mucronate, 1-3 cm. long; flowers solitary or 2-3 in the leaf axils; calyx-teeth pubescent on the margins; corolla 1-3 cm. long, banner purple and wings reddish-purple; legumes 2.5-7 cm. long, pubescent at first but soon becoming glabrous.

(Includes *Vicia angustifolia* L.)

Disturbed areas and fields; introduced from Europe.

Vicia tetrasperma (L.) Schreber — Slender vetch

Annuals; stems weak, very slender, branching, 1-7 dm. long; leaflets 4-12, 0.5-2.5 cm. long, oblong or very narrowly linear, somewhat widely spaced on the rachis; stipules very small, semi-sagittate; flowers nearly white to pale purple, 1 or 2 on slender peduncles shorter than the leaves, corolla 3.5-6 mm. long; legumes oblong, about 12 mm. long, glabrous, 3-5-seeded.

Sparingly established in disturbed areas; introduced from Europe.

Vicia villosa Roth — Hairy vetch

Pubescent annuals; stems slender, climbing, 0.5-2 m. long; leaflets 10-20, 1.5-3 cm. long, narrowly oblong to elliptic, mucronate; flowers in dense 1-sided long-peduncled racemes; calyx purplish, saccate at the base, lower teeth much longer than upper; corolla 10-18 mm. long, banner dark blue or dark to reddish-purple, wings somewhat lighter; legumes 1.5-4 cm. long, glabrous.

A European species often used as a cover crop, widely escaped and naturalized.

Vicia villosa

OXALIDACEAE
Oxalis Family

Herbs with acid juice (ours) to shrubs or trees; leaves alternate, compound, with 3 leaflets (ours); sepals and petals 5; ovary 5-loculed, superior; stamens 10 or 15; fruit a loculicidal capsule.

Oxalis

Small herbs (ours); leaves all basal or alternate with 3 leaflets; petals 5, early deciduous; flowers one to several on a peduncle; stamens 10, with filaments widened at the base, and slightly fused; styles 5; capsule 5-loculed.

1a Leaves all basal; flowers never yellow
  2a Peduncles 1-flowered; petals 15-25 mm. long .... *O. oregana*
  2b Peduncles 2-9-flowered; petals 6-12 mm. long .... *O. trilliifolia*
1b Leaves not all basal; flowers yellow
  3a Filaments pubescent; petals 1-2 cm. long .... *O. suksdorfii*
  3b Filaments glabrous; petals less than 1 cm. long
    4a Plants rhizomatous .... *O. stricta*
    4b Plants taprooted, often creeping and rooting at the nodes................ .... *O. corniculata*

Oxalis corniculata L. Creeping yellow wood-sorrel

More or less fleshy taprooted perennials; stems often creeping and rooting at the nodes; herbage usually purplish; leaves alternate, leaflets 1-4 cm. long, obcordate; peduncles 2-5-flowered; petals 4-8 mm. long, yellow; capsules oblong, 6-25 mm. long, pubescent.

A weedy species, common in lawns and disturbed habitats.

Oxalis oregana

Oxalis oregana Nutt. Oregon wood-sorrel

Rhizomes stout, scaly; leaves all basal; leaves and peduncles 1-1.5 dm. tall; leaflets 1.5-3.5 cm. long, petioles more or less pubescent, green or reddish,

succulent, jointed to a dilated base, breaking away at the joint, leaving old bases on the rhizomes; flowers solitary, white or pinkish-lavender, 15-25 mm. long, the corolla often veined with lavender; capsules 7-12 mm. long.
Common in moist woods.

Oxalis stricta L. — Upright yellow Oxalis

Rhizomatous perennials; stems prostrate to erect, to 5 dm. long; leaflets 1-5 cm. long, glabrous or the margins ciliate; peduncles up to 7-flowered; petals 4-9 mm. long, yellow; capsules pubescent.
Troublesome weed of western Washington.

Oxalis suksdorfii Trel. — Yellow wood-sorrel

Clustered stems erect to trailing, 0.5-2.5 dm. tall, often from slender stolons, somewhat pubescent; leaves alternate with 3 broad leaflets 1-2.5 cm. long, deeply notched at apex, sparsely pubescent, petioles 2-5 cm. long; peduncles 1-3-flowered, borne in the leaf axils; petals 1-2 cm. long, yellow, obtuse; capsules short-cylindrical, on jointed pedicels which turn backward in fruit, curving upward at summit to bring capsules erect.
Rather common in edges of cultivated fields and in wasteland.

Oxalis trilliifolia Hook. — Many-flowered wood-sorrel

Rhizomes stout, scaly; leaves all basal, leaflets obovate-cuneate and emarginate, petioles sparsely pubescent; scapes 1-2.5 dm. tall, 2-9-flowered; flowers 6-12 mm. long, white or cream-colored to purple-tinged; capsules slender, 2-3 cm. long.
Mountain woods, also associated with *O. oregana* near coast.

GERANIACEAE
Geranium Family

Herbs (ours); leaves lobed or compound; stipules present, membranous; sepals and petals usually 5; stamens usually 5 or 10; ovary superior, 3-5-lobed, the styles united to the axis of an elongated receptacle, separating (beginning at the base) at maturity into separate, 1-2-seeded, long-beaked dry fruits, the beaks (the persistent styles) coil and uncoil with changes in the humidity of the air, probably assisting in planting the fruits in the ground.

Leaves palmately lobed or divided; fertile stamens typically 10........ **Geranium**
Leaves (ours) pinnately compound; fertile stamens 5 .... **Erodium**

Erodium

Annuals or perennials with basal and opposite leaves, these often unequal, simple or pinnately compound; petals pinkish; anther-bearing stamens 5, alternating with 5 scale-like sterile filaments; fruit as described for the family.

Leaflets pinnatifid; petals fringed at base.................................. *E. cicutarium*
Leaflets merely toothed or the terminal one lobed; petals not fringed at base... .... *E. moschatum*

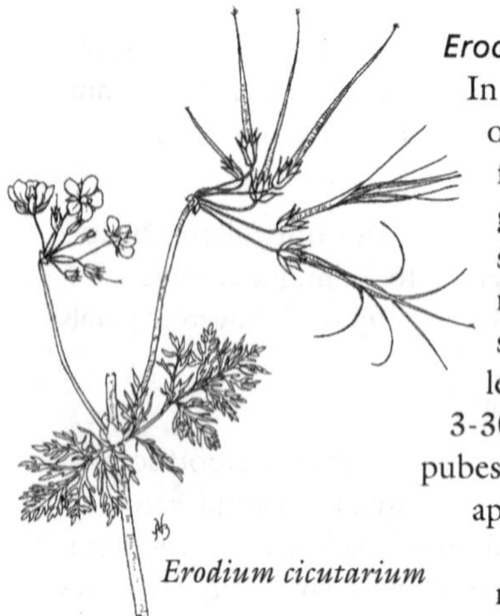
Erodium cicutarium

Erodium cicutarium (L.) L'Her. Filaree

In early spring or on hard ground, plants consist of a mostly flat rosette of leaves with short flowering stems, later or under more favorable growth conditions plants become large and spreading with long leaves and peduncles more or less erect; stems often reddish with swollen nodes; leaves pinnately compound, the leaflets pinnately lobed, more or less pubescent, 3-30 cm. long; flowers small, magenta; sepals pubescent, mucronate or with a stiff bristle at the apex; beak of fruit 2.5-5 cm. long.

Very common. Plants small along gravelled railroad beds; large and sturdy in gardens. Introduced from Europe.

Erodium moschatum L'Her. Large-leaved filaree

Plants, as in *E. cicutarium*, at first forming flat rosettes, later with stout ascending leafy stems 3-5 dm. long; leaves pinnately compound, the leaflets large, broad, coarsely toothed, the terminal leaflet generally 3-5-lobed; stipules large, conspicuous, silvery-papery, obtuse; sepals not bristly-tipped; beak of fruit 2-4 cm. long or longer.

Occasional, sometimes locally abundant; introduced from Europe.

Geranium

Annuals or perennials; stems with swollen nodes; leaves opposite or alternate, palmately lobed or divided; stipules membranous; flowers solitary, or in small axillary umbel-like cymes; stamens 10 (occasionally not all anther bearing), sometimes slightly united at base, alternately long and short, the longer with basal glands inside the filament; pistil 5-lobed with the 5 styles united nearly to the tip until maturity, when the carpels split apart and away from the central elongated axis of the receptacle, each segment long-beaked by the persistent twisted or coiled style.

1a Petals 15-24 mm. long; perennials.. *G. oreganum*
1b Petals 14 mm. long or less; annuals or biennials
  2a Ovary lobes glabrous
    3a Ovary lobes not wrinkled; petals 5-11 mm. long ..... *G. columbinum*
    3b Ovary lobes transversely wrinkled
      4a Petals 3-5 mm. long.. *G. molle*
      4b Petals 7-14 mm. long .... *G. robertianum*
  2b Ovary lobes pubescent
    5a Sepals acute; petals 2.5-5 mm. long; seeds smooth .... *G. pusillum*
    5b Sepals bristle-tipped or at least narrowed to a slender point; petals averaging over 4 mm. long; seeds usually not smooth
      6a Petals 7-14 mm. long with a distinct claw; sepals erect at time of flowering

7a Sepals keeled .... *G. lucidum*
7b Sepals not keeled, leaves deeply divided .... *G. robertianum*
6b Petals 4-8 mm. long without a distinct claw; sepals not erect
8a Peduncles 4-10 cm. long, slender; pedicels 1-3 cm. long.... *G. bicknellii*
8b Peduncles 3 cm. long or less; pedicels less than 1 cm. long
9a Seeds deeply pitted, the pits round to square; hairs of the fruits less than 1 mm. long .... *G. dissectum*
9b Seeds only slightly reticulate, the markings elongate; hairs of the fruits about 1 mm. long .... *G. carolinianum*

Geranium bicknellii Britt. Bicknell's Geranium

Annuals or biennials; stems 1-6 dm. tall, pubescence harsh, glandular; leaf blades cordate-orbicular in outline, 2-7 cm. broad, deeply 3-5-cleft; sepals bristle-tipped; petals 4-8 mm. long, pink or light purple; fruits minutely bristly; seeds reticulate-pitted.

Woods and fields.

Geranium carolinianum L. Carolina Geranium

Annuals; stems branching, spreading or ascending, sparsely or densely covered with white downward-turned hairs; leaves somewhat broadly triangular in outline, 5-7-parted, the segments irregularly lobed, the lobes mucronate, more or less pubescent, hairs of the petiole turning downward; flowers pink to purplish, about 4-7 mm. long; sepals white-hairy on the veins, abruptly narrowed at apex to a slender point or bristle-tipped; ovary lobes covered with long and short hairs, with or without glands, not wrinkled, beaks pubescent; seed surface netted.

Common in open woods and disturbed sites, although a native species, it tends to be weedy.

Geranium columbinum L. Long-stalked Geranium

Pubescent annuals; stems erect to decumbent, 1.5-8 dm. tall; leaf blades 5-9-lobed, these divisions again deeply divided into narrow segments, petioles much longer than the leaf blades; sepals bristle-tipped; petals 5-11 mm. long, purple; ovary lobes glabrous, not wrinkled; seeds finely netted.

Uncommon in our area except along the Columbia River; introduced from Europe.

Geranium dissectum L. Cut-leaved Geranium

Somewhat resembling *G. carolinianum*, but primary leaf segments generally parted nearly to the base, the lobes narrower; flowers magenta, 4-7 mm. long; ovary lobes both pubescent and transversely wrinkled, the seed surface deeply pitted and netted.

Common weedy species; introduced from Europe.

Geranium lucidum L. Shining crane's-bill

Annuals, shiny green and often tinged with red; stems 1-4 dm. tall, erect or ascending; leaves 2-6 cm. wide, divided into 5-7 obovate-cuneate lobes, these crenate or 3-toothed, the segments broad and mucronate; sepals 5-7 mm.

long, bristle-tipped, strongly keeled; petals pink, 8-10 mm. long; lobes of the ovary irregularly winkled and pubescent.

Native of Europe; relatively recently introduced and naturalized in our area.

Geranium molle L. — Dovefoot Geranium

Annuals; stems 1-8 dm. long, generally slender, weak, sometimes almost trailing, often strongly angular at the nodes; leaves pubescent, orbicular in outline, 5-7-lobed, the lobes again cut into somewhat narrow lobes, or those nearest summit of plant often with 5 narrow uncut lobes; pedicels widely spreading in fruit, and peduncles, also, often turned backward; petals 3-5 mm. long, notched, pink to purple; ovary lobes transversely wrinkled, glabrous; seed surface smooth.

Common in moist disturbed habitats; introduced from Europe.

Geranium oreganum Howell — Large Geranium

Perennials; stems stout, 4-8 dm. tall, usually pubescent, glandular at least above, branching; leaves 5-7-parted, the segments deeply lobed, scarcely mucronate; flowers 3.5-5 cm. in diameter, reddish-purple, the margins ciliate at the base; ovary lobes pubescent, sometimes glandular, not wrinkled, each segment 2-seeded.

Common in meadows and open woods.

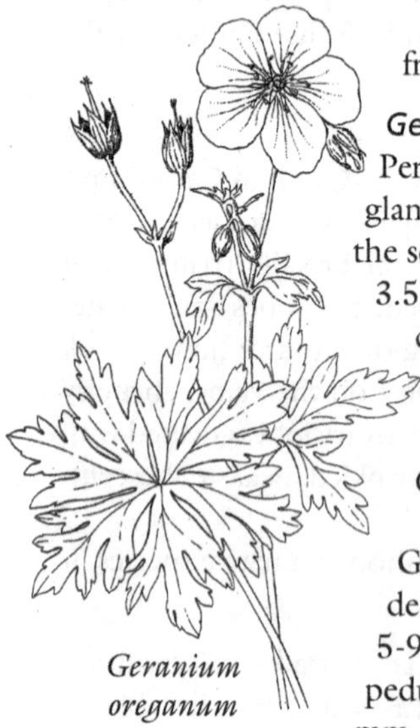

Geranium oreganum

Geranium pusillum L. — Small-flowered Geranium

Glandular pubescent annuals or biennials; stems decumbent to erect, 1-5 dm. long; leaf blades deeply 5-9-cleft, the segments irregularly cut and toothed; peduncles mostly 2-flowered; sepals actue; petals 2.5-5 mm. long, purple; ovary lobes keeled and pubescent, not wrinkled; seed surface smooth.

Common weedy species; introduced from Europe.

Geranium robertianum L. — Herb-Robert

Annuals or biennials; stems 1-5 dm. tall, erect to ascending, glandular-pubescent; leaves deeply dissected to compound into 3-5 segments, these again deeply lobed or dissected; sepals bristle-tipped, pubescent; petals 7-14 mm. long, red to pink or reddish-purple; ovary lobes keeled near the top, winkled, usually pubescent; seeds finely pitted.

In shady, often moist areas; introduced from Europe.

EUPHORBIACEAE
Spurge Family

Monoecious or dioecious herbs, shrubs or trees, often with milky sap; leaves opposite, alternate, or whorled, simple, entire or toothed; flowers much reduced, sometimes enclosed in a flower-like involucre; stamens 1 to many; ovary superior, 1-4-loculed, styles or stigmas as many or twice as many as the locules; fruit usually a capsule.

1a Involucre absent; calyx present in at least the staminate flowers; ovary 1-loculed .... **Eremocarpus**
1b Involucre present; calyx absent; ovary 3-loculed
  2a Plants erect; leaves usually alternate, at least below .... **Euphorbia**
  2b Plants prostrate; leaves opposite throughout .... **Chamaesyce**

Chamaesyce

Usually monoecious annual or perennial herbs with milky sap; stems prostrate (ours); leaves opposite, leaf bases typically asymmetrical; flowers reduced, borne in an involucre of 5 fused bracts, glands 4; staminate flowers clustered around the single pistillate flower, each staminate flower with a single stamen; pistillate flowers with a stalked ovary with 3 styles, ovary 3-loculed, 3-seeded; capsule round to 3-lobed.

Fruit less than 1.5 mm. long; leaf blades usually with a dark central blotch .... *C. maculata*
Fruit 1.5-2 mm. long; leaf blades without a dark central blotch .... *C. serpyllifolia*

Chamaesyce maculata (L.) Small — Spotted spurge

Annuals, freely branching from the base, forming more or less circular nearly prostrate mats; leaves 4-30 mm. long, oblong, asymmetrical at the base, the margins finely toothed, generally purple mottled and with a dark central blotch, pubescent at least when young; capsules pubescent; seeds wrinkled.

(=*Euphorbia supina* Raf. and *Euphoria maculata* L.)

Weedy species; sparingly introduced from the eastern United States.

Chamaesyce serpyllifolia

Chamaesyce serpyllifolia (Pers.) Small — Thyme-leaved spurge

Stems slender, often reddish at base, freely branched, prostrate, forming mats 1.5-9 dm. in diameter; leaves suborbicular to obovate, dull green above, paler beneath, 3-17 mm. long, minutely toothed toward apex, stipules fringed; involucres solitary in the leaf-axils; glands 4, horizontally elongated, depressed in the center, the wings white and generally toothed or lobed; seeds whitish, more or less pitted to nearly smooth.

(=*Euphorbia serpyllifolia* Pers.)

In dry ground, often in the bottom of dried pools.

Eremocarpus

Monoecious annuals; grayish-pubescent and strongly-scented; leaves alternate, entire, leaf blades with 3 main veins from base; staminate flowers in terminal clusters, the pistillate in the lower leaf axils; ovary with several basal glands; fruit a 2-valved 1-seeded capsule.

Eremocarpus setigerus (Hook.) Benth. — Turkey mullein

More or less prostrate mat-forming plants, grayish pubescent with bristles and minute stellate hairs; stems much-branched, 1-6 dm. long; leaves ovate and usually obtuse, petioled, the blades 6-60 mm. long; capsules 3-5 mm. long.

Dry ground.

Euphorbia

Usually monoecious annual or perennial herbs, or shrubs; flowers much reduced, but borne clustered in involucres which often resemble single flowers, these borne in the leaf axils or in terminal cymes; involucres 4-5-toothed, usually bearing glands in the sinuses, the glands often expanded or winged and may resemble small petals; within each involucre is a single pistillate flower and several staminate flowers; each staminate flower consists of a single stamen jointed to a minutely bracted pedicel; the pistillate flowers consist principally of a stalked ovary with 3 styles which protrude from the involucre, ovary 3-loculed.

1a Lower leaves obovate to nearly orbicular, petioled or at least narrowed at the base
  2a Ovary lobes crested; lower leaves distinctly petioled .... *E. peplus*
  2b Ovary lobes not crested; lower leaves narrow to the base but scarcely petioled .... *E. crenulata*
1b Lower leaves narrow, not petioled
  3a Leaves linear, less than 3 cm. long, suggesting fir needles ..*E. cyparissias*
  3b Leaves over 3 cm. long, not at all suggesting fir needles.........*E. lathyris*

Euphorbia crenulata Engelm. — Western wood spurge

Glabrous annuals or biennials 1-6 dm. tall; leaves 1.5-3.5 cm. long, obovate to spatulate, lower leaves not distinctly petioled though usually narrowed to the base; glands of the involucre 2-horned; capsules glabrous.

In dry habitats; not common.

Euphorbia cyparissias L. — Cypress spurge

Stems clustered, erect, 3 dm. or less tall, scaly below, leafy above, from woody creeping rhizomes, stems branching above, ending in many-rayed umbels; leaves numerous, linear, alternate below, lanceolate and whorled below umbels, 1-3 cm. long; bracts small, mostly broader than long, sessile; glands crescent-shaped; capsule nearly globose.

A cultivated species occasionally escaped and naturalized but not common.

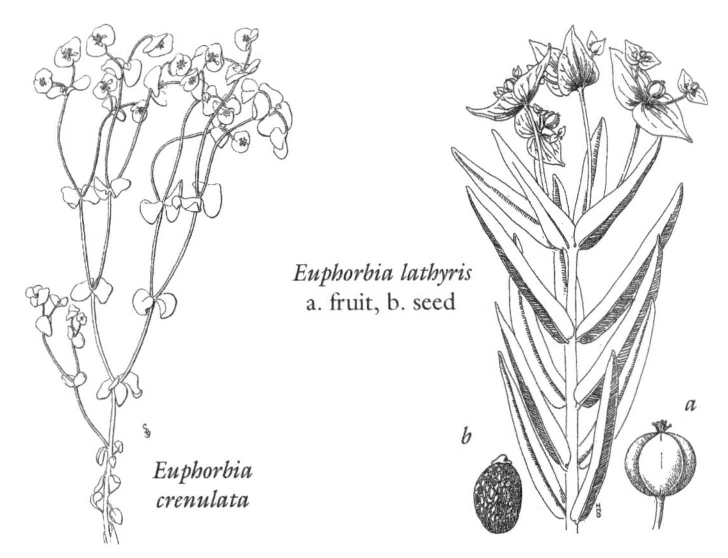

Euphorbia lathyris
a. fruit, b. seed

Euphorbia crenulata

Euphorbia lathyris L. — Caper spurge or Mole plant

Glabrous, glaucous stout leafy annuals or biennials, 2-10 dm. tall, somewhat branched above; lower leaves alternate, then opposite, narrow, turned downward, middle leaves narrowly lanceolate, 3.5-13 cm. long, sessile, generally whorled beneath inflorescence; bracts opposite, sessile, broadly ovate, acute at apex; glands crescent-shaped; capsule nearly globose with rounded lobes; seed-surface generally wrinkled.

Escaped from cultivation and well established, especially in the Willamette Valley. Children eating the fruits of this species are known to have become severely ill.

Euphorbia peplus L. — Petty spurge

Glabrous annuals; stems 1-5 dm. tall, nearly erect, branched above into a several-rayed leafy compound umbel; lower leaves alternate, slender-petioled, those subtending the inflorescence opposite or whorled, nearly to quite sessile; leaf blades obovate to ovate, 5-25 mm. long, generally entire; glands of involucre sharply crescent-shaped; capsule lobes 2-crested on the back.

A common weed; introduced from Europe.

CALLITRICHACEAE
Water-starwort Family

Submerged, floating or emergent and rooting in wet ground; stems slender; leaves opposite or rarely whorled, entire; flowers minute, usually unisexual, 1-3 in the leaf axils, perianth lacking; staminate flowers usually consisting of a single stamen; pistillate flowers of a single pistil; ovary deeply 4-lobed, separating at maturity into 4 1-seeded fruits.

Callitriche

Characters those of the family.

1a Leaf bases not joined by wing-like ridges, leaves all submerged and linear ... .*C. hermaphroditica*
1b Leaf bases joined by wing-like ridges, some leaves usually spatulate or oblanceolate, some leaves often floating, or plants stranded
 2a Margins of the fruit uniformly winged throughout............. *C. stagnalis*
 2b Margins of the fruit wingless or wings broader above the middle
  3a Fruit longer than broad, wings broader above the middle.... *C. verna*
  3b Fruit as long as broad, wings, if present, only above the middle .... *C. heterophylla*

Callitriche hermaphroditica L. — Northern water-starwort
Submerged; stems 5-15 cm. long; leaves linear, 1-veined, bifid at the tip, the base slightly clasping, but not joined by wing-like ridges; fruit 1-2 mm. long, narrowly winged above, irregularly and shallowly pitted.

Submerged in quiet water.

Callitriche heterophylla Pursh — Different-leaved water-starwort
Submerged, often with a floating rosette of leaves, or plants emergent; stems 5-40 cm. long; leaves joined at their bases by a minute wing-like ridge, submerged leaves linear, 1-veined, 5-25 mm. long, notched at the tip, floating leaves and those of emergent plants broadly obovate, up to 1 cm. broad, 3-veined; fruits wingless or narrowly winged only above, irregularly and shallowly pitted. Two varieties of this species occur in our area and they are separated on the basis of the fruit, with var. *heterophylla* having fruit 0.6-0.8 mm. long and var. *bolanderi* (Hegelm) Fassett having fruits 0.8-1.4 mm. long.

Ponds and slow moving water, often becoming stranded in mud.

Callitriche stagnalis Scop. — Pond water-starwort
Submerged, usually with floating rosettes of leaves or mat-forming when emergent; stems 3-15 cm. long; leaf bases joined by small winged ridges; leaves variable, from linear-lanceolate and 1-veined to oblanceolate and 3-5-veined, up to 1 cm. wide; fruit 1-1.8 mm. long, not quite as broad, winged-margined.

In quiet or slow moving water or stranded at the waters edge; introduced from Europe, but widely established in North America.

Callitriche verna L. — Spring water-starwort
Submerged or stranded; stems 5-20 cm. long; leaf bases joined by small winged ridges, submerged leaves linear. 0.5-2 cm. long, 1-veined, floating and emergent leaves spatulate, 3-veined, as much as 4 mm. in width; fruit about 0.9-1.4 mm. long, not quite as wide, wings broader above the middle, minutely pitted, with the pits tending be in vertical rows.

Still or slow-moving water; common.

EMPETRACEAE
Crowberry Family

Low evergreen shrubs, mostly dioecious or monoecious; leaves linear; flowers solitary or in terminal or axillary clusters; perianth segments 3-6; ovary superior; stamens 2-4; fruit drupe-like.

Empetrum

Flowers inconspicuous, generally both perfect and imperfect flowers found in the leaf axils; styles of pistillate flowers deeply 2-9-lobed; fruit black or red.

Empetrum nigrum L. — Crowberry

Stems prostrate at base, 1.5-4.5 dm. long, freely branched, forming dense cushions; leaves crowded, linear, thick, grooved on the lower surface, about 6 mm. long; flowers purplish; fruits 4-6 mm. in diameter, black.

In scattered stations along the coast.

LIMNANTHACEAE
Meadow-foam Family

Low annuals; leaves alternate, deeply pinnately divided to pinnately compound; flowers solitary in the leaf axils; sepals and petals 2-6; petals white or yellowish, sometimes pinkish-tinged; stamens usually of the same number as the petals or twice as many; pistils 2-5 or ovary deeply 2-5-lobed, separating at maturity into 1-seeded fruits.

Plants fleshy, stems weak, 2-20 cm. long; leaflets 3-5 .... **Floerkea**
Plants not fleshy; leaflets often more than 5 and plants usually over 20 cm. tall .... **Limnanthes**

Floerkea

A monotypic genus.

Floerkea proserpinacoides Willd. — False mermaid

Fleshy glabrous annuals; stems 2-20 cm. tall; leaves pinnately-compound, leaflets 3-5, oblanceolate to ovate, 2-20 mm. long; sepals and petals generally 3; sepals 2-4 mm. long; petals 1-3 mm. long, white; stamens 3-8; ovary deeply 2-3-lobed and separating at maturity into 1-seeded indehiscent fruits about 2.5 mm. long.

Wet areas.

Limnanthes

Annuals; leaves alternate, deeply pinnately lobed or compound; flowers solitary in the leaf axils; sepals and petals 4-6, petals white or yellowish; stamens 8 or 10; ovary 4-5-lobed.

Limnanthes alba Benth. Meadow-foam

Stems to 6 dm. tall; leaflets 5-9, about 15 mm. long, cut into 3 segments or entire; sepals 7-8 mm. long; petals 1-1.5 cm. long, white; fruit smooth to winkled or tuberculed.

Vernal pools, ditches and roadsides; native of California; cultivated and escaped and naturalized.

ANACARDIACEAE
Sumac Family

Trees, shrubs or vines with alternate leaves; flowers small, sometimes unisexual; sepals partly united about a disk; petals generally 5; stamens 5 or 10; ovary superior, 1-loculed, 1-ovuled; styles 1-3; fruit drupe-like, dry at maturity.

Toxicodendron

Vines, shrubs or small trees, usually with compound deciduous leaves of 3-9 leaflets; inflorescence axillary; flowers yellow to yellowish-green; stamens 5; stigmas 3; fruit globose.

Toxicodendron diversilobum (Torr. & Gray) Greene
Poison oak

Erect shrubs, reaching 4 m., sometimes tree-like, or becoming vine-like (to 35 m.) and climbing tall trees; leaves divided into 3 leaflets; leaflets ovate, roundish, or elliptical, the margins wavy or slightly lobed; petioles and young twigs covered with a rusty pubescence; panicles bearing very small greenish flowers; fruits whitish, corrugated, remaining on the shrub in dense clusters after the leaves have fallen.
(=*Rhus diversiloba* Torr. & Gray)

Toxicodendron diversilobum

Abundant in western Oregon; absent, or nearly so, on the immediate coast and at high altitudes in the mountains.

One of our most attractive and most widely distributed shrubs, but in sensitive individuals it causes severe allergic contact dermatitis. Quantities of the fruits are sometimes found, in winter, in the holes of small rodents, indicating that they are put to some use. Bees make an excellent honey from the nectar and it is not toxic.

CELASTRACEAE
Staff-tree Family

Shrubs or trees with simple alternate or opposite leaves; flowers usually regular, small, perfect, the parts in 4's or 5's, the pedicels jointed; calyx deeply lobed; petals distinct, spreading; stamens on a disk, alternate with the petals; ovary surrounded by the disk, free from or united with it, 2-5-loculed; fruit a capsule, berry, samara, drupe or nutlet; seeds often with a fleshy aril.

Low evergreen shrubs; flower parts in 4's..**Paxistima**
Tall deciduous shrubs or small trees; flower parts in 5's (ours)....... **Euonymus**

Euonymus

Large deciduous (ours) shrubs or small trees; twigs angled and often with corky ridges; leaves opposite, flowers small, purplish, in axillary cymes; flower parts generally in 5's (rarely 4's or 6's); fruit a 3-5-lobed capsule.

Euonymus occidentalis Nutt. ex Torr. Western wahoo or Burning bush
Shrubs 2-6 m. tall, with weak slender branches, the youngest stems 4-angled; leaves ovate or elliptic, thin, minutely serrate, acute or acuminate, short-petioled; peduncles axillary 2.5-5 cm. long, 1-4-flowered; flowers pediceled, 8-12 mm. wide, purplish, the petals whitish-margined; seeds in a red-orange aril.

Along shaded streams; not common.

Paxistima

Small evergreen shrubs with 4-angled twigs and leathery leaves; flowers small, greenish or reddish, solitary or in short cymes in the leaf axils; flower parts in 4's; capsule 2-loculed. (This genus is also spelled *Pachystima.*)

Paxistima myrsinites (Pursh) Raf. Oregon boxwood
Glabrous freely branching prostrate to ascending evergreen shrubs 2-9 dm. tall; leaves small, shiny, leathery, opposite, oblong or obovate, serrate, narrowed at base, with a very short petiole; flowers in small axillary cymes, reddish or purplish; seeds partially covered by a thin whitish aril.

Mostly in the mountains and foothills in shaded areas; occasionally at sea level. The shrub somewhat suggests evergreen huckleberry, but is readily distinguished by the opposite leaves.

ACERACEAE
Maple Family

Trees or shrubs with deciduous opposite leaves; flowers with both stamens and pistils, or sometimes unisexual; sepals and petals usually 5, or petals sometimes lacking; stamens 4-12, often borne on a fleshy glandular disk; ovary superior, 2-loculed with 1-2 styles, developing into 2-winged samaras, the wings facilitating dispersal by the wind.

Acer

Leaves simple, generally palmately lobed, or rarely pinnately compound; flowers drooping in axillary racemes, panicles or corymbs.

1a Large trees; flowers many in racemes or panicles; fruit stiff-hairy .... *A. macrophyllum*
1b Small trees or shrubs; flowers in few-flowered corymbs; fruit glabrous
  2a Leaves 7-9-lobed; wings of fruit widely spreading .... *A. circinatum*
  2b Leaves 3-5-lobed; wings of fruit not widely spreading......... *A. glabrum*

Acer circinatum Pursh — Vine maple

Small trees or straggling shrubs to 8 m. tall; leaves as broad as long, 4-12 cm. in diameter, cut into 7-9 sharp lobes less than half the distance to base, the lobes toothed; flowers 3-12 in a corymb, perfect or staminate, long-pediceled; sepals deep red; petals greenish-white; stamens 6-10, filaments pubescent below; wings of the fruit forming a straight or nearly straight line at right angles to the pedicel, deep red at maturity, glabrous.

Acer circinatum

Found along shaded streams, becoming more common with ascending altitude. One of our few scarlet-foliaged trees in the autumn.

Acer glabrum Torr. var. *douglasii* (Hook.) Dipp. — Douglas' maple

Small trees or shrubs, 1-10 m. tall, bark gray to reddish; leaves similar in size to those of vine maple, 3-5-lobed, paler beneath; flowers in corymbs; fruit glabrous, the wings not widely spreading.

Along streams, chiefly in the mountains.

Acer macrophyllum Pursh — Big-leaf maple

Round-topped tree reaching 30 m. in height; leaves as broad as long, 1-3 dm. in diameter, palmately 5-lobed to near or below the middle, the lobes further cut irregularly on the margins; flowers in long racemes or panicles, greenish or yellowish, staminate and perfect flowers appearing together; stamens generally 10, filaments densely white-woolly; ovary densely bristly with short stiff hairs; wings of fruits 3.5-5 cm. or more long; mature fruit dark brown.

Common. This is one of our most gorgeously colored trees in the autumn, the foliage turning a deep gold.

Acer macrophyllum

BALSAMINACEAE
Jewel-weed Family

Mostly succulent annual to perennial herbs, although sometimes woody at the base; leaves alternate, opposite or whorled; flowers borne in the leaf axils, irregular; sepals 3 or rarely 5, often petaloid, one usually spurred or saccate; petals 5; stamens 5; pistil compound; fruit typically a capsule which splits explosively when touched, or less often a berry-like drupe.

Impatiens

Succulent annuals or perennials; leaves simple, alternate, opposite or whorled; flowers showy, borne in axillary racemes; sepals 3, one saccate and usually spurred; petals 5, the upper petal notched at the tip, the lateral petals in unequal pairs; fruit a capsule which explosively splits into 5 valves at maturity, expelling the seeds.

1a At least some of the leaves opposite or whorled .... *I. glandulifera*
1b Leaves all alternate
  2a Saccate sepal not spurred .... *I. ecalcarata*
  2b Saccate sepal spurred
    3a Flowers usually paired .... *I. noli-tangere*
    3b Flowers in few-flowered axillary racemes .... *I. capensis*

Impatiens capensis Meerb. Orange touch-me-not or Orange Balsam
Succulent glabrous annuals to 1.75 m. tall; leaves alternate, ovate to elliptic, up to 9 cm. long, coarsely-toothed; flowers in few-flowered axillary racemes, drooping on slender pedicels, usually orange-yellow and reddish-brown spotted, spur to 2.5 cm. long, strongly incurved.

Introduced from eastern and central North America; escaped in moist to wet habitats, especially along rivers and streams.

Impatiens ecalcarata Blank. Spurless touch-me-not or Jewel-weed
Stems stout, 4-12 dm. tall, much-branched, often swollen at the lower nodes; leaves alternate, the blades 3-10 cm. long, elliptic-lanceolate to ovate, coarsely serrate with 6-14 teeth, on slender petioles 1-3 cm. long; flowers usually borne in pairs, 1-2 cm. long, yellow to orange; one sepal saccate, but not spurred.

In damp to wet, usually shady areas.

Impatiens glandulifera Royle Policeman's helmet
Stems 6-15 dm. tall; leaves alternate to opposite, but the uppermost whorled, leaf blades 6-15 cm. long, oblong-ovate to ovate-elliptic, sharply and closely serrate, on stout petioles 2-5 cm. long; flowers in clusters of 3 or more, 2-3 cm. long; lateral sepals purplish, the saccate sepal with a short recurved spur 4-5 mm. long, white to reddish and usually purplish-spotted; petals white to red; capsule drooping, 1-2.5 cm. long.

Escaped from cultivation and established in moist places, especially along stream banks; native of Asia.

Impatiens noli-tangere L. Touch-me-not or Jewel-weed

Succulent annuals 2-6 dm. tall, glabrous; stems freely branching; leaves alternate, leaf blades 3-12 cm. long, elliptic-ovate, coarsely crenate-serrate, the teeth with a short, sharp, slender point; flowers usually paired, yellowish and often with red to reddish-brown spots; saccate sepal 1-2 cm. long, narrowed gradually to a strongly recurved spur 6-10 mm. long.

Moist wooded areas.

RHAMNACEAE
Buckthorn Family

Shrubs or small trees; leaves simple; flowers small, solitary or in few-flowered umbels which are borne in racemes or panicles; sepals, petals and stamens in 5's or 4's; petals curved at tips in the form of a hood, or lacking; stamens alternate with sepals; ovary superior or inferior, 2-5-loculed; fruit drupe-like or a capsule.

Fruit fleshy, drupe-like; flowers greenish, inconspicuous .... **Rhamnus**
Fruit a capsule; flowers white, pink or blue, showy .... **Ceanothus**

Ceanothus

Shrubs or small trees, often with stiff spreading branches and opposite or alternate leaves; flowers very small but borne in showy umbels or panicles, mostly heavy-scented; calyx 5-lobed; petals 5, hooded, long-clawed; stamens 5, longer than petals or sepals; capsule somewhat thick-walled, more or less globose, 3-lobed and 3-loculed.

1a Leaves opposite or appearing whorled, grayish-green .... *C. cuneatus*
1b Leaves alternate, not grayish-green
  2a Leaves evergreen, sticky and fragrant; panicles broad.......... *C. velutinus*
  2b Leaves deciduous, usually not sticky
    3a Leaves serrate; flowers white or pinkish .... *C. sanguineus*
    3b Leaves entire; flowers white or generally blue.......... *C. integerrimus*

Ceanothus cuneatus (Hook.) Nutt. ex Torr. & Gray
Greasewood or Common buckbrush

Shrubs 1-4 m. tall, with stiff branches and branchlets spreading at nearly right angles, older stems with gray bark, young twigs with soft grayish-velvety pubescence; leaves opposite, but sometimes appearing whorled, oblong, oblanceolate or obovate, obtuse, entire or slightly toothed at apex, 0.5-2 cm. long, bluish-green and conspicuously veined on under surface, minutely pubescent, more densely so on the lower surface, short-petioled; umbels many-flowered, borne on short stiff branchlets; flowers creamy-white; capsule slightly elongated, with 3 horns near apex, at maturity upper part of capsule separating from lower, leaving receptacle and base of ovary attached to shrub.

Found principally on rocky or gravelly locations and along old stream beds.

Ceanothus integerrimus Hook. & Arn. Deerbrush

Stems widely branched, 1-4 m. tall; leaves deciduous, alternate, oblong to broadly oval, acute at apex, mucronate, obtuse at base, entire, with 3 main veins, usually pubescent; panicles 8-10 cm. long, narrow, on peduncles as long; flowers white to more often blue. Dry hillsides.

Ceanothus sanguineus Pursh

Redstem Ceanothus, Oregon tea or Buckbrush

Slender shrubs 1-3 m. tall, with reddish branches; leaves deciduous, 3-10 cm. long, alternate, broadly ovate to elliptic, obtuse at apex, glabrous or usually pubescent at least on the veins of the under surface, glandular-toothed, with 3 veins from the base, these with lateral branches; panicles dense; flowers white or pinkish; capsules deeply 3-lobed.

Rather common in wooded areas.

Ceanothus sanguineus

Ceanothus velutinus Dougl. Cinnamon bush

Bushy shrubs 0.5-3 m. tall, rarely slender and taller; leaves alternate, ovate to elliptic, shining and sticky above with a spicy odor, velvety beneath with thick minute hairs or glabrous beneath in var. ***hookeri***, finely toothed and glandular on the margin, 3-9 cm. long, strongly 3-veined from base, with many lateral veins; panicles generally dense and broad; flowers small, white; capsules 3-lobed, glandular, the upper part splitting off the base at the calyx line.

Variety ***velutinus*** is found in the Cascades and eastward with var. ***hookeri*** M. Johnston occurring in the Willamette Valley and the Coast Range.

Ceanothus velutinus

Rhamnus

Shrubs or trees; leaves deciduous or evergreen, alternate; flowers greenish, inconspicuous, bearing both stamens and pistils, or sometimes unisexual, produced in umbels or solitary; sepals partly united, forming a 4-5-lobed calyx, the upper part falling after flowering; petals 4-5, small, incurved, or absent; stamens 4 or 5; ovary 2-4-loculed; fruit fleshy.

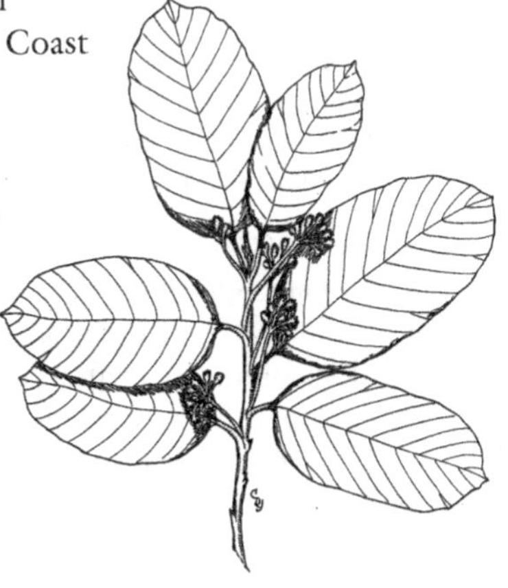

Rhamnus purshiana

Rhamnus purshiana DC. Cascara or Chittam

Shrubs or small trees, reaching 12 m. in height; leaves deciduous, broadly elliptic-oblong, abruptly pointed at apex, obtuse or slightly heart-shaped at base, the lateral veins conspicuously parallel, margins minutely toothed, glabrous to sparsely pubescent, petioles and young twigs velvety pubescent; sepals, petals and stamens 5; fruit purplish-black, with a sickly sweet taste.

Common in open woods. The bark yields a cathartic.

MALVACEAE
Mallow Family

Herbs, shrubs or trees with alternate simple palmately veined or lobed leaves and mucilaginous juice; flowers often perfect, sometimes pistillate, rarely with stamens and pistils on different plants; calyx 5-lobed, often with a circle of bractlets at the base; petals 5, generally conspicuously veined, twisted in the bud; stamens many, united in 1 or more series in a tube around the pistil, the corolla base united with base of tube; ovary superior, pistil composed of several to many carpels, with as many style-branches and stigmas; carpels separating at maturity, or fruit a berry or a several-loculed capsule splitting to allow escape of seeds.

Bractlets 3, distinct, borne on the calyx .... **Malva**
Bractlets none or rarely 1 borne below calyx...................................... **Sidalcea**

Malva

Annuals, biennials or perennials; calyx 5-lobed, subtended by 3 distinct bractlets united to the base of the calyx; petals white, pink or purple, broad, generally notched at apex; fruit 6-15-sectioned, somewhat compressed, the sections separating at maturity.

1a Petals 1.5-3 cm. long; plants usually erect
  2a Upper cauline leaves deeply dissected.............................. *M. moschata*
  2b Upper cauline leaves merely 5-7-lobed ....*M. sylvestris*
1b Petals less than 1.5 cm. long; plants more or less prostrate
  3a Petals 8-14 mm. long, claws of the petals pubescent...........*M. neglecta*
  3b Petals 4-5 mm. long, claws of the petals glabrous ....*M. parviflora*

Malva moschata L. Musk mallow

Perennials, sparsely pubescent, branching, 3-6 dm. tall; leaves orbicular in outline, 3-5-cleft, those of the upper leaves divided almost to base, the divisions pinnately cleft and toothed; petals rose-colored or white, 2-3 cm. long; carpels densely pubescent.

A European species, escaped from gardens and naturalized in some localities.

Malva neglecta Wallr. Dwarf cheese-weed

Annuals or biennials, more or less prostrate, spreading, 3-6 dm. or more in diameter, usually stellate-pubescent; leaves nearly circular in outline, generally

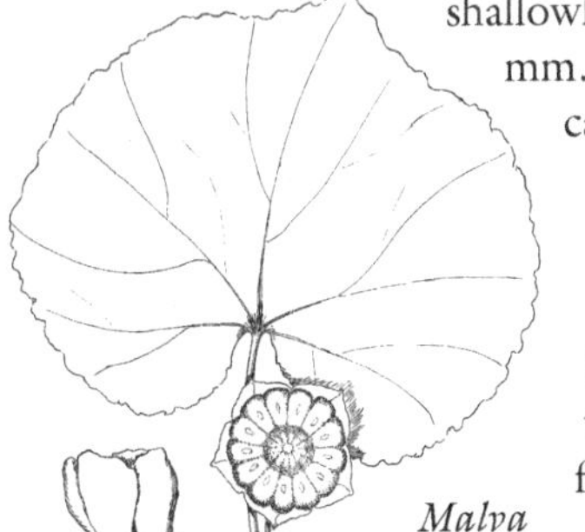

shallowly 5-7-lobed, the lobes crenate; corollas 8-14 mm. long, whitish to pink, lavender or bluish; carpels about 15, puberulent.

Waste places and neglected gardens.

Malva parviflora L. Cheese-weed

Somewhat resembling M. *neglecta,* but flowers smaller, the corollas 4-5 mm. long, the claw of the petals glabrous and matured fruit surfaces wrinkled.

A weedy species; introduced from Europe.

Malva sylvestris L. Common or High mallow

Stems erect, 5-30 dm. tall; leaf blades 3-7-lobed, the lobes crenate; petals 1.5-3 cm. long, notched, purple or pink with darker veins, the claws pubescent; matured fruit surfaces wrinkled.

Introduced from Europe, escaped from cultivation and established in disturbed habitats.

Sidalcea

Annual or perennial herbs with nearly orbicular palmately-veined and/or cleft leaves; flowers pink, purplish-lavender, or white, in terminal spikes or racemes; flowers dimorphic, perfect flowers on 1 plant and pistillate on another with those of the perfect flowers the largest; bractlets none, or 1 and attached below the base of the calyx; stamens numerous, united in 2-3 series; carpels 5-10, 1-seeded.

1a Plants strictly coastal
  2a Calyx purplish, glabrous except for the ciliate lobes; stems hollow .... *S. hendersonii*
  2b Calyx with simple and/or stellate pubescent .... *S. hirtipes*
1b Plants not on the immediate coast
  3a Petals white to pale pink; racemes loosely flowered .... *S. campestris*
  3b Petals deep pink to reddish-purple; racemes often spike-like
    4a Pubescence of the lower stem simple; petals 5-15 mm. long .... *S. nelsoniana*
    4b Pubescence of the lower stem of forked or stellate hairs
      5a Calyx prominently veined, the lobes widest above the base; petals 8-19 mm. long .... *S. cusickii*
      5b Calyx not prominently veined, the lobes widest at the base; petals 9-25 mm. long .... *S. malvaeflora*

Sidalcea campestris Greene Tall wild hollyhock

Perennials; stems slender, 7.5-18 dm. tall, smooth below, slightly pubescent near summit; foliage grayish-green; leaves orbicular in outline, palmately cut nearly to the base into 7-9 slender divisions, these pinnately cut into narrow divisions, divisions of lower leaves slightly broader than those above, leaves somewhat stiffly pubescent, hairs of the leaf blade mostly simple, those of

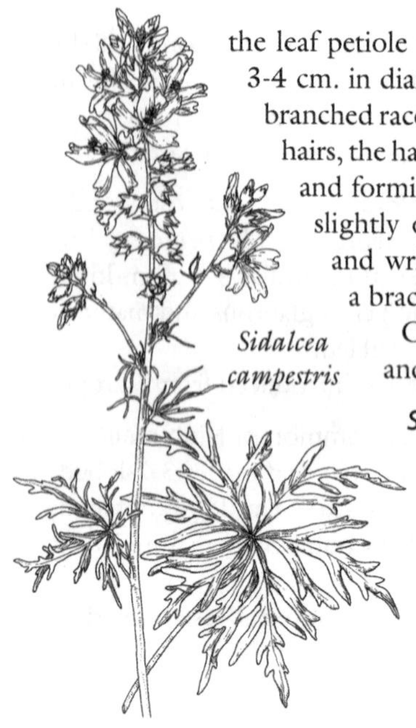
Sidalcea campestris

the leaf petiole forked or stellate; flowers white to pale pink, 3-4 cm. in diameter, in a long terminal simple or somewhat branched raceme; calyx densely covered with minute stellate hairs, the hairs on the margins and veins of the sepals longer and forming a fringe; petals 9-25 mm. long, oblong or slightly obovate, notched at apex; carpels pubescent and wrinkled at maturity, each flower subtended by a bractlet.

Common along road sides; flowering in June and July.

Sidalcea cusickii Piper
Cusick's checker-mallow

Rhizomatous perennials; stems 4-18 dm. tall, often hollow, finely stellate pubescent; basal leaves nearly orbicular, deeply crenate, the cauline leaves deeply palmately cleft; racemes spike-like; calyx 6-10 mm. long, purplish and finely stellate pubescent to nearly glabrous, prominently veined; petals 8-19 mm. long, deep pink; fruits nearly smooth on the back but netted on the sides.

Road sides and other open habitats, mostly south of our range.

Sidalcea hendersonii Wats. Henderson's checker-mallow

Nearly glabrous perennials; stems hollow, mostly simple, usually purplish, 6-15 dm. tall; basal leaves orbicular, 5-7-lobed, the lobes toothed, ciliate, cauline leaves deeply divided; flowers bright rose-colored, in dense terminal and axillary racemes; fruits glabrous, smooth.

Along the coast, often in marshes.

Sidalcea hirtipes C.L. Hitchc. Hairy-stemmed checker-mallow

Perennials; densely pubescent with simple or branched hairs; racemes compound, spike-like, many-flowered; calyx with stellate pubescence as well as numerous short stiff hairs; petals deep pink to reddish-purple; carpels wrinkled on the sides.

Coastal mountains to coastal bluffs; northern Oregon to Washington.

Sidalcea malvaeflora (DC.) Benth. subsp. *virgata* (Howell) C. L. Hitchc.
Showy wild hollyhock

Rhizomatous perennials; stems slender, 2-10 dm. tall, pubescent with stellate hairs; foliage deep green; basal leaves orbicular in outline, palmately cut one-third of the distance to base or deeper into 5-7 oblong toothed lobes, long-petioled, upper leaves more deeply cut, short-petioled, blades and petioles covered with minute branched or stellate hairs; flowers in a loose raceme, deep pink with magenta tinge, showy; calyx densely covered with fine stellate hairs; petals broadest above the middle, notched.

(=*Sidalcea virgata* Howell)

Common along road sides; flowering in April and May.

Sidalcea nelsoniana Piper — Nelson's checker-mallow

Perennials; stems generally clustered, 4-12 dm. tall, glabrous to rough pubescent; basal leaves rounded, shallowly lobed, the cauline leaves deeply cleft; racemes more or less spike-like, many-flowered; calyx dark red to reddish-purple, nearly glabrous or stellate pubescent; petals 5-15 mm. long, reddish-purple; carpels nearly smooth.

Moist habitats.

HYPERICACEAE
St. John's Wort Family

Herbs (ours), shrubs or trees; leaves simple, opposite or whorled, exstipulate, usually with black dots or embedded glands; sepals generally 5; petals usually 5, rarely 4, yellow or white; stamens 15-100, filaments free or united into 3-8 groups; pistil 1, ovary superior; fruit a capsule or a berry.

Hypericum

Annuals, perennials or shrubs; leaves sessile, more or less gland-dotted and usually with black dots along the margins; sepals and petals 5, petals yellow; stamens numerous; ovary 1-5-loculed; fruit a capsule.

1a Stems decumbent and mat-forming; petals 2-4 mm. long . *H. anagalloides*
1b Stems erect; petals 6-12 mm. long
  2a Weedy species of fields and hillsides; capsule not lobed... *H. perforatum*
  2b Plants of wet places; capsule 3-lobed .... *H. formosum*

Hypericum anagalloides Cham. & Schldl. — Bog St. John's wort

Mat-forming annuals or perennials; stems 4-angled, 5-25 cm. long, rooting at the nodes; leaves 5-15 mm. long, elliptic to ovate or obovate, 5-7-veined, inconspicuously gland-dotted, slightly clasping at the base; flowers in bracteate cymes or solitary; sepals 2-3.5 mm. long; petals 2-4 mm. long, yellow or pinkish-orange; stamens usually 15-25; capsules about 3 mm. long.

In ditches, bogs and swamps and on marshy margins of slow moving streams.

Hypericum anagalloides

Hypericum formosum H.B.K. subsp. *scouleri* (Hook.) C.L. Hitchc. — Western or Slender St. John's wort

Erect perennials from rhizomes and stolons; stems 1-8 dm. in height; leaves 1-3 cm. long, ovate to oblong or obovate, black-dotted, especially along the margins, inconspicuously dotted on the lower surface, slightly clasping at the base; bracteate cymes few-flowered; flowers strongly musk-scented; sepals 4-5 mm. long, black-dotted; petals 6-12 mm. long, deep yellow, black-dotted on the margins; stamens 75-100 in 3 groups; capsules 6-9 mm. long, 3-lobed and 3-loculed.

Wet meadows, springs and other moist areas, widely distributed but generally not abundant.

Hypericum perforatum L. St. John's wort, Goat weed or Klamath weed
Rhizomatous perennials; stems stout, often reddish, more or less woody at base and producing slender stolons, sending out slender, short, sterile branches, flowering stems more or less branched above, the upper branches slender, ending in simple or compound cymes; leaves mostly ovate to oblong to linear, entire, the margins often slightly rolled backward, 1-2.5 cm. long, sparsely black-dotted and containing translucent glands, appearing (when held to the light) as if punctured by pin pricks; flowers deep yellow; sepals 4-5 mm. long, narrow, acute; petals 8-12 mm. long, narrow, black-dotted; stamens in 3 groups; capsule not lobed.

A very serious weed on the Pacific Coast, spreading rapidly. Ingestion by livestock can cause photosensitization. After fruiting the whole plant turns a rusty brown, as if burned, and colors many acres of waste land in Oregon, Washington, and California, in late summer.

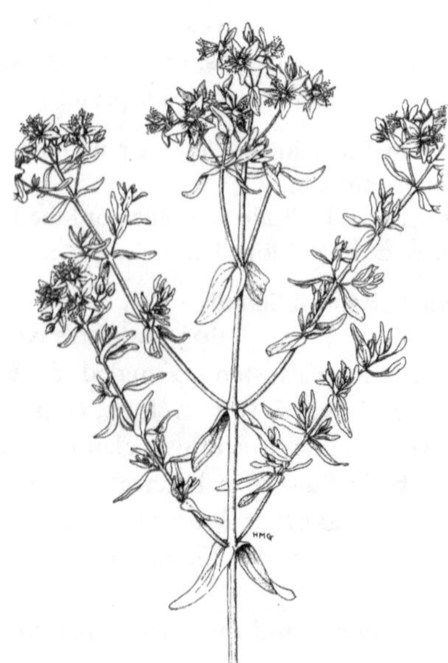

Hypericum perforatum

VIOLACEAE
Violet Family

Annuals, perennials or shrubs; leaves basal or usually alternate, simple or compound, stipules present; flower parts in 5's; corolla irregular, of 3 types of separate petals, upper (2), lateral (2), lower (1), this spurred or saccate at the base; filaments of stamens flattened and extending above the anthers which are borne on the inner faces, the anthers coming together, but not actually fused, around the pistil; pistil 1, ovary superior, 1-loculed, with 3 parietal placentae; fruit a 3-valved capsule or rarely a berry.

Viola

Annuals or perennials; leaves basal or alternate, simple or compound, stipulate; flowers showy, often solitary and axillary; sepals 5, subequal, usually auriculate near the base; the 2 lower stamens with appendages prolonged into the spur of the lower petal; capsule often explosively dehiscent.

1a Leaves evergreen; leaves cordate-ovate to orbicular; flowers yellow
  2a Leaves thick, purple or brown dotted; stoloniferous ..... *V. sempervirens*
  2b Leaves thin, not purple or brown dotted; not stoloniferous .... *V. orbiculata*
1b Leaves not evergreen
  3a Leaves all basal
    4a Leaves lanceolate to oblanceolate; plants limited, in our area, to coastal cranberry bogs .... *V. lanceolata*
    4b Leaves orbicular to reniform
      5a Flowers violet, blue or purplish, lower 3 petals white at the base... .... *V. langsdorfii*
      5b Flowers white or pale lavender
        6a Petals white, the lower 3 veined with purple ..... *V. macloskeyi*
        6b Petals tinged with lavender at least on the back ..... *V. palustris*
  3b Leaves not all basal
    7a Stipules leaf-like, pinnately lobed with 5-9 segments and one large terminal lobe; annuals .... *V. arvensis*
    7b Stipules not leaf-like; perennials
      8a Leaf blades deeply dissected; the 2 upper petals deep purple or blue, the laterals and lower yellow to nearly white .... *V. hallii*
      8b Leaves not deeply dissected
        9a Flowers predominately yellow or white, although sometimes purplish-tinged
          10a Leaves ovate to lanceolate, stems not naked below .... *V. nuttallii*
          10b Leaves broadly heart-shaped to reniform or orbicular, the lower stems often naked; petals yellow
            11a Leaves abruptly acute................................ *V. glabella*
            11b Leaves not acute.................................... *V. orbiculata*
        9b Flowers predominately blue, violet or purple, sometimes with the petals white at the base

12a Head of the style glabrous; coastal .... *V. langsdorfii*
12b Head of the style bearded
  13a Spur slender, longer than broad .... *V. adunca*
  13b Spur as broad as long
    14a Flowers 1.5-2.5 cm. long, pale blue, shading to white at the base.... .... *V. howellii*
    14b Flowers about 1.5 cm. long, purplish-violet, yellow at the base; subalpine .... *V. flettii*

Viola adunca Sm. — Western blue violet

Stems tufted, 2.5-20 cm. tall, with many leaves arising from a slender vertical rhizome; leaf blades 0.5-4 cm. long, 1-3 cm. wide, broadly ovate to somewhat elliptic, obtuse or slightly heart-shaped at base, margins crenate; peduncles longer than the leaves; flowers 5-16 mm. long, the lower petal with a long slender, curved to slightly hooked spur, petals blue to violet, the lower 3 usually whitish at the base, often dark-veined, the lateral 2 white-bearded; capsules 6-12 mm. long, glabrous.

This is a variable species with numerous named varieties, all of which appear to intergrade.

Wet meadows, damp woods and in drier sites.

Viola adunca

Viola arvensis Murr. — Wild pansy

Annuals; stems 0.5-3.5 dm. long, prostrate to erect, freely branching; leaf blades ovate to lanceolate, crenate-serrate, 1.5-3 cm. long, puberulent; stipules leaf-like, pinnately lobed with 5-9 segments and one large terminal lobe; flowers solitary in the leaf axils; sepals about equal to or longer than the petals, up to 12 mm. long; petals white, cream-colored or yellowish, often purplish- or bluish-tinged, the spur short; capsules 5-9 mm. long, glabrous.

Introduced from Europe and escaped from cultivation; weedy in lawns, gardens and fields.

Viola flettii Piper — Flett's violet

Plants glabrous, from stout creeping rhizomes; stems 3-15 cm. tall; leaves few, 1.5-4 cm. wide, broadly reniform, serrate-crenate, slender-petioled; stipules lanceolate; peduncles about the same length as the leaves; flowers violet, yellow at the base, about 1.5 cm. long, spur short, lateral petals bearded.

Rocky areas at high elevations in the Olympic Mountains of Washington.

Viola glabella Nutt. — Wood violet

Leaves and stems arising from a scaly horizontal rhizome; basal leaf blades 2-10 cm. long, long-

Viola glabella

petioled, broadly heart-shaped with acute apex, entire to finely crenate-serrate; stipules membranous, 5-10 mm. long; flowers 6-16 mm. long; petals yellow, the lower 3 dark-veined, the lateral 2 bearded, spur very short; capsules 7-17 mm. long.

Stream banks and moist woods.

Viola hallii Gray — Wild pansy

Stems several, arising from short erect rhizomes; leaves mostly palmately divided, the divisions again parted into narrow segments, the segments callous-tipped; flowers 10-15 mm. long, the 2 upper petals deep purple or blue, the lateral and lower yellow to nearly white with purple veins at the base, lateral petals slightly bearded at base; stigma bearded below the apex.

Uncommon in our limits, in open woods, usually on gravelly soil.

Viola howellii Gray — Howell's violet

Perennials from scaly rhizomes; stems slender, 0.5-2 dm. tall; leaves few, broadly cordate, crenate, ciliate, slender-petioled; peduncles as long as or longer than leaves, with minute bracts; flowers 1.5-2 cm. long, mostly pale blue, shading to white at the base, spur short, curved or merely saccate, lateral petals bearded.

Open woods, somewhat local.

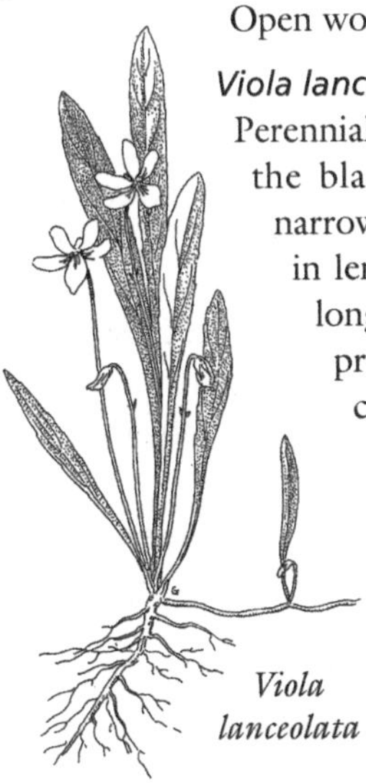

Viola lanceolata L. — Lance-leaved violet

Perennial by creeping rhizomes and stolons; leaves glabrous, the blades 4-12 cm. long, lanceolate to oblanceolate, narrowed to a long petiole; peduncles equalling the leaves in length or somewhat shorter; petals white, 6-11 mm. long, the lower and lateral petals purple-veined; capsules produced both by ordinary flowers and inconspicuous cleistogamous flowers borne on the stolons.

Found in cranberry bogs on the Pacific Coast, apparently introduced by way of cranberry stock. A native of eastern North America.

Viola lanceolata

Viola langsdorfii (Regel) Fisch. — Aleutian violet

Rhizomatous perennials; stems 2-10 cm. tall; leaf blades cordate to reniform, 2-5 cm. wide, crenate-serrate; stipules 3-10 mm. long, ovate to lanceolate; flowers 1.5-2.5 cm. long; petals violet, blue or purplish, lower 3 petals white at the base, the lateral 2 usually white-bearded, the spur saccate.

Bogs and other wet areas, along the coast.

Viola macloskeyi Lloyd — White violet

Rhizomatous and stoloniferous perennials 3-10 cm. tall; leaves all basal, leaf blades reniform to cordate to ovate-cordate, 1-5.5 cm. long, nearly entire to crenate; stipules lanceolate, membranous; flowers 5-12 mm. long; petals white, the lower 3 purple-veined, the lateral 2 bearded or not, spur short, saccate; capsules 5-9 mm. long.

Wet meadows, bogs and seepage areas in the mountains.

Viola nuttallii Pursh var. *praemorsa* (Dougl. ex Lindl.) Wats. **Upland yellow violet**

Stems and basal leaves arising from short vertical rhizomes; leaves broadly or narrowly ovate to lanceolate, 2-10 cm. long, obtuse or acute at apex, tapering at base or abruptly narrowed to a long petiole, margins entire or slightly toothed, usually pubescent; flowers yellow, the upper petals usually darker on the back, the lower and lateral petals dark-veined, the laterals bearded. (=*Viola praemorsa* Dougl. ex Lindl.)

In somewhat moist places, shaded by shrubbery or taller herbs.

Viola nuttallii

Viola orbiculata Geyer ex Hook. **Round-leaved violet**

Similar to *Viola sempervirens* and sometimes treated as a variety of that species, but leaves thin, only rarely evergreen and not purple or brown dotted, nearly orbicular, 2-4 cm. broad, the margins crenate.

In the mountains.

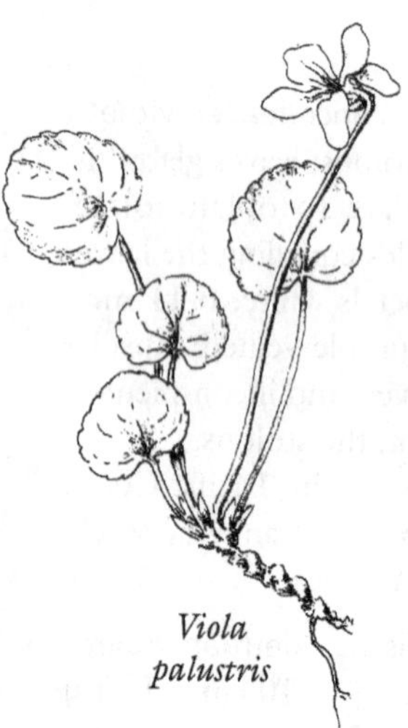

Viola palustris

Viola palustris L. **Swamp or Marsh violet**

Rhizomatous and stoloniferous perennials; leaves all basal, leaf blades 2-6.5 cm. long, 2.5-3.5 cm. wide, cordate to reniform, crenate, on petioles 2-17 cm. long; stipules lanceolate, entire, scarious; flowers 8-15 mm. long; auricles of the sepals prominent, 1-2 mm. long; petals white to pale violet, the lower 3 purple-veined, the lateral 2 bearded, spur short; capsules 6-7 mm. long.

Wet meadows, stream banks, swampy areas and moist shaded woods.

Viola sempervirens Greene **Evergreen violet**

Stems above ground stolon-like, arising with a cluster of leaves from the summit of a slender vertical rhizome; leaf blades evergreen, more or less leathery, 1-4 cm. long, orbicular or cordate-ovate, with low rounded teeth, purple or brown-dotted; peduncles erect, 5-10 cm. long; flowers 5-17 mm. long, appearing somewhat flattened, pale yellow, sometimes purple-veined, lateral petals bearded, spur short, saccate; capsules brown-mottled.

Common in moist open woods.

ELAEAGNACEAE
Oleaster Family

Evergreen or deciduous trees or shrubs, sometimes spiny, usually with rusty to silvery scale-like or stellate pubescent; leaves opposite or alternate, simple; flowers perfect or unisexual; calyx 4-lobed; petals lacking; stamens 4 or 8; pistil 1, simple; fruit 1-seeded, indehiscent, enclosed in the fleshy calyx and therefore appearing drupe-like.

Shepherdia

Dioecious shrubs or trees; leaves opposite, entire; flowers in the leaf axils, sometimes appearing before the leaves; staminate flowers with an 8-lobed disk inside the floral tube; calyx lobes 4, usually spreading; stamens 4-8; pistillate flowers with 4, short, erect calyx lobes.

Shepherdia canadensis (L.) Nutt. — Canadian buffalo berry

Shrubs 1-4 m. tall, not spiny, young branches brown-scurfy; leaf blades 1.5-7 cm. long, 1-3 cm. broad, ovate-oblong, green above, white-scurfy with brown scales on the lower surface, short-petioled; fruit elliptic, 6-8 mm. long, yellow to reddish.

In various habitats, often along stream banks.

LYTHRACEAE
Loosestrife Family

Annual or perennial herbs (ours) with simple entire leaves; flowers perfect, usually regular; calyx 4-8-lobed, the tube enclosing the ovary, but not united with the ovary; petals, if present, separate, as many as the calyx lobes and borne at the top of the floral tube; stamens of the same number as the petals, or more commonly twice as many, filaments usually unequal in length; pistil 1, ovary compound, superior; fruit a capsule.

Lythrum

Leaves commonly opposite, rarely whorled or alternate, simple and sessile; flowers axillary or in terminal spikes; floral tube cylindrical, striated; calyx 5-7-lobed, the lobes alternating with linear appendages; petals 5-7, white to pink or purple; capsule elongate, 2-loculed.

1a Rhizomatous perennials, 0.5-2 m. tall; petals 7-14 mm. long..... *L. salicaria*
1b Annuals or biennials, not rhizomatous, but sometimes rooting at the nodes; petals, if present, less than 4 mm. long
  2a Leaves opposite throughout, succulent, 0.5-1.5 cm. long....... *L. portula*
  2b Leaves, at least the upper, alternate, 1-3 cm. long.......... *L. hyssopifolia*

Lythrum hyssopifolia L. Hyssop loosestrife

Glabrous, glaucous annuals or biennials; stems 1-6 dm. tall, erect or spreading, slightly angled; leaves alternate or the lower opposite, 1-3 cm. long, 1-8 mm. wide, linear to oblong-elliptic; flowers solitary, sessile in the axils of the leaves; floral tube 4-6 mm. long; calyx lobes 0.5-1 mm. long, the alternating appendages about 1 mm. long, linear; petals 2-3.5 mm. long, white to rose; stamens 4-6; capsules 4-6 mm. long.

Marshes, vernal pools and ditches; introduced from Europe.

Lythrum portula (L.) Webb

Prostrate to decumbent annuals; stems 5-25 cm. long, rooting at the nodes; leaves succulent, opposite, 0.5-1.5 cm. long, oblong to obovate; flowers solitary, sessile in the leaf axils; floral tube 1-2 mm. long; calyx lobes triangular, about equal to the alternating appendages; petals, if present, about 1 mm. long, white to rose; capsules globose.

(=*Peplis portula* L.)

Vernal pools, margins of lakes and ponds and in marshy areas; introduced from Europe.

Lythrum salicaria L. Spiked or Purple loosestrife

Rhizomatous perennials; stems angled, erect, 0.5-2 m. in height; leaves predominantly opposite, but often some alternate and whorled, 3-15 cm. long, lanceolate, cordate at the base, pubescent; flowers in terminal, dense, interrupted spikes; floral tube 4-5 mm. long; calyx lobes triangular, about 1 mm. long, the alternating appendages 1.5-3 mm. long, linear; petals 7-14 mm. long, reddish-purple; stamens 8-12; capsules 3-4 mm. long.

Marshes, stream banks, ditches and other wet areas; introduced from Europe.

ONAGRACEAE
Evening Primrose Family

Annuals, perennials or shrubs; leaves opposite, alternate or whorled or all basal; calyx lobes and petals 2-5, rarely lacking; stamens usually of the same number as the petals or twice as many; ovary inferior; fruit usually a capsule, rarely indehiscent.

1a Sepals and petals minute, 2; fruit indehiscent, bristly....................... **Circaea**
1b Sepals and petals 4-6, or the petals absent; fruit a capsule
  2a Aquatic or subaquatic; stems prostrate to decumbent to floating; petals 5-6, or lacking .... **Ludwigia**
  2b Not prostrate or decumbent aquatics; petals 4
    3a Seeds with a tuft of hairs at the apex.................. **Epilobium** (in part)
    3b Seeds without a tuft of hairs
      4a Ovary and capsule 2-loculed .... **Gayophytum**
      4b Ovary and capsule 4-loculed
        5a Sepals erect in flower............................... **Epilobium** (in part)

5b Sepals reflexed in flower, sometimes remaining attached at the tips and turned sideways
6a Anthers attached near the middle; flowers yellow ....**Oenothera**
6b Anthers attached at or near base; flowers pink to purple or rarely white.. .... **Clarkia**

Circaea

Perennials with tuberous rhizomes or stolons; leaves opposite; flowers inconspicuous; calyx lobes 2; petals 2, white, notched; stamens 2; fruit indehiscent, pear-shaped, covered with hooked hairs.

Circaea alpina L. Enchanter's nightshade
Somewhat succulent from short rhizomes; stems 1-5 dm. tall, generally simple; leaves in several pairs, long-petioled, spreading at nearly right angles, the blades thin, 2-8 cm. long, broadly ovate, truncate or subcordate at base, abruptly acute or acuminate at apex, minutely and distantly toothed; flowers in bracted panicles or in terminal and axillary racemes; calyx lobes white, 1-2 mm. long; petals white, 1-1.5 mm. long; fruit about 2 mm. long.

Stream banks and moist woods, mainly in the mountains.

Circaea alpina
a. flower; b. fruit

Clarkia

Annuals, the epidermis of the stem often exfoliating near the base; leaves simple; inflorescence a leafy spike or raceme; flowers usually showy; sepals 4, reflexed individually, in pairs or united and reflexed to one side; petals 4; stamens 4 or 8; ovary 4-loculed; fruit usually a capsule.

1a Calyx lobes separate or in pairs and turned back in flower .... *C. purpurea*
1b Calyx lobes remaining united above and turned to one side in flower or split only on 1 side
2a Petals usually with a crimson spot; stigmas linear; flower buds erect........ .... *C. amoena*
2b Petals unspotted; stigmas short and broad; flower buds nodding............ .... *C. gracilis*

Clarkia amoena (Lehm.) Nels. & Macbr. Farewell-to-spring
Stems usually erect, slender, simple or branching, 2-9 dm. tall, puberulent, somewhat grayish-green, often reddish below; leaves 1-7 cm. long, linear or lanceolate, sometimes narrowed at base, the lower often having disappeared at flowering time, leaving the stem bare below, the upper leaves often folded; flowers showy; the calyx generally splitting on only one side, but the lobes free at the base; petals obovate, obtuse or truncate, 1-6 cm. long, 1-1.5 cm. wide, violet or rose-lavender, often with a crimson spot near the base; stamens 8, about 1/2 the length of petals, filaments sometimes with a few long

Clarkia amoena
a. stamen
b. style & stigma

hairs, anthers slender, pubescent, curved at the tips; style slightly exceeding the stamens; stigmas linear, pale yellow, minutely pubescent; ovary more or less indistinctly ribbed, tapering at each end.

Common along roadsides in Oregon in early summer. There is much variation in this species, and there are perhaps a number of distinct varieties. A dwarf form often occurs somewhat abundantly in dried silty- clay-loam soil in the Willamette Valley, either alone or with the typical species.

Clarkia gracilis (Piper) Nels. & Macbr. — Slender godetia

Stems erect, slender, generally simple, 3-9 dm. tall, usually with short appressed hairs; leaves 1-8 cm. long, linear to narrowly lanceolate, scattered; buds somewhat nodding; calyx splitting on only one side at maturity; petals 8-23 mm. long, obovate, rose-colored; capsule sessile, 2.5-5 cm. long, cylindrical, not conspicuously ribbed.

Dry open areas.

Clarkia purpurea (Curtis) Nels. & Macbr. — Purple godetia

Stems up to 7.5 dm. tall or rarely taller; leaves 1.5-7.5 cm. long, linear to ovate, entire or nearly so; inflorescence erect, congested or loose-flowered; sepals reflexed individually or in pairs; petals 5-25 mm. long, pink, purple to maroon or rarely white, with or without a purple spot; stamens 8; capsule 1-3 cm. long.

Meadows or woodlands.

Three subspecies occur within our area and can be separated as follows:

1a Leaves broadly lanceolate to ovate .... subsp. *purpurea*
1b Leaves linear to narrowly lanceolate
  2a Petals 15-25 mm. long . subsp. *viminea* (Dougl.) H. Lewis & M. Lewis
  2b Petals less than 15 mm. long .... subsp. *quadrivulnera* (Dougl.) H. Lewis & M. Lewis

Epilobium

Annual or perennial herbs, rarely shrubs, with alternate or opposite leaves; flowers axillary or terminal, solitary, or in racemes; calyx generally produced beyond the ovary, 4-lobed; petals 4, generally obovate, often notched; stamens 8; ovary 4-loculed; seeds usually bearing a tuft of white hairs.

1a Flowers yellow; stigmas 4-cleft.. *E. luteum*
1b Flowers purplish or pink, rose-colored or white
  2a Flowers generally 3 cm. or more in diameter
    3a Leaf blades averaging well over 6 cm. long, veins conspicuous on lower surface; flowers many .... *E. angustifolium*

3b Leaf blades 6 cm. or less in length, leaf veins not conspicuous; flowers few .... *E. latifolium*

2b Flowers generally smaller than 2.5 cm. in diameter

4a Annuals, lower stems often peeling

5a Seeds without a tuft of hairs

6a Petals 3-10 mm. long, white to rose to purple; fruits not beaked .... *E. densiflorum*

6b Petals 1-3 mm. long, white to pink; fruits beaked........ *E. torreyi*

5b Seeds with a tuft of hairs

7a Plants 3-20 dm. tall, often glandular; petals 2-15 mm. long; floral tube 1.5-15 mm. long .... *E. brachycarpum*

7b Plants 0.5-4 dm. tall, not glandular; petals 2-5 mm. long; floral tube 0.5-1 mm. long .... *E. minutum*

4b Perennials

8a Stems erect, producing offset rosettes or fleshy turions

9a Veins of the leaves obscure; flowers nodding........ *E. halleanum*

9b Veins of the leaves evident; flowers rarely nodding...... *E. ciliatum*

8b Stems usually not strictly erect, more or less ascending; plants rhizomatous and/or stoloniferous

10a Stems pubescent with inconspicuous lines of hairs or glabrous but not glaucous; stolons usually leafy........ *E. alpinum*

10b Stems glabrous and glaucous; stolons short, scaly........ *E. glaberrimum*

Epilobium alpinum L. — Alpine willow-herb

Rhizomatous and stoloniferous; stems 1-3.5 dm. tall, nearly erect to weakly ascending, glabrous or with inconspicuous lines of pubescence above and glandular-puberulent in the inflorescence; leaves mostly opposite, 0.5-1.5 cm. long, bright green, thin, glabrous or nearly so, linear to ovate, entire or slightly toothed, abruptly narrowed at the base and usually short-petioled; petals 3-14 mm. long, notched, white or rose-colored; capsule 2-7 cm. long, pediceled.

(includes *E. clavatum* Trel., *E. hornemannii* Reichb., *E. lactiflorum* Hausskn. and *E. oregonense* Hausskn.)

Wet places in the mountains, often subalpine or alpine.

Epilobium angustifolium L. — Fireweed

Perennials; stems erect, often reddish, mostly simple, 6-30 dm. tall; leaves alternate, long-lanceolate, deep green above, paler beneath, minutely and distantly toothed or nearly entire, the lower narrowed to short petioles; flowers in long terminal racemes, or some in the axils of the upper foliage leaves, long--pediceled; petals 8-25 mm. long, rose-colored; capsules slender, spreading.

Common in edges of open woods and in logged and burned-over lands.

Epilobium brachycarpum Presl — Tall willow-herb

Annuals; stems peeling below, 3-20 dm. tall, much-branched above, the branches slender; leaves 2.5-5 cm. long, alternate, linear or narrowly lanceolate, often folded and somewhat curved, sparingly denticulate, narrowed to a short petiole; inflorescence a loose much-branched panicle, often minutely glandular;

petals 2-15 mm. long, pink or pale magenta or whitish, spreading, deeply notched; stigma club-shaped; capsules slender, with 4 conspicuous ridges and 4 smaller ridges between, beaked, 2 cm. long or less, on short slender pedicels; seeds very minutely roughened, the tuft of hairs white.

(=*Epilobium paniculatum* Nutt. ex Torr. & Gray)

In open dry ground, common.

Epilobium ciliatum Raf.

Perennials producing offset rosettes or fleshy turions; stems less than 2 m. tall; leaves lanceolate to ovate, 1-15 cm. long, petioles absent or up to 8 mm. long; inflorescence pubescent and usually glandular; calyx lobes 2-7.5 mm. long; petals 2-14 mm. long, white to rose-colored to purple; fruit 1.5-10 cm. long, pubescent. This is an extremely variable species with three subspecies in our area: subsp. ***ciliatum*** which has lanceolate leaves, petals 2-9 mm. long and produces basal rosettes; subsp. ***glandulosum*** (Lehm.) P. Hoch & Raven which has ovate leaves, petals 4-14 mm. long and produces fleshy turions; and subsp. ***watsonii*** (Barbey) P. Hoch & Raven with ovate leaves, petals 5-14 mm. long and produces basal rosettes.

Epilobium ciliatum

(includes *E. adenocaulon* Hausskn., *E. brevistylum* Barbey, *E. glandulosum* Lehm. and *E. watsonii* Barbey)

Stream banks, wet meadows and other drier sites.

Epilobium densiflorum (Lindl.) P. Hoch & Raven — Dense spike-primrose

Annuals to 15 dm. tall; stems pubescent, peeling below; leaves alternate, except near the base where they may be opposite, 1-8.5 cm. long, narrowly lanceolate, subsessile, at least the upper pubescent; inflorescence densely leafy and sometimes glandular; floral tube 1.5-4 mm. long; calyx lobes 2-9 mm. long; petals 3-10 mm. long, white to purple; capsules 4-11 mm. long.

[=*Boisduvalia densiflora* (Lindl.) Wats.]

Stream banks and other moist areas.

Epilobium glaberrimum Barbey — Smooth willow-herb

Glabrous and glaucous perennials with short scaly stolons; stems to 8.5 dm. tall, more or less ascending; leaves 1-7 cm. long, lanceolate to ovate, clasping the stem; calyx lobes 1.5-7.5 mm. long; petals 2.5-12 mm. long, white to rose to purple; fruit 2-7.5 cm. long, sometimes pubescent. Two subspecies occur in our area: subsp. ***glaberrimum*** with stems 2-8.5 dm. tall, leaves 2-7 cm. long and capsules 4.5-7.5 cm. long; and the much less common subsp. ***fastigiatum*** (Nutt.) P. Hoch & Raven which is less than 3.5 dm. tall, leaves 1-3.5 cm. long and capsules 2-5.5 cm. long.

Stream banks, moist rocky areas and drier sites.

Epilobium halleanum Hausskn.
Perennials producing underground turions; stems to 9 dm. tall, usually pubescent with lines of stiff hairs; leaves 0.5-4.5 cm. long, ovate to lanceolate, ciliate, sessile or short-petioled; flowers typically nodding; floral tube less than 2 cm. long; calyx lobes 1-3 mm. long; petals 1.5-5.5 mm. long, white to pink; capsules 2.4-6 cm. long, often pubescent.
Moist areas.

Epilobium latifolium L. Red fireweed
Stems 1-6 dm. tall, more or less erect, leafy, generally branched; leaves alternate, appearing glabrous but mostly microscopically pubescent, broadly lanceolate, thick, 1-6 cm. long, narrowed at base but scarcely petioled; flowers showy, purple, about 3 cm. or more in diameter, axillary or forming a short terminal raceme.
Wet places in the northern mountains.

Epilobium luteum Pursh Yellow Willow-herb
Perennials with short leafy stolons; stems 1-8 dm. in height; leaves 2.5-8 cm. long, spatulate, to ovate, sharply but minutely toothed, the upper sessile, the lower short-petioled; calyx lobes 1-1.2 cm. long; petals 1.2-2.2 cm. long, pale yellow; capsules 3.5-7.5 cm. long, pubescent.
Margins of streams and wet meadows in the mountains.

Epilobium minutum Lindl. Small-flowered willow-herb
Slender annuals; stems 0.5-4 dm. tall, peeling below, simple or branched, minutely pubescent; leaves mostly alternate, lanceolate or oblong, narrowed to a short petiole, entire or somewhat toothed, the upper leaves narrower; petals 2-5 mm. long, pinkish or white, notched; capsules slender, slightly curved.
In open woods and dry places.

Epilobium torreyi (Wats.) P. Hoch & Raven Stiff spike-primrose
Annuals; stems 1-6.5 dm. in height, with grayish-pubescence, peeling below; lowest leaves opposite and often glabrous, 0.5-4.5 cm. long, linear-lanceolate, nearly sessile; flowers often cleistogamous; calyx lobes 0.7-2 mm. long; petals 1-3 mm. long, white to pink or purplish; capsules 8-13 mm. long, beaked.
[=*Boisduvalia stricta* (Gray) Greene]
Stream banks and other moist areas.

Gayophytum

Slender annuals with alternate linear or narrowly lanceolate leaves and axillary flowers; calyx lobes and petals 4 each, the latter very small; stamens 8, alternating long and short; ovary 2-loculed; fruit a capsule.

Gayophytum diffusum Torr. & Gray Spreading groundsmoke
Stems up to 6 dm. tall, glabrous or pubescent; leaves 1-6 cm. long, 1-5 mm. wide, reduced upwards; calyx pubescent; petals 1.2-7 mm. long; ovary pubescent; capsule 3-15 mm. long, constricted between the seeds.
Usually in dry open areas.

Ludwigia

Aquatic or sub-aquatic annuals, perennials to subshrubs; leaves alternate or opposite, simple; calyx lobes 4-7; petals 4-7, or occasionally lacking; stamens 4-12; fruit an irregularly dehiscing capsule.

1a Leaves opposite; petals lacking...*L. palustris*
1b Leaves alternate; petals present
  2a Leaves narrowly elliptic to lanceolate or broadly obovate, petioles 3-23 mm. long; petals 15-30 mm. long ....*L. hexapetala*
  2b Leaves oblong to orbicular, petioles 3-60 mm. long; petals 7-24 mm. long..*L. peploides*

Ludwigia hexapetala **(Hook. & Arn.) Zardini, Gu & Raven**
Perennials, often somewhat woody at the base; stems 2-20 dm. long, often floating, rooting at the nodes, glabrous to pubescent; leaves alternate, 1-11 cm. long, narrowly elliptic to lanceolate to broadly obovate, petioles 3-23 mm. long; calyx lobes 5-6, 8-20 mm. long; petals 5-6, 1.5-3 cm. long, yellow; stamens 10-12; capsules reflexed, cylindrical, 1-3 cm. long on pedicels 1-6 cm. long.
(=*Jussiaea uruguayensis* Camb.)
Swamps, lake margins, ditches and sloughs.

Ludwigia palustris **(L.) Ell.** **Water purslane**
Mat-forming prostrate perennials; stems 1-5 dm. long, sometimes submerged, floating or rooted in the mud, rooting at the nodes, much-branched, subglabrous; leaves opposite, 2-6 cm. long, fleshy, elliptic to obovate, abruptly acute, entire; calyx lobes 4, 1-2 mm. long; petals lacking; stamens 4; capsules erect, 1.5-5 mm. long, with 4 green stripes.
Marshes, ponds, wet meadows and ditches.

Ludwigia peploides **(Kunth) Raven**
Mat-forming perennials, stems floating or creeping, 1-30 dm. long; leaves alternate, less than 1 dm. long, glabrous or pubescent on the upper surface, oblong to orbicular, subentire; calyx lobes 5-6, 3-12 mm. long; petals 5-6, 7-24 mm. long, yellow; stamens 10-12; capsules 1-2.5 cm. long.
Marshes, stream banks, ponds and ditches.

Oenothera

Prostrate or erect herbs with alternate or basal leaves; flowers usually yellow, white or purple; calyx tube elongated beyond the ovary, the lobes 4, their tips free or united, at length reflexed; petals 4; stamens 8; ovary 4-loculed, stigma deeply-lobed; capsules 4-valved.

1a Petals 3.5-5 cm. long ....*O. glaviovana*
1b Petals less than 3.5 cm. long
  2a Longest hairs with a red blister-like base; inflorescence open, few-flowered..*O. villosa*
  2b Longest hairs without a red blister-like base; inflorescence dense...........
  ....*O. biennis*

Oenothera biennis L. Common evening primrose
Biennials; stems stout, erect, mostly simple, often reddish, 3-20 dm. tall, more or less pubescent with stiff hairs and glandular-pubescent at least in the inflorescence; leaves 5-20 cm. long, oblanceolate to oblong, toothed to subentire, pointed, upper sessile, lower short-petioled; flowers in a dense spike; petals 1-3 cm. long, yellow; capsules 2-4 cm. long, nearly straight.
Disturbed areas.

Oenothera biennis
a. basal leaf

Oenothera glaviovana Micheli
Biennials; stems erect, 5-15 dm. tall, at least the longest hairs with reddish blister-like bases; cauline leaves 5-15 cm. long, lanceolate, the margins often wavy , toothed to nearly entire; petals 3.5-5 cm. long, yellow but fading reddish; capsules 2-3.5 cm. long.
A common escape from cultivation.

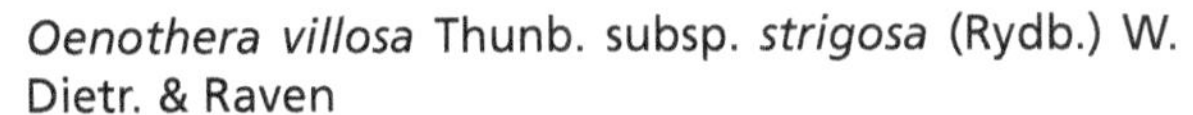
Oenothera villosa Thunb. subsp. *strigosa* (Rydb.) W. Dietr. & Raven
Biennials; stems erect 5-20 dm. tall; some hairs with reddish blister-like bases; leaves lanceolate, entire or nearly so, 1-3 dm. long; inflorescence few-flowered; petals 0.5-2 cm. long, yellow; capsules 2-3.5 cm. long.
Usually in moist areas.

HALORAGIDACEAE
Water-milfoil Family

Mostly perennial aquatics with whorled (rarely alternate or opposite), pinnately-dissected submerged leaves and emergent simple to divided leaves or bracts; flowers small, usually unisexual, the species monoecious or with a few perfect flowers, or rarely dioecious; calyx 2-4-lobed; petals, if present, 2 or 4, small; stamens 4 or 8; ovary inferior; fruit fleshy or nutlet-like and usually separating at maturity.
This family is sometimes spelled **Haloragaceae**.

Myriophyllum

Rhizomatous perennials; submerged leaves 3-6 per whorl, filiformly pinnatifid, emergent leaves or bracts variously arranged, but often opposite, entire to pinnately dissected; inflorescence typically emergent; flowers unisexual and axillary, usually with the staminate flowers above the pistillate ones, the middle flowers sometimes perfect; calyx lobes 4; petals 4, minute, soon deciduous or sometimes lacking on the pistillate flowers; stamens 4 or 8; pistil compound, the four carpels separating into nutlet-like segments at maturity.

1a Dioecious species
2a Stems stout; leaves in whorls of 4-8, uniformly pinnatifid with 21-35 filiform divisions.. *M. aquaticum*
2b Stems delicate; leaves opposite or in whorls of 3, with less than 15 segments.. *M. ussuriense*
1b Monoecious species; leaves or bracts subtending the flowers smaller and usually much less dissected than the lower leaves
3a Bracts linear to oblong, 0.5-1.5 cm. long, serrate or entire; stamens 4.... .... *M. hippuroides*
3b Bracts usually ovate, if more than 0.5 cm. long, then pinnately divided; stamens typically 8
4a Leaf whorls from the middle and below usually less than 1 cm. apart; bracts pinnate, longer than the fruit........................... *M. verticillatum*
4b Leaf whorls from the middle and below usually 1 cm. or more apart
5a Lowermost bracts usually shorter than the flowers; leaves in the middle of the stem usually with less than 26 filiform segments .... *M. sibiricum*
5b Lowermost bracts longer than the flowers and fruit; leaves in the middle of the stem usually with more than 26 linear segments........... .... *M. spicatum*

Myriophyllum aquaticum (Vell.) Verdc. Parrot's feather
Dioecious species; stems stout, 2-4 mm. thick and up to 2 m. in length; leaves in whorls of 4-8, 1.5-4 cm. long, pinnatifid with 21-35 filiform divisions, the leaves subtending the flowers the same as the others; staminate flowers with 8 stamens and 4 small, early deciduous petals; pistillate flowers apetalous, the stigma plumose.

(=*M. brasiliense* Camb. in St. Hilaire)

Usually mostly submerged; common in ponds, lakes, backwaters of streams and rivers; introduced from South America.

Myriophyllum aquaticum

Myriophyllum hippuroides Nutt. ex Torr. & Gray Western Water-milfoil
Stems weak, much-branched, up to 1 m. in length; leaves in whorls of 4-5, the submerged ones 1.5-3 cm. long and pinnately-dissected into 13-21 (rarely more) segments, these transitional to the serrate or entire bracts; stamens 4; stigma short, not plumose.

Quiet and slow-moving water.

Myriophyllum sibiricum V. Komarov Common Water-milfoil
Stems weak, simple or branched, often over 1 m. in length, purplish, but becoming whitish when dry; submerged leaves 3-4 per node, 1-3 cm. in length, pinnately dissected into 13-25 filiform segments; bracts 1-3 mm. long, narrowly to broadly ovate, serrate to entire, usually shorter than the flowers and the fruit.

(=*M. exalbescens* Fern.)

In quiet or slow-moving, often brackish water.

Myriophyllum spicatum L. Eurasian Water-milfoil

Stems weak, often more than 2 m. in length, pinkish when dry; submerged leaves 3-5 in a whorl, 1-2.5 cm. in length, those in the middle of the stem usually with more than 25 segments; bracts 1-3 mm. long, the lowest longer than the fruit and serrate, the middle and upper entire and usually shorter than the fruit; stamens 8.

In quiet or slow-moving water; introduced from Eurasia.

Myriophyllum ussuriense (Regel) Maxim.

Dioecious species; stems delicate, light green, up to 2.5 dm. long; leaves opposite or the lower usually in whorls of 3, 1-8 mm. long, entire or divided into 2-14 segments, those of the upper leaves usually much-reduced and bract-like, red-tipped; winter buds present; staminate flowers with 8 stamens.

Semi-terrestrial or on muddy or sandy soil in shallow water of lakes and streams; introduced from Asia.

Myriophyllum verticillatum L. Whorled Water-milfoil

Stems weak, usually much-branched, up to 2.5 m. in length; winter buds of crowded leaves produced in the fall; submerged leaves in whorls of 4-5, up to 4.5 cm. long with 18-34 linear divisions; bracts 3-10 mm. in length and pinnately divided; fruit 2-2.5 mm. long.

Quiet or slow-moving water.

HIPPURIDACEAE
Mare's-tail Family

Rhizomatous perennials; stems erect and unbranched, hollow; leaves in whorls of 4-12, simple, entire and sessile; flowers mainly perfect, inconspicuous, solitary in the axils of the upper leaves; calyx minute; petals lacking; stamen one; pistil simple, ovary inferior; fruit indehiscent, 1-seeded.

Hippuris

The Hippuridaceae is considered to be monogeneric and monotypic.

Hippuris vulgaris L. Mare's-tail

Stems 3-6 dm. in height; leaves 4-12 to a whorl, entire, linear to narrowly elliptic.

A variable species, previously split into two or three species.

In shallow water, usually one-half or more of the stem submerged.

ARALIACEAE
Ginseng Family

Trees, shrubs, vines or perennial herbs; leaves opposite, alternate or whorled, mostly palmately lobed or compound; flowers generally in umbels, these solitary or more often racemose or paniculate; ovary inferior; sepals minute, usually 5; petals and stamens usually 5; ovary 1-15-loculed; fruit a berry or a drupe.

1a Shrubs densely armed with spines .... **Oplopanax**
1b Plants unarmed
  2a Leaves pinnately compound; stems erect .... **Aralia**
  2b Leaves simple; a woody vine .... **Hedera**

Aralia

Shrubs, small trees or perennial herbs; leaves alternate, compound or decompound; flowers perfect, greenish-white; sepals, petals and stamens 5; ovary 5-6-loculed; fruit a berry.

Aralia californica Wats.

Stems stout, 2-3 m. in height; leaves 1-3 ternate or pinnate, the leaf 1-2 m. long, the leaflets 6-30 cm. long, ovate to oblong, serrate, usually asymmetrical at the base, glabrous or nearly so; inflorescence much branched; flowers greenish; berry globose, purplish-black.

Moist woods and stream banks.

Hedera

Woody vines climbing by aerial rootlets; leaves evergreen, alternate, simple, sometimes lobed, usually stellate pubescent; flowers small; sepals, petals and stamens 5; fruit a berry.

Hedera helix L.
English ivy

Vines creeping on the ground or climbing to as high as 30 m.; leaves leathery, dark green, leaves on sterile stems palmately 3-5-lobed, those of the fertile stems nearly entire, more or less obovate to broadly ovate, stellate pubescent on the under surface or nearly glabrous, long-petioled; flowers inconspicuous; berry purplish-black.

Widely cultivated and often escaped and naturalized.

Oplopanax

A monotypic genus.

Oplopanax horridum
a. pistil

Oplopanax horridum (Sm.) Miquel — Devil's club

Densely spiny deciduous shrubs, 1-4 m. tall, with weakly ascending branches, leafy only near the tips; leaves palmately 5-15-lobed, cordate, orbicular in outline, 1-3.5 dm. broad, pubescent and stiff-spiny on the veins and the petioles; flowers greenish; inflorescence a raceme or panicle of dense umbels; sepals and petals small, the stamens usually longer than the petals; berry scarlet.

In moist woods.

APIACEAE
Parsley Family

Herbs or rarely shrubs with flowers in simple or compound umbels or heads; stems generally hollow, mostly rank-scented or fragrant; leaves usually alternate or all basal with sheathing petioles, but without stipules; calyx tube united with the ovary, the teeth small or absent; petals usually 5, generally long-clawed, the blade turned inward or somewhat hooded; stamens 5, alternate with the petals, attached to a disk above the ovary; ovary inferior, styles 2, united at base and often swollen to form a stylopodium; ovules 2, pendulous; fruit a schizocarp consisting of 2 carpels united at first, generally separating later from base upward into mericarps The alternate name for this family is **Umbelliferae**.

Cultivated species of plants belonging in this family may sometimes be found in waste places as escapes from gardens but are rarely established in the wild. Among these are: Garden parsnip (*Pastinaca sativa* L.) a biennial with a stout taproot and large pinnnately compound leaves and yellow flowers; and Coriander (*Coriandrum sativum* L.) a strongly scented annual, growing 3-7.5 dm. tall, with upper leaves finely dissected, lower leaves once or twice pinnate, with broad segments; flowers white; fruits nearly globose, about 3 mm. in diameter.

1a Fruits linear or nearly so, measuring (with beak if present) slightly less than 1 cm. to much longer
  2a Beak of the fruit several times longer than the body .... **Scandix**
  2b Beak of fruit, if present, much shorter than the body .... **Osmorhiza**
1b Fruits not linear
  3a Fruits bristly, scaly, prickly, or warty
    4a Umbels dense, head-like; at least lower leaves opposite...... **Eryngium**
    4b Umbels not condensed (umbellets sometimes head-like); leaves mostly alternate or basal
      5a Flowers greenish, yellow or purple; fruits covered with tubercle-based hooked bristles .... **Sanicula**
      5b Flowers white
        6a Involucre conspicuous, of cleft bracts .... **Daucus**
        6b Involucre absent or of linear bracts
          7a Fruits ovoid, about 4 mm. long, short-beaked, covered with short hooked bristles except for the beak .... **Anthriscus**

7b Fruit 3 mm. long, elliptic-oblong, densely covered with roughened barbed bristles........ **Torilis**
3b Fruits not bristly, prickly, warty, or scaly (sometimes pubescent)
8a Fruits flattened front to back and usually winged
9a Flowers reddish-purple or yellow .... **Lomatium** (in part)
9b Flowers white, pinkish, or greenish
10a Prostrate or spreading seashore species; ribs of the fruit all broadly corky-winged........ **Glehnia**
10b Not prostrate or spreading seashore species
11a Leaves mostly basal; plants less than 8 dm. tall........ **Lomatium** (in part)
11b Stem leafy; mostly taller plants
12a Marginal flowers of the umbel enlarged; only lateral ribs of the fruit winged .... **Heracleum**
12b Marginal flowers of the umbel not enlarged; all ribs of the fruit winged or very conspicuous
13a Leaves finely dissected, the ultimate leaf segments small .... **Conioselinum**
13b Ultimate leaf segments large........ **Angelica**
8b Fruits either not flattened or flattened side to side
14a Leaves not compound; stems low, creeping; umbels simple
15a Leaves reduced to hollow apparently jointed petioles........ **Lilaeopsis**
15b Leaf blades orbicular to reniform, petioled........ **Hydrocotyle**
14b Leaves compound
16a Plants aquatic or semi-aquatic (see *Conium* if leaves finely dissected)
17a Leaves once pinnate; fruit laterally compressed, the ribs all corky .... **Sium**
17b At least some of the leaves twice or thrice-pinnate
18a Styles 1/2 as long as fruit; lateral veins of leaflets running to center of teeth .... **Oenanthe**
18b Styles shorter; lateral veins of leaflet closely following upper margins of teeth........ **Cicuta**
16b Plants not aquatic
19a Low plants, less than 2 dm. tall
20a Seashore plants, densely pubescent; leaves thick; all the wings of the fruit thick and corky .... **Glehnia**
20b Not seashore plants; slender with thin leaves; only lateral wings of the fruit thick........ **Orogenia**
19b Plants averaging at least 2 dm. tall (generally much taller)
21a Ribs of fruit inconspicuous; leaves few, leaflets linear........ **Perideridia**
21b Ribs of fruit very prominent, at least when dry
22a Plants with spotted stems; odor of dried plants distinctly mouse-like .... **Conium**
22b Stems not spotted; odor not mouse-like ....**Ligusticum**

Angelica

Stout taprooted perennials with hollow leafy stems and ternately or pinnately compound leaves; umbels compound, large and showy, terminal on the main stem or long axillary peduncles; flowers whitish; involucre generally lacking; involucel none or of a few slender bracts; stylopodium conical; fruit oblong to orbicular and usually compressed front to back (dorsally).

1a Seaside plants
  2a Leaves usually densely woolly beneath; lateral wings of the fruit broader than the dorsal .... *A. hendersonii*
  2b Leaves glabrous; ribs of the fruit subequal, thick and corky..... *A. lucida*
1b Not seaside plants
  3a Primary leaflets sharply reflexed; involucels conspicuous... *A. genuflexa*
  3b Primary leaflets not sharply reflexed; involucels inconspicuous or absent .... *A. arguta*

Angelica arguta Nutt. — Sharp-tooth Angelica

Stems sometimes purplish, 0.5-2 m. tall, glabrous and shining or minutely pubescent towards the inflorescence; leaves ternate, then once or twice pinnate, the leaflets narrowly ovate to elliptic, irregularly cleft and sharply serrate, 4-14 cm. long, glabrous or pubescent on the veins of the lower surface, pale green; umbels reaching 1.5 dm. or more in diameter; involucre none; involucels none or sometimes of a few small bracts; fruit oblong, 4-9 mm. long, veins prominent, lateral wings as wide as the body, not corky.

Stream banks, wet meadows and marshy places in the mountains and foothills.

Angelica genuflexa Nutt. — Kneeling Angelica

Stems 1-2 m. tall, glabrous below inflorescence; leaves ternately-compound, 1-8 dm. long, triangular to ovate in outline, the primary leaflets sharply reflexed, leaflets ovate to lanceolate, serrate, 4-10 cm. long, 1.5-5 cm. wide, acute to acuminate; flowers white to pink; fruit nearly orbicular, 3-4 mm. long, notched at each end, lateral wings broader than the body.

Stream banks, swamps and other wet areas in the woods; west of the Cascades.

Angelica hendersonii Coult. & Rose — Henderson's Angelica

Plants strong-scented; stem hollow, 0.5-2 m. tall; leaves petioled, sheathing at the base, divided into 3's or 5's, then pinnate, leaflets broadly ovate, 4-10 cm. long, usually densely woolly beneath, green and minutely and profusely veined above, 5 cm. or more long, toothed, upper leaves reduced to sheaths; compound umbels terminal and often axillary on long peduncles, compact, densely short white-hairy; involucre none; involucel of a few slender bracts; umbellets head-like; flowers yellowish-white, the pedicels and ovaries densely pubescent; fruit 6-10 mm. long, broadly oblong, lateral wings broad, thick and corky.

(=*Angelica tomentosa* Wats. var. *hendersonii* DiTomaso)

Common along the seashore, especially where beach and bluff meet.

Angelica lucida L. Sea-watch

Stems 1-1.5 m. tall; leaves smooth and shining, leaflets 3-10 cm. long, lanceolate or broadly ovate, toothed or lobed; umbels pubescent, 5-15 cm. broad; involucre absent or much reduced; involucels conspicuous; fruit 4-9 mm. long, oblong, corky-ribbed.

Coastal bluffs and beaches.

Anthriscus

Taprooted annuals with pinnately dissected or compound leaves; inflorescence a few-rayed compound umbel; flowers white; involucre none; involucel of a few reflexed bracts; fruit short-beaked, often covered with hooked bristles.

Anthriscus caucalis M. Bieb. Bur-chervil

Pubescent, rank-smelling annuals; stems erect 2-9 dm. tall; leaves several times pinnate, the ultimate segments toothed; umbels short-peduncled, small, with 3-7 rays; fruits ovoid, about 4 mm. long, short-beaked, covered with short hooked bristles except for the beak.

[=*Anthriscus scandicina* (Weber) Mans.]

A European weed, sparingly introduced, but often locally abundant.

Cicuta

Glabrous rhizomatous perennials with the base of the stem thickened and hollow with cross partitions containing a yellow to brownish sap; leaves pinnately-dissected to compound; flowers in compound umbels; involucre usually lacking; involucels inconspicuous; petals white to greenish; calyx teeth small; fruit ovate to globose, only slightly compressed laterally, with low corky ribs.

Cicuta douglasii

Cicuta douglasii (DC.) Coult. & Rose Western water hemlock

Stems stout, hollow, somewhat branched, striate, often reddish, 0.5-2 m. tall, sometimes reaching 2.5 cm. in diameter, growing from rhizomes 7.5-15 cm. long and often reaching 4 cm. broad, the internodes of the stem base and rhizome greatly shortened, the interior of the structure thus appearing penetrated by low broad chambers separated by thin plates; leaves mostly twice-pinnate, the petioles sheathing, at least at the base, those of the lower leaves long, leaflets 1-15 cm. long, lanceolate, sharply toothed, the lateral veins appearing to end in the notches between the teeth; involucre none or of a few narrow bracts; involucel of several narrow reflexed bracts; rays 2-8 cm. long; fruit 2-4 mm. long, brownish with lighter broad low ribs, roundish, tipped by the 2 short styles and the somewhat conspicuous calyx teeth.

This species is extremely poisonous to both humans and livestock.

Wet areas, including marshes, ditches and stream banks.

Conioselinum

Tall leafy perennials with glabrous ternately decompound leaves; umbels compound; flowers white; calyx teeth obsolete; fruit oblong, dorsally flattened, glabrous, with prominent ribs, some or all of which are winged; stylopodium somewhat conspicuous, conical.

Conioselinum pacificum (Wats.) Coult. & Rose

Stems stout, 2-15 dm. tall, glabrous or sometimes minutely pubescent above; leaves large, somewhat fern-like, much divided, first ternately, then pinnately, the leaflets 1-5 cm. long, narrowly ovate, the margins generally deeply toothed or cleft; umbels 6-30-rayed; involucral bracts few, linear; bracts of the involucel generally conspicuous, sometimes cleft and longer than the flowers; fruit 5-8.5 mm. long, ovate or oblong, dorsal and intermediate ribs conspicuous, but hardly winged, the lateral ribs broadly winged.

Margins of the coastal marshes and on high sea beaches.

Conium

Tall taprooted biennials with red- or purple-spotted hollow stems; leaves pinnately many times divided; flowers white, in compound umbels; involucre and involucel present, of small narrowly ovate bracts; fruit broadly ovate, its wavy margined ribs very prominent when dry.

Conium maculatum L. — Poison hemlock

Tall biennial herbs, usually much branched toward the top, sometimes reaching a height of 3 m. and having a distinct mouse-like odor; stems coarse, green and red- or purplish-mottled; leaves finely dissected into toothed leaflets, the lower leaves petioled and 2-6 dm. long, usually having disappeared at flowering time, the upper leaves sessile and gradually becoming smaller toward top of plant; umbels numerous, compound, 2.5-5 cm. in diameter, rays 8-20, 1-5 cm. long; involucre of several small ovate bracts, more or less lobed at the base and spreading or later turning downward; umbellets about 15-rayed, consisting of perfect and sterile flowers; involucel present, similar to involucre but smaller; flowers very small; calyx-teeth obsolete; petals white, nearly equal, obovate, with an in-turned apex; stamens white or anthers slightly pinkish; stylopodium conspicuous, flattened, consisting of two white, scalloped lobes; styles very short, white; fruits about 3 mm. long, somewhat flattened laterally, conspicuously ribbed, the ribs wavy.

A very poisonous plant. This is the species used by the ancient Greeks for capital punishment.

Road sides, gardens, stream banks and other moist disturbed habitats; introduced from Eurasia.

Daucus

Taprooted annuals or biennials with pinnately dissected decompound leaves; flowers usually white, the petals unequally 2-lobed; umbels compound; involucre and involucel usually present and conspicuous; calyx teeth evident to obsolete; fruits oblong to ovate, ribs bearing bristles and prickles.

Biennials; umbels mostly 5-10 cm. in diameter; central flower of the umbel usually purple .... *D. carota*
Annuals; umbels mostly 2.5 cm. or less in diameter; central flower of the umbel white .... *D. pusillus*

Daucus carota L. — Wild carrot

Biennials from fleshy taproots with a carrot-like odor, the plant represented the first year by a rosette of finely dissected leaves; second year stem branched, erect, 1.5-12 dm. tall, pubescent with stiff hairs; leaves many times divided, the ultimate segments linear or lanceolate; involucre of pinnately parted bracts, the segments long-linear; umbels 5-10 cm. or more broad, the rays turned inward in fruit; flowers small, white (rarely pinkish), generally 1 or several purple flowers in center of each umbel; fruits 3-4 mm. long, widest at the middle, somewhat flattened, the ribs bristly, the bristles mostly barbed.

A very abundant and troublesome weed introduced from Europe. All varieties of garden carrot are considered derived from this species.

Daucus pusillus Michx. — Rattlesnake weed

Differs from *D. carota* in being annual, usually much shorter, the umbel usually 2.5 cm. or less broad, the involucral segments short and the fruit widest below the middle.

Native of North America, common in dry habitats.

Eryngium

Biennials or perennials; leaves often pinnately or palmately dissected and usually spinulose-serrate, young leaves linear and segmented; petioles segmented; inflorescence dense, surrounded by an involucre; calyx lobes obvious; petals white, blue or purple; fruit slightly to strongly compressed, usually covered with scales or tubercules.

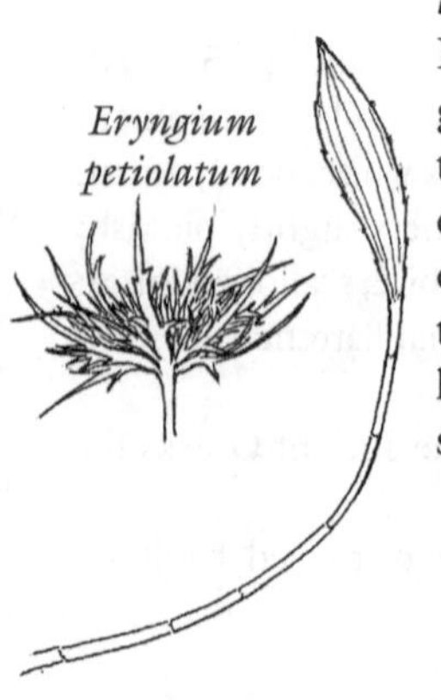
Eryngium petiolatum

Eryngium petiolatum Hook. — Oregon coyote-thistle

Erect perennials; stems sometimes much-branched, grayish-green, 1.5-5 dm. tall; first leaves often reduced to tubular petioles with conspicuous cross walls, bladeless or with small narrow sharply-toothed blades, upper leaves opposite, lanceolate, blades up to 1 cm. wide, nearly entire to spinulose-serrate; inflorescence head-like, the bracts long, conspicuous, stiff, needle-pointed; petals white; fruit scaly, about 2 mm. long.

Vernal pools and other wet areas.

Glehnia

A monotypic genus.

Glehnia littoralis (Gray) Miq. subsp. *leiocarpa* (Mathias) Hulten Beach silver-top

Perennial low-growing herbs with very short aerial stems from a long thick taproot; leaves thick, densely white-woolly except usually on the upper surface of leaflets, ternate then ternate-pinnate, petioles to 2.5-15 cm. long, sheathing at the base, leaflets broad, more or less lobed, 1-7 cm. long, toothed or lobed, conspicuously veined above; peduncle usually much shorter than the leaves, bearing a compound umbel of whitish blossoms; bracts of involucre few or none, of the involucel, several, linear; inflorescence becoming in fruit a compact cluster of 5-15 globose umbels each consisting of globose corky winged fruits, the wings of each fruit usually 10.

(=*Glehnia leiocarpa* Mathias)

On sandy beaches along the coast

Glehnia littoralis

Heracleum

Tall, stout biennials or perennials; leaves large, ternately to pinnately compound, petioles strongly sheathing; flowers small, white, in large flat-topped compound umbels, the outer flowers larger and often sterile; fruits broad, strongly dorsally flattened.

Heracleum lanatum Michx. Cow parsnip

Large perennials; stems 1-3 m. tall, thick, hollow, strongly ribbed, roughened on the ribs, and more or less woolly; leaflets 3, the terminal one rarely again divided, the central leaflet larger than the lateral ones, all lobed and irregularly toothed, 0.5-4 dm. broad, the petioles broadly sheathing, generally densely woolly; umbels 1-4 dm. broad; petals white; fruits 7-12 mm. long, 5-9 mm. wide with several reddish marks running down from the apex.

Moist, usually shady habitats, often along streams.

Heracleum mantegazzianum Somm. & Levier, a coarser and taller species has apparently escaped from cultivation and become established on the Olympic Peninsula.

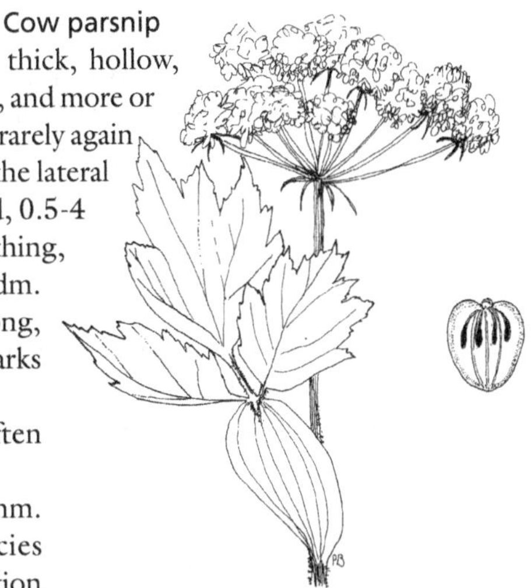

Heracleum lanatum

Hydrocotyle

Perennial herbs with creeping stems giving rise to leaves and peduncles; leaf blades more or less orbicular to reniform, often peltate, long-petioled; calyx lobes minute; petals white, yellowish, purple or greenish; fruit nearly round to slightly laterally compressed, the dorsal ribs often forming prominent margins, the lateral ribs narrow.

Hydrocotyle ranunculoides L. — Water pennywort

Glabrous perennials with floating or creeping stems, rooting and sending up leaves and flowers at the nodes; leaves long-petioled, the blades 1.5-6.5 cm. wide, orbicular-reniform with a deep basal sinus, with 3-7 rounded lobes; peduncles much shorter than the leaves, erect in flower, turned downward in fruit; umbels simple, few-rayed; fruit nearly round, 1-3 mm. long and 2-3 mm. wide, the ribs obscure.

Marshes, lake margins, ponds and other low wet areas

Ligusticum

Taprooted perennials; leaves ternately to pinnately compound, leaflets entire to deeply pinnately lobed; flowers in large compound umbels; calyx lobes evident or obsolete; petals white to pinkish or purple; fruit nearly terete or oblong, slightly laterally compressed, the ribs prominent, sometimes winged.

Ribs of the fruit wingless; lowland plants; flowers white.............. *L. apiifolium*
Ribs of the fruit wing-like; mountain plants; flowers usually pinkish.... *L. grayi*

Ligusticum apiifolium (Nutt.) Gray — Celery-leaved lovage

Stems 0.5-1.5 m. or more tall, slender, few-leaved, whole plant glabrous except for the inflorescence and the leaf margins; leaves at first several times ternate, then pinnate, the leaflets ovate, deeply and narrowly lobed or cleft; rays of umbel many, unequal; bracts of involucel linear or obsolete; fruits broadly oblong, 3-5.5 mm. long, the ribs conspicuous, but not winged.

Meadows and open woods and road sides at low elevations.

Ligusticum grayi Coult. & Rose — Purple lovage

Stems slender, 2-8 dm. tall, nearly leafless above, glabrous; leaves mostly basal, once or twice ternately divided, then pinnate; leaflets ovate, 1-3.5 cm. long, toothed and more or less lobed; umbels few-rayed, the rays unequal; bracts of involucel few, linear-lanceolate; flowers usually pinkish; fruits oblong, 4-6 mm. with narrow wing-like ribs.

Subalpine meadows and open or wooded slopes in the Cascades.

Lilaeopsis

Glabrous rhizomatous perennials; leaves arising directly from the rhizome, narrow, elongate, hollow and segmented, without a defined blade; umbels simple, borne in the leaf axils; calyx teeth small; petals white or pinkish-tinged; fruit glabrous, subglobose, the lateral ribs thick, the dorsal slender or all the ribs alike, stylopodium flat.

Lilaeopsis occidentalis Coult. & Rose
Rhizomes prostrate, creeping; leaves solitary or arising in tufts along the rhizome, 2.5-30 cm. long, 0.8-4.5 mm. wide, hollow but segmented; axillary peduncles usually recurved, shorter than the leaves, 5-15-flowered; fruit 1-2.5 mm. long.

Salt marshes, muddy lake shores or wet brackish areas near the coast and along the Columbia River.

Lomatium

Perennials from a taproot or a thickened and tuber-like base; leaves several times divided, often mostly basal; umbels compound; calyx lobes minute or obsolete; fruits dorsally flattened, the lateral ribs often with thin wings. Some species inclined to be wholly or partly dioecious.

1a Leaves decompound, dissected into numerous small divisions
 2a Herbage gray from the fine pubescence .... *L. macrocarpum*
 2b Herbage glabrous or sparingly pubescent, but not gray
  3a Involucels of filiform bracts; plants 4-18 dm. tall ....*L. dissectum*
  3b Involucels usually of short broad dissected or toothed bractlets; plants 6.5 dm. or usually much shorter
   4a Wings of the fruit thin; common .... *L. utriculatum*
   4b Wings of the fruit corky-thickened; uncommon .....*L. bradshawii*
1b Leaves with several large divisions forming more or less definite leaflets
 5a Peduncles conspicuously swollen below the umbel............*L. nudicaule*
 5b Peduncles not swollen
  6a Leaflets narrowly linear, entire or shallowly toothed .. *L. triternatum*
  6b Leaflets ovate to oblong in outline, variously cleft or parted
   7a Wings of the fruit nearly as wide as the body ....
   .... *L. martindalei* var. *martindalei*
   7b Wings of the fruit only about 1/2 as wide as the body or less
    8a Fruit 8-15 mm. long; involucels absent or inconspicuous.........
    .... *L. martindalei* (in part)
    8b Fruit 5-9 mm. long; involucels present.......................*L. hallii*

Lomatium bradshawii (Rose) Math. & Const. Bradshaw's Lomatium
Taprooted perennials; stems erect, 2-6.5 dm. tall; leaves mostly all basal, ternately-pinnately dissected into narrow segments; flowers pale yellow in compound umbels, the rays of the umbel up to 12 cm. long, usually unequal; bracts of the involucel short and broad and dissected into 2-3 segments, these again lobed and toothed; fruit 5-7 mm. long, with thick corky wings about the width of the body.

Uncommon in moist meadows and prairies in the Willamette Valley.

Lomatium dissectum (Nutt. in Torr. & Gray) Math. & Const. var. *dissectum* Purple parsley
Stems thick, finely ribbed, hollow, 4-18 dm. tall; leaves with sheathing petioles, the blades many times divided into narrow ultimate segments, minutely pubescent to glabrous and glaucous; rays of umbels 10-30, subtended by several narrow bracts, each ray 3-10 cm. long, ending in a compact pom-

pom about 1 cm. in diameter, of minute dark reddish-purple or rarely yellow flowers, subtended by a few bracts; fruits woody or corky, 1-1.5 cm. long, with thick corky lateral wings.

Common along the road sides and wooded areas from the valleys to the mountains.

Lomatium hallii (Wats.) Coult. & Rose — Hall's Lomatium

Stems 2-4 dm. tall; leaves mostly basal, pinnately to ternately-pinnately dissected, the segments deeply pinnatifid or toothed; flowers yellow; fruits elliptic, glabrous, 5-9 mm. long, the wings about 1/2 as broad as the body.

Usually in rocky areas on the west slope of the Cascades.

Lomatium macrocarpum (Nutt. in Torr. & Gray) Coult. & Rose — Large-fruited Lomatium

Stems several, stout, 1-5 dm. tall; whole plant more or less gray-green from the fine pubescence; leaves all at or near the base, once or twice ternate then pinnate, the segments linear, about 2-10 mm. long; umbels stout-rayed, rays 1-8 cm. long; involucre none or of a few bracts; involucel of conspicuous slender bracts principally longer than the pedicels; flowers small, white or slightly yellowish or purplish; fruits variable, 1-2 cm. long, oblong to elliptic with wings usually narrower than the body.

Willamette Valley, not abundant.

Lomatium martindalei Coult. & Rose — Few-leaved Lomatium

Stems generally several, 1-4 dm. tall, bearing 1 or 2 cauline leaves; foliage usually glabrous and somewhat glaucous; leaves thick, often twice pinnate, the ultimate segments toothed or pinnatifid; flowers white to yellowish-white or rarely pure yellow; fruits glabrous, 8-15 mm. long, the wings as broad as the body of the fruit in var. *martindalei* or both the fruits and the wings more narrow in var. *angustatum* Coult. & Rose in which the leaves are generally only once pinnate with the segments deeply pinnatifid or toothed. Variety *flavum* (G.N. Jones) Cronq. of the Olympic Mountains has yellow flowers and narrow fruits.

Rocky places in the Cascades and high peaks of the Coast Range.

Lomatium nudicaule (Pursh) Coult. & Rose — Naked-stemmed hogfennel or Pestle parsnip

Plants 2-8 dm. tall; foliage glabrous and glaucous; leaves all basal, divided first in 3's, each branch pinnately compound with 5 or more (rarely fewer) leaflets, the leaflets oblong to lanceolate, 2-9 cm. long; main rays of umbel 10-20, borne on enlarged summit of stem, rays unequal in length, 2-20 cm. long, ending in dense clusters of small yellow flowers on short thread-like pedicels; fruits reaching 1.4 cm. in length, oblong, or elliptic, with thin lateral wings.

Low hillsides, open or wooded areas, often in gravelly soil.

Lomatium triternatum (Pursh) Coult. & Rose Hogfennel

Plants 1.5-8 dm. tall, finely puberulent throughout or the leaves nearly glabrous; leaves mostly basal, several times divided in 3's, then pinnate, the leaflets narrowly linear; rays of umbel 5-30, unequal in length, 1-10 cm. long in same umbel, without subtending bracts, rays ending in head-like clusters of small yellow flowers; bractlets lanceolate; fruits oblong to elliptic, 7-20 mm. long, winged.

Meadows and open areas, low hillsides to mid elevations in the mountains.

Lomatium utriculatum (Nutt. in Torr. & Gray) Coult. & Rose Spring gold

Stems 0.3-6 dm. tall, erect, or in early spring their stems very short, scarcely rising above the ground; leaves 3 or 4 times divided into very small narrow segments, the petioles conspicuously inflated and veined, often reddish; umbels 2-4.5 cm. in diameter, nearly flat-topped, rays 5-7 (sometimes more), without subtending bracts, the rays unequal in length, ending in head-like clusters of small yellow flowers subtended at the base of their short pedicels by several short broad bractlets; fruits 5-14 mm. long, elliptic, brown and yellowish striped, the wings broad, thin, and yellowish.

Hillsides. Among our earliest appearing spring flowers; flowering when the plants are scarcely above the ground surface, later plants are taller, and are found flowering throughout the spring.

Lomatium utriculatum

Oenanthe

Glabrous perennials; leaves pinnately-compound or dissected, the leaflets toothed, with veins running through the center of each tooth to its tip; inflorescence a compound umbel, outer flowers of the umbel sometimes staminate; calyx teeth evident; petals white or reddish; fruit terete or laterally compressed, the ribs corky.

Oenanthe sarmentosa Presl ex DC. Wild celery or Water parsley

Plants growing mostly in shallow water; rhizomatous; stems freely branching, 4.5-15 dm. long, decumbent to ascending; leaves mostly 2-pinnate, or at first ternate then pinnate, the leaflets 1-6 cm. long, ovate, coarsely toothed or sometimes somewhat lobed or cleft, the midrib of each extending through center of leaflets to the tip; flowers white; fruit 2.5-3.5 mm. long, ovoid-cylindrical, with long styles.

Common in marshes, stream banks, ditches and other wet habitats.

This plant, because of its similar habitat, is often confused with the poisonous water hemlock (**Cicuta**). However, the two can be separated by the shape of fruit, length of style (very short in **Cicuta**), and particularly by venation of the leaflets.

Orogenia

Dwarf glabrous and glaucous perennials with fleshy tuberous roots and ternately once or twice compound leaves with linear segments; flowers white in umbels of unequal rays; fruit oblong or ovate, nearly terete or only slightly compressed laterally, the lateral ribs much thickened and corky, the dorsal slender or obscure; stylopodium flattened.

Orogenia fusiformis Wats.

Stems 5-15 cm. tall arising from elongated tuberous roots; leaves all basal, the outermost reduced to sheaths, the others once or twice ternate into linear segments, these entire or narrowly cleft; umbels 2-8-rayed, umbellets head-like; flowers white; fruit about 3-4 mm. long, lateral ribs thickened into inflexed wings, the dorsal ribs obscure.

Open, often gravelly areas from the valleys to mid elevations in the mountains.

Osmorhiza

Perennials from thickened aromatic roots; leaves mostly basal, ternately decompound; flowers white, greenish-yellow or purple, in compound umbels; calyx teeth obsolete; fruit linear to oblong or club-shaped.

1a Fruits glabrous; flowers yellow .... *O. occidentalis*
1b Fruits bristly; flowers white, purple or greenish
  2a Flowers purple or rarely greenish; mostly found in the mountains ....
  .... *O. purpurea*
  2b Flowers white; a widespread species .... *O. berteroi*

Osmorhiza berteroi DC. — Common sweet cicely

Stems erect, branched above, 2-10 dm. tall, nearly glabrous or with some spreading white hairs; leaves twice-ternate, the lobes ovate, coarsely and rather regularly toothed, with a few hairs; umbels 2-10-rayed, the rays slender, long; fruit 1-2.5 cm. long, beaked, long-tapered at base, bristly, particularly below, borne on pedicels of equal or greater length.

(=*Osmorhiza chilensis* Hook. & Arn.)

Common in open woods and in disturbed habitats.

Osmorhiza occidentalis (Nutt.) Torr. — Smooth-fruited sweet cicely

Stems branched, 2-12 dm. tall, generally pubescent at first, often becoming nearly glabrous; leaves several times ternate, leaflets ovate, coarsely toothed and lobed, upper sheaths veiny and generally long-ciliate; rays of umbel 4-12, often varying from 1-9 cm. long in the same umbel; fruits glabrous, 1-2 cm., narrowed at base, very short-beaked, truncate at apex.

In the mountains but also in the Willamette Valley.

Osmorhiza purpurea (Coult. & Rose) Suksd. — Purple sweet cicely

Stems 1-6 dm. tall, nearly glabrous; leaves once or twice ternate, leaflets 2-7 cm. long, mostly long-ovate, the margin deeply and irregularly cut; rays of umbel 2-6, 3-10 cm. long; fruit 8-15 mm. long, long-tapered at base, short-beaked, ribs bristly at base.

Usually in moist habitats in the Cascade and Olympic Mountains.

Perideridia

Slender glabrous biennials or perennials from fibrous or tuberous roots; leaves once or twice ternately to pinnately compound, the leaflets linear; flowers white or pink, borne in compound umbels; calyx teeth present; stylopodium conical; fruit globose to oblong sometimes slightly laterally compressed, the ribs unequal, not winged.

Fruits nearly globose; rays of umbel usually 7-16.......................... *P. gairdneri*
Fruits oblong; rays of umbel usually 10-30.................................... *P. oregana*

Perideridia gairdneri (Hook. & Arn.) Math. Wild false caraway
Roots spindle-shaped, often clustered; stems 3-14 dm. tall, slender, few-leaved, simple or branched above; uppermost leaves simple, all other leaves once or rarely twice pinnate, leaflets usually 3-7, those of the upper leaves long-linear, of the lower leaves narrowly oblong, the leaves often dried by the flowering season; involucels of 8-13 linear bracts, rays of umbel usually 7-16; flowers white; fruit 1.5-3.5 mm. long, nearly globose to slightly flattened, styles 1/2 as long as fruit.

In moist low ground, often in open woods and meadows. The roots were eaten by the Native Americans.

Perideridia oregana (Wats.) Math. Oregon yampah
Resembling *P. gairdneri* but shorter, usually less than 7 dm. in height and rays of the umbel usually 10-30 and the fruits oblong, 2.5-6 mm. long.

Open woods and meadows.

Sanicula

Biennials to perennials, often with mostly basal leaves, these pinnately or palmately cleft or compound; flowers both perfect and staminate in the same inflorescence or all staminate, white, greenish, yellow or purple, in irregularly umbellate heads; involucre of toothed bracts; involucel bracts usually small, entire; calyx teeth evident; fruit not ribbed, covered with prickles, scales or bristles.

1a Prostrate yellow-green plants of bluffs or dunes along the coast.................
....*S. arctopoides*
1b Plants not prostrate, foliage green or purplish
  2a Leaves pinnately divided.. *S. bipinnatifida*
  2b Leaves mostly ternately lobed or divided
    3a Lower leaf blades deeply palmately 3-5-lobed or cleft.. *S. crassicaulis*
    3b Lower leaves ternately compound.............................. *S. graveolens*

Sanicula arctopoides Hook. & Arn. Footsteps-of-spring or Yellow mats
Plant growing low against the ground, more or less rosette-like, 1-3 dm. broad, stems and foliage yellowish; stems very short, bearing several spreading branches, each ending in a few-rayed umbel or a single umbellet; leaves palmately cleft into 3 segments, these lobed, the lobes all sharply toothed and cut, or the lowermost leaves merely lobed with toothed margins; involucre of several toothed leaf-like bracts; involucel of conspicuous entire bracts; flowers

yellow, about 1/2 perfect and 1/2 staminate; fruit 2-5 mm. long, bristly above, the bristles hooked at the tip.

On exposed bluffs along the coast.

Sanicula bipinnatifida Dougl. ex Hook. Purple snake-root
Stems erect, 1.5-6 dm. tall, naked above, the whole plant green or sometimes purplish; basal leaves several, long-petioled, the blade 3.5-20 cm. long, pinnately several lobed or divided, the 2 basal segments also generally again lobed, the lobes all narrowly and deeply toothed, the teeth mucronate and continuing down the rachis; cauline leaves few, shorter-petioled; umbels several rayed, the rays ending in purple or rarely yellow globose heads; bracts of involucre more or less leaf-like, toothed, those of involucel somewhat hidden, entire or slightly toothed, the rays of the umbel often with a pair of narrow bracts; fruits covered with slender hooked bristles.

Dry ground, common.

Sanicula crassicaulis Poepp. ex DC. Snake-root
Stems erect, 1-12 dm. tall, widely branching above, glabrous; basal leaves several, long-petioled, the blades 3-12 cm. nearly orbicular in outline, cordate, deeply 3-5-lobed, the lobes toothed, the teeth ending in bristles; cauline leaves few, somewhat similar but shorter-petioled, often more deeply divided, the lobes narrower; flowers very small, yellow or rarely purple, in irregular umbels; bracts of involucre more or less leaf-like, those of involucel generally entire; fruits usually short-stalked, covered with hooked bristles.

Rather common in open woods and in waste places.

Sanicula graveolens Poepp. ex DC. Sierra snake-root
Stems slender, erect, 5 dm. or less tall; basal leaf blades orbicular in outline, 1.5-4 cm. long, with 3 leaflets, these generally again deeply 3-5-cleft or lobed, toothed; cauline leaves longer; umbels scattered along the stem, generally of only a few slender rays, ending in small pale yellow flower clusters; staminate flowers more numerous than the pistillate; involucre of small somewhat leaf-like bracts; involucel bracts inconspicuous, more or less united; fruits covered with stout curved prickles.

Prairies, open woods from the valleys to the low mountains.

Scandix

Slender annuals with much-dissected leaves and white flowers in compound umbels; both sterile and fertile flowers in the same umbel, the staminate with stamens and a green disk, the fertile either perfect or pistillate with long styles and a purple disk; rays few, involucre absent or of one bract, umbellets of several rays and with several bracts; fruit slender with a long narrow beak, the body minutely warty.

Scandix pecten-veneris

Scandix pecten-veneris L. Shepherd's-needle or Venus'-comb

Taprooted annuals; leaves several times pinnately dissected into fine linear segments; flowers white in few-rayed compound umbels, umbellets compact; petals white; body of the fruit 5-15 mm. long, the beak 2-7 cm. long.

An introduced species, more or less weedy.

Sium

Perennials; leaves pinnately-compound, serrate; inflorescence a compound umbel; calyx lobes obsolete; petals white; fruit glabrous, nearly round to slightly laterally compressed, ribs corky, the intervals reddish brown.

Sium suave Walt. Hemlock water-parsnip

Glabrous perennials; stems 4.5-18 dm. tall, conspicuously ribbed; lower leaves on long-winged petioles, reduced upwards, once-pinnate with 7-13 leaflets, these serrate or less often, pinnately-lobed, leaflets 2-9 cm. long, 1.5-20 mm. wide; bracts of involucre and involucel linear, reflexed; umbel rays many; petals white; fruit 2-3 mm. long, nearly round.

Usually in shallow water, or very wet ground in swamps, marshes, stream banks and ponds.

Torilis

Hispid taprooted annuals; leaves pinnately dissected; involucre and involucel usually of slender bracts, or involucre sometimes wanting; both perfect and staminate flowers present in most umbels; petals white or reddish; fruit bristly, the bristles often rough-surfaced and barbed at the tip.

Torilis arvensis

Torilis arvensis (Huds.) Link

Hispid annuals 1.5-9 dm. tall; stems slender, branching above the base, often purplish, slightly ridged; leaves once- or twice-pinnate, the leaflets sharply toothed, petioles flattened and sheath-like but sheathing the stem only at the bases; umbels compound, long-peduncled, arising from the stems opposite the leaves; involucre generally consisting of one linear bract, rarely several linear bracts or some of these reduced to minute scales; rays of umbel 3-10, 1-2.5 cm. long; involucel of several slender bracts; umbellet compact, a few sterile flowers scattered among the fertile; calyx-lobes slender; petals white, 2-lobed with an incurved tooth between, pubescent on the back; stamens shorter than the petals; fruit 3 mm. long, elliptic-oblong, densely covered with purplish (later whitish) roughened barbed bristles.

Sparingly introduced, probably by way of Europe, from its probable original home in South Africa.

Other species of **Torilis** may occasionally enter our area.

CORNACEAE
Dogwood Family

Trees or shrubs, or low perennial herbs; leaves opposite, alternate or whorled, simple; flowers small, in cymes or head-like glomerules; calyx united with the ovary, the 4-5 teeth small or obsolete; petals 4, rarely 5, or occasionally lacking; stamens 4; ovary 1-4-loculed; fruit fleshy, a drupe or a berry.

Cornus

Perennial herbs, shrubs or trees; leaves simple, exstipulate, usually opposite, rarely alternate or whorled; inflorescence sometimes surrounded by large bracts, giving the appearance of a single flower with white petals; flowers usually small, sepals, petals and stamens usually 4; pistil 1, compound; ovary inferior; fruit a 2-seeded drupe.

1a Flowers not surrounded by white bracts; shrubs .... *C. sericea*
1b Flowers surrounded by white or pinkish bracts
  2a Trees .... *C. nuttallii*
  2b Low herbs 7.5-20 cm. tall .... *C. unalaschkensis*

Cornus nuttallii Aud. — Western flowering or Common dogwood

Trees 3-15 m. tall; leaves 5-14 cm. long, narrowly to broadly elliptic or obovate, tapering to the petiole, generally abruptly acute at apex; flowers in a compact cluster surrounded by 4-6 large conspicuous petal-like bracts, the bracts white or cream-colored, velvety, 4-7 cm. long, broadest at the center, abruptly acute; fruits scarlet, 15 mm. or less long.

Common throughout forests in the Cascades and westward.

Flowers appear with or before the leaves and often again in autumn with the fruits. An attractive tree, the foliage turning deep red in the autumn.

Cornus sericea L. — Creek dogwood

Shrubs 1-6 m. tall, young branches reddish turning grayish-green; leaf blades 4-12 cm. long, ovate usually acuminate, paler and often pubescent at least on the under surface; inflorescence a cyme; flowers numerous; petals white or

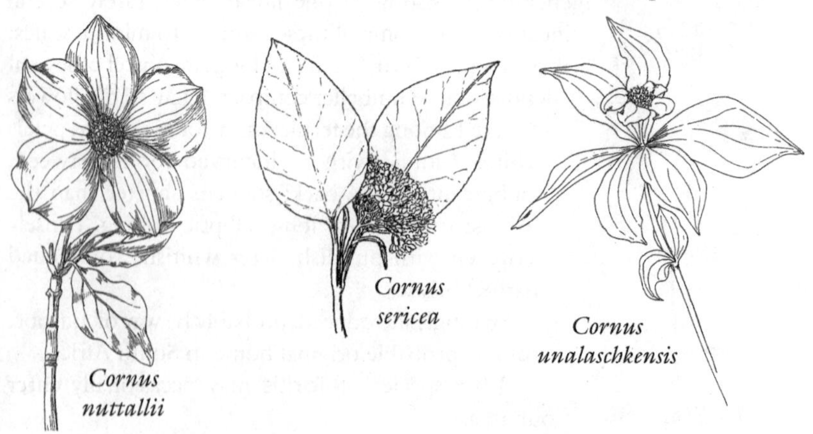

Cornus nuttallii — *Cornus sericea* — *Cornus unalaschkensis*

cream-colored, 2-4.5 mm. long; drupe 7-9 mm. long, white to bluish-tinged. Two subspecies ocurr in our area: subsp. *occidentalis* (Torr. & Gray) Fosb. with leaves that are densely pubescent beneath and petals 3-4.5 mm. long, and subsp. *sericea* with leaves glabrous to sparingly strigose beneath and petals 2-3 mm. long.

[=*Cornus stolonifera* Michx. and *Cornus occidentalis* (Torr. & Gray) Cov.]

Moist habitats, often along streams.

Cornus unalaschkensis Ledeb. Bunchberry

Rhizomatous perennials 7.5-20 cm. tall from slender creeping rhizomes; leaves 4 -7 in a whorl at summit of aerial stem, several pairs of reduced leaves below, leaves near base of plant scale-like, whorled leaves obovate, 2.5-8 cm. long, conspicuously veined; flower cluster surrounded by 4 broad white or pinkish-tinged bracts; tips of the flower petals pinkish; fruits clustered, scarlet.

(*Cornus canadensis* L. misapplied)

Margins of bogs and marshes and in moist forests.

ERICACEAE
Heath Family

Trees, shrubs, or perennial herbs; leaves simple, mostly alternate, often evergreen, and leathery; flowers regular, usually more or less bell-shaped, sepals and petals in 5's or 4's; petals usually united, rarely lacking; stamens 4-12, not attached to the corolla, or only slightly so at the base; anthers dehiscent by pores or slits, often awned; ovary superior or inferior; fruit a capsule, berry or a drupe. In the subfamily Monotropoideae the plants are fleshy, non-green saprophytes and the leaves are reduced to bracts.

1a Plants herbaceous; stamens 8-12
- 2a Plants without green leaves, leaves reduced to scales and nongreen; saprophytes or parasites (see also *Pyrola picta*)
  - 3a Petals united, forming a bell-shaped or urn-shaped corolla
    - 4a Plants white or pinkish; anthers not awned ....**Hemitomes**
    - 4b Plants reddish-brown; anthers awned .... **Pterospora**
  - 3b Petals free from each other or nearly so, or none
    - 5a Flowers in a dense spike-like raceme
      - 6a Plant white to brownish; both sepals and petals present ....**Pleuricospora**
      - 6b Stems red and white striped; only 1 set of perianth parts present ....**Allotropa**
    - 5b Flowers solitary or in a 1-sided or recurved raceme
      - 7a Flower solitary at end of the stem; plants white, drying black... ....**Monotropa**
      - 7b Flowers in a 1-sided raceme; plants yellowish or reddish. ....**Hypopitys**

2b Plants with green leaves (except *Pyrola picta* in some cases)
8a Flowers in corymbs or umbels; filaments dilated at middle or at the base........ **Chimaphila**
8b Flowers solitary or in racemes; filaments not dilated
9a Flowers solitary........ **Moneses**
9b Flowers not solitary
10a Flowers in 1-sided racemes; petals with tubercules.... **Orthilia**
10b Racemes not 1-sided; petals without tubercules.......... **Pyrola**
1b Plants woody, trees or shrubs (some prostrate or vine-like); stamens 4-12
11a Ovary inferior or appearing so because of the fleshy persistent calyx
12a Ovary truly inferior; fruit a berry........ **Vaccinium**
12b Ovary only appearing inferior due to the fleshy persistent calyx which encloses the fruit........ **Gaultheria**
11b Ovary superior
13a Petals free from each other or nearly so
14a Leaves deciduous; flowers copper-red, axillary, usually solitary........ **Cladothamnus**
14b Leaves evergreen; flowers white, in terminal corymbs..... **Ledum**
13b Petals united into a toothed, lobed, or cleft corolla
15a Corolla nearly cup-shaped or rotate containing 10 "pockets" which hold the young anthers........ **Kalmia**
15b Corolla without "pockets" which hold the young anthers
16a Flowers showy, 1.5-4 cm., or more, wide..... **Rhododendron**
16b Flowers smaller
17a Leaves deciduous; flower parts mostly in 4's .... **Menziesia**
17b Evergreen trees or shrubs (some low); flower parts generally in 5's (rarely 4's)
18a Leaves up to 5 mm. broad; fruit a dry capsule
19a Anthers awned at the tip........ **Cassiope**
19b Anthers not awned........ **Phyllodoce**
18b Leaves well over 5 mm. broad; fruit berry-like
20a Trees (ours) with thin peeling bark............ **Arbutus**
20b Shrubs, sometimes prostrate or creeping
21a Sepals united, becoming fleshy and enclosing the fruit........ **Gaultheria**
21b Sepals free, not becoming fleshy........ **Arctostaphylos**

Allotropa

A monotypic genus.

Allotropa virgata Torr. & Gray ex Gray — Sugar stick

Fleshy saprophytic perennials 1-4.5 dm. tall with red and white striped stems; leaves reduced and scale-like, linear-lanceolate, yellowish-brown to pinkish; flowers in a spike-like raceme; only 1 set of 5 perianth parts present, variously interpreted as sepals or petals; stamens 10 or sometimes only 8; ovary superior, 5-loculed; stigma flattened, 5-lobed; fruit a loculicidal capsule.

Coniferous forests.

Arbutus

Evergreen trees or shrubs; leaves leathery, shining; flowers in a large dense panicle; corolla urn-shaped, 5-toothed, much larger than the 5-parted calyx; stamens 10, shorter than corolla, filaments pubescent, anthers bearing awns; ovary superior, 5-loculed, borne on a disk; fruit a berry.

Arbutus menziesii

Arbutus menziesii Pursh   Pacific madrone

Trees reaching heights of 30 m. or more; bark reddish or golden, shining, the old bark peeling in long strips; leaves ovate or oblong, dark green and glossy above, lighter beneath, smooth, minutely but sharply toothed on margin in first year shoots, but leaves of older growth generally entire, 7.5-15 cm. long; corolla waxy white, inflated, with 5 minute teeth curved back from apex and with 10 translucent glands at the base; berries subglobose, granular, orange to scarlet, 8-13 mm. in diameter.

Wooded areas.

An attractive tree with beautiful flowers and fruit.

Arctostaphylos

Evergreen shrubs with dark red or red-brown smooth and polished bark; leaves simple, generally entire, often standing edge upward by the twisting of the petiole; flower parts generally in 5's; sepals not united; flowers borne in racemes or panicles; corollas urn-shaped, white or pink; stamens 10, each anther bearing a pair of awns and opening by pores; ovary situated on a disk, 2-10-loculed; fruit fleshy, berry-like.

1a Erect shrubs
  2a Young stems pubescent with long hairs; leaves gray-green .... *A. columbiana*
  2b Stems without long hairs; leaves yellow-green........................ *A. patula*
1b Prostrate, decumbent or creeping shrubs, often rooting at the nodes
  3a Leaves obtuse or retuse at apex; corolla mostly pinkish; fruit scarlet........ .... *A. uva-ursi*
  3b Leaves apiculate at apex; corolla white or pinkish; fruit brownish-red..... .... *A. nevadensis*

Arctostaphylos columbiana Piper   Hairy manzanita

Shrubs 0.5-3 m. tall, rarely taller, much-branched above, the young twigs densely covered with long and short often glandular hairs; leaf blades 2-6 cm. long, gray-green, finely puberulent, especially on the under surface, oblong to oblong-ovate, acute at apex, obtuse at base, entire or with very minute teeth; inflorescence densely pubescent; flowers white or pinkish in short compact clusters; fruit about 1 cm. in diameter.

Often in gravelly soil, from the coast to the west slope of the Cascades.

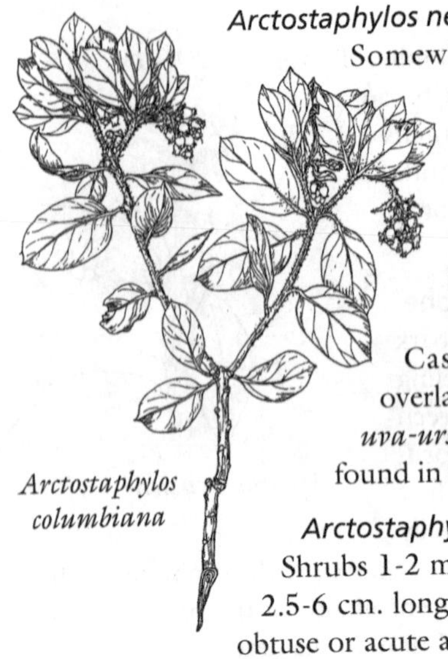

Arctostaphylos nevadensis Gray Pinemat manzanita

Somewhat resembling *A. uva-ursi*, but stems decumbent, not trailing, often forming dense low cushions rather than mats; leaves thick, ovate, oval, or somewhat obovate, distinctly apiculate; flowers white or pinkish, few to many in a raceme; fruit globose, brownish-red.

Often in rocky areas in the forests of the Cascades, beginning at low altitudes, and overlapping at its borders the area covered by *A. uva-ursi* and *A. columbiana*. Hybrids are frequently found in such situations.

Arctostaphylos columbiana

Arctostaphylos patula Greene Green-leaf manzanita

Shrubs 1-2 m. tall; older bark reddish-brown; leaf blades 2.5-6 cm. long, broadly ovate to nearly orbicular, the apex obtuse or acute and usually apiculate; inflorescence a panicle; flowers mostly pink; fruit glabrous, 1 cm. or less in diameter, brownish-orange.

Open forested areas and road sides, mostly south of our range.

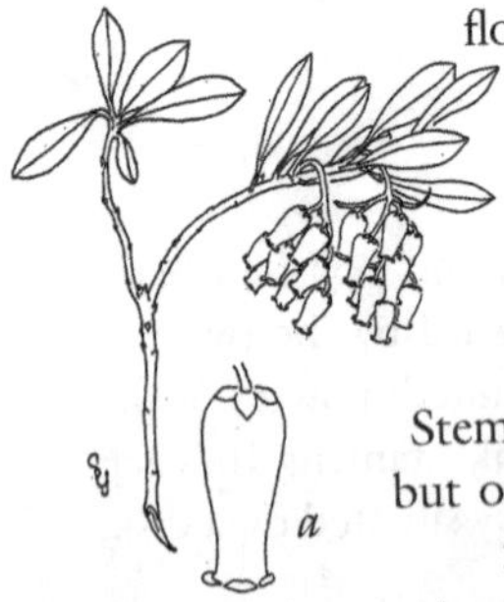

Arctostaphylos uva-ursi Spreng. Kinnikinnic or Bearberry

Stems to 0.5 m. long, prostrate, trailing, forming dense mats, but only 7.5-15 cm. high; bark dark reddish-brown, peeling from the stems in large flakes; leaves 1-3 cm. long, generally obovate, mostly obtuse at apex, netted veins conspicuous, smooth or finely pubescent when young; flowers pinkish, in short racemes, slightly fragrant; corolla 6 mm. long; fruit scarlet, smooth, 6-12 mm. in diameter.

On the coast and in the mountains.

Arctostaphylos nevadensis
a. flower

Cassiope

Low, branched evergreen shrubs; leaves opposite or alternate, appressed or spreading; flowers small, solitary at apices of shoots, or several and axillary; corolla pink or white, the small lobes 5 or rarely 4; stamens 10 or rarely 8, the anthers awned and opening by pores; fruit a subglobose 5- (rarely 4-) valved capsule.

Leaves alternate, spreading, short-petioled .... *C. stelleriana*
Leaves opposite, scale-like, appressed, sessile .... *C. mertensiana*

Cassiope mertensiana (Bong.) G. Don White heather

Tufted freely branched evergreen shrubs with small overlapping scale-like leaves; flowers small, axillary, nodding, borne near the tips of branches; corolla white, the lobes 4-5; capsule valves splitting at the tips.

Spreading over large areas at timberline in the mountains.

Cassiope stelleriana (Pall.) DC. Moss heather

Low, mat-forming shrubs, very leafy; leaves small, persistent, linear to narrowly oblanceolate; flowers usually terminal and solitary, white to pink; anthers long-awned.

Mount Rainier near timberline, and northward.

Chimaphila

Low semi-herbaceous evergreens; leaves alternate, or irregularly opposite to whorled, leathery, toothed, short-petioled; flowers in terminal umbel-like corymbs; calyx 5-lobed; petals 5, spreading; stamens 10; ovary superior, style short, stigma disk-shaped with radiating lobes; capsule 5-loculed.

Leaves usually oblanceolate; flowers 3-15 .... *C. umbellata*
Leaves lanceolate to elliptic; flowers 1-3 .... *C. menziesii*

Chimaphila menziesii (D. Don) Spreng Little prince's-pine

Stems 8.5-15 cm. tall, often branched; leaves lanceolate to elliptic, the margin sharply toothed or entire, 1-5 cm. long, short-petioled; flowers white or pinkish, 1-3 on terminal and sometimes lateral peduncles; filaments flattened at the base and pubescent.

Coniferous forests in the mountains.

Chimaphila umbellata (L.) Bart. Prince's-pine or Pipsissewa

Stems 1-3 dm. tall, generally simple, from creeping rhizomes; leaves leathery, short petioled, 3-7.5 cm. long, usually oblanceolate, sharply serrate, especially toward the apex, borne in several false whorls; flowers pinkish, 3-15 in a terminal corymb; filaments flattened and ciliate at the base.

Coniferous woods.

Cladothamnus

A monotypic genus.

Cladothamnus pyrolaeflorus Bong. Copper-bush

Erect deciduous shrubs 0.5-3 m. tall; leaves alternate, 1.5-5 cm. long, oblong-oblanceolate, thin, nearly glabrous, paler beneath, margin entire, veins few, somewhat conspicuous, petioles very short; flowers axillary, usually solitary; petals free, narrow, 1-1.5 cm. long, copper-colored; ovary superior, becoming a somewhat depressed globose capsule.

Moist areas, especially along stream banks; Saddle Mountain, Oregon and northward.

Gaultheria

Erect to prostrate subshrubs or shrubs; leaves alternate, evergreen, leathery; inflorescence a raceme or the flowers solitary and axillary; calyx usually 5-lobed; corolla urn-shaped or bell-shaped; stamens 5-10; ovary superior, but appearing inferior in fruit; fruit a capsule enclosed in the fleshy calyx forming a pseudo-berry.

1a Flowers in racemes; shrubs; leaves 3-10 cm. long, 2.5-5 cm. wide *G. shallon*
1b Flowers solitary in the leaf axils; subshrubs; leaves 1-4 cm. long, 0.5-3 cm. wide
2a Stems densely brown-pilose; calyx glandular-pubescent..... *G. ovatifolia*
2b Stems glabrous to merely puberulent; calyx glabrous......... *G. humifusa*

Gaultheria humifusa (Grah.) Rydb. Alpine wintergreen

Trailing subshrubs up to 2 dm. long, but rarely more than 5 cm. in height; herbage glabrous or puberulent; leaves thin, broadly ovate to elliptic, 1-2.5 cm. long, 0.5-1.5 cm. wide, entire to finely serrulate; flowers solitary in the leaf axils; corolla 3-4 mm. long, bell-shaped, pinkish; anthers unawned; fruit red, 5-7 mm. wide, spicy.

In wet to moist coniferous forests in the mountains at 1,600 m. and above.

Gaultheria ovatifolia Gray Oregon wintergreen

Low spreading subshrubs, branches to 3.5 dm. long, densely brown-pilose; leaves ovate, 1.5-4 cm. long, 1-3 cm. wide, margins serrulate, tips acute; flowers solitary in the leaf axils, 3.5-5 mm. long; corolla bell-shaped, white to pinkish; anthers unawned; fruit red, 6-8 mm. wide.

Subalpine bogs to wet coniferous forests.

Gaultheria shallon

Gaultheria shallon Pursh Salal

Creeping to erect shrubs from 1-20 dm. tall; leaves ovate, 3-10 cm. long, 2.5-5 cm. wide, the tip acute, serrate; flowers borne in glandular-pubescent bracteate racemes; corolla urn-shaped, 7-10 mm. long, pinkish; anthers awned; fruit dark purple, 6-10 mm. wide.

Moist areas, but also in drier sites; from the coast to the Cascades mountains. Widely cultivated and the leaves commonly used in the floral trade.

Hemitomes

A monotypic genus.

Hemitomes congestum Gray Gnome-plant

White to cream-colored or pinkish saprophytic herbs 0.5-2 dm. tall with a stout scaly stem usually mostly underground; flowers flesh-colored, densely clustered; sepals 2 or 4; corolla more or less bell-shaped, 4-5-lobed, the lobes pubescent on the inner surface; stamens typically 8 or 10, filaments pubescent above the middle; ovary 1-loculed, but sometimes appearing more than 1-loculed by the intrusion of the 4-10 placenta.

In woods of the mountains and the along coast.

Hypopitys

Succulent saprophytic herbs with stems slender or thick, more or less pubescent; leaves scale-like; flowers several to many, white, pink, red, or yellow, in a one-sided raceme, nodding at first but becoming erect; terminal flower parts in 5's, lateral flower parts generally in 3's or 4's; sepals generally as many as petals; petals saccate at the base; stamens 6-10; ovary superior, 3-5-loculed below, but 1-loculed above, stigma funnel-form; disk 8-10-toothed; capsule 3-5-valved.

Hypopitys monotropa Crantz — Hairy pinesap

Whole plant reddish or yellowish, more or less pubescent, 0.5-4 dm. tall; leaves scale-like, ovate, entire or slightly fringed, closely overlapping below, scattered above; flowers few to many, at first nodding, later erect; sepals pubescent on the margins; petals usually pubescent; ovary, style, and stigma pubescent, a circle of hairs turning downward from the rim of the funnel-like stigma; fruit subglobose.

(=*Monotropa hypopitys* L.)

In coniferous woods.

Kalmia

Low evergreen shrubs with opposite, alternate or whorled leathery leaves, and flowers in terminal clusters; calyx and corolla segments 5 each; corolla nearly cup-shaped or rotate containing 10 "pockets" in which the anthers are borne before maturing; ovary superior, 5-loculed; fruit a septicidal capsule.

Kalmia microphylla (Hook.) Heller — Swamp Laurel

Low shrubs 1-6 dm. tall; leaves glabrous and deep green above, grayish and glandular-puberulent on the under surface, entire, the margins sometimes revolute, leaf blades oblong to lanceolate, 0.5-6 cm. long, 3-25 mm. wide, short petioled; flowers deep pink to purplish, 7-11 mm. long; capsules 2-5 mm. wide.

Kalmia microphylla

At high altitudes in the Cascades this species may have smaller flowers and leaves.

In *Sphagnum* bogs and wet meadows. Very beautiful when in flower, but poisonous.

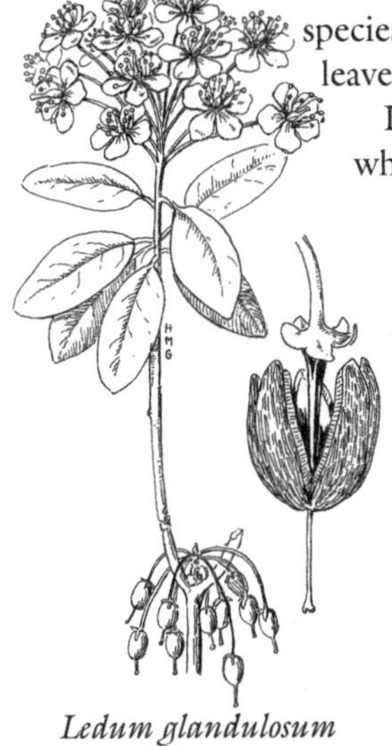

Ledum glandulosum

Ledum

Shrubs with fragrant alternate entire evergreen leaves; flowers white, in corymbs, cymes or racemes; calyx lobes usually 5, small, nearly separate; petals usually 5; stamens 5-12, anthers dehiscent by pores; ovary superior, 5-loculed; fruit a septicidal many-seeded capsule.

Lower surface of leaves whitish, not woolly .... *L. glandulosum*

Lower surface of leaves rusty-woolly .... *L. groenlandicum*

Ledum glandulosum Nutt. var. *columbianum* (Piper) C.L. Hitchc.
Labrador tea

Shrubs 6-20 dm. tall, bark smooth and brown, branches gland-dotted and puberulent; leaves oblong, 1-7 cm. long, dark green and smooth above, whitish and glandular beneath, with margins strongly revolute; flowers white to cream-colored with glandular pedicels; ovary pubescent and glandular; capsules globose, 3-6 mm. long.

Bogs and other wet areas near the Columbia River and along the coast.

Ledum groenlandicum Oedr. Rusty Labrador tea

Similar to *L. glandulosum*; shrubs 0.5-2 m. in height; branches pubescent; leaves linear-elliptic, dark green and glabrous to pubescent above, rusty-pubescent on the under surface; petals white; capsules ovoid, 4-5 mm. long.

In bogs and swamps along the coast in Washington.

Menziesia

Shrubs with alternate deciduous leaves; flowers small, in umbel-like clusters from terminal buds; calyx shallowly 4-5-lobed; corolla urn-shaped to bell-shaped, 4-5-lobed; stamens 5, 8 (ours) or 10; ovary superior, 4-5-loculed; fruit a septicidal capsule.

Menziesia ferruginea

Menziesia ferruginea Smith
Fool's huckleberry

Erect shrubs 0.5-4 m. tall with reddish shreddy bark, branches often more or less grouped, appearing whorled, pubescent and glandular when young, often becoming glabrous; leaves thin, dark green above, pale beneath, 1.5-6.5 cm. long, sparsely pubescent above, obovate, the margins crenate-serrate, the apex apiculate, short-petioled; flowers clustered at beginning of new year's growth; petals 4, united almost to apex into an urn-shaped corolla 6-8 mm. long, pinkish-orange to reddish-yellow; stamens 8; capsule ovoid, 5-7 mm. long.

Stream banks and moist coniferous woods; along the coast and in the Cascades.

Moneses

A monotypic genus.

Moneses uniflora (L.) Gray
Single beauty

Moneses uniflora

Evergreen perennials; leaves in single or double whorls or the lowest paired; flowering stems 5-15 cm. high; leaves thick, ovate or nearly round, serrate, 8-25 mm. broad; flower solitary, 1-2 cm. broad, white to pinkish, fragrant; ovary superior, stigma circular with 5 lobes; capsules erect.

(=*Pyrola uniflora* L.)

Moist forests.

Monotropa

Succulent white saprophytic herbs with leaves modified into scales and each stem with a single large nodding flower, the capsule becoming erect at maturity; sepals 1-4, deciduous (probably actually calyx-like bracts with the sepals lacking); petals 4-6, not saccate at the base; stamens usually 10, the anthers peltate; ovary 5-loculed, stigma funnel-form; capsule 5-valved; seeds many, minute.

Monotropa uniflora L. — Indian pipe or Ghost flower

Whole plant glabrous, waxy white, drying black; stems generally clustered, 0.5-2.5 dm. tall, growing up among the remains of the last year's shoots; leaves scale-like, lanceolate, entire; flower bell-shaped, terminal, nodding, reaching 2.5 cm. in length; capsule erect, somewhat elongated, obtusely angled.

In the humus of deep woods.

Orthilia

A monotypic genus.

Orthilia secunda (L.) House — One-sided wintergreen

Stems from long slender, often branched, rhizomes; leaves ovate, entire or minutely toothed, pale green, blades 1-4 cm. long on petioles 6-12 mm. long; peduncles and inflorescence 7.5-18 cm. tall, the flowers white to greenish, nodding, in a slender 1-sided raceme; petals 2 tubercled at base; stamens 10; ovary superior, 5-loculed, style straight; fruit a nodding capsule.

(=*Pyrola secunda* L.)

In deep mountain woods.

Phyllodoce

Low alpine shrubs with crowded alternate linear evergreen leaves with margins inrolled below, the leaf shape, arrangement and scars resembling those of conifers; flowers solitary or in shortened clusters at the ends of branches, each flower 2-bracted at the base; corolla 5-toothed or lobed; stamens usually 10, the anthers opening by short oblique terminal slits; ovary superior; capsule 5-loculed, 5-valved.

Corolla yellowish, densely glandular .... *P. glanduliflora*
Corolla pinkish or magenta, not glandular .... *P. empetriformis*

Phyllodoce empetriformis (Smith) D. Don — Mountain red heather

Low, much branched shrubby plants, the stems prostrate at the base, the tips rising 1.5-4.5 dm. in height; leaves 8-16 mm. long, somewhat resembling conifer needles; flowers crowded at the ends of the branches in the axils of bracts, and appearing to be in umbels; corolla bell-shaped; pinkish or magenta, about 6 mm. long, the lobes shorter than the tube.

In the mountains in moist areas at about timberline; very showy.

Phyllodoce glanduliflora (Hook.) Coville — Yellow heather

Stems woody, ascending, 1-4 dm. tall, very leafy; leaves thick, 6-12 mm. long, resembling conifer needles; flowers several in terminal umbels, pale yellow to greenish-yellow, the corolla urn-shaped, entire flower and pedicel densely glandular.

Abundant at timberline in the mountains.

Pleuricospora

A monotypic genus.

Pleuricospora fimbriolata Gray — Fringed pinesap

White or brownish saprophytic herbs; lower leaves scale-like, entire; upper leaves and bracts fringed, the bracts as long as the petals and sepals; flowers in a dense spike-like raceme; sepals generally 4-5, fringed; petals separate, 4-5; stamens twice as many; ovary compound, 1-loculed.

Deep woods in the Cascades and Coast Range; uncommon.

Pterospora

A monotypic genus.

Pterospora andromedea Nutt. — Pine drops

Brownish or reddish perennial glandular saprophytes; stems unbranched, reddish, 1 to several, 1-15 dm. tall, densely glandular-pubescent and bearing many reflexed flowers in a long spike or raceme; calyx deeply 5-lobed or the sepals free; corolla urn-shaped, 5-toothed; stamens 10, anthers awned; ovary superior, stigma 5-lobed; capsule 5-lobed, 5-loculed.

Coniferous woods in the mountains.

Pyrola

Rhizomatous evergreen perennials; leaves mostly basal; flowers in a terminal raceme; calyx 5-lobed; petals 5; stamens 10, the anther opening by pores usually borne on tubes; ovary superior, styles 5-lobed; capsule 5-loculed, typically reflexed.

1a Floral bracts longer than the pedicels, ovate; flowers white or pinkish, drying purplish .... *P. asarifolia*

1b Floral bracts shorter than the pedicels

  2a Plants nearly leafless or leaves conspicuously veined with white or mottled; flowers greenish, white, or pinkish-purple edged with white; sepals acute .... *P. picta*

  2b Leaves not white-veined or mottled; flowers greenish-white; sepals obtuse .... *P. chlorantha*

Pyrola asarifolia Michx. — Alpine Pyrola

Leaves obovate to oblong or nearly orbicular, usually purplish on the under surface, rounded at the apex, entire to crenate or sharply toothed; peduncles and inflorescence 2-6 dm. tall, with 1 thin scale near the middle; flowers white or pinkish, drying purplish; style turned downward at base, curved and exserted. Two subspecies occur in our area: subsp. ***asarifolia*** with entire to crenate margins and sepals 2-3.5 mm. long and subsp. *bracteata* (Hook.) E. Haber with sharply toothed leaves and sepals 3-6 mm. long.

Moist coniferous woods.

Pyrola chlorantha Sw. Green-flowered wintergreen
Leaf blades thick, not shining, 2-3.5 cm. long, ovate to obovate or reduced and bract-like, shorter than the petioles, margin entire or wavy; few-flowered; sepals less than 2 mm. long; petals longer, greenish-white; style turned downward, curved, exserted.

Deep coniferous woods.

Pyrola picta

Pyrola picta Smith
White-veined wintergreen
Leaves several at base to nearly leafless, leaves usually ovate, leathery, deep green, white-veined, blades 2-8.5 cm. long, margins entire or toothed, about the same length as the petioles; flowering stems 1.5-4 dm. tall; bracts small, triangular; flowers few to many, greenish, white, or pinkish-purple edged with white; style turned downward at base, exserted.

Common in open coniferous woods.

Rhododendron

Shrubs (ours), with alternate leaves; flower clusters terminal or lateral; calyx 5-toothed, small; corolla more or less bell-shaped to funnel-form, showy, often somewhat irregular; stamens 5-10 (ours); ovary superior, style 1, slender, topped by a knob-like stigma; fruit a septicidal capsule.

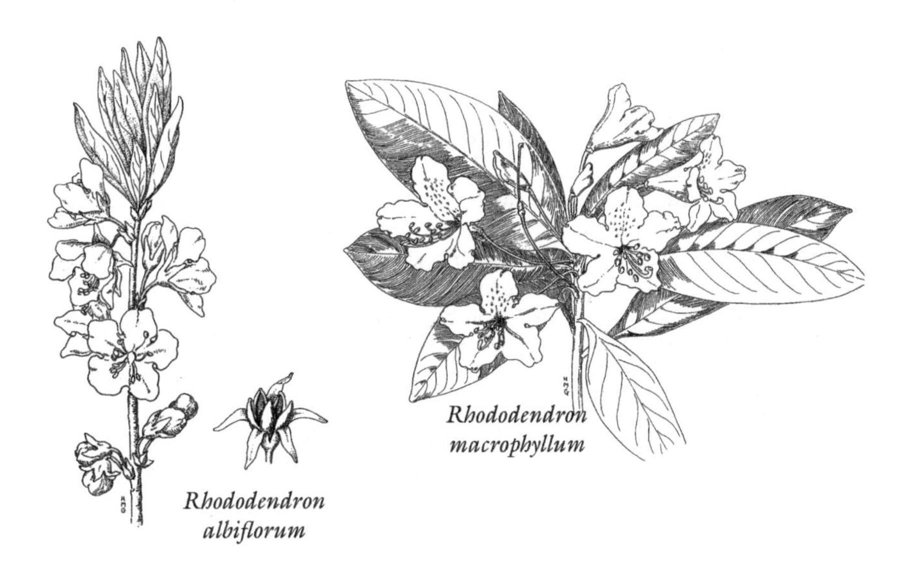

Rhododendron albiflorum

Rhododendron macrophyllum

Flowers usually rose-colored, many in a terminal inflorescence; lowland and foothill evergreen shrubs .... *R. macrophyllum*
Flowers creamy-white in few-flowered lateral clusters; alpine or subalpine deciduous shrubs .... *R. albiflorum*

Rhododendron albiflorum Hook. White-flowered Azalea
Deciduous shrubs 1-2 m. tall, slightly glandular and pubescent; leaves 4-9 cm. long, thin, bright green, oblong, narrowed at each end, margin wavy; flowers creamy-white, 1.5-2.5 cm. wide, generally in 2's or 3's from lateral buds.
Stream banks and wet areas at timberline and below, in the mountains.

Rhododendron macrophyllum D. Don ex G. Don
Western Rhododendron
Erect shrubs, compact when in the open or with long straggling branches when crowded, reaching 2-3.5 m. in height, or becoming a small tree 8 m. tall; leaves leathery, evergreen, stiff, oblong or elliptic-oblong, narrowed at both ends or obtuse at base, green above, often rusty beneath, margins sometimes somewhat in-rolled; calyx small, 5-lobed; corolla spreading bell-shaped, 3.5 cm. or more wide, varying from pale pink to deepest rose-colored (rarely white), the lobes wavy-margined, the upper lobe dotted with green, stamens 10; ovary covered with reddish down, style curved downward; capsule woody.
Along the coast to the foothills of the Coast Range and the Cascades.

Vaccinium

Trailing to erect shrubs or subshrubs with alternate deciduous or evergreen leaves; flowers solitary or in racemes; calyx lobes obsolete or 4-5; corolla urn-shaped to cup-shaped, 4-5-lobed; stamens 8 or 10, anthers opening by pores, often 2-awned; ovary inferior, compound; fruit a berry.

1a Leaves evergreen
 2a Erect shrubs; leaves serrate, 2-5 cm. long; corollas pink, urn-shaped .... *V. ovatum*
 2b Trailing vines; leaves entire, 1.5 cm. long or less; petals fused only at the base, typically reflexed
  3a Bracts 2-5 mm. long, foliaceous .... *V. macrocarpon*
  3b Bracts less than 2 mm. long, not foliaceous .... *V. oxycoccus*
1b Leaves deciduous
 4a Berries red; stems green, sharply angled .... *V. parvifolium*
 4b Berries blue, purple, or black; stems angled or not
  5a Calyx deeply parted into nearly separate sepals; flowers 1-4 in the leaf axils; calyx and corolla segments often 4
   6a Leaves thick, conspicuously veined beneath .... *V. uliginosum*
   6b Leaves thin, obscurely veined beneath .... *V. occidentale*
  5b Calyx less deeply lobed to merely wavy; flowers solitary in the leaf axils; calyx and corolla segments mostly 5
   7a Flowers appearing before the leaves are fully expanded; leaves entire or nearly so; twigs yellow-green, angled .... *V. ovalifolium*

7b Flowers appearing after the leaves are expanded; leaves minutely toothed
8a Corolla long-ovoid, nearly twice as long as wide .... *V. caespitosum*
8b Corolla nearly globose or at least much less than twice as long as broad
9a Low shrubby plants 3 dm. or less tall; leaves thin but not membranous .... *V. deliciosum*
9b Erect shrubs 5-15 dm. tall; leaves membranous.. *V. membranaceum*

Vaccinium caespitosum Michx. Dwarf huckleberry

Low shrubs 4.5 dm. or less tall; first year twigs green, minutely and closely pubescent, older twigs gray, bark peeling in broad patches; leaves 1-4.5 cm. long, bright green or sometimes bluish-green, shining, and sometimes glandular, especially beneath, glabrous or the teeth tipped with cilia, petiole 3 mm. or less long; flowers solitary and axillary, corolla long-ovoid, whitish to pinkish, 6 mm. or less long; berries small, often flattened, dark maroon to purple-black, glaucous, thus appearing bluish.

Wet meadows, bogs and rocky slopes in the mountains.

Vaccinium caespitosum

Vaccinium deliciosum Piper Blue huckleberry

Deciduous shrubs more or less prostrate, woody at base and often rooting at the nodes, with short erect branches 1-3 dm. tall, young twigs grayish, glaucous; leaves dull green, almost glaucous, microscopically white-dotted, blades 1.5-3.5 cm. long, broadly ovate or elliptic, acutish or obtuse at apex, narrowed at base, minutely toothed; flowers solitary and axillary; corolla nearly globose, pinkish; berry up to 1 cm. in diameter, purplish black, glaucous, thus appearing bluish.

Meadows in the high Cascades and Olympics.

Vaccinium macrocarpon Ait. Cranberry

Vine-like, stems to 4 dm. long, creeping and rooting at the nodes; leaves evergreen, 7-15 mm. long, oblong, the margins rolled backward, under surface of leaf whitish; flowers lateral on leafy branches; sepals 4, fused; petals 4, pink, 6-10 mm. long, fused only at the base, reflexed; anthers unawned; berry nearly globose, deep red, 1-2 cm. in diameter.

Occasionally escaped from cultivation and found in boggy areas on the coast. Native of the eastern United States.

Vaccinium membranaceum Dougl. ex Torr. Mountain huckleberry

Erect or spreading shrubs 5-15 dm. tall; leaves thin, 1.5-5 cm. long, pale green but scarcely glaucous, paler beneath, ovate or obovate to elliptic, glabrous, or slightly pubescent on the veins, obtuse or narrowed at base, acute at apex, minutely toothed, the teeth sometimes tipped by deciduous cilia or glandular; flowers solitary and axillary, more or less nodding, pedicels up to 12.5 mm. long; corolla nearly globose to urn-shaped, pinkish; fruit slightly flattened, sometimes reaching 12.5 mm. in diameter, dark purplish-red to black, the pedicels erect in fruit.

Wet meadows and slopes at 900 m. and above in the mountains. The fruits have a good flavor.

Vaccinium occidentale Gray — Western huckleberry

Low shrubs 3-10 dm. tall; leaves oval or obovate-oblong, 1-2.5 cm. long, rather thin, somewhat glaucous, with inconspicuous veins, margins entire; flowers mostly solitary, but up to 4 per leaf axil; corollas urn-shaped, pinkish; berry 9 mm. or less long, blue-black, glaucous.

Swamps or wet meadows in the mountains.

Vaccinium ovalifolium Smith — Tall blue huckleberry

Tall shrubs reaching 2 m. or more, but beginning to fruit at 6 dm. or less; leaves thin, 2-5 cm. long, oval or somewhat ovate, entire or rarely minutely and irregularly toothed near base, short-petioled; flowers slightly elongated, pinkish, solitary and axillary, appearing before the leaves; berries 6-9 mm. in diameter, blue to black, glaucous.

Deep woods near the coast, also low in the mountains. The fruits have a good flavor and are widely used, where abundant.

Vaccinium ovatum
a. stamen

Vaccinium ovatum Pursh — Common evergreen huckleberry

Evergreen shrubs 0.5-4 m. tall; leaves thick, leathery, shining above, ovate-oblong, serrate, 2-5 cm. long; flowers in small axillary racemes; corollas pink, urn-shaped to bell-shaped, 4-10 mm. long; berries purplish-black or glaucous, thus sometimes appearing blue, 6-8 mm. in diameter.

Along the coast to the west slope of the Cascades. Frequently used as an ornamental shrub and the foliage is used in the florist trade.

Vaccinium oxycoccus L. — Wild cranberry

Trailing vines, the stems cord-like with slender short ascending branches; leaves usually persistent, ovate to lanceolate, 5-15 mm. long, the margins rolled backward, under surface of leaf whitish; flowers 1-10 on slender terminal or lateral stems, nodding; petals 4, fused only at the base, typically reflexed, 5-8 mm. long, pale pink or purplish; berries dark red, globose, about 8 mm. in diameter.

In sphagnum bogs.

Vaccinium parvifolium Sm. Red huckleberry

Shrubs 0.5-4.5 m. tall, with green sharply angled slender branches; leaves thin, deciduous, ovate to elliptic, lighter green beneath, 1-3 cm. long, margins entire or toothed on young growth; solitary and axillary; corollas nearly globose, greenish-white to yellowish-pink; berries bright red, 6-10 mm. in diameter, reflexed, mildly acid.

Rather common from the coast to the west slope of the Cascades.

Vaccinium parvifolium

Vaccinium uliginosum L. Bog blueberry

Low deciduous shrubs rarely higher than 6 dm.; leaves obovate, whitish beneath, 1-3 cm. long, margins rarely toothed; flowers solitary or 2-4 in the leaf-axils; corollas white to pink, urn-shaped; berry deep bluish-black but glaucous which makes it appear bright blue, 6-10 mm. in diameter, drooping.

In swamps and bogs and wet meadows from the coast to the mountains.

PLUMBAGINACEAE
Leadwort Family

Annuals, perennials or shrubs, usually with basal leaves and regular perfect flowers with all parts in 5's; sepals united into a 5-angled persistent calyx; petals united, at least at base; stamens opposite the petals and attached to them at the base; ovary superior, usually 5-angled above, 1-loculed, 1-ovuled; fruit an achene, utricle or a capsule.

Armeria

Perennials to shrubby; stems reduced, bearing dense basal tufts of linear leaves, and naked peduncles terminating in flower heads; heads consisting of crowded clusters of nearly sessile flowers, each cluster subtended by a papery bract, the whole head surrounded by an involucre of bracts, the 2 outermost united to form a sleeve-like sheath extending down the peduncle for an inch or more below the head; calyx papery, funnel-shaped; corolla of 5 petals slightly united at the base and each bearing a stamen.

Armeria maritima (Mill.) Willd. Thrift or Sea-pink

Perennials; leaves tufted, numerous, linear, 5-12 cm. long; peduncles 1-4.5 dm. tall; calyx tube pubescent and with 10 dark ribs, 5 of these projecting through the papery limb and continuing as short teeth, the margin of the limb erose between the teeth; calyx and corolla both pink.

Common on sandy bluffs along the ocean and occasionally inland.

PRIMULACEAE
Primrose Family

Herbs with simple leaves; flowers regular with the parts generally in 4's or 5's; sepals united at least at base; petals united at least at base, into a tube or a ring; stamens attached to corolla tube, opposite the lobes; ovary 1-loculed, with basal or free-central placentation, usually superior, style 1, stigma 1; fruit a capsule.

1a Tufted perennials forming mats; leaves in rosettes .... **Douglasia**
1b Not tufted, mat-forming perennials
 2a Leaves all basal; flowers in umbels (ours)
  3a Flowers minute, 4 mm. or less long................................ **Androsace**
  3b Flowers showy, 10 mm. long or more; calyx and corolla lobes reflexed in flower... **Dodecatheon**
 2b Leaves not all basal
  4a Flowers sessile in the leaf axils, leaves often alternate, at least above
   5a Annuals; petals present; leaves mostly alternate........ **Centunculus**
   5b Perennials; corolla none; calyx corolla-like.......................... **Glaux**
  4b Flowers, if in the leaf axils, not sessile; leaves mostly opposite or whorled
   6a Leaves in a whorl at top of stem, with scattered scales or leaves below; the flowers arising from center of whorl .... **Trientalis**
   6b Leaves opposite; flowers axillary or terminal
    7a Flowers yellow (ours); capsule splitting longitudinally.............
    .... **Lysimachia**
    7b Flowers reddish-salmon with blue center (rarely white); capsule circumscissile .... **Anagallis**

Anagallis

Small annual or perennial herbs with opposite or whorled leaves; flowers solitary in the leaf axils; calyx and corolla 5-lobed; stamens 5, opposite the corolla lobes; filaments pubescent; fruit a circumscissile capsule.

Anagallis arvensis L.
Scarlet Pimpernel or Poor man's weather-glass

Low decumbent to erect annuals; leaves 0.5-2 cm. long, broadly ovate, acute at apex, palmately veined, sessile, glabrous; corolla 5-7 mm. long, orange or salmon, with deep blue center, the margins minutely glandular-fringed; pedicels reflexed at maturity; top of capsule separating from base at maturity, to allow escape of seeds.

In disturbed ground; somewhat uncommon in our limits.

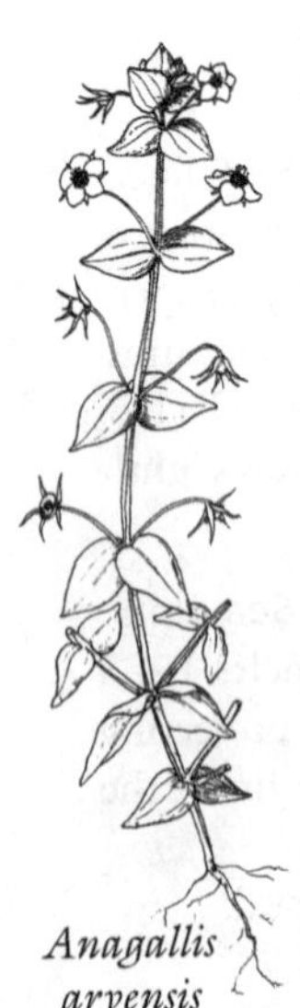
Anagallis arvensis

Androsace

Annuals or perennials with leaves in a basal rosette and small flowers solitary or more often in umbels; corolla funnel-shaped or abruptly widened above the tube, 5-lobed; capsule 5-valved.

Androsace filiformis Retz. Fairy-candelabra

Annuals; peduncles 3-12 cm. tall; leaves in a basal rosette, the blades 0.5-3 cm. long, ovate, coarsely toothed, obtuse at apex, on petioles as long as or longer than the blade; flowers minute, white, on hair-like pedicels 1-3.5 cm. long in umbels; calyx lobes shorter than or about equal to the tube; fruit globose.

Wet meadows; uncommon in our limits.

Centunculus

Annuals; leaves entire, alternate or the lower ones opposite; flowers small, solitary and sessile or subsessile in the leaf axils; calyx lobes 4-5; corolla urn-shaped, white to pinkish, with 4-5 acute lobes; capsule circumscissile.

Centunculus minimus L.

Glabrous annuals; stems 2-10 cm. long, decumbent, rooting at the nodes; leaves oblanceolate, 3-10 mm. long, decurrent on the stem; flowers solitary; calyx 2-3 mm. long; corolla about 1.5 mm. long, pink, urn-shaped, the lobes acute; capsule globose, about 2 mm. long.

Vernal pools and other moist areas.

Dodecatheon

Perennial herbs with all basal leaves; flowers few or many in an umbel rising on a long peduncle; flowers parts in 4's or 5's, often on the same plant; corolla with a short tube and long strongly reflexed lobes; anthers erect, surrounding the style; fruit a 1-loculed capsule.

1a Stamens united by filaments into a tube
- 2a Leaves gradually narrowed to the petiole; capsule opening by valves at the apex .... *D. pulchellum*
- 2b Leaves abruptly narrowed to the petioles; capsule opening by a cap .... *D. hendersonii*

1b Stamens free
- 3a Leaves abruptly narrowed to the petioles; corolla lobes white or cream-colored .... *D. dentatum*
- 3b Leaves gradually narrowed to the petiole; corolla usually reddish-purple, sometimes yellowish to white .... *D. jeffreyi*

Dodecatheon dentatum Hook. White bird-bill or shooting star

Rhizomatous perennials; leaf blades ovate to oblong-lanceolate, 3-10 cm. long, the margins undulate to dentate, rounded to cordate at the base and abruptly narrowed to a long petiole; scapes 1-4 dm. tall, 2-12-flowered; flower parts in 5's; corolla 12-20 mm. long, the tube yellow with a purple-red ring at the base, the lobes white to cream-colored; anthers 6-7 mm. long, dark red to purple, filaments free; capsules 6-10 mm. long.

Stream banks and moist to wet shaded slopes; in the Cascades and along the Columbia Gorge.

Dodecatheon hendersonii Gray Shooting star or Bird-bill
Perennials with bulblet bearing roots; leaves borne in a flat rosette, 2.5-15 cm. long, broadly ovate to elliptic, somewhat fleshy, narrowed abruptly to a slender petiole; scapes 1.5-5 dm. long, green below, reddish above, bearing 3-17 flowers in a spreading umbel; flower parts generally in 5's (sometimes in 4's); calyx lobes long-ovate, sharp-pointed; corolla 15-25 mm. long, the lobes reflexed, purplish-black in the center, bordered by yellow, margined with white, with the lobes magenta, oblong; tube of filaments purplish black; capsule opening by a dehiscent cap, then splitting longitudinally.

Dodecatheon hendersonii

Open woods and thickets; common.

Dodecatheon jeffreyi Van Houtte
Tall mountain shooting star
Rhizomatous perennials; leaf blades oblanceolate to lanceolate, 0.5-5 dm. long, 1-6 cm. wide, entire to serrulate, gradually narrowed to the long petiole; scapes 1-6 dm. tall, 3-20-flowered; flower parts in 4's or 5's; corolla tube yellow to cream, usually banded just below the lobes, the lobes 1-2.5 cm. long, reddish-purple, sometimes pale yellow to white, anthers 6-10 mm. long, filaments short, free; capsules circumscissile, 6.5-11 mm. long.

Wet meadows and stream banks in the mountains.

Dodecatheon pulchellum (Raf.) Merrill Few-flowered shooting star
Glabrous to glandular-pubescent perennials; leaf blades 2-25 cm. long, ovate, oblanceolate or spatulate, entire to denticulate, gradually narrowed to the winged petiole; scapes 0.5-4.5 dm. tall, 1-25-flowered; flower parts in 5's; corolla tube short, yellow, usually with a purplish ring, corolla lobes 9-14 mm. long, magenta to lavender or rarely white; anthers 3-7 mm. long, filaments united into a tube; capsules 5-15 mm. long.

This is an extremely variable species, but the variation appears to depend on conditions such as moisture and elevation.

[=*Dodecatheon pauciflorum* (Durand) Greene]

Wet saline areas to wet mountain meadows and stream banks, from the coast to the high mountains.

Douglasia

Low tufted perennials, often more or less woody at the base; leaves in rosettes; flowers solitary or in umbels; calyx tube keeled; corolla tubular with spreading lobes, the tube narrowed and 5-crested at the throat; stamens separate, shorter than the corolla tube; capsules 1-3-seeded.

Douglasia laevigata Gray
Low tufted mat-forming plants, nearly glabrous; leaves 6-20 mm. long, oblanceolate, glabrous to ciliolate; flowers bright pinkish-red, 10-12.5 mm. long, pediceled to subsessile, 2-10 in involucrate umbels.

In the Cascades and Olympics and on Saddle Mountain.

Glaux

A monotypic genus.

Glaux maritima L. Sea-milkwort
Succulent rhizomatous perennials; stems erect, 3-40 cm. tall; leaves sessile, opposite below and alternate above, 4-25 mm. long, 1-10 mm. wide, linear to oblong, entire; flowers solitary in the leaf axils; calyx 4-5 mm. long, 5-lobed, petaloid, white to pink; corolla lacking; stamens 5; capsules about 2.5 mm. long.

Coastal salt marshes, tide flats, inland saline marshes and wet meadows.

Lysimachia

Perennials; leaves opposite or whorled, finely gland-dotted; flowers solitary in the leaf axils or racemose to paniculate; flower parts 5-7; corolla yellow, lobed nearly to the base; stamens 5; fruit a capsule.

Flowers axillary: stems creeping...*L. nummularia*
Flowers in a terminal raceme; stems erect.................................... *L. terrestris*

Lysimachia nummularia L. Moneywort
Stems leafy, slender, creeping, forming conspicuous green mats; leaves nearly orbicular, 12.5 mm. or more long, short-petioled; flowers yellow, 1.8-2.5 cm. wide, axillary.

In moist places in the Willamette and Columbia River bottom lands. Sparingly introduced, but locally sometimes covering considerable areas.

Lysimachia terrestris (L.) B.S.P. Swamp candle
Stems erect, 1-9 dm. tall; leaves generally opposite, narrowed at the base, but sessile, dark green above, paler with dark dots beneath; flowers borne in an elongated raceme; corolla lobes narrow, yellow, streaked with dark red lines.

Cranberry bogs along the coast; introduced from the eastern United States.

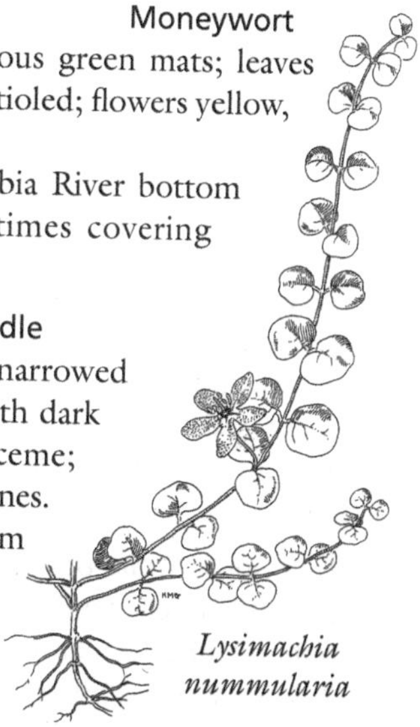

Lysimachia nummularia

Trientalis

Small perennials from tuberous roots; stem unbranched, generally crowned at apex by a whorl of leaves; one or more pedicels, each bearing a single flower, arise from the center of leaf whorl; corolla rotate, deeply 5-9-lobed; stamens united at base, generally equal in number to corolla lobes and opposite them; fruit a globose 5-valved capsule.

Main leaves clustered at top of stem; flowers pink; woodland plants. .... *T. latifolia*

Some leaves scattered below the terminal cluster; flowers white; plants of bogs or other wet areas.. *T. arctica*

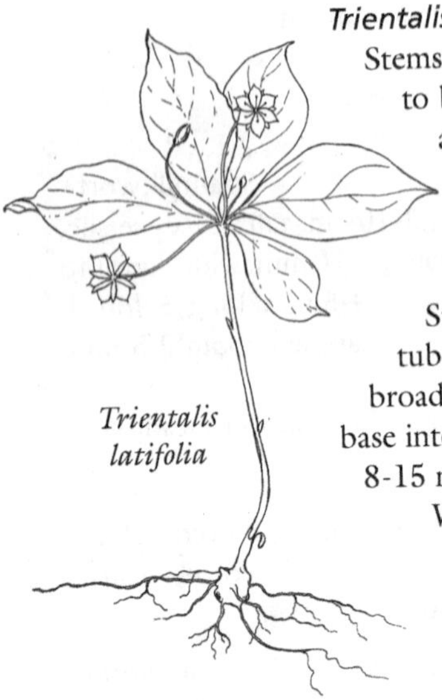

Trientalis arctica Fisch. Arctic starflower

Stems 5-20 cm. tall; leaves 1.5-5 cm. long, obovate to broadly elliptic, smaller foliage leaves scattered along stem below the terminal cluster; flowers white, 1-1.5 cm. wide.

Bogs or swampy areas.

Trientalis latifolia Hook. Western starflower

Stem 7-25 cm. tall from somewhat elongated tubers; leaves of the whorl 3-8, 2.5-10 cm. long, broadly obovate, acute or obtuse at apex, tapering at base into a short petiole, lower leaves scale-like; corolla 8-15 mm. wide, pale to deep pink, star-shaped.

Wooded areas; common.

Trientalis latifolia

OLEACEAE
Ash Family

Trees, shrubs or vines; leaves usually opposite, rarely alternate, simple or pinnately-compound, deciduous or evergreen, exstipulate; flowers often unisexual; calyx small, 4-15-lobed; corolla lobes 4-6 or the corolla sometimes lacking; stamens usually 2; pistil 1, ovary superior, compound; fruit a drupe, capsule or a samara (ours).

Fraxinus

Trees or shrubs with opposite pinnately compound deciduous leaves; flowers perfect, or imperfect with stamens and pistils on different plants, or with both perfect and staminate or pistillate flowers on the same plant; flowers sometimes appearing before the leaves; calyx 4-lobed, small; corolla consisting of 2 or 4 petals, or absent; stamens generally 2; fruit a 1-winged samara containing 1 seed.

Fraxinus latifolia Benth. Oregon ash

Dioecious trees 10-35 m. tall, with thick rough grayish-brown bark; herbage puberulent to tomentose when young, but becoming nearly glabrous; leaves 2-3 dm. long with 3-9 (mostly 5-7) leaflets; leaflets oblong, ovate to obovate, tapering at both ends, sometimes toothed near apex; inflorescence of dense panicles; corolla lacking; staminate flowers with a vestigial calyx and 2 stamens; pistillate flowers with a small 4-lobed calyx; samara oblanceolate, slightly notched at apex, 2.5-5 cm. long.

Common along streams and in river bottoms usually in wet areas, but also found in drier sites.

GENTIANACEAE
Gentian Family

Annuals or perennials; leaves simple, opposite or whorled, exstipulate; flowers usually perfect, calyx and corolla usually 4-5-lobed, but occasionally up to 12-lobed; corolla often with scales or nectary pits within, usually convolute in the bud; stamens attached to the corolla, as many as the corolla lobes and alternate with them; pistil 1, compound, ovary superior; fruit a capsule.

1a Corolla tube slender, the lobes spreading, pink (rarely white).. **Centaurium**
1b Corolla funnel-form or bell-shaped, blue (rarely white or yellowish)
  2a Annuals; corolla without teeth in the sinuses, but with an inner circle of hairs.......... **Gentianella**
  2b Perennials; corolla with teeth in the sinuses (sinuses rarely truncate), without an inner circle of hairs .... **Gentiana**

Centaurium

Glabrous annuals or biennials with opposite entire leaves; calyx deeply 4-5-lobed; corolla 4-5-lobed, generally white to pink with a slender tube and and abruptly spreading lobes; stamens attached to throat of the corolla, anthers twisting with age; ovary 1-loculed, but the placentae in such close contact as to simulate a partition; style slender; capsule 2-valved.

Lower leaves forming a rosette; corolla lobes 5-7 mm. long........ *C. erythraea*
Lower leaves not in a rosette; corolla lobes 2-5 mm. long..... *C. muhlenbergii*

Centaurium erythraea Raf. — Common centaury

Annuals or biennials; stems erect, simple or somewhat branched above, 4-angled, 1-6 dm. tall; lower leaves in a basal rosette, 2.5-4 cm. long, broadly obovate, with 3 veins more conspicuous than the others; upper leaves becoming smaller, elliptic to lanceolate, slightly clasping; stem terminating in a rather dense compound cyme, the central flower in each simple cyme sessile or nearly so, the lateral flowers short-pediceled, 2-bracted; calyx to 1 cm. long; corolla lobes bright pink to magenta, 5-7 mm. long, ovate, obtuse; capsules about 1 cm. long.

(=*Centaurium umbellatum* Gilib.)

Road sides and other waste places; introduced from Europe.

Centaurium muhlenbergii (Gris.) Wight — Muhlenberg's centaury

Annuals, 3-30 cm. tall; lower leaves oblong to obovate, 5-25 mm. long, rounded, cauline leaves 1-2 cm. long, 2-9 mm. wide, linear-oblanceolate, acute, sessile; calyx 6-12 mm. long, deeply divided; corolla 1-1.5 cm. long, the lobes 2-5 mm. long, pink; capsules 6-7 mm. long.

Wet areas and moist forests.

Gentiana

Annuals, biennials or perennials; leaves opposite; flowers solitary or in cymes, 4-6-merous; corolla tube narrowly bell-shaped to funnel-form, the lobes spreading; capsules 1-loculed.

1a Appendages between the corolla lobes not cleft; principally lowland plants; in coastal and foothill bogs .... *G. sceptrum*
1b Appendages between the corolla lobes cleft; boreal or alpine plants
2a Plants with a definite basal rosette of leaves; found in the southern Cascades .... *G. newberryi*
2b Plants without a definite basal rosette of leaves; found throughout the Cascades .... *G. calycosa*

Gentiana calycosa

Gentiana calycosa Griseb. — Mountain bog gentian
Glabrous tufted, more or less decumbent, perennials, 0.5-4.5 dm. tall; leaves mostly 7-9 pairs, the lower ones fused and strongly sheathing, middle and upper cauline leaves 1-5 cm. long, 0.5-3 cm. wide, broadly ovate; flower parts in 5's; corolla 2-5 cm. long, dark blue and mottled with green, appendages between the lobes cleft; ovary long-stalked.

Wet alpine or subalpine meadows, swamps or stream banks.

Gentiana newberryi Gray — Newberry's gentian
Alpine perennials; stems decumbent at base, the tips erect, 5-10 cm. tall or taller, 1-5-flowered; basal leaves tufted, broadly spatulate, obtuse, narrowed at base, 2-5 cm. long, cauline leaves smaller, sometimes nearly orbicular, sometimes short-petioled; flowers 2-5.5 cm. long, blue outside, with a brownish band from each corolla lobe to base, white inside, greenish-dotted, the appendages between the lobes divided into 2 triangular lobes, these often fringed.

High Cascades, barely within our southern limits.

Gentiana sceptrum Griseb. — King's or Common gentian
Tufted perennials, 2-12 dm. tall; leaves 10-15 pairs with the lower ones much reduced and fused for most of their length, upper cauline leaves oblong-lanceolate, 3-8 cm. long, 7-25 mm. wide; flower parts in 5's; corolla 3-5 cm. long, bluish, often mottled with green, scarcely or not at all toothed at the angles between lobes; ovary stalked.

Wet meadows, marshes and bogs, west of the Cascade Mountains, especially along the coast.

Gentianella

Glabrous annuals; leaves basal and cauline, the cauline opposite; flowers 4-5-merous; corolla lacking appendages or pleats between the lobes, funnelform; ovary sessile; fruit a capsule.

Gentianella amarella (L.) Boerner
Annuals; stems erect, 0.5-8 dm. tall, slightly angled; basal leaves oblanceolate, 0.5-4 cm. long, soon withering, cauline leaves mostly in 5-8 pairs, up to 6 cm. long and 3 cm. wide, lanceolate to narrowly oblanceolate, the base clasping; flowers numerous, terminal and in axillary clusters; corolla appearing cylindrical, blue to purple or lavender to nearly white, 1-2 cm. long.

(=*Gentiana amarella* L.)

Wet meadows, bogs and other moist areas; uncommon in western Oregon; but found in Washington from the coastal mountains to the Cascades.

MENYANTHACEAE
Buckbean Family

Glabrous rhizomatous perennials; leaves alternate, simple or trifoliolate from long sheathing petioles; flower parts in 4's-6's, but usually in 5's; corolla rotate to funnel-form; stamens 5; ovary superior or inferior; fruit a capsule or indehiscent.

Menyanthes

A monotypic genus.

Menyanthes trifoliata L. Buckbean

Rhizomes thick, covered with bases of old leaves; stems prostrate or the flowering branches ascending; leaves with 3 fleshy, sessile or very short-stalked leaflets, 2-12 cm. long, 1-5 cm. wide, on petioles 1-3 dm. long, the leaflets ovate to obovate, entire to dentate; inflorescence of long pedunculate racemes; calyx 2-5 mm. long, deeply 5-6-lobed; corolla 1-1.5 cm. long, the lobes spreading, whitish and usually pink- to purplish-tinged, covered with short scale-like hairs on the inner surface; ovary inferior; fruit 1-loculed, indehiscent or irregularly bursting.

Bogs, ponds, shallow lakes, swamps and wet meadows.

APOCYNACEAE
Dogbane Family

Perennial herbs (ours), shrubs or trees with milky juice and simple entire opposite (ours), alternate or whorled leaves; calyx deeply 5-lobed; stamens 5, borne on the corolla, alternate with its lobes; ovaries 2, free at base but united above into a single style and stigma; fruit follicles (ours).

Corolla greenish, white or pink, 1 cm. or less in length ....**Apocynum**
Corolla blue, over 1.5 cm. long; stems more or less trailing.................. **Vinca**

Apocynum

Rhizomatous perennials; flowers tubular to urn-shaped or bell-shaped, borne in cymes; stamens alternating at base of corolla with 5 small appendages, the anthers more or less fused at the tips to the stigma; follicles long; seeds with a tuft of long hairs.

The following two species hybridize.

Corolla 4-10 mm. long, bell-shaped, the lobes spreading or recurved; leaves drooping or spreading.. *A. androsaemifolium*
Corolla 2.5-5 mm. long, tubular to urn-shaped, the lobes nearly erect; leaves ascending.. *A. cannabinum*

Apocynum androsaemifolium L. Dogbane

Stems to 6 dm. tall, freely branched; leaves drooping or spreading, ovate or oblong to nearly orbicular, usually heart-shaped at base, obtuse or acute and mucronate at apex, dark green above, paler beneath, 2.5-6 cm. long, with short petioles; flowers in small cymes; calyx small, the teeth slender; corolla bell-shaped, 4-10 mm. long, white to pink, sometimes with deeper pink veining, the lobes spreading to recurved; follicles 7-15 cm. long.

(Also spelled *androsimifolium*)

Generally found where somewhat shaded by other vegetation.

Apocynum androsaemifolium

Apocynum cannabinum L. Indian hemp

Stems to 12 dm. tall, freely branching, especially above; leaves ascending, the blades 3-12 cm. long, ovate to oblong, tapered or more or less cordate at base, short-petioled or the upper sessile; corollas 2.5-5 mm. long, tubular to urn-shaped, white to greenish, the lobes erect; follicles 6-9 cm. long.

Usually in moist areas, but also weedy in cultivated ground.

Vinca

Erect or trailing perennials with opposite leaves and large axillary solitary flowers; corollas with a slender tube and an abruptly spreading limb; stamens attached above the middle of the corolla tube; fruit of follicles; seeds unappendaged.

Vinca major L. Common or Greater periwinkle

More or less trailing and forming large mats; leaves evergreen, short-petioled, the blades 3-9 cm. long, ovate to broadly lanceolate, ciliate on the margins; calyx lobes linear, ciliate; corollas 3-5 cm. wide, the tube about 2 cm. long, blue; follicles 3-5 cm. long, rarely produced.

Commonly cultivated as a ground cover and occasionally escaped and naturalized or persistent in areas long after the buildings around which it was planted are gone.

A second species, *Vinca minor* L., may also occur in similar situations. It has smaller flowers and lacks cilia on the calyx lobes.

ASCLEPIADACEAE
Milkweed Family

Shrubs, vines or herbs (ours) with milky juice; leaves simple, entire, opposite or whorled (rarely alternate); flowers regular; calyx and corolla each 5-lobed; stamens 5, attached to the base of corolla and united into a tube which is further united with the united style tips of the 2 superior ovaries, anthers generally appendaged at the top or base, pollen grains coherent in masses, and suspended from the summit of slits between the anthers, so arranged that an insect in walking over the flower draws them out in pairs when its foot is caught in the slits and drawn upward, and deposits them upon the stigma of other flowers when its foot is again caught; fruit of follicles; seeds flattened and bearing a tuft of silky hairs.

Asclepias

Annual or perennial herbs or shrubs; calyx divisions turned downward, persistent; stamen tube bearing 5 hooded appendages, each generally horned; follicles ovoid or lanceolate, one often abortive; seeds flat, silky-pubescent at apex, closely overlapping on the placenta which separates at maturity from the line of attachment.

Stems slender; leaves mostly whorled; pedicels not woolly .... *A. fascicularis*
Stems stout; leaves opposite; pedicels densely woolly .... *A. speciosa*

Asclepias fascicularis Dcne. — Narrow-leaved milkweed

Rhizomatous perennials; stems stiff, slender, sometimes tufted, 2-8 dm. tall, leafy; leaves in whorls or 3-6 or some opposite, narrowly oblong to linear,

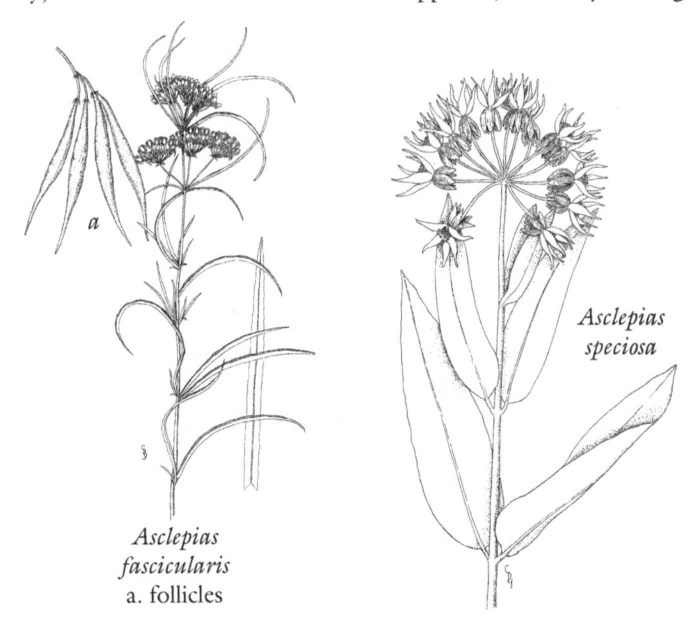

Asclepias fascicularis
a. follicles

Asclepias speciosa

5-15 cm. long, short-petioled; umbels several to many, terminal and lateral; flowers many, greenish or purplish, 2.5-5 mm. long; follicles erect, 5-12 cm. long, broadest near the base.

Occasionally found in large patches but not common in our range.

Asclepias speciosa Torr. Showy milkweed

Rhizomatous perennials; whole plant soft grayish-pubescent, or sometimes the stems nearly glabrous below, the inflorescence densely woolly; leaves opposite, somewhat heart-shaped to ovate or oblong, or the upper lanceolate; corolla lobes purplish, about 1 cm. long; stamen hoods somewhat longer, horned, with a long appendage; follicles densely woolly, 6-12 cm. long, erect on recurved pedicels.

Road sides and along streams; more common east of the Cascades.

CONVOLVULACEAE
Morning glory Family

Herbs (ours), mostly twining or trailing, with alternate leaves; sepals 5, separate, or united at base; corolla bell- or trumpet-shaped, generally large and showy, folded and twisted in the bud; stamens 5, attached to and shorter than corolla; pistil generally compound and 1-2-loculed; ovary superior; fruit usually a capsule.

Calyx closely subtended by 2 large bracts .... **Calystegia**
Bracts 2, narrow, borne near the middle of the peduncle .... **Convolvulus**

Calystegia

Perennials or subshrubs; stems usually twining, sometimes high climbing; leaves often arrowhead-shaped; flowers usually solitary in the leaf axils, large and showy; calyx with 2 closely associated bracts; ovary superior, typically 1-loculed; capsule often inflated.

1a Plants somewhat fleshy; leaves reniform .... *C. soldanella*
1b Plants not fleshy; leaves arrowhead-shaped or heart-shaped
  2a Plants less than 4.5 dm. tall, often trailing, but not twining....................
  .... *C. atriplicifolia*
  2b Stems twining, up to 3 m. long .... *C. sepium*

Calystegia atriplicifolia Hallier f. Hedge-row morning glory

Glabrous herbs; stems more or less erect or trailing up to 4.5 dm. long; leaves mostly broader than long, obtuse and mucronate at apex, more or less arrowhead-shaped or heart-shaped at base, 3.5-9 cm. long; flowers mostly in the lower leaf axils, subtended by 2 broad overlapping bracts; corolla broadly tubular, spreading slightly at the mouth, 3-6.5 cm. long, white or cream.

(=*Convolvulus nyctagineus* Greene)

In rocky places in wooded areas or in meadows.

Calystegia sepium (L.) R. Br. — Large wild morning glory

Rhizomatous perennials; stems reaching 3 m. long, climbing over other vegetation; leaves 4-12 cm. long, the blades generally broadly arrowhead-shaped, on petioles as long; flowers 3-7 cm. long, spreading, white or pinkish, the calyx closely subtended by the 2 broad bracts, pedicel longer than the flower.

(=*Convolvulus sepium* L.)

Showy, but not widely distributed in our region. A serious weed.

Calystegia soldanella (L.) R. Br. — Beach morning glory

Plants somewhat fleshy, prostrate, spreading; leaves reniform, broader than long, 2.5-5 cm. broad, on petioles at least as long as blades; flowers pink to purplish, about 5 cm. long, broadly spreading, the calyx closely subtended by 2 broad bracts.

(=*Convolvulus soldanella* L.)

Sandy ocean beaches.

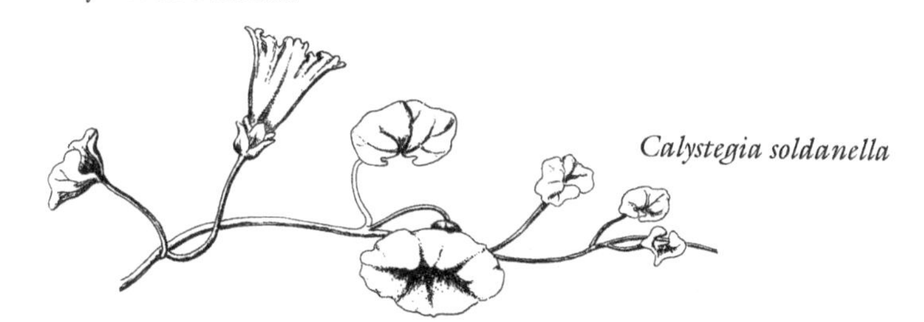

Calystegia soldanella

Convolvulus

Twining or spreading prostrate plants; flowers usually showy, bracts usually 2, situated well below the calyx; ovary 2-loculed; capsule globose, usually more or less inflated.

Convolvulus arvensis L.

Field or Orchard morning-glory or Bindweed

Stems trailing or twining, 3-12 dm. long, from deep-seated rhizomes; leaves narrow, arrowhead-shaped to hastate, with narrow acute lobes, somewhat pubescent to nearly smooth; corolla spreading, bell-shaped, white or deep pink outside, less than 25 mm. long; bracts 2, narrow, near the middle of peduncle.

An introduced weed and one of our worst field and garden pests.

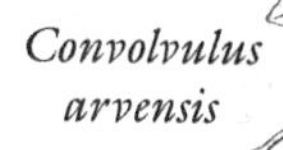

Convolvulus arvensis

CUSCUTACEAE
Dodder Family

Parasitic herbs, lacking chlorophyll; stems twining, thread-like, yellow to orange; leaves lacking or reduced to small scales; flower parts in 3's to 5's, perfect; calyx deeply-lobed; corolla white to yellowish, tubular to bell-shaped, the tube usually with scale-like appendages alternating with the stamens; ovary superior, 2-3-loculed; fruit a capsule or berry-like. This family is sometimes included in the **Convolvulaceae**.

Cuscuta

Characters those of the family.

1a Stigmas filiform; capsule circumscissile near the base
  2a Stamens extending beyond corolla .... *C. epithymum*
  2b Stamens included with in the corolla .... *C. approximata*
1b Stigmas capitate
  3a Corolla appendages inconspicuous, 0.1 mm. long or absent..................
  .... *C. californica*
  3b Corolla appendages conspicuous, over 0.5 mm. long
    4a Corolla tube shorter than wide .... *C. pentagona*
    4b Corolla tube longer than wide
      5a Calyx and corolla lobes obtuse .... *C. cephalanthi*
      5b Calyx and corolla lobes acute
        6a Corolla lobes about equalling the tube; salt-marsh species........
        .... *C. salina*
        6b Corolla lobes about 1/2 the length of the tube. *C. subinclusa*

Cuscuta approximata Bab. — Clustered dodder

Stems twining, yellowish; flowers sessile or nearly so, several in a head-like cluster; calyx yellowish, fleshy, the lobes usually 4, triangular-ovate, the tips slightly recurved; corolla 2-4 mm. long, bell-shaped to globose, 5-lobed, lobes slightly shorter than the tube, oblong-ovate, keeled; stamens included in the corolla, appendages fringed on the apex, nearly equalling the tube and united for 1/4 their length; capsule globose, circumscissile near the base.

Parasitic on crops, especially alfalfa.

Cuscuta californica Hook. & Arn. — California dodder

Stems twining, yellowish; inflorescence spike-like; calyx 2-3 mm. long, the lobes triangular to lanceolate; corolla 4-7 mm. long, tubular to bell-shaped, gland-dotted, 5-lobed, the lobes about as long as the tube, acute, entire, usually reflexed, appendages in the corolla inconspicuous; stamens exserted; capsule globose.

On a variety of hosts along road sides and in grasslands and woods.

Cuscuta cephalanthi Engelm. — Bush dodder

Stems dark yellow, somewhat coarse; flowers sessile or on short thick pedicels, in loose clusters; calyx lobes 3-5, oblong and obtuse; corolla 2-4 mm. long, lobes 3-4, obtuse, shorter than the tube, erect to spreading, appendages oblong, fringed principally near the apex; stamens usually slightly exserted; capsule globose; usually one seed maturing, this rounded, orange, scar a short depressed slit.

Parasitic on a variety of hosts, particularly golden-rod and other tall herbs and shrubs.

Cuscuta epithymum Murr. Small clover dodder

Stems slender, reddish; flowers sessile in dense clusters; calyx 5-lobed, lobes acute; corolla white or pinkish, 5-lobed, the lobes acute, remaining on the capsule, the tube urn-shaped, slightly longer than the lobes, appendages broad, fringed; stamens extending beyond the corolla; capsule globose, circumscissile near the base; seeds nearly spherical, brown or gray, scar a small paler-colored rounded spot.

Parasitic on clover, thyme, tomato, and other plants.

Cuscuta pentagona Engelm. Field dodder

Stems slender, pale yellow green; flowers sessile or nearly so, in dense globose clusters; calyx lobes 5, obtuse, broad, the calyx rarely longer than the corolla tube; corolla 1.5-3 mm. long, tube short and broad, with 5 acute lobes, appendages obovate, deeply fringed all around, sometimes longer than the corolla tube; stamens exserted; capsule globose; seeds lemon-yellow, gray, or light brown, broadly ellipsoid, one face rounded, the other flattened in 2 planes from a central ridge, scar a short white elevated line.

Parasitic on alfalfa, clovers, and other herbaceous plants.

Cuscuta pentagona on clover

Cuscuta salina Engelm.
Marsh dodder

Stems thread-like, golden-brown, twining; flowers 2-5 mm. long, narrowly bell-shaped; calyx lobes acute, as long as the corolla tube; corolla lobes broad at base, entire or fringed, acute or acuminate, nearly the length of tube, appendages inside the corolla 3-5 mm. long; capsule globose-ovoid; only 1 seed usually maturing, this yellow, orange or dark brown, scar a short white line, a pronounced protrusion not common to other dodder seeds occurring at the scar end.

Parasitic on *Salicornia* and other salt marsh plants, the stems often so abundant as to make conspicuous golden-brown patches over the marsh.

Cuscuta subinclusa Dur. & Hilg. Canyon dodder

Stems slender, yellowish; flowers sessile or nearly so, in dense globose clusters; calyx 4-5-lobed, the lobes acute to acuminate; corolla 4.5-5.5 mm. long, tubular, the lobes shorter than the tube, slightly fringed, appendages fringed on the margins; capsule ovoid; seed usually 1.

Often on shrubs, in moist areas, especially near streams.

POLEMONIACEAE
Phlox Family

Herbs or rarely shrubs; leaves alternate or opposite, simple or compound; flower parts typically in 5's; sepals partly united; petals partly united, corolla regular; stamens attached to the corolla and alternate with the lobes; ovary superior, 3-loculed (or rarely 2-loculed); stigmas 3 (rarely 2); fruit a capsule, opening by 3 (or rarely 2) valves.

1a Leaves opposite or appearing whorled or only the upper rarely alternate
  2a Leaves entire .... **Phlox**
  2b Leaves palmately divided into narrow segments .... **Linanthus**
1b Leaves alternate
  3a Calyx uniform in texture, not papery between the lobes, not bursting with the developing capsule
    4a Leaves simple, although often deeply lobed or cleft .... **Collomia**
    4b Leaves pinnately compound........ **Polemonium**
  3b Calyx more or less papery between lobes, often bursting with the developing capsule
    5a Inflorescence spiny-bracted; calyx lobes unequal, needle-pointed as are the leaves .... **Navarretia**
    5b Inflorescence not spiny-bracted; calyx lobes equal
      6a Flowers generally blue, not dotted .... **Gilia**
      6b Flowers scarlet to nearly white, finely dotted .... **Ipomopsis**

Collomia

Glandular-pubescent herbs; leaves alternate; flowers (in ours) in head-like clusters, each flower subtended by a leaf-like bract; calyx broadly funnel-shaped, not splitting between the teeth with the development of the capsule, but widening to form out-turned folds; corolla narrowly trumpet-shaped, or the lobes spreading saucer-shaped; stamens often unequal in length; capsule 3-loculed, containing 1 to several seeds in each locule.

1a Rhizomatous perennials; plants alpine........ *C. larsenii*
1b Annuals; plants not alpine
  2a Leaves mostly more or less divided........ *C. heterophylla*
  2b Leaves all entire or the lower occasionally slightly lobed, but not deeply so
    3a Corolla 1.5-3 cm. long, yellow to salmon-colored; anthers blue .... *C. grandiflora*
    3b Corolla 1-1.5 cm., pink to bluish-purplish or even white; anthers usually white .... *C. linearis*

Collomia grandiflora Dougl. ex Lindl. — Large-flowered Collomia

Erect annuals; stems 3-9 dm. tall; lower leaves lanceolate, 2-8 cm. long, leaves immediately beneath the flowers broader, minutely and stiffly pubescent; flowers trumpet-shaped, 1.5-3 cm. long, yellow to salmon-colored, in a dense terminal head, with smaller heads sometimes in the leaf axils; bracts subtending the flowers sticky-glandular; anthers blue.

Common in fields and on open hillsides.

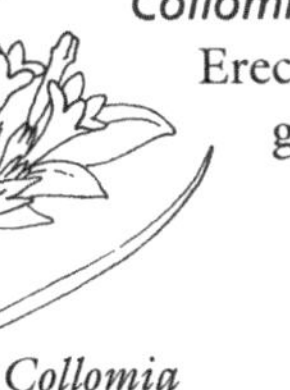
Collomia grandiflora

Collomia heterophylla Hook. Vari-leaved Collomia

Erect usually freely branched annuals 0.5-4.5 dm. tall, glandular-pubescent at least above; leaves varying much in shape and size, those beneath the flower clusters commonly broad, nearly entire; other leaves generally cleft or divided, the segments again cleft and toothed, all more or less glandular-pubescent; flowers in bracted clusters at the ends of branches; corolla slender, 8-16 mm. long, the lobes spreading, reddish-magenta or occasionally white; calyx less than 1/2 as long.

Margins of woods and on open slopes, sometimes along stream banks.

Collomia larsenii (Gray) Payson Talus Collomia

Rhizomatous perennials, tufted from a freely branched crown and spreading out for several dm.; leaves palmately once or twice cleft into 3-7 lobes; heads dense; corolla purplish to pink or even white, about 1.5 cm. long, at least twice as long as the calyx.

[=*Collomia debilis* (Wats.) Greene var. *larsenii* (Gray) Brand]

On talus slopes in the high Cascades.

Collomia heterophylla

Collomia linearis Nutt. Narrow-leaf Collomia

Erect annuals; stems simple or sometimes widely branching, whole plant pubescent or glabrous below, glandular above; leaves lanceolate or narrowly oblong, entire, narrowed at both base and apex, 1-8 cm. long; flowers in dense terminal heads and sometimes smaller lateral ones; corolla slender, pink to bluish-purplish or even white, 10-15 mm. long, twice as long as calyx.

Low ground; widespread.

Gilia

Annual or perennial herbs; leaves often in a basal rosette with the cauline leaves reduced upwards; calyx-tube semi-transparent and papery at the junction of the teeth, splitting between teeth or expanding as the capsule develops; corolla trumpet- shaped or the lobes spreading saucer-shaped, the tube generally longer than the lobes; capsule 3-loculed.

Gilia capitata

Gilia capitata Dougl. Field gilly-flower

Annuals; stems slender, scarcely branching, 2-10 dm. tall, pubescent, glandular, or glabrous; leaves 5-10 cm. long, forming a basal rosette and the cauline leaves alternate, gradually decreasing in size upward, pinnately compound, leaflets often again cut into linear segments; flowers in dense terminal or axillary heads; flowers blue, rarely white; corolla small, funnel-shaped, with long narrow lobes.

Common in fields and open sites.

Ipomopsis

Annuals or perennials; stems usually branched from the base; leaves alternate, simple, entire to deeply lobed and cleft; calyx tubular with membranous areas between the lobes; corolla bell-shaped or the lobes spreading. Sometimes included in the genus *Gilia*.

Ipomopsis aggregata (Pursh) V. Grant Skyrocket

Perennials 2-10 dm. tall; basal leaves pinnately lobed with 9-13 segments, cauline leaves with fewer segments; inflorescence long, loose; flowers trumpet-shaped, 2 cm. or more long, usually scarlet, but sometimes nearly white, finely flecked with white or yellow.

[=*Gilia aggregata* (Pursh) Spreng.]

A plant typically of eastern and southern Oregon, but abundant on peaks in eastern Linn and Lane Counties.

Linanthus

Annuals or perennials with opposite (the upper sometimes alternate) leaves palmately divided into narrow segments, or rarely entire; inflorescence in head-like clusters or flowers solitary; calyx tube membranous between the teeth; corolla bell-shaped to funnel-form or with the lobes spreading.

Annuals; flowers purplish, pink or yellow.. *L. bicolor*
Perennials; flowers white.. *L. nuttallii*

Linanthus bicolor (Nutt.) Greene Baby stars

Pubescent annuals; stem simple, 2.5-15 cm. tall; leaves very small at base, becoming larger upward, palmately cut into very narrow divisions, with 2 much-divided leaves at a node, therefore appearing as a whorl of simple leaves; flowers in a head at the apex, subtended by 1 or more pairs of leaves; calyx lobes narrow, somewhat stiff and sharp, pubescent; corolla 15-25 mm. long, much longer than the calyx, the tube thread-like then abruptly flaring, the lobes short, spreading, purplish or yellow, lobes pink to magenta.

Common on hill sides and open low ground.

Linanthus bicolor

Linanthus nuttallii (Gray) Milliken

Perennials; stems 1-3 dm. tall, clustered, usually pubescent; leaves palmately divided to base into 5-9 linear segments; flowers numerous, white; corolla tube woolly-pubescent.

[=*Gilia nuttallii* Gray and *Linanthastrum nuttallii* (Gray) Ewan]
Rocky cliffs and dry slopes in the Cascades.

Navarretia

Small stiff annual herbs with alternate mostly pinnatifid leaves; flowers in a close bracted head-like inflorescence; calyx tube scarious between the ribs, not bursting with the developing capsule, the ribs generally prolonged into bristles; corolla funnel-form or abruptly widened above the tube; capsule 2-3-loculed, rarely 1-loculed due to imperfect partitions.

Plants dark green, glandular, with a skunk-like odor .... *N. squarrosa*
Plant with a white pubescence, not glandular, without a skunk-like odor .... *N. intertexta*

Navarretia intertexta (Benth.) Hook. — Needle-leaf Navarretia

Stems erect, simple or branching below, 1-3 dm. tall, somewhat white-pubescent but not glandular; leaves once or twice pinnately divided, the segments narrow, rigid, sharply needle-pointed; heads dense, bracts conspicuously widened at base and white-woolly; corolla 4-12 mm. long, white or pale lavender or bluish, not much exceeding than calyx; ovary 2-loculed.
Common in open, often moist, areas.

Navarretia squarrosa (Esch.) Hook. & Arn. — Skunk weed

Stem erect, simple or branching, 1-4 dm. tall, whole plant sticky-pubescent, with a skunk-like odor; leaves once or twice pinnately divided, the segments narrow, weakly needle- pointed; heads dense; corolla 9-12 mm. long, blue, the tube slender, not much exceeding the calyx; ovary 3-loculed.
Common in open areas, more or less weedy.

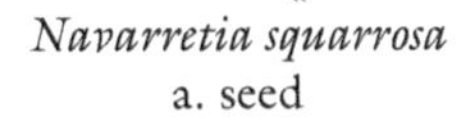

Navarretia squarrosa
a. seed

Phlox

Herbs, sometimes woody at base, with opposite entire leaves (or upper leaves alternate); flowers in cymes or solitary; calyx tube narrow, membranous between lobes, split by the developing capsule; corolla tube slender, the lobes spreading at right angles to the tube; stamens attached at different levels within the corolla tube.

1a Annuals; corolla about 4 mm. wide.. *P. gracilis*
1b Perennials: corolla 1 cm. wide or more
  2a Plants tufted or cushion-like .... *P. diffusa*
  2b Plants trailing, not tufted .... *P. adsurgens*

Phlox adsurgens Torr. — Trailing Phlox

Perennials; stems slender, trailing, minutely glandular near apex; leaves more or less broadly ovate, elliptic, or obovate, glabrous, sessile or short-petioled; flowers large, showy, in few-flowered cymes; corolla tube 12.5-25 mm. long, the lobes one-half as long, bright pink, broad, spreading.

Open woods, particularly southward.

Phlox adsurgens

Phlox diffusa Benth. — Spreading Phlox

Stems much branched at base, the branches prostrate, forming dense mats, the erect branches 2.5-5 cm. high; leaves linear-lanceolate, 6-20 mm. long, glabrous except near the base; flowers usually solitary at the ends of the branches, 12-20 mm. wide, pink, bluish, lavender, or whitish; calyx more or less white-pubescent; corolla tube slender, the lobes spreading.

Open rocky areas in the mountains, including the higher peaks of the Coast Range.

Phlox gracilis Greene

Annuals; stems slender, simple or branched, 6-25 cm. tall, somewhat sticky-glandular, at least above; leaves 1-3 cm. long, lanceolate to oblong, pubescent and glandular; corolla about 4 mm. wide and about 1 cm. long, scarcely longer than the glandular calyx, the lobes pink to white or magenta, the tube generally yellow.

[=*Microsteris gracilis* (Hook.) Greene]

Common in open areas, sometimes along streams.

Phlox gracilis

Polemonium

Annuals or perennials with alternate pinnately compound leaves with entire leaflets; flowers showy in corymbs, racemes, compound cymes, or panicles; calyx green throughout, the tube short; corolla funnel-form to saucer-shaped; stamens attached at nearly equal levels in the corolla tube.

1a Plants densely tufted; leaves mainly basal; flowers 7-13 mm. long... *P. pulcherrimum*

1b Plants not densely tufted

2a Corolla lobes blue; flowers 1-1.5 cm. long...................... *P. occidentale*

2b Corolla lobes cream-colored at first, turning pinkish to lavender; flowers 1.5-3 cm. long.. *P. carneum*

Polemonium carneum Gray Great Polemonium

Perennials; stems 3-9 dm. tall, generally much branched, spreading or weakly ascending, whole plant nearly glabrous or finely glandular-pubescent in the inflorescence and sometimes on the stems; leaflets 7-21, thin, ovate to lanceolate, acute, 1-4.5 cm. long; flowers 1.5-3 cm. long, generally cream at first, then pinkish, to lavender.

Western Washington, and scattered stations in western Oregon.

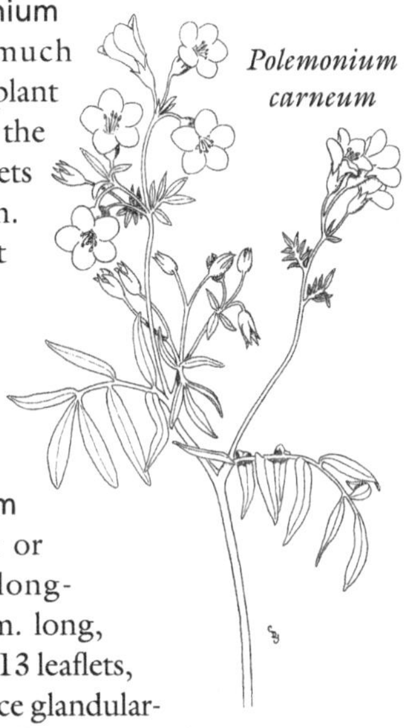

Polemonium carneum

Polemonium occidentale Greene
Western Jacob's ladder or
Western Polemonium

Rhizomatous perennials, 1.5-9 dm. tall, erect or decumbent, basal and lower cauline leaves long-petiolate, with 11-27 lanceolate leaflets 0.5-4 cm. long, middle and upper leaves much reduced and with 9-13 leaflets, the 3 terminal leaflets sometimes fused; inflorescence glandular-pubescent, the flowers in several cymose clusters; flowers 1-1.5 cm. long; corolla lobes blue, longer than the tube, the tube usually yellow.

Stream banks, damp meadows, swamps and other wet places, mostly in the mountains.

Polemonium pulcherrimum Hook. Showy Polemonium

Perennials; stems densely tufted at base, 0.5- 4.5 dm. tall, spreading or ascending, glandular-pubescent in the inflorescence; leaflets of basal leaves generally 11-23, cauline leaves few and with fewer leaflets, oblong or obovate; flowers 7-13 mm. long, pale or deep blue to purplish with yellow centers, borne in somewhat close cymes.

A very showy plant often about timberline in the mountains, sometimes lower; frequently abundant.

HYDROPHYLLACEAE
Waterleaf Family

Annuals, perennials or rarely shrubs; leaves alternate or opposite, simple to pinnately-compound, exstipulate; flowers solitary or in modified, often helicoid, cymes; flowers usually 5-parted, calyx deeply-lobed; corolla regular or nearly so, rotate to funnel-form or cylindric; stamens borne on the corolla tube; ovary superior, 1-loculed or appearing 2-loculed by the intrusion of the parietal placentae; fruit a capsule.

1a Flowers (in ours) axillary, or solitary at ends of branches .... **Nemophila**
1b Flowers not borne as above
  2a Leaves reniform to nearly orbicular, lobed........................ **Romanzoffia**
  2b Leaves not reniform to nearly orbicular
    3a Inflorescence sometimes helicoid but not conspicuously so; stems more or less succulent; fibrous rooted perennials ....**Hydrophyllum**
    3b Flowers in conspicuously helicoid cymes, these sometimes head-like; stems not succulent; taprooted annuals, biennials or perennials **Phacelia**

Hydrophyllum

Perennial herbs; leaves alternate, mostly basal, pinnately parted or compound; flowers in loose or compact head-like clusters; corolla bell-shaped, with narrow internal scales, the edges joining in pairs to form nectar cavities; stamens exserted; style 1, stigmas 2.

Leaves usually with 7-15 leaflets; calyx lobes bristly on margins and pubescent on back.. *H. occidentale*
Leaves with 3-5 leaflets, the terminal one 3-5-lobed; calyx teeth bristly on margins but rarely pubescent on back...................................... *H. tenuipes*

Hydrophyllum occidentale Gray — Western waterleaf

Plant branching from the base, erect, somewhat succulent, 1.5-6 dm. tall; stems mostly pale-green, usually bearing coarse down-turned hairs; leaves to 2 dm. wide, with 7-15 leaflets, the terminal much longer, generally 3-parted, all the leaflets coarsely toothed to lobed, pubescent; calyx stiff-hairy; corolla open bell-shaped, bluish or lavender; stamens much longer than corolla.

Common in shaded woods, along stream banks and in moist meadows.

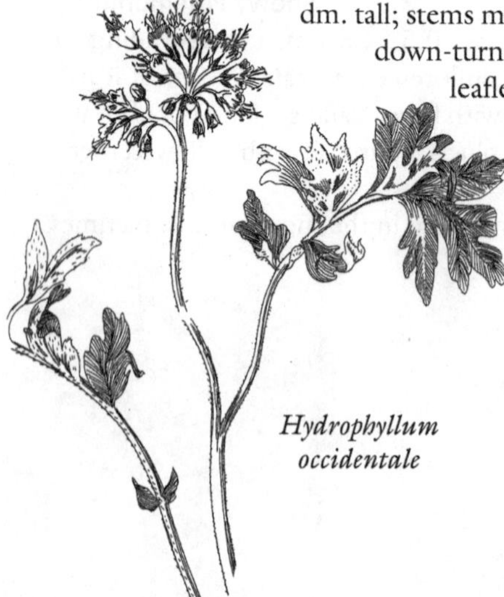

Hydrophyllum occidentale

Hydrophyllum tenuipes Heller — Pacific waterleaf

Plants somewhat succulent, erect, 2-8 dm. tall, herbage pubescent; leaves with 3-5 leaflets, the upper 3-5-lobed, all the leaflets coarsely toothed; flowers pale violet, bluish or greenish-white.

In moist woods and along streams.

Nemophila

Annuals with opposite or alternate leaves, sometimes both on the same plant; flowers solitary; calyx 5-lobed, with a small backward-turned appendage in each notch; stamens 5, short, attached near base of corolla, alternating with 10 small scales in pairs; style 2-cleft; capsule 1-loculed.

1a Flowers 1-3.5 cm. in diameter; peduncles much longer than leaves .... *N. menziesii*

1b Flowers 8 mm. or less in diameter

  2a Calyx appendages nearly as long as calyx lobes; stems mostly spreading .... *N. pedunculata*

  2b Calyx appendages much shorter than the calyx lobes; stems mostly weakly ascending .... *N. parviflora*

Nemophila menziesii Hook. & Arn. var. *atomaria* (Fisch. & Mey.) Chandler — Pale baby blue-eyes

Nemophila menziesii

Stems several, branched, spreading or more or less ascending, angular, roughened by short backward turned hairs; leaves pinnately 5-13-parted, the segments oblong, rough pubescent at least on the margins, opposite, on long petioles; flowers long-stalked, broadly cup- or saucer-shaped, 1-3.5 cm. in diameter, white, more or less blue-veined, generally black dotted within, pubescent at the center.

Fields and open hillsides; not widely distributed, but abundant where it occurs.

Nemophila parviflora Dougl. ex Benth. var. *parviflora* — Wood Nemophila

Stems decumbent or weakly ascending, usually rough with down-turned hairs, internodes long; leaves opposite or alternate, the lower mostly pinnately 5-parted or lobed, the segments broad, toothed; calyx appendages much smaller than the calyx lobes; corolla bell-shaped, 1-5 mm. in diameter, white or bluish.

Moist shady places.

Nemophila pedunculata Dougl. ex Benth. — Small Nemophila

Stems several, somewhat stout, spreading, angular, branched, minutely roughened by down-turned hairs, 6-30 cm. (rarely more) long; leaves opposite or rarely some alternate, 5-9-parted, the segments entire

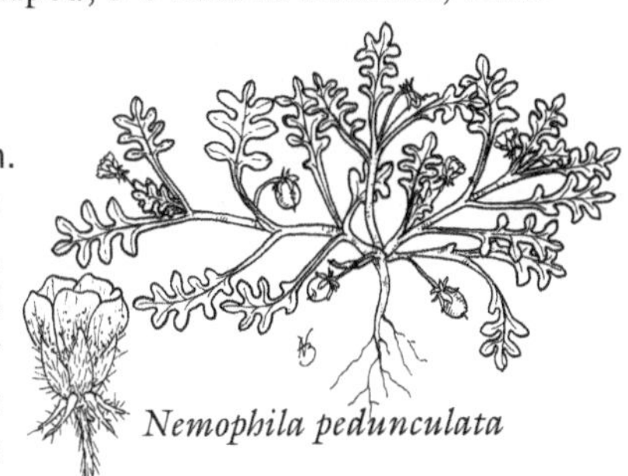
Nemophila pedunculata

or toothed, pubescence rough; calyx lobes alternating with reflexed appendages and about equal to them in length; corolla cup-shaped, 3-8 mm. in diameter, white or bluish, sometimes dotted or veined with blue or purple.

Open places and stream banks.

Phacelia

Annual, biennial or perennial herbs with alternate (lower sometimes opposite) leaves; flowers mostly showy, borne in helicoid cymes; calyx deeply 5-parted; corolla rotate to funnel-form, often with internal scales; stamens usually attached at the same level in the corolla; capsule 1-loculed or sometimes 2-loculed below the middle.

1a Flowers white or yellowish
  2a Dry-land plants; pubescence mostly appressed .... *P. heterophylla*
  2b Plants of moist places; pubescence spreading .... *P. nemoralis*
1b Flowers deep-violet or blue, very showy
  3a Perennials.. *P. sericea*
  3b Annuals
    4a Leaves linear, entire or with a few linear lobes .... *P. linearis*
    4b Leaves ovate or oblong in outline, deeply lobed to compound.........
    .... *P. tanacetifolia*

Phacelia heterophylla Pursh — Varileaf Phacelia

A highly variable species, of which a number of varieties have been described. Enumeration of these is not attempted here, but the following description is largely inclusive: Biennials or perennials; stems 1-12 dm. tall, typically stout; leaves entire or with 1 pair (or more) of smaller lobes or leaflets at base of blade, petioles of lower leaves slender, pubescence of usually appressed, stiff hairs, whole plant grayish, or hairs few and plants green; helicoid cymes generally numerous, terminal and lateral, peduncled, pubescent; flowers white or yellowish, 4-6 mm. long, stamens exserted, the filaments more or less pubescent with long hairs.

Dry sandy or gravelly exposed slopes.

Phacelia linearis (Pursh) Holz. — Threadleaf Phacelia

Annuals; stems simple or branched, erect, 1-5 dm. tall, whole plant more or less pubescent, young stems sometimes covered with short white hairs; leaves 1-8 cm. long, linear, entire, or cut into several linear, entire lobes; inflorescence of terminal and often lateral dense helicoid cymes; flowers showy, violet-blue, the corolla lobes spreading to about 1-2 cm. in diameter.

Not common in our limits.

Phacelia nemoralis Greene — Woodland Phacelia

Perennials resembling *P. heterophylla,* but plants more robust and occurring in shady moist places; plants green; the pubescence bristly and spreading; helicoid cymes slender, and often more open than in *P. heterophylla.*

Occasional in shaded places along streams.

Phacelia sericea (Graham) Gray — Silky Phacelia

Perennials; stems more or less tufted from a woody base, 1-3 dm. tall, whole plant silvery-silky to nearly glabrous and bright green; leaves clustered at the base, becoming fewer and shorter up stem, deeply pinnately-parted, the segments often again narrowly lobed; flowers 6 mm. or more broad, violet, showy, in helicoid cymes arranged in a dense spike.

In the high mountains.

Phacelia tanacetifolia Benth.

Annuals 1.5-10 dm. tall; stems glandular-pubescent; leaves deeply lobed to more often compound, leaflets toothed or lobed; inflorescence dense, usually 2-4-branched; calyx lobes densely pubescent; corolla 6-9 mm. long, blue.

Native of the southwestern United States. Apparently distributed in packets of wild flower seed; escaped from cultivation and becoming established.

Romanzoffia

Annuals or perennials; herbage more or less succulent; leaves simple, reniform to orbicular, basal leaves long-petiolate, cauline leaves alternate and reduced; inflorescence of loose cymes; calyx divided nearly to the base; corolla bell-shaped to funnel-form, white with a yellow center.

1a Annuals; some of the lower leaves opposite .... *R. thompsonii*
1b Tuberous or bulbous perennials; leaves alternate
  2a Coastal species; inflorescence shorter than or scarcely equalling the leaves .... *R. tracyi*
  2b Inland species or, if coastal, inflorescence greatly exceeding the leaves
    3a Stems from a cluster of brown-woolly tubers, and commonly with bulbils in the leaf axils .... *R. californica*
    3b Stems not from tubers, but from a bulb-like structure formed by thick overlapping petiole bases; bulbils not present in leaf axils ... *R. sitchensis*

Romanzoffia californica Greene — California mistmaiden

Perennials 1-4 dm. tall, bearing brown-tomentose tubers; leaf blades 1-5 cm. broad, reniform-orbicular, shallowly-lobed, palmately-veined, cauline leaves reduced and often bearing axillary bulbils; calyx lobes 2-5 mm. long; corolla 5-12 mm. long; ovary glandular-puberulent; capsules 6-10 mm. long.

(*Romanzoffia suksdorfii* Greene misapplied)

Wet rocky cliffs in the Coast Range and rarely in the Cascades; also on ocean bluffs.

Romanzoffia sitchensis Bong. — Sitka mistmaiden

Perennials 0.5-3 dm. tall; petioles widened and overlapping to form a bulbous base, but not tuber-bearing, leaf blades reniform to orbicular, 0.5-4 cm. wide, coarsely toothed to shallowly lobed, cauline leaves, if present, few and much reduced; calyx lobes 2-4 mm. long; corolla 5-11 mm. long; capsule 4-7 mm. long.

Wet cliffs in the mountains.

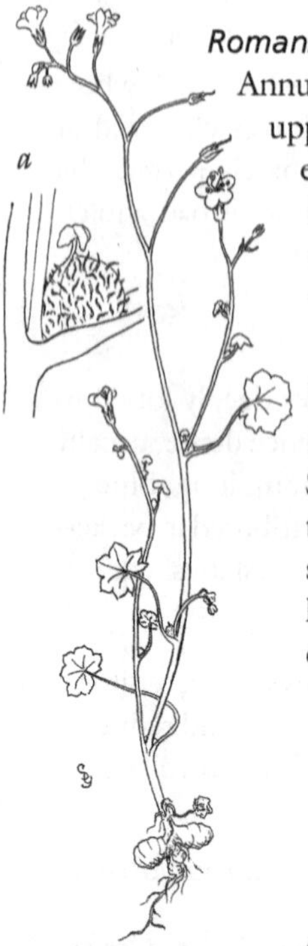

Romanzoffia thompsonii Marttala Thompson's mistmaiden

Annuals 1.5-18 cm. tall; at least some of the lower leaves opposite, upper leaves alternate, the blades nearly orbicular to obovate, entire to 3-5-lobed, 1-10 mm. long and 5-9 mm. wide, the base truncate to attenuate; inflorescence of few-flowered cymes; calyx lobes 1-3 mm. long; corolla long funnel-form, 4-5.5 mm. long.

Seasonally wet, usually rocky areas. Cascades Mountains and western foothills from Marion County, Oregon south.

Romanzoffia tracyi Jeps. Tracy's mistmaiden

Stems erect, arising from clusters of brown-tomentose tubers; leaf blades succulent, reniform or nearly round, shallowly 7-9-lobed, 1-5 cm. wide, on petioles 2-8 cm. long; calyx lobes 2-5 mm. long; corolla 6-8 mm. long; capsule 5-8 mm. long.

On moist cliffs along the coast.

Romanzoffia californica

a. bulbil in leaf axil

BORAGINACEAE
Borage Family

Annuals, perennials or shrubs, generally with a rough pubescence; leaves simple, alternate or rarely opposite, or both alternate and opposite on same plant; flowers usually in 1-sided coiled cymes, rolled inward, and unrolling during flowering; sepals 5, united at base; petals united at base into a long or short tube, forming a 5-lobed corolla and often bearing crests (fornices) at the apex of the tube opposite the lobes; stamens 5, attached to corolla alternate with its lobes; ovary superior, generally 4-lobed, usually dividing at maturity into 4 1-seeded nutlets; nutlets smooth or more often roughened or prickly.

1a Corolla without conspicuous swellings in throat
- 2a Flowers white .... **Plagiobothrys** (in part)
- 2b Flowers not white
  - 3a Flowers yellow or orange; plants with rough pubescence .. **Amsinckia**
  - 3b Flowers blue (often pink at first); plants without conspicuously rough pubescence .... **Mertensia** (in part)

1b Corolla with a circle of conspicuous swellings or crests in throat at bases of lobes
- 4a Mature nutlets with stout barbed prickles .... **Cynoglossum**
- 4b Nutlets not barbed, sometimes warty
  - 5a Corolla lobes erect or only slightly spreading; flowers blue or pink .... **Mertensia** (in part)
  - 5b Corolla lobes abruptly and widely spreading
    - 6a Receptacle flat or nearly so; nutlets smooth and shining, attached at exact base .... **Myosotis**

6b Receptacle more or less elongated; nutlets wrinkled or warty, mostly attached above exact base
   7a Receptacle much elongated; nutlets attached from base to center along a furrow generally forked at base .... **Cryptantha**
   7b Receptacle not as elongated; nutlets without furrows .... **Plagiobothrys** (in part)

Amsinckia

Bristly pubescent annuals, the hairs often with enlarged, hardened bases; flowers yellow or orange in elongating helicoid cymes, resembling 1-sided racemes; calyx-lobes generally 5; corolla with a long tube, the lobes spreading at right angles to the tube, the tube with inner folds but without crests; nutlets smooth or roughened.

1a Leaves finely and irregularly toothed; nutlets black .... *A. spectabilis*
1b Leaves entire; nutlets brown or gray
   2a Stamens inserted below the middle of the corolla tube; throat of the corolla obstructed .... *A. lycopsoides*
   2b Stamens inserted above the middle of the corolla tube; throat of the corolla open .... *A. intermedia*

Amsinckia intermedia
a. nutlet

Amsinckia intermedia Fisch. and Mey.
Fireweed fiddle-neck

Plants erect, somewhat branching above, 3-8 dm. tall, bristly pubescent throughout; lower leaves oblong, narrowing slightly downward toward base, but abruptly dilated at base, upper leaves long ovate, sessile, with entire or irregular margins; corolla tube very slender, lobes pale or deep yellow; pubescence of the inflorescence tawny; nutlets angled, transversely wrinkled, granular, brown.

A common weed.

Amsinckia lycopsoides Lehm.
Tarweed fiddle-neck

Stems erect to procumbent, very bristly; leaves lanceolate, or the upper broader; corollas 7-10 mm. long, deep yellow; nutlets brownish, granular.

Waste ground and open areas.

Amsinckia spectabilis Fisch. and Mey.
Seaside fiddle-neck

Stems erect to prostrate or spreading; leaves bright green, lanceolate, finely and irregularly toothed, bristly, especially on the margins; corollas orange to yellow; nutlets blackish, roughened and sometimes ridged.

Bluffs and dunes along the coast.

Cryptantha

Small generally hispid annual, biennial or perennial herbs; corolla small, white, abruptly widened above the tube; nutlets typically 4, attached by one-third their length to the greatly elongated receptacle.

Cryptantha intermedia (Gray) Greene Common white forget-me-not

Annuals; stems erect 1-3 dm. (rarely to 6 dm.) tall, slender-branched; leaves linear to oblong, 1-5 cm. long, hispid; cyme branches often paired, flowers scattered, bracts none; corolla about 5 mm. wide, white with yellow crests; nutlets about 2 mm. long, brownish, mottled with small and large white papillae, inner face grooved from apex to base, the groove widening at base.

[=*Cryptantha hendersonii* (Nels.) Piper]

Dry ground, often in wooded areas.

Cynoglossum

Coarse annual, biennial or perennial herbs with entire leaves; calyx deeply 5-cleft; corolla crested at the throat; nutlets more or less divergent, covered with stout barbed prickles.

Corollas pink, becoming deep blue; inflorescence terminal on a naked peduncle; bracts lacking or scale-like..*C. grande*

Corollas reddish-purple; inflorescence of numerous false racemes from the axils of the upper leaves ....*C. officinale*

Cynoglossum grande

Cynoglossum grande Dougl. Western hound's-tongue

Coarse erect perennials 2-9 dm. tall; stems usually smooth; leaves long-ovate, hispid at least beneath, the blades 1-2 dm. long, the petioles of the lower leaves longer, leaves reduced upwards; flowers in a loose branching inflorescence; corolla lobes spreading, 8-15 mm. broad, pink at first, becoming deep clear blue, with conspicuous white crests in the throat; nutlets prickly.

Open woods.

Cynoglossum officinale L. Common hound's-tongue

Biennials 3-12 dm. tall, villous throughout; lower leaves oblanceolate, tapering to the petiole, blades 8-20 cm. long, upper leaves sessile, oblong to lanceolate; flowers reddish-purple, about 1 cm. broad; nutlets prickly.

Weedy in disturbed sites.

Mertensia

Perennial herbs with erect or weak stems; leaves large; flowers blue or pink and blue; stamens attached to the corolla and included within the tube; style and stigma 1; nutlets roughened, attached slightly above base.

1a Stems from subterranean corm-like base; corolla not sharply divided between the tube and lobes .... *M. bella*
1b Stem not arising from a corm-like base; corolla sharply divided between the tube and the lobes
  2a Anthers 3.5-5 mm. long; calyx lobes obtuse or scarcely acute .... *M. platyphylla*
  2b Anthers 2-3.5 mm. long; calyx lobes narrow, acute or acuminate .... *M. paniculata*

Mertensia bella Piper — Oregon lungwort

Stems simple, generally less than 5 dm. tall, solitary from a more or less globose corm-like root; lower leaves broadly elliptical or nearly round, upper leaves narrower; flowers many in branching raceme-like cymes; calyx longer than narrowed portion of corolla tube; corolla 6-10 mm. long with short broad lobes, bright blue, crests not well developed.

Wet meadows in the mountains; at Tombstone Prairie on the South Santiam, and sporadic southward.

Mertensia paniculata (Ait.) G. Don var. *borealis* (Macbr.) Wms. — Tall bluebells

Stems from a branched caudex or a stout rhizome, 2-15 dm. tall; leaf blades 3-14 cm. long, thin, broadly ovate, acute or acuminate at the apex, broadly obtuse to subcordate at base, the lower petioled, the petioles fringed with hairs, and the under surface sparsely pubescent; calyx lobes fringed with hairs, also pubescent within; corolla 9-16 mm. long, the tube broad, shorter than the expanded portion, the crests evident.

Open woods, wet meadows and stream banks in Olympics and Cascades and along the Columbia River.

Mertensia platyphylla Hel. — Bluebells or Lungwort

Stems weakly erect from a stout branching rhizome; leaves sparsely pubescent, broadly ovate, acute or acuminate at apex, obtuse to truncate or even slightly heart-shaped at base, 5-12 cm. long, the lower long-petioled; flowers in loose few-flowered cymes at summit of stem, peduncle generally bearing 1 or 2 leafy bracts, pedicels long; calyx short; corolla tubular, abruptly widened at middle, lobes scarcely spreading, pink at first, becoming a clear blue at maturity.

Shady, moist margins of woods.

Mertensia platyphylla

Myosotis

Annuals to perennials, often short-lived; inflorescence an elongating helicoid cyme; corolla with a slender tube and abruptly spreading lobes, blue, white or yellow, bearing prominent crests in the throat; nutlets smooth, shining.

1a Stems creeping and stoloniferous, perennials; corolla 5-10 mm. broad .... *M. scorpioides*
1b Stems not stoloniferous, usually annuals or biennials or sometimes short-lived perennials; corolla 5 mm. or less broad
 2a Corollas white; calyx lobes somewhat different in size .... *M. verna*
 2b Corollas yellowish-white to blue or both on the same plant
  3a Hairs of the calyx closely appressed, not hooked at the tip; flowers all blue .... *M. laxa*
  3b Hairs of the calyx spreading, some of them minutely hooked at the tip; flowers yellowish-white at first, changing to blue .... *M. discolor*

Myosotis discolor Pers. Small blue forget-me-not

Annuals to biennials, simple or branched, 0.5-5 dm. tall; leaves oblong or obovate-oblong or the upper ovate, upper sessile, lower narrowed toward base, hispid; flowers small; calyx densely clothed with stiff hairs mostly minutely hooked at tips; corolla yellowish-white at first, changing to deep blue, the base of corolla tube remaining yellowish; nutlets dark brown to black.

Fields and hillsides.

Myosotis discolor

Myosotis laxa Lehm. Small forget-me-not

Annuals to short-lived perennials; stems weak, 1-4 dm. long, more or less decumbent, but not stoloniferous, glabrous or usually with appressed pubescent; lower leaves oblanceolate, gradually changing upward to lanceolate; flowers in long weak helicoid cymes; corolla 2-5 mm. broad, deep blue with a yellow center.

In shallow water or moist ground.

Myosotis scorpioides L. Common forget-me-not

Perennials; stems 2-6 dm. long, creeping and stoloniferous; leaves 2.5-8 cm. long, lower leaves oblanceolate, upper leaves elliptic to oblong; corolla blue, 5-10 mm. broad; nutlets blackish.

In shallow water or wet ground; introduced from Europe.

Myosotis verna Nutt. White forget-me-not

Annuals or biennials; stems erect or spreading, often freely branching, 0.5-4 dm. long, hirsute; leaves 1-5 cm. long, obovate to oblong, decreasing gradually upwards, lower most tapering at base into a broad petiole, upper sessile, mucronate; upper part of stem forming a long helicoid cyme, becoming 10-13 cm. in length, lowest flowers borne in leaf axils, upper flowers bractless, lateral cymes present on larger plants; calyx clothed with long hairs slightly hooked at tips; corollas 1-2 mm. broad, white; nutlets light brown, convex on outer face, angled on inner face, keeled between faces.

Found in low ground which is very wet in winter and dry in summer.

Plagiobothrys

Annuals or perennials; leaves narrow, lower leaves often opposite, upper usually alternate; inflorescence tightly coiled in flower; calyx cleft to the middle or below; corolla white, the crests sometimes yellow, the lobes spreading; nutlets often keeled on both dorsal and ventral and sometimes on lateral faces, but not grooved on inner keel, often wrinkled, attached at or above base.

1a Lower cauline leaves opposite, the upper alternate
  2a Corolla 5-10 mm. wide; nutlets broadly ovoid .... *P. figuratus*
  2b Corolla 1-4 mm. wide; nutlets narrowly ovoid .... *P. scouleri*
1b All but the lowest pair of leaves alternate
  3a Nutlet cruciform, ridged and tuberculate horizontally ...... *P. tenellus*
  3b Nutlet beaked but not cruciform, coarsely and reticulately ridged, minutely granular between ridges .... *P. nothofulvus*

Plagiobothrys figuratus (Piper) Johnst. Scorpion grass

Erect annuals; stems several to many, erect or spreading, 1-4 dm. long with appressed pubescence; leaves 6-15 cm. long, linear to narrowly lanceolate or oblong, pubescent, the lower opposite; flowers in helicoid cymes, closely inrolled in bud; calyx pubescent; corolla, 5-10 mm. in diameter, the lobes spreading, white, tube yellow, crests well developed, yellow; nutlets wrinkled and tuberculate.

Common on low ground which is very wet in winter and dry in summer.

Plagiobothrys figuratus

Plagiobothrys nothofulvus Gray Popcorn flower

Stems erect or nearly so; roots, petioles, sap and mid-ribs of basal leaves usually purplish; lower leaves long, narrowed to base, then dilated, upper leaves lanceolate; pubescence of stems and leaves white, rusty on pedicels and calyx-lobes; flowers 4-9 mm. in diameter; calyx-lobes falling away in fruit, leaving the tube as a cup enclosing the nutlets; nutlets grayish-green, transverse wrinkles distant, the scar large.

Vernally wet areas.

Plagiobothrys scouleri (Hook. & Arn.) Johnst. Scouler's popcorn flower

Annuals; stems one to several, slender, prostrate to erect, 5-20 cm. or more long, these and the leaves more or less strigose; leaves linear; flowers small, 1-4 mm. wide, borne in the axils of bracts, or the lowermost sometimes in leaf axils; nutlets narrowly ovoid, rugose, scar small.

Vernally wet areas.

Plagiobothrys tenellus (Nutt.) Gray — Small popcorn flower

Annuals; stems generally several from the base, branching, usually erect, 5-25 cm. tall; leaves ovate to lanceolate, mostly in a basal rosette, cauline leaves smaller; pubescence white or tawny in the inflorescence; flowers small, 2-4 mm. broad, white; nutlets grayish green, shining, closely wrinkled transversely, the wrinkles warty.

On land which is very wet in winter and baked in summer or sometimes on dry slopes in open woods.

VERBENACEAE
Verbena Family

Herbs (ours) to shrubs and trees with opposite or whorled leaves; flowers complete; calyx persistent, 4-5-toothed; corolla 4-5-lobed, slightly or quite irregular; stamens 4, in 2 pairs or 5; stigma entire or 2-lobed, style simple; ovary superior, 2-4-loculed, splitting at maturity into 2-4 1-seeded indehiscent nutlets (ours) or a capsule or drupe-like.

Verbena

Annual or perennial herbs usually with 4-angled stems and opposite leaves; flowers in simple or panicled bracteate spikes; calyx narrow, 5-angled, unequally 5-toothed; corolla 5-lobed, slightly irregular; stamens 4, included, one pair sometimes sterile; fruit enclosed by the calyx, at maturity splitting into nutlets.

Bracts of the inflorescence inconspicuous, 2-2.5 mm. long .... *V. hastata*
Bracts of the inflorescence conspicuous, 4-15 mm. long .... *V. bracteata*

Verbena bracteata Lag. & Rodr. — Bracted vervain

Stems usually prostrate or decumbent, sparingly pubescent, 1-6 dm. long; leaves oblong to obovate, variously toothed, cleft or pinnately divided, tapering to a winged petiole; inflorescence terminal, bracteate; corolla 5-6 mm. long, nearly hidden by the bracts, blue, pinkish or rarely white; nutlets reticulate above.

Occasional in waste places.

Verbena hastata

Verbena hastata L. — Tall wild Verbena or Blue vervain

Stems slender, unbranched except at top, 3-15 dm. tall; herbage hispid; leaves 7-15 cm. long, lanceolate, tapering to a slender tip, conspicuously veined, margins toothed, the teeth becoming larger toward the base, and the bases sometimes lobed; inflorescence often much branched into slender stiffly erect spikes, the individual spikes peduncled and leafless; bracts shorter than the flowers;

corollas 2.5-5 mm. long, purplish-blue, opening upward on the spike, only a few flowers of a spike open at one time; nutlets smooth.

Moist to wet areas; abundant locally but not widespread.

LAMIACEAE
Mint Family

Herbs or shrubs with 4-angled stems and simple opposite leaves, often aromatic; flowers borne singly or in cymes in the leaf axils, often so densely that they appear whorled; leaves sometimes becoming bract-like above; sepals partly united, the lobes equal or unequal; corolla tubular below, irregular and often 2-lipped; stamens 4, in 2 pairs, all anther-bearing, or 2 anther bearing and 2 sterile, or 2 absent, all attached to the corolla; ovary superior, 4-lobed or parted, dividing at maturity into 4 nutlets; style 1, 2-lobed at summit. The alternate name for this family is **Labiatae**.

1a Ovary merely 4-lobed; nutlets attached by inner margins .... **Trichostema**
1b Ovary deeply 4-parted; nutlets attached by their bases
  2a Corolla scarcely 2-lipped, the lobes nearly equal; corolla usually 4-lobed
    3a Plants very aromatic; stamens 4 .... **Mentha**
    3b Plants scarcely or not at all aromatic; fertile stamens 2 .... **Lycopus**
  2b Corolla distinctly 2-lipped, the lobes unequal
    4a Fertile stamens 2 .... **Salvia**
    4b Fertile stamens 4
      5a Calyx teeth equal or nearly so
        6a Flowers solitary or 2 in the leaf axils; plants trailing.... **Satureja**
        6b Flowers several in leaf-axils, or in spikes or racemes
          7a Calyx teeth spinulose-tipped; stems retrorsely-bristly on the angles .... **Stachys**
          7b Calyx teeth not spinulose-tipped although sometimes acuminate
            8a Calyx inconspicuously veined, not becoming greatly enlarged in fruit .... **Lamium** (in part)
            8b Calyx prominently 10-veined, becoming enlarged in fruit .... **Physostegia**
      5b Calyx 2-lipped or the teeth clearly unequal
        9a The 2 calyx lips entire, calyx crested on upper side. **Scutellaria**
        9b The calyx lips 2-3-toothed or lobed, calyx not crested.
          10a Inflorescence dense, terminal .... **Prunella**
          10b Inflorescence axillary or sometimes also terminal
            11a Lower lip of the corolla constricted at the base of the enlarged central lobe .... **Lamium** (in part)
            11b Lower lip of the corolla not constricted at the base of the central lobe
              12a Stems creeping; corolla blue or purplish . **Glechoma**
              12b Stems erect; corolla white, yellow or pinkish .... **Melissa**

Glechoma

Glechoma hederacea

Creeping perennial herbs; leaves long-petioled, the blades cordate to reniform, crenate or toothed; flowers in small clusters in the leaf axils; calyx slender, shorter than the corolla tube, unequally 5-toothed, 15-veined; corolla strongly 2-lipped; stamens 4.

(Also spelled *Glecoma*)

Glechoma hederacea L.
Ground ivy or Creeping Charlie

Stems 1-5 dm. long, leafy, creeping, rooting at the nodes and with ascending branches bearing flowers in the leaf axils; leaf blades 1-5 cm. long; calyx narrow, irregularly toothed; corolla blue to purplish, tube much longer than the calyx, the limb 2-lipped.

Escaped from cultivation and widely naturalized, most often in moist areas.

Lamium

Pubescent annual or perennial herbs; calyx 5-lobed, the lobes acute or acuminate; corolla strongly 2-lipped, the upper lip in bud enclosing the lower, like a cap, the lower lip constricted at the base of the enlarged central lobe; stamens 4; plants not aromatic.

1a Perennials; leaves generally white blotched.......................... *L. maculatum*
1b Annuals; leaves not white blotched
  2a Upper leaves sessile.. *L. amplexicaule*
  2b Leaves all petioled, uppermost generally purplish............ *L. purpureum*

Lamium amplexicaule

Lamium amplexicaule L. Dead-nettle

Annuals; stems 1-4 dm. long, generally several from base, prostrate to ascending, pubescent; upper leaves fan shaped, palmately veined, coarsely lobed or toothed, sessile, lower leaves smaller, rounded to heart-shaped, petioled, pubescent; flowers in dense whorls in upper axils; calyx pubescent; corolla narrow, tubular, 1-2 cm. long, purplish-red, upper lip cap-like, erect, velvety pubescent without, lower lip spreading at right angles, paler, purple-dotted.

A common garden weed, often a winter-annual.

Lamium maculatum L. Spotted dead-nettle

Perennials; stems 1-7 dm. long, creeping at the base, then more or less ascending; leaf blades more or less triangular to heart-shaped, generally with a whitish blotch in middle, the margins usually deeply toothed, all petioled, calyx lobes unequal; corollas 2-2.5 cm. long, purplish or rarely white.

Occasionally escaped from cultivation and becoming weedy.

Lamium purpureum L. Red dead-nettle

Annuals; stems weakly ascending, plants somewhat pubescent, lower leaves cordate and obtuse, upper ovate, acute, all petioled, the uppermost leaves generally purplish, the blades turned downward; calyx lobes long and slender, unequal; corollas 1-1.8 cm. long, pinkish-purple.

Introduced; weedy and locally abundant in gardens and moist places.

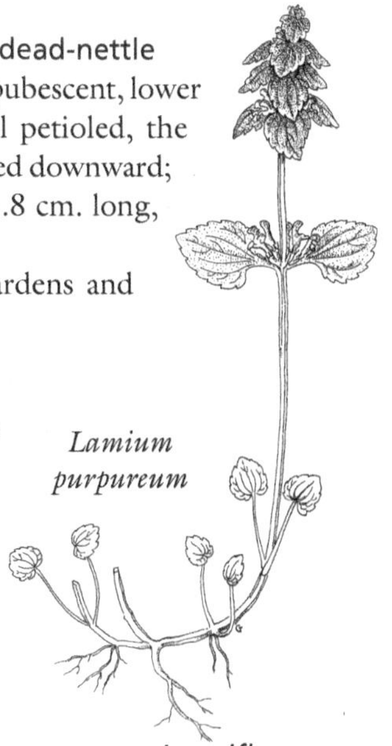

Lamium purpureum

Lycopus

Rhizomatous perennials, scarcely aromatic; leaves sharply toothed; flowers small, whitish, in dense axillary clusters resembling whorls; calyx 4-5-toothed; corolla 4-lobed, the upper lobe broader than the other 3; the 2 upper stamens without anthers, or absent; the 2 lower stamens fertile; nutlets wedge-shaped, thickened or corky at the angles.

Rhizomes tuber-bearing; leaves irregularly toothed .... *L. uniflorus*
Rhizomes not tuber-bearing; leaves deeply lobed or cleft at least on the lower half of the blades .... *L. americanus*

Lycopus americanus Muhl. Cut-leaved water horehound

Stems sharply 4-angled, erect, from a creeping rhizome, smooth or somewhat pubescent at least at the nodes, 2-8 dm. tall; leaves lanceolate or the lower oblanceolate in outline, deeply toothed or often cut into narrow lobes, especially at the base, the lower 3-10 cm. long, reduced upwards, long-acuminate, narrowed at base to a short petiole, light green, somewhat paler beneath; flowers small in dense axillary clusters, appearing whorled; calyx teeth longer than the nutlets, nearly as long as the corolla; corolla 2-3 mm. long, white, with 4 lobes, the upper lobe broader and slightly notched, the tube lined with a circle of hairs near the stamen bases; the 2 upper sterile stamens very small, white, somewhat knobbed at the apex; nutlets very granulose on upper inner angles.

Common in marshy ground and other moist areas.

Lycopus americanus

Lycopus uniflorus Michx. Northern bugleweed

Rhizomes tuber-bearing; stems 1-7 dm. tall, puberulent to finely strigose; leaves 2-8 cm. long, elliptic to lanceolate, margins irregularly toothed, short-petiolate; flowers in compact clusters in the leaf axils; calyx lobes obtuse to acute; corolla white to pinkish, 2.5-4 mm. long.

Stream banks, bogs, swamps, lake margins and marshes.

Melissa

Lemon-scented perennials with broad coarsely toothed leaves; flowers borne in axillary clusters, all the flowers of each node generally turning in the same direction; calyx 2-lipped, the upper lip flat, 3-toothed, the lower 2-lobed, the calyx tube bent upward, 13-veined, turning downward in fruit; corolla longer than the calyx, the upper lip entire or slightly notched, hood-like, the lower lip 3-lobed; stamens 4; the anther sacs widely separated; nutlets smooth.

Melissa officinalis L. — Lemon balm

Herbage lemon-scented; stems, 2-15 dm. tall, commonly much-branched, finely glandular-pubescent; leaf blades ovate, the margins coarsely toothed from middle to apex, 2-14 cm. long, gradually reduced upwards; flowers in axillary clusters; calyx flattened and bent upward about the middle, ciliate on veins and lower teeth, the 2 lower teeth deeply cut, the 3 upper short; buds yellowish, corolla 1-1.5 cm. long, white to yellow or becoming pinkish.

Introduced from Europe and escaped from cultivation and well established in moist ground, ditches and wet meadows, also in somewhat drier sites.

Mentha

Aromatic rhizomatous perennials; flowers in whorls in the axils of leaves or bracts or in spike-like inflorescences; calyx 10-veined, the 5 teeth equal or nearly so; corolla 4-lobed, the upper lobe formed by the fusion of the 2 upper lobes and broader than the other 3 lobes; stamens 4, nearly equal; nutlets smooth.

1a Flowers in terminal spikes or heads; stems glabrous or nearly so
  2a Cauline leaves on petioles 3-14 mm. long .... *M.* x*piperita*
  2b Cauline leaves sessile or subsessile, or the petioles 1 mm. or less .... *M. spicata*
1b Flowers in whorls in the leaf axils; stems puberulent to pubescent, often densely so
  3a Floral leaves much reduced, usually less than 1 cm. long; leaves velvety pubescent at least on the lower surface .... *M. pulegium*
  3b Floral leaves little reduced, 1.5-3 cm. long; leaves not velvety pubescent .... *M. arvensis*

Mentha arvensis L. — Field mint

Perennials from slender creeping rhizomes, rooting at the lower nodes and often sending up suckers at the base; stems 1-12 dm. tall, puberulent to pubescent; leaves 2-10 cm. long, ovate to elliptic, serrate, acute, the lower leaves short-petiolate, the upper subsessile, pubescent at least on the veins of the under surface, both surfaces with scattered minute embedded glands; flowers in dense whorls in the axils of the middle and upper leaves; calyx nearly regular, slightly constricted near base, dilated above, glandular-pubescent; corolla 4-7 mm. long, white to pink, lavender to purplish; stamens extending beyond the corolla.

Stream banks, lake shores, wet fields and marshy areas.

Mentha xpiperita L. Peppermint

Rhizomatous perennials; stems 3-10 dm. tall, usually glabrous; leaves lanceolate to ovate or elliptic, serrate, acute, 2.5-7 cm. long, dull dark green, minute glands embedded in the leaf surface seen with a lens, the leaf pubescent only on the veins beneath, or glabrous; whorls of flowers in dense spike-like inflorescences, the lateral spikes generally overtopping the terminal at maturity, bracts ciliate, shorter than the flowers except the lower which are sometimes leaf-like and longer; corolla 3.5-6 mm. long, white, lilac, pink or purple. Plants often sterile.

[A hybrid between *Mentha aquatica* L. and *Mentha spicata* L.]

Moist areas including fields, stream banks and ditches; widely cultivated and escaped and naturalized.

Mentha pulegium L. Pennyroyal

Rhizomatous perennials with decumbent to ascending stems 1.5-6 dm. tall, rooting at the lower nodes; herbage somewhat velvety pubescent; leaves short-petioled to nearly sessile, elliptic to ovate, toothed or entire, 5-25 mm. long, much reduced on basal sterile shoots and on the flowering branches; flowers in dense distant whorls in the axils of much reduced leaves on long branches; calyx and corolla glandular-pubescent; corolla 5-8 mm. long, purplish; stamens extending beyond the corolla.

An introduced plant which has become naturalized, especially in ditches, marshes and along stream banks.

Mentha spicata L. Spearmint

Stems erect or spreading from a creeping rhizome, generally branched, smooth, 3-10 dm. tall; leaves shining green, lanceolate, acuminate, 1.5-7 cm. long, slightly narrowed to truncate or subcordate at the base, sessile or nearly so, the midrib, primary and secondary veins all conspicuous, making the leaf appear reticulated, margins somewhat sharply serrate, the minute glands embedded in the leaf surface visible with a lens, the leaves pubescent only on the veins beneath, or glabrous; spikes very slender, interrupted, long, terminating the main stem and branches, those of the branches normally shorter than that of the main stem; bracts slender, ciliate, some of them generally as long as the flowers; flowers pale bluish or lavender to nearly white; calyx gland-dotted, the teeth ciliate, nearly as long as the tube; corolla 2-4 mm. long, sparsely gland-dotted, the upper lip slightly notched; stamens shorter than corolla.

Introduced from Europe; common in moist areas, lake shores, stream banks and ditches.

Physostegia

Erect perennial herbs with sessile leaves, and 2-lipped flowers in terminal spikes or racemes; calyx equally 5-toothed, becoming inflated in fruit; corolla much longer than calyx, upper lip entire, lower 3-lobed, spreading; stamens 4, the lower pair longer; nutlets smooth.

Physostegia parviflora Nutt. Purple dragon-head

Stems erect or ascending, 3-10 dm. tall, rarely branched, entire plant glabrous; leaves lanceolate to linear-oblong, sharply toothed to nearly entire, sessile or nearly so; flowers purple, 1-2.5 cm. long, many in terminal and lateral racemes.

(=*Dracocephalum nuttallii* Britt.)

Along streams; uncommon.

Prunella

Perennials; stems nearly simple, bearing a thick spike-like or head-like inflorescence at apex; calyx broad, with a wide obscurely 3-toothed upper lip, and a deeply and sharply 2-lobed lower lip; corolla longer than the calyx, the tube dilated above an inner oblique circle of hairs, upper lip of corolla 2 lobed, lower lip 3-lobed, the middle lobe the longest; stamens 4, paired, the filaments branched near the apex, the anthers borne on the lower branch.

Prunella vulgaris L. Heal-all or Self-heal

Stems often tufted at the base, prostrate or decumbent to erect, 1-5 dm. tall, smooth, or sparsely pubescent above; leaves long-petioled, glabrous, oblong-lanceolate or ovate, broadly angled or rounded at base, acute at apex, obscurely toothed; inflorescence a dense spike, flowers in whorls of 6's, each whorl subtended by 2 broad bracts; flowers blossoming progressively in the spike, the lower gone when the upper open; corolla showy, 8-18 mm. long, purple to violet or rarely white, twice as long as the calyx, the upper lip arching over the stamens, the broad middle lobe of the lower lip fringed on the margin.

Prunella vulgaris

Variety *vulgaris* is apparently an introduced European variety and var. *lanceolata* (Barton) Fern. is native. They differ with var. *vulgaris* being nearly prostrate and having cauline leaves twice as long as wide and var. *lanceolata* decumbent to erect with cauline leaves about 3 times as long as wide.

Road sides, gardens and lawns, stream banks, and margins of moist woods.

Salvia

Annual or perennial herbs or shrubs; flowers in interrupted spikes or in racemes or panicles; calyx 2-lipped, 10-15-veined; corolla distinctly 2-lipped, the lower lip 3-lobed, the 2 lobes of the upper lip often reduced; stamens 2.

Salvia aethiopis L. Mediterranean sage

Taprooted biennials; stems freely branching, 2-7 dm. tall, loosely woolly at least when young; leaves at first felt-like from the dense pubescence, but sometimes becoming nearly glabrous above, 6-25 cm. long, blades ovate to cordate or elliptic, irregularly shallowly lobed to coarsely toothed, the lower leaves petiolate; inflorescence large, open; flowers borne in whorls subtended by a pair of sessile bracts; calyx loose-woolly, purplish; corolla 1.5-2.5 cm. long, yellowish white; stamens exserted.

Introduced from the Mediterranean region; only occasional west of the Cascades.

Satureja

Fragrant evergreen perennials to vines or shrubs; flowers 1 to several in the leaf axils; calyx 5-toothed; corolla 2-lipped; stamens 4, in 2 pairs.

Satureja douglasii (Benth.) Briq. Yerba buena or Oregon tea

Rhizomatous perennials; stems slender, prostrate, or at first erect, 1.5-10 dm. long, rooting at the nodes, more or less pubescent; leaves very broadly ovate to nearly round, irregularly toothed, smooth, or somewhat pubescent, gland-dotted, blades 1-3.5 cm. long; calyx and corolla short-hairy without, the former conspicuously ribbed; flowers white, generally in pairs at a node, 1 in each axil of the opposite leaves.

Common in open woods. Very fragrant; it is one of our most-loved Pacific Coast herbs. Leaves and stems often purplish in winter.

Scutellaria

Annuals, perennials or sometimes low shrubs; flowers axillary or in bracteate racemes; calyx 2-lipped, the lips entire, the upper with a crest-like swelling, the 2 lips splitting apart in fruit; corolla 2-lipped, upper lip hood-like, lower lip 3-lobed; stamens 4 in two dissimilar pairs.

1a Flowers 6-8 mm. long, in slender racemes .... *S. lateriflora*
1b Flowers 12-32 mm. long, solitary in the leaf axils
  2a Flowers 12-22 mm. long .... *S. antirrhinoides*
  2b Flowers 22-32 mm. long .... *S. angustifolia*

Scutellaria angustifolia Pursh

Resembling *S. antirrhinoides,* but somewhat larger and with narrow scarcely-petioled leaves and a corolla 2.2-3.2 cm. long.

Only occasionally entering our limits.

Scutellaria antirrhinoides Benth.
Large skullcap

Rhizomatous perennials; stems branching, leafy, 1-3.5 dm. tall, whole plant finely puberulent; leaves ovate to oblong, obtuse, short-petioled, entire or nearly so, less than 3.5 cm. long; flowers 1.2-2.2 cm. long, dark blue, borne in the upper leaf axils, the corolla tube somewhat abruptly inflated above the calyx.

Moist shady places.

Scutellaria antirrhinoides

Scutellaria lateriflora L.
Common skullcap

Stems arising singly from a creeping rhizome, 2-15 dm. tall; leaf blades ovate to broadly lanceolate, coarsely-toothed, the tips acute,

Scutellaria lateriflora

3-10 cm. long, at least the lower and middle cauline leaves petiolate; flowers in slender axillary and often terminal racemes; corolla 6-8 mm. long, blue.

Moist woods, marshes, wet meadows, stream banks and other moist places.

Stachys

Annuals or perennials; herbage pubescent; flowers in whorls in an interrupted or continuous bracteate spike; calyx bell-shaped, 5-10-veined, 5-toothed, the teeth sharp; corolla 2-lipped; stamens 4; plants mostly with a disagreeable odor, especially when bruised.

1a Corolla tube 15-25 mm. long; flowers deep red or magenta .... *S. cooleyae*
1b Corolla tube less than 15 mm. long; flowers pink to purplish or white
  2a Leaves subtending at least the lowest 2 flower clusters long-petiolate .... *S. mexicana*
  2b Leaves subtending the lowest 1 or 2 flower clusters sessile or very short-petiolate... *S. rigida*

Stachys cooleyae **Heller** — **Giant hedge-nettle**
Rhizomatous perennials; stems 6-18 dm. tall, retrorsely-bristly on the angles; leaf blades 5-15 cm. long, only gradually reduced upwards, cordate-ovate, coarsely crenate, on petioles 1.5-4.5 cm. long; flowers showy, borne in loose spike-like arrangements; calyx 8-13 mm. long; corolla deep red or magenta, lower lip often white-spotted, the tube 15-25 mm. long.

Swamps, marshes, stream banks and other moist areas.

Stachys mexicana **Benth.** — **Great or Mexican hedge-nettle**
Rhizomatous perennials; stems 3-8 dm. tall, retrorsely-bristly on the angles; leaves only gradually reduced upwards, cordate-ovate, coarsely crenate, on petioles over 1 cm. long; calyx 5-9 mm. long; corolla pink to purplish, the tube 8-14 mm. long

Swampy areas and moist woods along the coast and in the Coast Range.

Stachys rigida **Nutt. ex Benth.** — **Rigid hedge-nettle**
Rhizomatous perennials; stems 1-10 dm. tall, retrorsely-bristly on the angles; leaf blades 1.5-9 cm. long, nearly glabrous to densely pubescent, sometimes glandular, oblong to lanceolate or ovate, lower leaves petioled, middle and upper leaves sessile or nearly so; calyx 5-6 mm. long; corolla white to pink or purplish, the tube 6-14 mm. long.

(=*Stachys ajugoides* Benth. var. *rigida* Jeps. & Hoover)

Moist to wet areas, but also in drier sites.

Trichostema

Annuals, perennials or subshrubs; leaves usually entire; inflorescence of axillary cymes or racemes; calyx 5-toothed, 10-veined, regular or irregular; corolla tube often abruptly bent or curved, 5-lobed, the lower lobe larger than the upper 4 lobes; stamens 4, the filaments purplish, generally much longer than the corolla, and curved outward and downward, the anther cells widely separated; nutlets attached by the inner margins.

Leaves with 3-5 conspicuous veins from below the middle; corolla tube much longer than calyx .... *T. lanceolatum*
Leaves pinnately veined; corolla tube nearly obscured by the calyx .... *T. oblongum*

Trichostema lanceolatum Benth. Blue-curls or Vinegar or Camphor weed
Stems 1-4 dm. tall, erect, simple or branched, loosely or densely leafy, the whole plant pubescent with both glandular and non-glandular hairs; leaves oblong to lanceolate, sessile or short-petioled, with 3-5 conspicuous veins from below the middle; flowers 18 mm. or less long, in dense or loose axillary clusters; calyx deeply lobed, about 1/2 as long as corolla tube; corolla light purplish or blue, the slender tube bent sharply backward about the middle, the 2 lobes of the upper lip shorter than the 2 lateral lobes of the lower, about equalling the lower middle lobe, the latter minutely purple-spotted.

Odor of the plant strong, generally considered disagreeable.

Dry ground, particularly where water earlier stood; widely distributed but usually not abundant.

Trichostema oblongum Benth. Mountain blue-curls
Glandular-pubescent annuals; stems erect, simple or often branched, 0.5-5 dm. tall, generally pubescent with both long and short hairs; leaves broadly oblong, obovate or elliptic, obtuse or acute at apex, narrowed at base, with short and long hairs; cymes in axils of foliage leaves; calyx very pubescent, nearly as long as the tube of the corolla; corolla blue, the tube slightly upturned, but not abruptly bent.

Moist areas in meadows and stream banks and in sandy ground.

SOLANACEAE
Nightshade Family

Herbs, shrubs, vines or rarely trees; leaves usually alternate and simple, exstipulate; flowers regular, solitary or borne in terminal or axillary umbels, panicles, or cymes; calyx lobes usually 5; corolla lobes generally 5, regular or nearly so; stamens 5, borne on the corolla tube and alternate with the lobes; ovary superior, usually 2-loculed; fruit a berry or a capsule.

1a Corolla rotate or the lobes turned backwards; fruit a berry .... **Solanum**
1b Corolla funnel-shaped, tubular, or with a slender tube and an abruptly spreading limb; fruit a capsule.
  2a Flowers solitary in forks of the stem; capsules large, prickly .... **Datura**
  2b Flowers in racemes or panicles; capsules smooth .... **Nicotiana**

Datura

Coarse herbs or shrubs, often ill-smelling; leaves entire to lobed, mostly petioled; flowers showy, borne singly in the axils of branches; calyx 5-toothed, tubular, generally persistent in part at base of fruit; corolla trumpet-shaped, white to purple; fruit a spiny capsule.

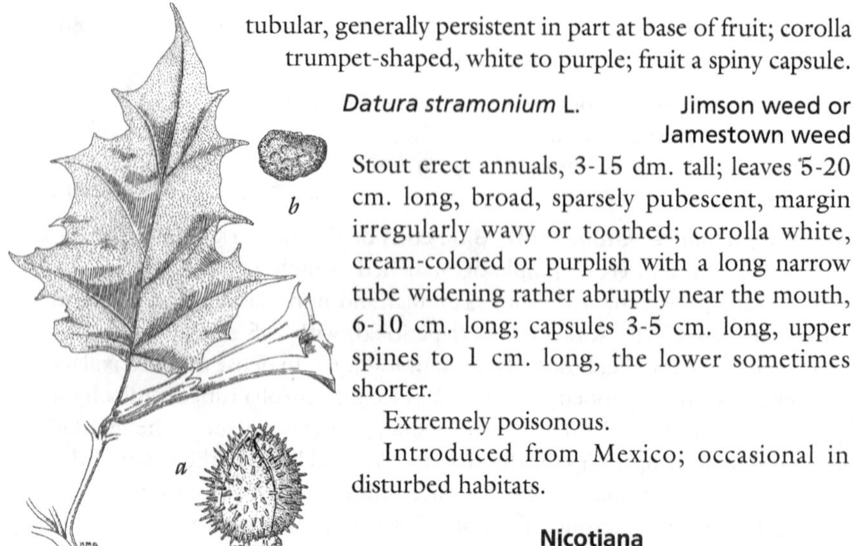

Datura stramonium
a. capsule b. seed

Datura stramonium L. Jimson weed or Jamestown weed

Stout erect annuals, 3-15 dm. tall; leaves 5-20 cm. long, broad, sparsely pubescent, margin irregularly wavy or toothed; corolla white, cream-colored or purplish with a long narrow tube widening rather abruptly near the mouth, 6-10 cm. long; capsules 3-5 cm. long, upper spines to 1 cm. long, the lower sometimes shorter.

Extremely poisonous.

Introduced from Mexico; occasional in disturbed habitats.

Nicotiana

Strong-scented narcotic herbs to trees; leaves entire; flowers in panicles or racemes; calyx persistent; corolla funnel-shaped or with a slender tube and an abruptly spreading limb, 5-lobed; capsule 2-4-valved.

Nicotiana quadrivalvis Pursh Wild tobacco

Annuals; stems stout, 3-20 dm. tall, branching, whole plant glandular-pubescent; leaves ovate-lanceolate, to oblanceolate, the upper narrowed at base; flowers few, white, often tinged with green, the lobes rounded; capsule globose.

Formerly reported along streams in western Oregon, but no present locality in our range is known. Said to be the only plant cultivated by the Native Americans of the Northwest.

Solanum

Annuals, perennials, shrubs or vines; inflorescence umbel-like or of panicles; calyx bell-shaped; corolla rotate, the lobes usually becoming reflexed; anthers 5, coming together around the style; fruit a berry or a capsule.

1a Rhizomatous perennials or becoming woody at the base, shrub-like or more often vine-like in riparian habitats; stems to 3 m. long; fruit bright red at maturity .... *S. dulcamara*
1b Not as above
  2a Calyx enlarging and enclosing the lower half of the brownish-green fruit .... *S. sarrachoides*
  2b Calyx not enlarging and enclosing the lower half of the fruit
    3a At least some of the hairs glandular .... *S. nigrum*
    3b None of the hairs glandular .... *S. americanum*

Solanum americanum Miller — American black nightshade

Stems 3-8 dm. tall; leaves ovate, entire to wavy-margined; calyx short, the lobes reflexed in fruit; corolla white, 3-6 mm. in diameter; berry 5-8 mm. in diameter, green to black.

A common garden and wayside weed of our area.

Solanum dulcamara L. — Bittersweet nightshade

Rhizomatous perennials or becoming woody at the base, shrub-like or more often vine-like in riparian habitats; stems to 3 m. long; some leaves simple, broadly ovate, 2.5-12 cm. long, other leaves with a pair of small basal lobes or leaflets, rarely with only 1 or 4 lobes; corolla purplish-blue, 8-12 mm. long, the lobes becoming reflexed; stamens deep yellow, projecting beak-like from the corolla; berries 8-12 mm. long, ovoid, changing from green through yellow and orange to bright red.

Moist, often disturbed habitats, especially along streams and lake margins; native of Eurasia, apparently escaped from cultivation.

Solanum nigrum L. — European black nightshade

Stems spreading or erect, glandular-pubescent, sometimes sparingly so; leaves ovate, entire or wavy-toothed; flowers to 1 cm. in diameter, white or purplish, in small umbels, the calyx lobes sometimes reflexed in fruit; berry 6-8 mm. in diameter, black at maturity.

Native of Europe; an occasional weed in gardens and disturbed habitats.

Solanum sarrachoides Sendtn. ex Mart. — Hairy nightshade

Annuals; stems more or less decumbent, densely pubescent and more or less glandular; calyx lobes broad and rounded, the calyx enlarging and enclosing the lower half of the brownish-green fruit; corolla white, 3-5 mm. in diameter.

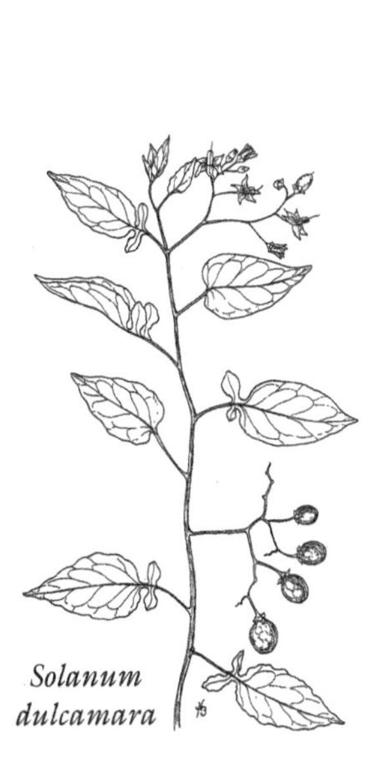

Solanum dulcamara

a. *Solanum nigrum*
b. fruit of *Solanum sarrachoides*

An occasional weed; introduced from South America.

Of this species and the other nightshades, the herbage and at least the unripe fruits are reported poisonous, in S. *dulcamara* the fruits are still poisonous when ripe.

SCROPHULARIACEAE
Figwort Family

Annuals, perennials or shrubs; leaves alternate or opposite, simple to variously dissected; calyx 4-5-lobed; corolla 4-5-lobed, often 2-lipped, rarely absent; stamens usually 4, rarely 2 or 5, a sterile stamen (staminodium) or 2 often present; ovary superior, usually 2-loculed; fruit a capsule.

1a Corolla not distinctly 2-lipped, though lobes varying somewhat in size
  2a Anther bearing stamens 5 .... **Verbascum**
  2b Anther bearing stamens 4 or 2
    3a Anther bearing stamens 4
      4a Leaves alternate and basal; corolla 4 cm. long or more.... **Digitalis**
      4b Leaves opposite; corolla less than 3 cm. long
        5a Stigmas 2-lobed .... **Mimulus** (in part)
        5b Stigma unexpanded ....**Tonella**
    3b Anther bearing stamens 2
      6a Corolla rotate; leaves mostly cauline................................**Veronia**
      6b Corolla bell-shaped or inflated and tubular
        7a Leaves opposite and sessile...........................**Gratiola** (in part)
        7b Leaves mostly basal, petioled...................................**Synthyris**
1b Corolla distinctly 2-lipped
  8a Corolla spurred or saccate at the base
    9a Corolla with a long narrow spur
      10a Stems erect; perennials (ours).. **Linaria**
      10b Stems prostrate or decumbent; annuals ....**Kickxia**
    9b Corolla merely saccate
      11a Sepals distinct; capsule asymmetrical......................**Antirrhinum**
      11b Sepals united at the base; capsule symmetrical .... **Collinsia**
  8b Corolla not spurred or saccate
    12a Upper lip of the corolla forming an elongated beak or rounded and helmet-shaped
      13a Anther sacs equal and parallel
        14a At least the lower leaves opposite..................... **Parentucellia**
        14b Leaves alternate and basal ....**Pedicularis**
      13b Anther sacs either only 1 per anther or unequally set; upper lip of the corolla forming a beak
        15a Tip of the corolla beak closed; stigma unexpanded................ ....**Orthocarpus**
        15b Tip of the corolla beak open; stigma expanded or 2-lobed
          16a Anther sac 1 per stamen................................ **Triphysaria**
          16b Anther sacs 2 per stamen, unequally set ....**Castilleja**

12b Upper lip of the corolla not elongated into a beak nor rounded and helmet-shaped
 17a Anther bearing stamens 2
  18a Anther sacs of each stamen parallel and separate; staminodia present as 2 knobbed ridges on the lower lip of the corolla .... **Lindernia**
  18b Anther sacs of each stamen not parallel and not separate; staminodia absent or represented by lines without knobs .... **Gratiola** (in part)
 17b Anther bearing stamens 4
  19a Sterile filament present
   20a Fertile filaments attached to the corolla at the same level, their bases densely pubescent .... **Nothochelone**
   20b Fertile filaments attached to the corolla at different levels, not densely pubescent at the very base .... **Penstemon**
  19b Sterile filament lacking, although staminodia often represented by knobs, glands or scales
   21a Calyx deeply 5-lobed or the sepals free or nearly so
    22a Leaves opposite; corolla 1.5 cm. long or less..... **Scrophularia**
    22b Leaves alternate and basal; corolla 4-6 cm. long ...... **Digitalis**
   21b Sepals united into a definite calyx tube
    23a Stigmas 2 or 1 and 2-lobed .... **Mimulus** (in part)
    23b Stigma unexpanded
     24a Stamens exserted .... **Tonella**
     24b Stamens included within the corolla .... **Collinsia**

Antirrhinum

Annual or perennial herbs; leaves usually opposite below and alternate above; corolla 2-lipped, saccate on the lower side of the base; fertile stamens 4; capsule asymmetrical, opening by 2 or 3 pores.

Antirrhinum orontium L. Lesser snapdragon

Annuals 2-5 dm. tall; leaves linear, 2-5 cm. long; flowers axillary or raceme-like; sepals linear, unequal; corolla 2-lipped, pink, 1-1.5 cm. long; capsule asymmetrical, pubescent, 6-10 mm. long.

Introduced species; occasional in gravelly areas.

Castilleja

Annuals, perennials to subshrubs, more or less parasitic on roots of other plants; leaves alternate, entire to dissected, sessile; inflorescence spike-like, bracts similar to the leaves, or more dissected, often with colored tips; calyx 4-lobed, usually unequally so; corolla 2-lipped, the upper lip beak-like, lower lip 3-toothed or sac-like and usually much-reduced; stamens 4; fruit a loculicidal capsule.

1a Annuals
 2a Bracts green; leaves linear-lanceolate, pubescent and glandular .... *C. tenuis*
 2b Bracts tipped with purple, yellow or white; leaves lanceolate to ovate
  3a Plants of salt marshes and coastal bluffs; lower lip with inflated pouches.. *C. ambigua*
  3b Plants not maritime; lower lip scarcely inflated .... *C. attenuata*

1b Perennials
4a Leaves all (or nearly all) entire
5a Calyx lobes and lobes of the bracts typically obtuse
6a Bracts yellow .... *C. levisecta*
6b Bracts red- to orange-red-tipped .... *C. affinis*
5b Calyx lobes and lobes of the bracts acute to acuminate
7a Bracts mostly scarlet; stems several; plants of lowlands and mountains; common .... *C. miniata*
7b Bracts green with a yellow band and scarlet tips; stems arising singly; plants alpine; uncommon .... *C. suksdorfii*
4b All but the lowest leaves with narrow lobes (see also *C. suksdorfii*)
8a Bracts yellow .... **C. levisecta**
8b Bracts red or rose (rarely whitish)
9a Upper lip of the corolla 1/2 the length of the tube; bracts rose to carmine, never scarlet .... *C. parviflora*
9b Upper lip of the corolla about as long or longer than the tube; bracts scarlet to reddish-orange
10a Lateral lobes of the leaves usually about equal in width to the central lobe; plants alpine .... *C. rupicola*
10b Lateral lobes of leaves usually much narrower than the central lobe
11a Leaves nearly glabrous .... *C. chambersii*
11b Leaves hispid to villous .... *C. hispida*

Castilleja affinis Hook. & Arn. subsp. *litoralis* (Penn.) Chuang & Heckard
Pacific or Oregon-coast paintbrush

Perennials 1.5-6 dm. tall; finely hispid-villous or rarely nearly glabrous; leaves 3-8 cm. long, oblong, mostly entire; corolla 2.5-4 cm. long, the beak about 3/4 the length of the tube or longer; bracts broad and obtuse, entire or with 2-4 short lateral lobes, red- to orange-red-tipped.

(=*Castilleja litoralis* Penn.)

Coastal dunes and bluffs.

Castilleja ambigua Hook. & Arn.

Annuals, 1-3 dm. tall; leaves 1-5 cm. long, lanceolate to ovate, entire to 5-lobed; bracts of the inflorescence tipped with white, yellow or purple, 3-9-lobed, the central lobe rounded; calyx 12-20 mm. long, unevenly divided into linear lobes; corolla 14-25 mm. long, pale yellow to purplish; capsules 8-12 mm. long.

(=*Orthocarpus castillejoides* Benth.)

Salt marshes and coastal bluffs.

Castilleja attenuata (Gray) Chuang & Heckard

Annuals; stems slender, erect, generally not clustered, 1-5 dm. tall, pubescent above; leaves long-lanceolate, gradually tapering at apex, upper leaves generally with narrow lobes; bracts 3 or more lobed, white- or yellow-tipped; calyx lobes 4, narrow; corolla white or yellowish, with shallow lobes, lower lip scarcely inflated, purplish-red-dotted, the lobes linear.

(=*Orthocarpus attenuatus* Gray)

Common in meadows and pastures.

Castilleja chambersii M. Egger & Meinke

Low perennials; stems 1.5-3.5 dm. tall, ascending, usually branched from the base; leaves 2-4 cm. long, divided into 5-11 segments, these often cleft or divided; bracts unevenly divided, scarlet to reddish-orange; corolla 3-4.5 cm. long, the beak equal to or longer than the tube.

Known only from near the summits of a few Coast Range peaks in Clatsop County, Oregon.

Castilleja hispida Benth. Indian paintbrush

Erect perennials; stems little branched, 1.5-5 dm. tall, more or less villous to hispid, at least the upper bracts glandular-pubescent; lower leaves generally undivided; upper leaves with 3-5 pinnate lobes with 3-5 main veins from the base; bracts deeply 3-5-lobed, the central lobe broader, lateral lobes of bracts and leaves widely spreading; tips of bracts and corolla brilliant scarlet; spikes dense.

In this exceedingly variable species, a number of varieties or subspecies have been recognized and, by some botanists, treated as species.

Variability occurs in amount of pubescence, and shape and proportions of leaves.

Dry fields and open woods; uncommon in our range.

Castilleja levisecta Greenman Golden Indian paintbrush

Glandular-pubescent perennials; stems 1.5-5 dm. long, more or less decumbent at the base; lower leaves mostly entire, upper leaves usually with 3-7 narrow lobes; bracts entire or with several short lobes, yellow.

Meadows and moist places; uncommon in our limits.

Castilleja miniata Dougl. Scarlet Indian paintbrush

Perennials; stems slender to stout, 3-8 dm. tall, principally glabrous, but sometimes villous above; leaves 3-6 cm. long, lanceolate to linear, entire, sessile, with 3 prominent veins from the base; inflorescence a dense terminal spike, sometimes smaller lateral spikes from upper axils; bracts red-tipped (rarely yellowish), entire to toothed or cleft at apex into 3-5 (or rarely up to 9) linear lobes; corolla 1.5-4 cm. long, the beak greenish-yellow with red margins, the lower lip dark green, much-reduced.

A highly variable species.

Our most common *Castilleja*; usually occurring in moist to wet ground.

Castilleja parviflora Bong.

Tufted perennials from a woody base, sometimes forming large clumps; stems 1.5-4.5 dm. tall, glabrous to pubescent; leaves mostly cleft half their length into 3-5 narrow spreading lobes, the terminal lobe broader at the base; bracts mostly 3-5-cleft or lobed, rose-colored to carmine, rarely white; corolla 12-28 mm. long, yellowish-green, upper lip slender, about equalling the tube, lower lip very short.

Common in meadows of the Cascades and high Coast Range peaks at about timberline.

Castilleja rupicola Piper — Cliff paintbrush

Perennials; stems less than 3 dm. tall, somewhat pubescent; leaves and bracts mostly with 3-7 narrow lobes; bracts scarlet-tipped; upper corolla-lip longer than tube; lower lip short, thickened, dark green.

Rocky areas in the northern Cascades.

Castilleja suksdorfii Gray

Perennials; stems 3-8 dm. tall; herbage glabrous to somewhat pubescent; leaves linear-lanceolate, the lower entire, upper sometimes narrowly 3-5-lobed, the lobes erect; bracts deeply 3-5-lobed, the lobes often toothed, green with a yellow band and scarlet tips; corolla 2.5-5 cm. long, the upper lip longer than the tube.

Moist to wet subalpine meadows of the Cascades.

Castilleja tenuis (Heller) Chuang & Heckard

Annuals, 1-4.5 dm. tall; herbage pubescent and glandular, at least above; leaves linear-lanceolate, 1-4 cm. long, entire or 3-5-lobed; bracts of the inflorescence 3-7-lobed, green; calyx 8-12 mm. long; corolla 12-20 mm. long, white or yellow, the lower lip with 3 pouches.

(=*Orthocarpus hispidus* Benth.)

Wet meadows and other moist sites.

Collinsia

Annuals with opposite or whorled leaves and axillary flowers; calyx 5-lobed; corolla turned downward at a sharp angle with base of tube, swollen or spurred above, 2-lipped, upper lip erect, 2-lobed, lower lip 3-lobed, the central lobe folded to enclose stamens and style; fertile stamens 4 in 2 dissimilar pairs, the 5th stamen represented by an inconspicuous gland at the base of corolla.

Corolla 9-18 mm. long, the tube twice as long as calyx tube .... *C. grandiflora*

Corolla 4-8 mm. long, the tube little longer than calyx tube .... *C. parviflora*

Collinsia grandiflora Dougl. — Large innocence

Stems erect or somewhat spreading, stout or slender, 1-3.5 dm. tall; leaves opposite below, mostly whorled above; lower leaf blades nearly orbicular to short-oblong, long-petioled, several toothed to entire, upper leaves long-lanceolate, scarcely petioled, entire or somewhat toothed; flowers in upper leaf axils; pedicels pubescent; corolla 9-18 mm. long, much longer than calyx, corolla tube lavender, lobes deep blue.

Rather common, in fields and the margins of woods. One of our most beautiful spring flowering plants.

Collinsia grandiflora

Collinsia parviflora Dougl. — Small innocence

Stems 1 to several from the base, 0.5-3 dm. tall (rarely taller); leaves lanceolate or ovate to oblanceolate or obovate, short-petioled or the upper nearly sessile, entire or toothed, minutely pubescent, opposite or the upper leaves several in a whorl, green, or whole plant sometimes reddish; flowers borne in the leaf axils; corolla 4-8 mm. long, pale blue and lavender with a whitish upper lip; pedicels elongating in fruit and sometimes turned downward.

Common in fields and open woods.

Digitalis

Biennials or perennials with basal and alternate leaves; flowers large, tubular, borne in a long terminal bracteate raceme; calyx deeply 5-lobed; stamens 4, in 2 pairs, shorter than corolla tube.

Digitalis purpurea L. — Foxglove

Biennials; stems erect, 1-3 m. tall, puberulent and glandular at least above; leaves in large basal tuft the first year, blades 1-3.5 dm. long, ovate, narrowed to a long petiole, the veins of the blade conspicuous, upper leaves becoming gradually smaller with shorter petioles, margins crenate or toothed, under surface densely pubescent, less so above; bracts of the inflorescence ovate; corolla 4-6 cm. long, conspicuously inflated, particularly on lower side, white to pink or purple with lower side of throat whitish and purple-dotted; flowers somewhat drooping and more or less turned to one side of the raceme.

Introduced from Europe; escaped from cultivation and especially common near the coast.

This species is the source of the drug digitalis and the plant is poisonous.

Gratiola

Annuals or perennials; leaves opposite, sessile; flowers solitary in axils of the upper leaves or racemose; calyx of 5 free sepals, often subtended by a pair of bractlets; corolla 2-lipped, upper lip entire or 2-lobed, lower lip 3-lobed; fertile stamens 2, often with 2 rudimentary stamens.

Pedicels without bracts; leaves acuminate .... *G. ebracteata*
Pedicels with a pair of bracts below the sepals; leaves obtuse or barely acute.....
.... *G. neglecta*

Gratiola ebracteata Benth. — Bractless hedge-hyssop

Somewhat fleshy annuals to 25 cm. tall; leaves 0.5-2.5 cm. long, ovate to lanceolate, acuminate, tapered to the clasping base, entire or dentate towards the apex; sepals 8-11 mm. long, without bracts below; corolla inflated, tubular with short lips, 5-7 mm. long, the tube yellowish, the lips white or pinkish with delicate lavender veining; fertile stamens 2; capsules 4-5 mm. long.

In shallow water or wet, muddy ground and vernal pools.

Gratiola neglecta Torr. — American hedge-hyssop

Annuals, 1-3 dm. tall; usually glandular at least above; leaves 0.5-5 cm. long, lanceolate to obovate, entire or with a few teeth, the tip obtuse or barely acute; sepals 3-7 mm. long, subtended by a pair of bracts; corolla 5-10 mm. long, the tube white or yellowish, the lobes usually white with a purple tinge; capsules 5-7 mm. long.

In shallow water or wet, usually muddy or sandy ground and vernal pools.

Kickxia

Annuals or perennials with prostrate to decumbent stems; leaves alternate or the lower opposite; flowers solitary in the leaf axils; calyx 5-lobed; corolla 2-lipped, spurred, the lower lip inflated and closing the throat; capsule circumscissle.

Some of the leaves arrowhead-shaped...*K. elatine*
Leaves cordate at the base, none of them arrowhead-shaped.............*K. spuria*

Kickxia elatine (L.) Dumort. Sharppoint Fluellin

Procumbent glandular-pubescent annuals; leaves alternate, or the lower opposite, at least the middle cauline leaves arrowhead-shaped, but sometimes with more than one pair of basal lobes; flowers borne in the leaf axils on long pedicels; corollas 0.5-1.5 cm. long, including the long spur, pale yellow with a violet upper lip; capsule globose, about 4 mm. in diameter.

Introduced from Europe; usually in moist sandy soil in disturbed habitats; becoming a problem in nursery stock.

Kickxia spuria (L.) Dumort.

Similar to *Kickxia elatine* but none of the leaves arrowhead-shaped, the corollas 1-1.5 cm. long, yellow with a purple upper lip.

Introduced from Europe, only occasional in our area.

Linaria

Annual or perennial herbs with opposite or whorled leaves, or the upper alternate; flowers in terminal spikes or racemes; calyx 5-parted; corolla 2-lipped, spurred at the base, the throat often closed or nearly so; stamens of 2 lengths, paired; capsules opening below the apex by pores or slits.

Flowers blue...*L. canadensis*
Flowers yellow and orange..*L. vulgaris*

Linaria canadensis (L.) Dum.-Cours. Blue toad-flax

A slender, blue-flowered annual or biennial, rarely found in our area.

Linaria vulgaris Mill. Toad-flax or Butter and eggs

Perennials; stems erect, 3-9 dm. tall, leafy, simple or with slender basal branches; leaves linear or somewhat wider, more or less glaucous, lower leaves 2-10 cm. long, becoming shorter above, gradually reduced to bracts; inflorescence a long spike-like raceme; calyx-lobes small, nearly regular; corolla 2-3.5 cm. long including the long slender spur, upper lip and spur yellow, lower lip orange, with 2 bearded orange lines extending into the throat; seeds many, winged.

Naturalized from Europe and established in many parts of the Northwest as a troublesome weed.

Linaria vulgaris

Lindernia

Low annual or biennial herbs with sessile opposite leaves; flowers solitary in the leaf-axils, or in terminal racemes; sepals 5, united only at the base; upper lip of corolla 2-lobed, lower lip 3-lobed; fertile stamens 2, inserted near base of corolla; the other pair reduced to filaments which form pubescent ridges with knobs high in corolla throat; style thread-like; stigma 2-lobed; capsule septicidal, many-seeded.

Lindernia dubia (L.) Pennell — False pimpernel

Glabrous annuals; stems 5-35 cm. tall; leaves 5-30 mm. long, lanceolate to ovate, entire to finely toothed; sepals 2.5-7 mm. long; corolla 7-10 mm. long, white to lavender. Two varieties occur in our area: var. *dubia* with leaves only slightly reduced above and with the pedicels surpassed by or only slightly surpassing the subtending leaves and var. *anagallidea* (Michx.) Cooperr. with the pedicels much surpassing the subtending leaves.

Wet areas, especially on muddy banks along rivers and margins of ponds.

Mimulus

Annual or perennial herbs or shrubs with opposite leaves; flowers generally showy, in racemes or solitary in the leaf axils; calyx 5-lobed, 5-angled; corolla more or less 2-lipped, the tube 2 ridged on the lower side; stamens 4 in 2 unequal pairs; stigmas 2; capsule loculicidal.

1a Flowers red, orange or purplish
  2a Plants 15 cm. or less in height; annuals
    3a Lower lip reduced to 1 (rarely 3) minute lobe(s) .... *M. douglasii*
    3b Lobes of lower lip about equal to the lobes of the upper lip .... *M. tricolor*
  2b Plants 20 cm. or more in height; perennials
    4a Flowers pinkish-purple, marked with yellow; stamens enclosed by the corolla .... *M. lewisii*
    4b Flowers red to orange; stamens exserted .... *M. cardinalis*
1b Flowers predominantly yellow
  5a Plant slimy-glandular, musk-scented .... *M. moschatus*
  5b Plant not musk-scented, rarely slimy
    6a Leaf blades distinctly pinnately veined .... *M. dentatus*
    6b Leaf blades (or the large terminal lobe) palmately veined
      7a Calyx broadly inflated, the upper tooth larger than the others .... *M. guttatus*
      7b Calyx narrow, the teeth about equal or the lower longer
        8a Plants with obvious stems, generally branching, longer than peduncles .... *M. alsinoides*
        8b Stems short, the peduncles longer than the stems .... *M. primuloides*

Mimulus alsinoides Dougl. ex Benth. — Baby monkey-flower

Annuals; stems weakly ascending to rather stiffly erect, simple or branching, 2.5-30 cm. tall, rarely reaching 40 cm., glabrous; leaf blades broadly ovate to deltoid, finely toothed, tapering to the petioles, palmately veined, 1-2.5 cm.

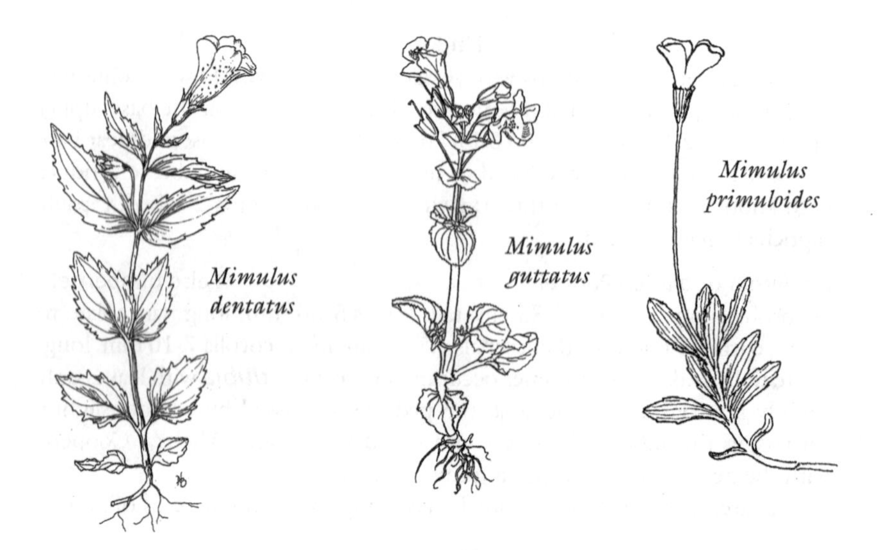

long (or shorter in low specimens); calyx 4-8 mm. long, the 2 lower teeth longer than the upper 3; corollas 8-15 mm. long, 2-lipped, yellow, with a deep red or purplish-red spot at the base of the lower lip; capsule 5-7 mm. long.

Moist to wet, usually moss covered cliffs, lake shores, banks of streams, and on rocks exposed to spray.

Mimulus cardinalis Dougl. ex Benth. Cardinal monkey-flower
Pubescent rhizomatous perennials; stems 2.5-8 dm. tall; leaves 2-11 cm. long, 1-6 cm. wide, oblong to spatulate, sharply dentate, 3-5-veined from the base, sessile, the upper leaves clasping; calyx 2-3 cm. long; corolla red to orange, strongly 2-lipped, 3-6 cm. long; stamens exserted; capsule 12-18 mm. long.

Stream banks, wet cliffs and seeps, scarcely reaching our southern limits.

Mimulus dentatus Nutt. Woods monkey-flower
Pubescent rhizomatous perennials; stems 1-4 dm. tall; leaves sessile or short-petiolate, leaf blades lance-ovate, 1.5-7 cm. long, 1-3.5 cm. wide, serrate, 5 or more veined from below the middle; calyx 8-18 mm. long; corolla 2.5-4 cm. long, yellow or the lobes tinged with red or purple, usually red-spotted within; capsule 8-9 mm. long.

On wet banks in shady woods.

Mimulus douglasii (Benth.) Gray
Pubescent annuals; stems 1-5 cm. tall, simple or somewhat branched; leaves ovate to oblong, 0.5-4 cm. long, entire or toothed; corolla up to 4 cm. long, the tube very slender below, broader above, 2 or 3 times as long as the calyx, reddish-purple, mottled or striped with yellow and purple, the upper lip broadly 2-lobed and showy, the lower reduced generally to a small lobe, the lateral lobes wanting or minute.

Low wet areas; mostly south of our area.

Mimulus guttatus DC. Common monkey-flower

Annuals to perennials, sometimes stoloniferous or rhizomatous; stems sometimes only 2 cm. tall, other times up to 1 m. tall; leaves up to 1 dm. long in large plants, sometimes purplish beneath, glabrous to pubescent, ovate to reniform, crenate to dentate or even with a few basal lobes, lower leaves petiolate, reduced upwards, upper leaves becoming sessile and clasping, 3-7-veined from near the base; corolla 1-4 cm. long, strongly 2-lipped, yellow, spotted or streaked with red, base of the lower lip pubescent.

This is an extremely variable species, but the infraspecific taxa are not well-defined.

Wet areas from the mountains to the coast.

Mimulus lewisii Pursh Pink monkey-flower

Rhizomatous perennials; stems several, simple, erect, 2.5-10 dm. tall, viscid-pubescent; leaves 1-3.5 cm. long, or the lower smaller, lanceolate to ovate, acute, sessile, often somewhat clasping, entire to irregularly toothed, 3-5-veined from the base; calyx 1.5-3 cm. long; corolla strongly 2-lipped, 3-5.5 cm. long, pinkish-purple, marked with yellow; stamens enclosed by the corolla; capsule 12-15 mm. long.

Stream banks and other wet areas in the mountains.

Mimulus moschatus Dougl. in Lindl. Musk monkey-flower

Viscid-pubescent slimy musk-scented rhizomatous perennials; stems prostrate to ascending, 0.5-7 dm. long; leaves 1-8 cm. long, 7-35 mm. broad, more or less ovate, irregularly and sometimes obscurely toothed, pinnately-veined, sessile or short-petiolate; calyx 7-13 mm. long, angled, the lobes nearly equal; corolla 1.5-3 cm. long, yellow and often reddish-brown streaked or dotted, the lobes nearly equal.

A variable species.

Stream banks, moist woods and other wet areas, from the coast to the mountains.

Mimulus primuloides Benth. Primrose monkey-flower

Mat-forming rhizomatous and/or stoloniferous perennials; stems 0.5-12 cm. long; leaves 0.5-5 cm. long, 3-11 mm. broad, glabrous to viscid-pubescent, crowded at the base of the stem, oblanceolate, entire or obscurely toothed, sessile or subsessile, 3-5-veined from near the base; peduncles one or few, slender, much longer than the stems; calyx 4-12 mm. long, the teeth nearly equal; corolla 1-2.5 cm. long, the lobes nearly equal, widely spreading, yellow, dotted with red or maroon; capsule 6-7 mm. long.

Wet mountain meadows, stream banks and seepage areas.

Mimulus tricolor Hartw. ex Lindl. Tricolored monkey-flower

Glandular-puberulent annuals, 1-15 cm. tall; leaves 1.5-4.5 cm. long and up to 1 cm. broad, oblanceolate to elliptic, at least the upper sessile, pinnately-veined; calyx 0.5-2.5 cm. long, the lobes unequal; corolla 3-5 cm. long, purple with a maroon spot at the base of each lobe and mottled with white, yellow and dark spots in the throat.

Vernal pools and wet meadows, usually in clay soil; Willamette Valley and southward.

Nothochelone

A monotypic genus.

Nothochelone nemorosa (Dougl.ex Lindl.) Straw Turtle-head

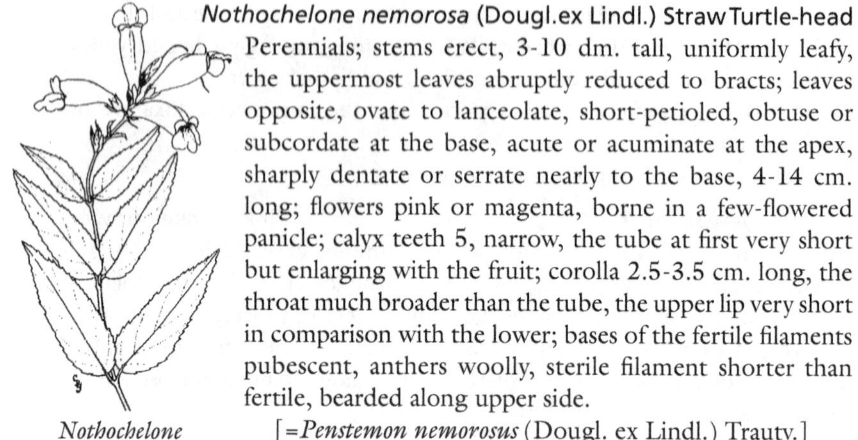

Nothochelone nemorosa

Perennials; stems erect, 3-10 dm. tall, uniformly leafy, the uppermost leaves abruptly reduced to bracts; leaves opposite, ovate to lanceolate, short-petioled, obtuse or subcordate at the base, acute or acuminate at the apex, sharply dentate or serrate nearly to the base, 4-14 cm. long; flowers pink or magenta, borne in a few-flowered panicle; calyx teeth 5, narrow, the tube at first very short but enlarging with the fruit; corolla 2.5-3.5 cm. long, the throat much broader than the tube, the upper lip very short in comparison with the lower; bases of the fertile filaments pubescent, anthers woolly, sterile filament shorter than fertile, bearded along upper side.

[=*Penstemon nemorosus* (Dougl. ex Lindl.) Trautv.]

Along streams and in wooded areas.

Orthocarpus

Annuals; leaves mostly alternate, sessile, the bracts subtending the flowers often bright-colored; flowers in spikes or spike-like racemes; calyx unequally 4-cleft; upper lip of corolla narrow, beak-like, enclosing the stamens, lower lip with 3 sac-like inflations; stamens 4; capsule loculicidal.

Bracts differing from the upper leaves; mountain plants .... *O. imbricatus*
Bracts, at least the lower, similar to the upper leaves; lowland plants ....
.... *O. bracteosus*

Orthocarpus bracteosus Benth. Purple owl-clover

Stems erect, little branching, 1-4 dm. tall, minutely pubescent and upper part of plant somewhat glandular; lower leaves long-lanceolate, upper 3-5-lobed, the lobes spreading; bracts purplish, broad, with 3-5 narrow spreading lobes; inflorescence compact; flowers showy, purple or lavender, exceeding the bracts; upper lip of corolla slender, slightly hooked at apex; lower lip with 3 much inflated sacs; anthers pubescent.

Fields, moist meadows and road sides, often forming conspicuous purple patches.

Orthocarpus imbricatus Torr. Mountain owl-clover

Stems slender, erect, 1-4 dm. tall, sometimes branched above; leaves entire, lanceolate to linear; bracts broadly overlapping, rounded at apex; corolla purple to rose-colored, lower lip slightly inflated, scarcely wider than upper lip, nearly as long.

In mountain meadows.

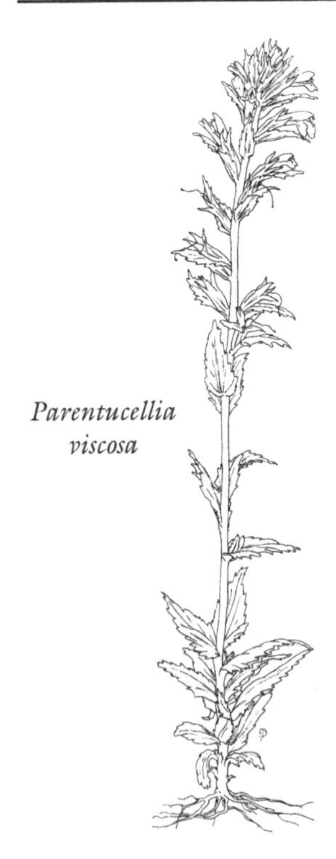
Parentucellia viscosa

Parentucellia

Sticky-pubescent annual or biennial herbs with flowers in a leafy spike-like raceme; leaves usually opposite, at least in part; upper lip of corolla narrow and elongate as in *Castilleja* and *Orthocarpus*, but differing from those genera in having anther sacs equal and parallel.

Parentucellia viscosa (L.) Caruel — Eyebright

Annuals; stems erect, generally simple, 1-7 dm. tall; leaves 1-4 cm. long, long-lanceolate, sessile, crenate-dentate; flowers nearly sessile, 1.5-2 cm. long, bright yellow.

A European species; weedy; frequently found in moist open areas.

Pedicularis

Perennial herbs with alternate toothed or pinnately compound leaves; flowers in a bracted spike or raceme; calyx 2-5-cleft or toothed; corolla conspicuously 2-lipped, the upper lip arched, sometimes beaked, the lower 3-lobed; stamens 4, concealed by the arched upper lip.

1a Leaves lanceolate, merely toothed; upper corolla-lip curved inward .... *P. racemosa*

1b Leaves pinnately deeply lobed to divided into narrow segments
- 2a Upper lip of the corolla hood-like, not beaked; corolla whitish to yellowish or rarely purplish .... *P. bracteosa*
- 2b Upper lip of the corolla beaked
  - 3a Upper lip of the corolla with a downward turned beak nearly concealed by the lower lip .... *P. contorta*
  - 3b Upper lip of the corolla with straight or up turned beak, not concealed by the lower lip
    - 4a Beak of the upper lip of the corolla straight .... *P. ornithorhyncha*
    - 4b Beak of the upper lip of the corolla upturned
      - 5a Pubescent, at least in the inflorescence; lower lip of the corolla well developed .... *P. attollens*
      - 5b Glabrous or nearly so; lower lip of the corolla much shorter than the beak .... *P. groenlandica*

Pedicularis attollens Gray — Little elephant's head

Stems 0.5-6 dm. tall, pubescent at least above; basal leaves divided into 17-41 segments, the segments linear and toothed, the leaves reduced upwards; corolla 4.5-7 mm. long, pink to purple and with darker markings, upper lip curved upwards, more or less trunk-like; capsules 6-10 mm. long.

Wet meadows, bogs and stream banks in the mountains.

Pedicularis bracteosa Benth. Bracted lousewort

Stems stout, erect, 3-10 dm. tall, often purplish; leaves thin, somewhat lax, with a slender rachis, the lower usually petioled, blades pinnately divided into narrow segments, these sharply toothed or lobed; lower bracts toothed, upper entire; inflorescence densely flowered; corolla whitish to yellowish or rarely purplish, upper lip hood-like, somewhat curved downward.

A variable species with several named varieties.

(=*Pedicularis flavida* Penn.)

Moist woods in the mountains.

Pedicularis contorta Dougl. Coiled-beak lousewort

Stems 1.5-6 dm. tall; basal leaves 3-18 cm. long, 1-3.5 cm. wide, the 25-41 segments linear, entire or more often finely serrate, the basal leaves slender-petioled; calyx 5-lobed, the lobes unequal, 6-10 mm. long; corolla 10-13 mm. long, the upper lip down turned, pale yellow to white and usually lightly marked with purple; capsule 6-10 mm. long.

Wet meadows, bogs, stream banks or drier forest sites.

Pedicularis groenlandica Retz. Elephant's head

Stems slender or stout, 2-8 dm. tall; leaves lanceolate in outline, basal leaves 3-25 cm. long, 0.5-4 cm. wide, segments 25-51, linear to oblong, sharply toothed, reduced upwards; spikes dense; corollas 1-1.5 cm. long, pinkish-purple to purplish-red, upper lip with a trunk-like beak, 6-13 mm. long, turned upwards; capsule 6-9 mm. long.

[=*Pedicularis surrecta* Benth.]

Wet meadow, bogs and stream banks in the mountains.

Pedicularis ornithorhyncha Benth. in Hook. Bird's-beak lousewort

Stems erect, 0.5-3 dm. tall; leaves mostly basal, divided into narrow segments, these often again cleft and toothed; flowers purple, upper corolla lip short, straight.

Moist areas in the high mountains of Washington.

Pedicularis racemosa Dougl. ex Hook. Leafy lousewort

Stems generally tufted, slender, simple or somewhat branched, 1-6 dm. tall; herbage nearly glabrous; leaves lanceolate, short-petioled, the blades 2.5-10 cm. long, veins conspicuous, the margins doubly toothed; flowers in loose terminal racemes, the lower bracts leafy; corolla whitish to yellowish or pink-tinged, the upper lip curved downward to the lower.

Mountain woods.

Penstemon

Perennial herbs, sometimes shrubby at the base, usually with opposite leaves, the upper sessile, reduced to small bracts above; flowers generally showy, in racemes, panicles, or cymes; calyx 5-parted nearly or quite to the base; corolla tubular, often inflated, more or less 2-lipped, the upper lip 2-lobed, the lower usually 3-cleft; anther bearing stamens 4, the filaments generally somewhat curved, the anther sacs attached only near the tip, the fifth stamen reduced to a filament, often flattened and/or bearded; capsule many-seeded.

Certain species of this genus are widely variable, particularly those of high mountain habitats. It seems undesirable to enumerate here with diagnostic characters, all the varieties which have been given names.

Other species than those described below, may sometimes enter our region from the east side of the Cascades.

1a Anthers densely woolly; leaves leathery; flowering branches arising from clumps or mats of woody stems
  2a Flowers deep pink to reddish-lavender .... *P. rupicola*
  2b Flowers purple
    3a Leaf blades mostly lanceolate to elliptic, 1.5-4 cm. long; flowering stems 10-50 cm. tall .... *P. cardwellii*
    3b Leaf blades ovate to oblong, 0.5-2 cm. long; flowering stems 5-20 cm. tall .... *P. davidsonii*
1b Anthers not woolly; leaves not leathery; stems erect, herbaceous or woody at the base
  4a Anther sacs splitting nearly the entire length, widely spreading *P. ovatus*
  4b Anther sacs opening only at the apex, or less than 1/2 their length; remaining horseshoe-shaped
    5a Corollas glandular-pubescent on the outside; inflorescence loose .... *P. richardsonii*
    5b Corollas glabrous; inflorescence rather compact .... *P. serrulatus*

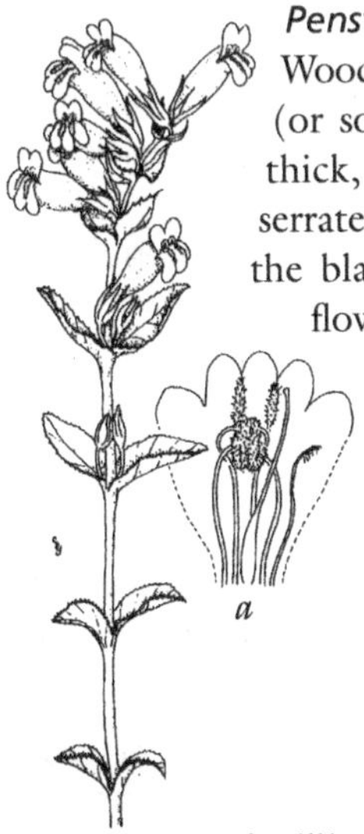

Penstemon cardwellii Howell   Cardwell's Penstemon

Woody stems 1-5 dm. tall, much-branched, bearing racemes (or sometimes panicles) at apices of the branches; leaves thick, leathery, lanceolate, or narrowly elliptic, obtuse, serrate to subentire, or those of the flowering stems entire, the blades 1.5-4 cm. long, the lower generally petioled; flowers deep purple, 2-4 cm. long; pedicels, bracts, and calyx glandular-pubescent; anthers woolly, the sterile stamen bearded.

In the Cascades and on high peaks of the Coast Range.

a

Penstemon cardwellii
a. open flower

Penstemon davidsonii Greene   Davidson's Penstemon

Flowering stems 5-20 cm. tall, arising from matted much-branched woody stems; leaves of mats thick, ovate to oblong, entire to minutely toothed, narrowed to a short-petiole, leaves of the flowering stems few, smaller; flowers purplish, 2-3.5 cm. long, in racemes, the inflorescence more or less glandular; stamens with woolly anthers; sterile stamen bearded.

Usually in rocky areas in the mountains.

Penstemon ovatus Dougl. ex Hook.   Broad-leaved Penstemon

Stems 2-10 dm. tall from a slightly woody base; leaf blades broadly or narrowly ovate, the upper sessile and clasping, the lower petioled, usually pubescent at least below, serrate; corolla glandular-pubescent, blue, the tube somewhat abruptly widened or sometimes tapering, 12-25 mm. long, about 1 cm. wide at mouth; sterile stamen bearded, sometimes protruding from the corolla.

Moist rocky places and open woods.

Penstemon richardsonii Dougl. ex Lindl. Richardson's Penstemon
Stems ascending, 2-8 dm. tall, often several from a woody base, glabrous to canescent, sometimes reddish, often branched; leaves opposite, lanceolate, often deeply and irregularly toothed, 2.5-7 cm. long, thick but not leathery, often glabrous; panicle generally large, loose, often with minute glandular hairs; calyx lobes ovate, acute; corolla glandular-pubescent on the outside, lavender to reddish, 18-32 mm.; anthers of fertile stamens horseshoe-shaped, splitting only at the apex, ciliate, the long sterile filament extending beyond the corolla, long-bearded.

Dry, rocky places along the Columbia River and eastward in the mountains.

Penstemon rupicola (Piper) Howell Cliff or Rock Penstemon
Stems woody, spreading mat-like; ascending flowering stems 7-15 cm. tall, pubescent; leaves of the mats glaucous, rigid, ovate, oval to orbicular, 6-18 mm. long, finely and often irregularly toothed to entire, the lower short-petioled, leaves of the flowering stems reduced; inflorescence glandular, racemes usually few-flowered; corolla deep pink to reddish-lavender, 2.5-4 cm. long; anthers long-woolly; sterile filament glabrous or sparsely bearded.

Rocky places in the mountains and in the Columbia River Gorge.

Penstemon serrulatus Menz. Cascade Penstemon
Stems slightly woody at the base, erect, 2-8 dm. tall, slender, glabrous or nearly so; leaf blades lanceolate to ovate-oblong, 2.5-10 cm. long, upper sessile, lower petioled, sharply toothed; calyx lobes sometimes fringed; corollas blue or generally pinkish-violet 1.5-3 cm. long, glabrous; anthers permanently horseshoe-shaped; sterile filament glabrous.

Moist areas; in the mountains and to the coast.

Scrophularia

Perennials; stems square in cross-section; leaves opposite; inflorescence a panicle; calyx 5-lobed; corolla 2-lipped, the upper lip 2-lobed, the lower lip shorter, 3-lobed; fertile stamens 4, the fifth stamen represented by a club-shaped knob or a scale; capsule septicidal.

Staminodium wider than long, fan-shaped.................................*S. lanceolata*
Staminodium longer than wide, club-shaped to obovate.............*S. californica*

Scrophularia californica Cham. & Schlecht. California Figwort
Glandular perennials; stems 6-18 dm. tall; leaf blades 7-20 cm. long, ovate to more or less triangular, truncate to cordate at the base, acute at the apex, sharply toothed or sometimes deeply cleft, petioles 2-7 cm. long; calyx 3-4 mm. long; corolla 8-12 mm. long, the upper lip maroon, the lower lip greenish-yellow; staminodium longer than wide, purplish-brown; capsule 6-8 mm. long.

Wet or moist habitats, west of the Cascades.

Scrophularia californica

Scrophularia lanceolata Pursh

Lance-leaved figwort

Glandular perennials; stems 5-15 dm. tall; leaf blades 5-15 cm. long, 2-7 cm. broad, ovate to triangular, cuneate, truncate, subcordate or rounded at the base, acute or acuminate at the apex, sharply-toothed, on petioles 1-4 cm. long; calyx 2-3 mm. long; corolla 8-14 mm. long, yellowish-green or tinged with maroon; staminodium wider than long, greenish-yellow; capsule 6-8 mm. long.

Wet or moist shaded habitats.

Synthyris

Perennial herbs; basal leaves and stems arising from rhizomes; flowers borne in bracteate racemes; calyx deeply 4-parted; corolla unequally 4-lobed, the upper lobe the largest; stamens 2, attached to corolla tube on upper side; capsule somewhat flattened.

1a Corolla deeply and irregularly cut into narrow lobes .... *S. schizantha*

1b Corolla not deeply cut into narrow lobes

2a Corolla lobes spreading, longer than the tube; leaf margins sharply lobed and toothed .... *S. stellata*

2b Corolla bell-shaped, the lobes slightly shorter than the tube; leaf margins with rounded lobes and teeth .... *S. reniformis*

Synthyris reniformis (Dougl.) Benth. Spring or Snow queen

Plants 5-20 cm. tall; leaves glabrous to pubescent, reniform to heart-shaped, 1-8 cm. broad, coarsely lobed, with 3-5 conspicuous veins from the base, long-petioled; flowers 4-9 mm. long, bell-shaped, blue (rarely white) borne in a short raceme.

Shady moist woods.

This is one of our longest-season spring flowers, the first blossoms appearing in January or even as early as November in a mild season, and continue until May or June. The color of the flowers is an unusually pure blue, making the plant very desirable for borders in the wild garden.

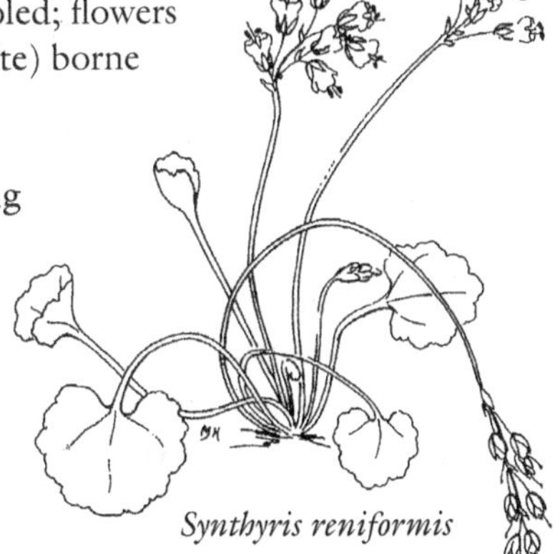

Synthyris reniformis

Synthyris schizantha Piper — Fringed Synthyris

Plants 1-3 dm. tall; basal leaves reniform to heart-shaped, 4-12 cm. broad, long-petioled, coarsely lobed, the lobes toothed, somewhat pubescent at least on the veins below, cauline leaves generally 2, nearly sessile; bracts much reduced and not at all leaf-like; corolla 4-6 mm. long, deeply and irregularly cut into narrow lobes.

In the mountains; uncommon in our limits.

Synthyris stellata Penn.

Stems 1-6 dm. tall, bearing reduced but somewhat conspicuous bracts; basal leaves, long-petioled, the blades nearly orbicular, deeply cordate at base, the margin evenly and shallowly lobed, the lobes generally 3-5-toothed, 2.5-9 cm. wide, lower bracts fan-shaped, narrowed at base, the margin sharply toothed, 6 mm. or more long below, becoming smaller, but still conspicuous, above; racemes generally longer than the leaves, somewhat densely flowered; flowers 4-7 mm. long, purplish-blue.

Moist slopes; in the Columbia River gorge.

Tonella

Annuals; plants similar in general appearance to *Collinsia,* but the corolla tube nearly straight, not bent at a sharp angle as in *Collinsia*, and some of the cauline leaves usually 3-parted or divided.

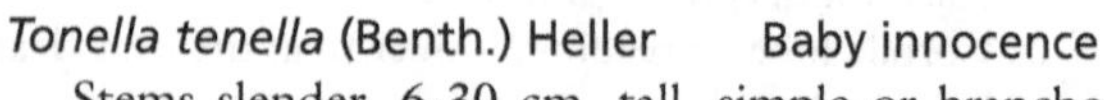

Tonella tenella (Benth.) Heller — Baby innocence

Stems slender, 6-30 cm. tall, simple or branched at base; first leaves orbicular in outline, palmately 3-5-lobed or entire, about 6 mm. broad, long-petioled, upper cauline leaves short-petioled to nearly sessile, usually divided into 3 narrow lobes, these toothed or entire; flowers minute, pale blue, the corolla about 2.5 mm long, somewhat 2-lipped, the tube nearly straight.

Moist shady areas; often found growing in moss on stumps and tree trunks.

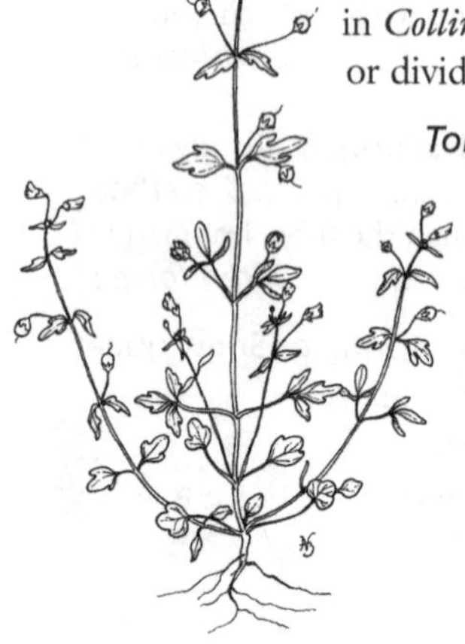

Tonella tenella

Triphysaria

Annuals; leaves sessile, the upper finely divided; inflorescence a bracteate spike; calyx 4-lobed; corolla 2-lipped, the upper lip beak-like, the lower lip deeply 3-pouched; stamens 4, anther sac 1 per filament; capsule loculicidal.

Corollas conspicuous, 10-25 mm. long .... *T. eriantha*
Corollas inconspicuous, 7 mm. or less long .... *T. pusilla*

Triphysaria eriantha (Benth.) Chuang & Heckard

Stems 1-3.5 dm. tall; plants more or less purplish, glandular and puberulent; leaves 3-7-lobed; corollas 1-2.5 cm. long, deep yellow to white with slender purple upper lip and conspicuous 3-pouched lower lip.

(=*Orthocarpus erianthus* Benth.)

Occasionally enters our range from the south.

Triphysaria pusilla (Benth.) Chuang & Heckard — Dwarf owl-clover

Stems 1 or several from base, decumbent to erect, 3-20 cm. tall, whole plant purplish; first leaves minute, entire, becoming increasingly pinnately cleft above, the divisions again cleft into 3-9 narrow segments; bracts longer than the minute reddish, yellow or purple flowers.

(=*Orthocarpus pusillus* Benth.)

Common on hillsides and in lawns and pastures, making marked masses of color, but inconspicuous individually.

Verbascum

Usually biennials with tall wand-like stems and basal and alternate leaves; flowers generally yellow or white, in terminal bracteate spikes, racemes or panicles; calyx 5-parted, the divisions nearly equal; corolla deciduous, rotate, 5-lobed, the upper lobes a little shorter than the lower; stamens 5, all fertile, at least the 3 upper filaments bearded; capsule 2-valved, many-seeded, the seeds pitted or rough-surfaced.

Plants densely woolly; upper leaves forming wing-like extensions down the stem .... *V. thapsus*

Plants not woolly; leaves not continued as wing-like extensions down the stem .... *V. blattaria*

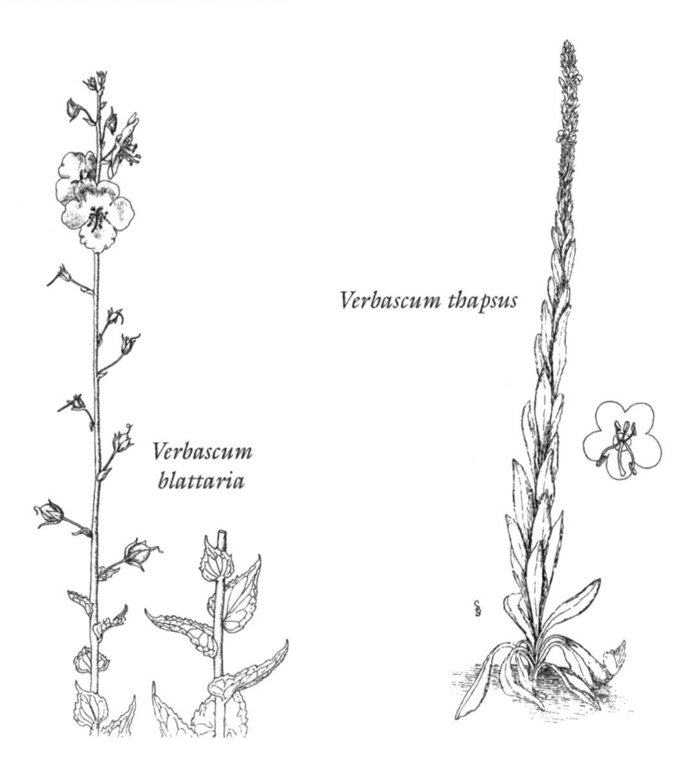

Verbascum blattaria L. Moth mullein

Plants glabrous or nearly so, the dark green stems slender, simple or sometimes branched above, 3-15 dm. tall; basal leaves oblong, somewhat narrowed at the base, veins prominent, margins crisped and toothed, cauline leaves ovate or lanceolate, toothed, cordate at base, somewhat clasping; inflorescence glandular; flowers 2.5 cm. in diameter, yellow or white, in a long raceme, the buds flattened vertically, the exposed corolla in the older buds orange or purplish; anthers orange, filaments purple, bearded with long silky purple hairs; seeds roughened.

Common in disturbed habitats; introduced from Europe.

Verbascum thapsus L. Common mullein

Plants densely woolly, the stems stout, spire-like, unbranched, except sometimes at the top, 5-18 dm. tall; leaves all felt-like, entire or toothed, yellowish-green, the basal leaves 1-3 dm. or more long, oblong or elliptic, petioled, cauline leaves oblong, crowded, the lower narrowed at the base, the upper broadest at the base, but extending wing-like down the stem below the point of attachment; flowers yellow in a dense woolly spike 3-9 dm. long, with shorter spikes sometimes branching from below; corolla slightly irregular; the upper stamens short with densely woolly filaments, the lower longer with filaments less woolly, anthers all orange; style slender, lying against the lower corolla lobe, somewhat curved; ovary woolly; seeds with horizontally elongated pits, the pits in rows.

Introduced from Europe; a rather common weed of roadsides and waste places.

Veronica

Annuals or perennials; leaves all opposite to alternate in the inflorescence; inflorescence a raceme or a spike or the flowers solitary in the leaf axils; sepals 4-5; corolla rotate, 4-lobed, the upper lobe the largest, the lower lobe the smallest; stamens 2; capsule usually notched or lobed at the apex, somewhat compressed perpendicular to the septum.

1a Flowers borne in axillary racemes; leaves all opposite
  2a Plants weedy; not aquatic; pubescent .... *V. chamaedrys*
  2b Aquatic or semi-aquatic species; glabrous or nearly so
    3a Leaves sessile .... *V. scutellata*
    3b Leaves short-petioled .... *V. americana*
1b Flowers solitary or in terminal racemes; upper leaves or well developed bracts often alternate.
  4a Rhizomatous perennials
    5a Flowers solitary in the axils of the leaves; plants mat-forming .... *V. filiformis*
    5b Flowers in the axils of reduced leaves or bracts or in terminal racemes
      6a Capsules wider than high; lower leaves often short-petioled .... *V. serpyllifolia*
      6b Capsules higher than wide; leaves all sessile

7a Style 5-10 mm. long; leaves usually glabrous.......................... *V. cusickii*
7b Style less than 2 mm. long; leaves pubescent................ *V. wormskjoldii*
4b Annuals
8a Leaves pinnately veined .... *V. peregrina*
8b At least the lower leaves with 3 or more veins from the base
9a Flowers nearly sessile in the leaf and bract axils................ *V. arvensis*
9b Flowers on long pedicels much exceeding the leaves or bracts
10a Leaves longer than broad .... *V. persica*
10b Leaves nearly orbicular or broader than long
11a Leaves 3-5-lobed.. *V. hederifolia*
11b Leaves merely toothed ....*V. filiformis*

Veronica americana (Raf.) Schwein. Common or American speedwell
Glabrous rhizomatous perennials; stems decumbent to nearly erect, 1-10 dm. long, rooting at the lower nodes; leaves all opposite, 1-8 cm. long, 0.5-3 cm. broad, lanceolate to ovate, serrate to nearly entire, short-petioled; flowers in axillary racemes; corolla 5-10 mm. wide, blue with darker lines; capsules notched.

Common in still or slow-moving water, often in ditches and also in wet meadows and springy areas.

Veronica arvensis L. Small speedwell
Annuals; stems weakly ascending or spreading, 2.5-40 cm. tall, slender; leaves alternate above, opposite below, the lower broadly ovate, palmately veined, crenately toothed, becoming narrower and smaller upward; flowers 2-3 mm. long, blue-violet, solitary in the axils of bracts; capsule heart-shaped at apex, pubescent.

Introduced from Europe; weedy in fields, gardens, and lawns.

Veronica chamaedrys L. Germander speedwell
Rhizomatous perennials; stems 0.5-4.5 dm. long, decumbent to ascending and rooting at the nodes; leaves 1-3 cm. long, ovate, obtuse, the lower most short-petioled, the others sessile, coarsely crenately toothed, all opposite; flowers borne in axillary racemes; corolla bright blue with a white center, about 1 cm. broad; capsule broadly triangular.

Introduced from Europe; weedy, especially in lawns.

Veronica cusickii Gray Cusick's speedwell
Rhizomatous perennials; stems 0.5-2 dm. tall, glandular-pubescent; leaf pairs crowded, usually glabrous, 5-25 mm. long, oval or elliptic, sessile, entire; flowers deep blue, showy, about 1 cm. wide; capsule deeply notched.

Meadows and openings in the forest, often in moist rocky areas; alpine.

Veronica filiformis Sm. Creeping speedwell
Annuals or more often perennials; stems slender and rooting at the nodes; leaves opposite, short-petioled, blades nearly orbicular or broader than long, the margins toothed, main veins palmate; flowers solitary in the leaf axils on long slender pedicels; flowers 1-1.5 cm. wide, deep blue with dark markings; capsules deeply notched, ciliate.

Weedy in lawns; escaped from cultivation.

Veronica hederifolia L. Ivy-leaf speedwell

Pubescent annuals; stems 0.5-3.5 dm. long, prostrate or nearly so; leaves 0.5-2 cm. long, wider than long, palmately veined and 3-5-lobed, the base truncate, lower leaves opposite, upper leaves alternate; flowers solitary, axillary; corolla pale blue; capsules only slightly notched.

(Also spelled *V. hederaefolia.*)

Introduced from Europe; weedy in disturbed habitats.

Veronica peregrina L. Purslane speedwell

Annuals 0.5-3 dm. tall; stems erect or somewhat spreading; leaves opposite below the inflorescence, linear to oblanceolate, 0.5-3 cm. long, 1-9 mm. wide, entire to irregularly toothed; bracts similar in shape to the leaves, but alternate, each subtending a solitary flower; corolla white, 2-3 mm. wide; capsule notched. Two varieties occur in our area, with the common variety being ***xalapensis*** (H.B.K.) St. Johns & Warren which is usually finely glandular-pubescent. Variety ***peregrina*** is glabrous.

Moist areas, wet meadows, swamps and stream banks.

Veronica persica

Veronica persica Poir. Birdseye or Winter speedwell

Annuals; stems more or less spreading, often rooting at the lower nodes; leaves broadly ovate to oblong, coarsely toothed, lower leaves short-petioled, the upper sessile; flowers bright blue with dark lines and a white center, borne on long slender axillary pedicels; capsule notched.

An introduced weed of waste areas, gardens and lawns.

Veronica scutellata L. Narrow-leaved speedwell

Rhizomatous perennials, usually glabrous, whole plant sometimes purplish or reddish; stems decumbent to erect, 1-6 dm. long; leaves opposite throughout, 2-8 cm. long, 2-15 mm. broad, linear-lanceolate, entire or with a few remote teeth, sessile; racemes axillary; corolla blue with darker lines, 5-10 mm. wide; capsules deeply notched.

Wet areas from meadows to ponds and stream banks.

Veronica serpyllifolia L. Thyme-leaf speedwell

Finely puberulent rhizomatous perennials; stems 5-30 cm. tall, sometimes producing long prostrate lower branches; leaves opposite or the much-reduced upper ones alternate, 1-2.5 cm. long, 0.5-1.5 cm. wide, elliptic to ovate, entire to crenate or toothed; flowers in terminal racemes; corolla white to blue, 4-8 mm. wide; capsules notched.

Two varieties occur in our area: var. ***humifusa*** (Dickson) Vahl which appears to be native and has bright blue flowers and var. ***serpyllifolia*** which is a native of Europe and is occasionally naturalized with white or pale blue flowers with darker lines.

Moist areas, especially wet meadows, stream banks and around lakes and ponds and sometimes as a weed in lawns.

Veronica wormskjoldii Roem. & Schult. American alpine speedwell
Villous rhizomatous perennials; stems 7-40 cm. long, rooting at the lower nodes; leaves opposite or the reduced upper ones alternate, 1-4 cm. long, 0.5-2 cm. broad, lanceolate to elliptic, entire to crenate, sessile; flowers in terminal racemes; corolla 6-10 mm. wide, deep blue; capsules notched.

Wet meadows, stream banks and other wet habitats in the mountains.

LENTIBULARIACEAE
Bladderwort Family

Small aquatic or sub-aquatic plants; leaves simple or dissected, basal, alternate or whorled, often dimorphic (these dissected leaves interpreted by some, perhaps rightly so, to be branching stem systems which are filamentous and leaf-like); calyx 2-5-lobed, 2-lipped; corolla 5-lobed, 2-lipped and spurred at the base; fertile stamens 2; ovary superior; pistil 1; fruit a capsule.

Leaves simple and entire, borne in a basal rosette .... **Pinguicula**
Leaves much-dissected, alternate; plants bearing bladder traps....... **Utricularia**

Pinguicula

Perennials; leaves simple and entire, borne in a basal rosette, fleshy, upper surface greasy, bearing stalked glands that secrete a viscous mucilage to capture prey and sessile glands that secrete the digestive enzymes; when prey is trapped the margins of the leaf slowly curl inward; scapes with a solitary flower; calyx 5-lobed, 2-lipped; corolla 5-lobed, 2-lipped, the spur straight.

Pinguicula vulgaris L. Butterwort
Leaves 2-5 cm. long and less than 2 cm. wide; scapes up to 15 cm. in height, obscurely glandular; calyx purple; corolla 1.5-3 cm. long, including the 5-9 mm. spur, pale violet or rarely nearly white, the throat pubescent; capsule 2-valved.

Subalpine in bogs and wet areas but at lower elevations on moist serpentine.

Utricularia

Usually submerged annuals or perennials; leaves mostly alternate, dissected into narrow segments; bladder traps borne on the leaves or on specialized branches, these bladders trap small organisms; flowers borne in racemes, rarely solitary; corolla yellow, upper lip nearly entire, the lower lip usually 3-lobed.

1a Scapes with a whorl of floating leaves with inflated petioles .... *U. inflata*
1b All the leaves alternate
  2a Leaf segments capillary, thread-like, more or less terete
    3a Plants robust; leaf margins with fine bristles; leaves with 20-150 ultimate segments.. *U. macrorhiza*
    3b Plants delicate; leaf margins usually smooth and glabrous; ultimate leaf segments usually not more than 4 .... *U. gibba*
  2b Leaf segments flattened, narrow, but not capillary

4a Leaf margins smooth and glabrous, leaves usually 2-parted at the base; bladder traps borne on the leaves.. *U. minor*
4b Leaf margins with fine bristles, leaves usually 3-parted at the base; bladder traps usually borne on specialized branches distinct from the leaves
5a Tip of the leaf segments obtuse but bearing a bristle, the ultimate leaf segments usually with 4 or more bristles on each side of the leaf margins.... .... *U. intermedia*
5b Tip of the leaf segments acute, gradually narrowed to a bristle, the ultimate leaf segments usually with less than 4 bristles on each side arising from small teeth on the leaf margins.................................. *U. ochroleuca*

Utricularia gibba L. Humped or Creeping bladderwort
Stems floating or creeping, often forming a tangled mass; leaves 2-15 mm. long, generally 2-parted at the base, each part sometimes forked again, the ultimate segments not more than 4, segments filiform; bladder traps few, borne on the stems and the leaves; flowers usually 2-6, but sometimes more; corolla yellow, 4-12 mm. long.
Shallow quiet or slow-moving water or occasionally stranded in mud.

Utricularia inflata Walt. Floating bladderwort
Plants free-floating; stems submerged with alternate leaves several times dichotomously branched into fine capillary segments, these bearing bladder traps; scapes with a whorl of 4-10 floating leaves with inflated petioles; flowers 8-14, yellow, up to 25 mm. long, the lower lip about twice as long as the spur.
Still or slow moving water; introduced in Washington.

Utricularia intermedia Hayne Flat-leaved or Northern bladderwort
Perennials; leaves numerous, polymorphic, those on stolons above the substrate imbricate, 1-20 mm. long, divided into as many as 15 segments, ultimate segments flattened, the apex obtuse, the margins bearing numerous bristles, or the leaves produced early or late in the growing season, sparsely denticulate, each tooth with one or more bristles; leaves borne on the stolons beneath the substrate much reduced; bladder traps few, produced on specialized branches distinct from the leaves; corolla yellow, 8-18 mm. long, the spur a little shorter than the lower lip.
Shallow water.

Utricularia macrorhiza Le Conte Common bladderwort
Large perennials; leaves numerous, 1.5-6 cm. long, two parted at the base then more or less pinnately-divided, the secondary segments dichotomously divided into 20-150 ultimate segments, these segments capillary and bearing minute bristles; bladder traps numerous; flowers 3-20, corolla yellow with reddish streaks on the base of the lower lip, 10-25 mm. long.
Quiet or slow-moving water.

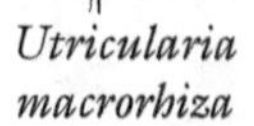

Utricularia macrorhiza

This may not actually be a distinct species, but rather a variant of the European species, *Utricularia vulgaris* L.

Utricularia minor L. Lesser bladderwort

Small perennials; leaves numerous, polymorphic, those on stolons above the substrate semi-orbicular in outline, 2-15 mm. long and divided into 7-22 segments, the ultimate segments flattened, the margins entire or nearly so, leaves borne on stolons beneath the substrate much reduced; bladder traps borne on the leaves; flowers 2-6; corolla pale yellow, 4-8 mm. long.

In shallow water of bogs and marshes.

Utricularia ochroleuca R. Hartman Buff bladderwort

Small perennials; leaves numerous, polymorphic, those on stolons above the substrate nearly orbicular in outline, divided into as many as 20 segments, the ultimate segments flattened, acute, the margins sparsely denticulate each terminating in 1-several bristles, leaves borne on the stolons beneath the substrate much reduced; bladder traps few and mostly borne on specialized branches distinct from the leaves; flowers 2-7; corolla 10-15 mm. long, pale yellow, upper lip usually about one-half the length of the lower lip.

Utricularia ochroleuca is not always recognized as a species distinct from *Utricularia intermedia.*

Usually in shallow water of bogs and marshes, but occasionally in deeper water in streams and lakes.

OROBANCHACEAE
Broom-rape Family

Herbs without green color, parasitic on roots; leaves reduced to alternate scales; flowers complete; petals united into a more or less 2-lipped tube; stamens 4, in pairs of 2 lengths, attached to the corolla-tube; ovary superior, compound, usually 1-loculed; fruit a loculicidal capsule, opening by 2 or 4 valves; seeds many, small.

Inflorescence glabrous; corolla with a ring of hairs at the base of the stamens... .... **Boschniakia**

Inflorescence pubescent; corolla without a ring of hairs at the base .... **Orobanche**

Boschniakia

Plants fleshy, ours parasitic on the roots of members of Ericaceae, developing a bulbous-like swelling at the point of attachment; plant spike-like above ground, bearing many flowers subtended by scale-like bracts, each flower sometimes also bearing 1 or 2 bractlets; corolla more or less inflated, somewhat 2-lipped, the upper lip generally slightly notched, with edges infolded, corolla with a ring of hairs at the base of the stamens; anthers pubescent; ovary 1-loculed, compound.

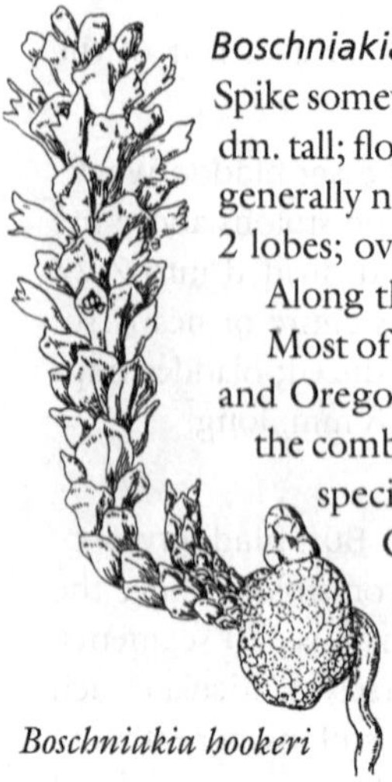

Boschniakia hookeri Walp. Small ground-cone

Spike somewhat slender, reddish or yellowish-brown, generally 1-1.5 dm. tall; flowers 1-1.5 cm. long; calyx usually with a few teeth; corolla generally not much curved; stigma typically somewhat 1-sided with 2 lobes; ovary usually with 2 (rarely 3-4) placentae.

Along the coast, parasitic on salal (*Gaultheria shallon*).

Most of the characteristics of this genus as it occurs in Washington and Oregon are somewhat variable, but the difference in host and the combination of characters appear stable enough to separate our species from the stouter form found in southern Oregon and California, *Boschniakia strobilacea* Gray, which is parasitic on other members of Ericaceae.

Boschniakia hookeri

Orobanche

Small yellow or purplish plants, usually glandular-pubescent; sepals 4-5, partly united, pointed at tips; petals 5, united to near the tips in a long curved glandular-pubescent tube; stamens shorter than tube; capsule opening by 2 valves.

1a Calyx divided into 2 segments, each segment commonly 2-lobed; flowers sessile .... *O. minor*
1b Calyx 5-lobed; flowers long-pedicellate
  2a Flowers 1-5; calyx lobes longer than the tube .... *O. uniflora*
  2b Flowers 4-20; calyx lobes equal to or shorter than the tube .... *O. fasciculata*

Orobanche fasciculata Nutt. Clustered broom-rape

Somewhat resembling 0. *uniflora* but with more flowers in the axils of bracts on stems 3-20 cm. long; flowers 1.5-3 cm. long, generally yellowish, sometimes purplish-tipped.

Dry open areas on various hosts.

Orobanche minor J.E. Smith Clover broom-rape

Stems simple, erect, glandular-pubescent, arising from a thickened base; scale-like leaves scattered along the entire length of the stem, small, acuminate, glandular; inflorescence a slender spike; corolla 1-1.5 cm. long, white or yellowish, with purple markings.

Parasitic on various crop plants; particularly those of the Fabaceae, Apiaceae and Asteraceae.

Orobanche uniflora

Orobanche uniflora L. Naked broom-rape

Stems mostly underground, scaly, bearing 1-5 flowers on long slender pedicels; calyx lobes tapering to a narrow point, longer than the calyx tube; whole plant often purplish-tinged, this color in the corolla deepening to violet or deep purple; corolla 1-3.5 cm. long, the lobes finely ciliate.

A variable species with several named subspecies and varieties. On various hosts.

PLANTAGINACEAE
Plantain Family

Annuals or perennials; leaves usually all basal; calyx deeply 4-parted; corolla 4-lobed, scarious; stamens 1-4; ovary superior, 2 or 4-loculed, or rarely only 1-loculed; fruit a capsule.

Plantago

Leaves usually all basal, the venation often appearing parallel; flowers in spikes, each flower subtended by a bract; sepals persistent in fruit; corolla with a short tube, the lobes spreading; stamens 4 (rarely only 2.), alternating with the corolla lobes; ovary 2-loculed, or falsely 4-loculed, capsule circumscissile, the top deciduous.

1a Leaves deeply pinnately lobed...*P. coronopus*
1b Leaves not deeply lobed
  2a Annuals
    3a Plant densely woolly throughout.............................. *P. patagonica*
    3b Plant glabrous or pubescent, but never woolly throughout
      4a Bracts conspicuous, much longer than the flowers .... *P. aristata*
      4b Bracts fleshy, about equal to the calyx; maritime species ....
      .... *P. elongata*
  2b Perennials
    5a Flowers of three kinds, pistillate, staminate and perfect...... *P. hirtella*
    5b Flowers all perfect
      6a Leaves elliptic to cordate-ovate..................................... *P. major*
      6b Leaves linear to narrowly oblanceolate or lanceolate
        7a Leaves linear to narrowly oblanceolate, succulent, spikes slender
        .... *P. maritima*
        7b Leaves lanceolate; spikes stout........................... *P. lanceolata*

Plantago aristata Michx. — Long-bracted plantain

Annuals 1-4.5 dm. tall; leaves 5-15 cm. long, narrow, ribbon-like; spikes stout, 2.5-15 cm. long, usually somewhat pubescent; bracts conspicuous, much longer than the flowers; stamens 4.

Occasional in disturbed habitats; introduced from the midwestern United States.

Plantago coronopus L.

Taprooted annuals or biennials; stems densely pubescent; leaves deeply pinnately lobed, narrowly lanceolate in outline, the lobes acute; spikes 1-6, 2-22 cm. long, cylindrical, densely flowered; bracts conspicuous, stamens 4.

Waste places; introduced from Europe.

Plantago elongata Pursh — Small or Slender plantain

Annuals 4-20 cm. tall; leaves 2-12 cm. long, linear or oblanceolate, fleshy, shorter than or equal to the peduncles; spikes 1-5 cm. long; bracts fleshy, about equal to the calyx; flowers often imperfect; stamens usually 2; seeds winged or not.

Along the coast in salt marshes, beaches and vernal pools.

Plantago hirtella H.B.K. Mexican plantain

Perennials; brownish-woolly at the crown of the caudex; leaves 5-35 cm. long, succulent, elliptic to elliptic-oblanceolate; scapes stout; spikes dense; bracts keeled; stamens 4.

Wet sandy ground along the coast of southwestern Washington.

Plantago lanceolata L. English or black plantain

Perennials, 1.5-9 dm. tall, brown-woolly at the base; leaves 5-40 cm. long, pubescent, long-lanceolate, 3-several-ribbed, gradually tapered to the short petiole, usually irregularly denticulate; spike 1-8 cm. long, dense; stamens 4, long-exserted.

A common garden and road side weed; native of Europe. Sheds large quantities of pollen and is said to be troublesome as a hayfever plant.

Plantago lanceolata

Plantago major L. Broad-leaved or Common plantain

Perennials from a short caudex, glabrous or nearly so, except in the inflorescence; scapes 0.5-8 dm. tall; rosette of leaves flattened against the ground to erect, leaf blades 4-18 cm. long, broadly elliptic to cordate-ovate, entire to irregularly toothed, strongly veined, abruptly narrowed to the petiole; bracts of the inflorescence about equal to the sepals; corolla lobes reflexed; stamens 4.

This is a highly variable species.

Weedy in many habitats, often in moist to wet sites, but grows readily in hard ground of paths and yards if moisture supply is sufficient. There are apparently native as well as introduced forms.

Plantago maritima L. Seaside plantain

Perennials, scapes 5-30 cm. tall; leaves 3-25 cm. long, linear to narrowly oblanceolate, succulent, usually entire; bracts of the inflorescence fleshy; corolla lobes somewhat conspicuous as whitish tufts above the calyx; bracts about as long as the flowers; stamens 4. Two varieties occur in our area, with var. *californica* (Fern.) Pilger occurring on coastal bluffs and cliffs and var. *juncoides* (Lam.) Gray occurring in wet, often saline areas.

(=*Plantago juncoides* Lam.)

A highly variable species.

Coastal.

Plantago patagonica Jacq. Indian-wheat

Annuals, 0.5-3 dm. tall, conspicuously woolly or the spikes silky; leaves 2-15 cm. long, narrow, ribbon-like, mostly 3-veined; spikes dense, slender, 1-12 cm. long; bracts narrow, slightly longer than the flowers; stamens 4.

Dry ground; uncommon west of the Cascades.

RUBIACEAE
Madder family

Herbs (ours), shrubs or trees with opposite or whorled entire leaves; flower parts in 4's or sometimes 5's in *Sherardia*; ovary inferior, 2-loculed (ours), separating into 2 (rarely 4) 1-seeded sections (ours), or fruit a capsule, drupe or berry.

Flowers in a close head-like inflorescence surrounded by an involucre.............
....**Sherardia**
Flowers in cymes or panicles or solitary, without an involucre.............**Galium**

Galium

Annual or perennial herbs or sometimes more or less woody; stems 4-angled; leaves typically in apparent whorls of 4 or more (rarely only 2); flowers white, yellow, purple, or greenish, rarely solitary in the leaf axils, usually in cymes or panicles; flowers perfect or less often unisexual; calyx lobes minute or obsolete; corolla usually 4-lobed (rarely 3); stamens of the same number as corolla lobes, attached alternate with them on the corolla tube; ovary 2-lobed, 2-loculed, 1 ovule in each cavity; styles 2, short, stigmas capitate, ovary consisting of paired carpels which separate at maturity; fruits glabrous or bristly.

1a Leaves in whorls or 4 (upper leaves sometimes reduced to 2)
  2a One pair of leaves in each whorl generally shorter than the other pair, or only 2 leaves at some of the upper nodes .... *G. bifolium*
  2b Leaves of a whorl essentially equal in length
    3a Flowers yellow or yellowish-green
      4a Fruit smooth; annuals .... *G. pedemontanum*
      4b Fruits with long hooked hairs; perennials ....*G. oreganum*
    3b Flowers white
      5a Leaves 1-veined .... *G. trifidum*
      5b Leaves 3-veined from the base .... *G. boreale*
1b Leaves in whorls of 5-8, rarely more.
  6a Fruit with hooked bristles
    7a Leaves ovate to spatulate; perennials .... *G. triflorum*
    7b Leaves linear to narrowly oblanceolate; annuals
      8a Stems and leaves with hooked bristles; fruits 1.5-5 mm. long........
      .... *G. aparine*
      8b Stems and leaves without hooked bristles; fruits about 1 mm. long
      ....*G. parisiense*
  6b Fruit glabrous and smooth or with small tubercules
    9a Flowers bright yellow; leaves narrowly linear ....*G. verum*
    9b Flowers white or pinkish
      10a Perennials (slightly woody at the base, at least in *G. mollugo*)
        11a Angles of the stems restrorsely hispid; apex of the leaves rounded... *G. trifidum*
        11b Angles of the stems smooth; apex of the leaves cuspidate.......
        .... *G. mollugo*

10b Annuals
12a Pedicels stout; inflorescence axillary, not surpassing the leaves .... *G. tricornutum*
12b Pedicels capillary; inflorescence much surpassing the leaves .... *G. parisiense*

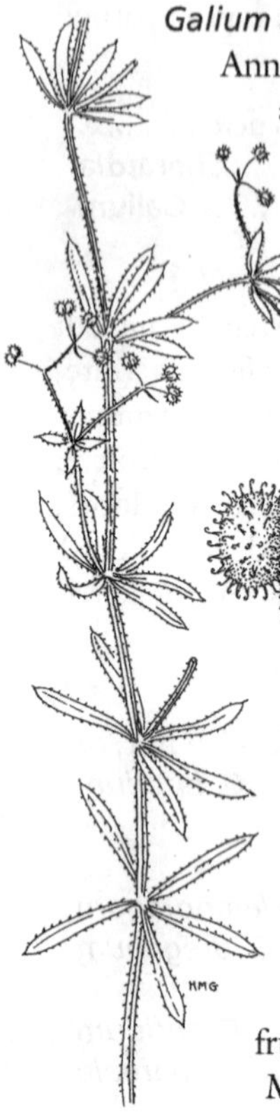

Galium aparine L. Goose-grass, Bedstraw or Cleavers
Annuals; stems weak, trailing or climbing, 3-9 dm. long, rarely short and erect, rough from short hooked bristles on the 4 angles; leaves 6-8 in a whorl, linear to narrowly oblanceolate, 1-4 cm. long, hooked-cilate on the margins and usually on the midrib; flowers few in simple or compound cymes; fruits covered with short stiff hooked hairs.
Common in somewhat shaded places and often weedy.

Galium aparine

Galium bifolium Wats. Twin-leaved bedstraw
Annuals; stems slender, erect, 4-20 cm. tall, mostly unbranched; lower leaves in whorls of 4 with 2 opposite leaves in each whorl often shorter than the other 2, upper leaves sometimes only 2 and opposite; flowers usually solitary, the pedicels becoming reflexed in fruit; fruits with hooked hairs.
Foothills and mountains; uncommon in our limits.

Galium boreale L. Northern bedstraw
Rhizomatous perennials; stems 2-8 dm. tall, usually glabrous except below the nodes; leaves in whorls of 4, 1.5-4.5 cm. long, linear to lanceolate, strongly 3-veined from the base, sessile; flowers white, in terminal paniculate cymes; fruit glabrous or pubescent, but the hairs not hooked.
Moist to wet areas.

Galium mollugo L. Wild madder
Perennials; stems 3-12 dm. tall, slightly woody at the base; leaves sessile, in whorls of 6, or rarely 8, linear or oblanceolate, scarious-ciliate on the margins; flowers minute, white, in a large terminal panicle of cymes; stamens exserted; fruit glabrous.
Waste ground and lawns; native of Europe.

Galium oreganum Britt. Oregon bedstraw
Erect perennials 1-4 dm. tall; leaves in whorls of 4, 2-5 cm. long, ovate to elliptic, scarious-ciliate at least on the margins, 3-veined; corolla yellowish-green; fruits with long hooked hairs.
Usually in shady woods.

Galium parisiense L. Lamarck's bedstraw

Annuals; stems slender, 1-2.5 dm. tall, much-branched, scarcely roughened on the angles; leaves 4-10 mm. long, linear, in whorls of 5-8, but usually 6; inflorescence in axillary and terminal compound cymes; corolla minute, white or pinkish; fruits small, granulate or with short-hooked hairs.

Grassy hillsides and fields, often weedy; native of Europe.

Galium pedemontanum All.

Annuals; stems simple, 0.5-3 dm. tall, retrorsely scarbous; leaves in whorls of 4; flowers small, yellow, in axillary cymes of 2-4; fruit smooth.

Recently introduced in our area; native of Europe.

Galium tricornutum Dandy Rough-fruit corn bedstraw

Annuals; stems 1-4 dm. tall, mostly simple; leaves narrow, in whorls of 6-8, 1-2 cm. long; flowers in axillary few-flowered cymes generally shorter than the leaves; pedicels curved downward in fruit; corollas white; fruit conspicuously roughened.

Rather frequently encountered in grain fields of the Willamette Valley and southward.

Galium trifidum L. Small bedstraw

Rhizomatous perennials; stems 0.5-6 dm. long, weak, more or less spreading to ascending and often rooting at the lower nodes, sharply 4-angled, hispid with barbs projecting downward; leaves commonly in whorls of 4, but sometimes 5-6, 0.5-2 cm. long, linear to narrowly ovate or spatulate, 1-veined, the apex obtuse; flowers in axillary and terminal clusters of 2-3 flowers, rarely solitary; corolla 3-4-lobed, white; fruit smooth. Two varieties occur in our area: var. *pacificum* Wieg. (=*Galium cymosum* Wieg.) has stems 1-5 dm. tall, while var. *pusillum* Gray occurs in mountain meadows and has stems 0.5-1 dm. tall.

Wet areas.

Galium triflorum Michx. Fragrant bedstraw

Perennials; stems weak, decumbent or trailing, 2-4.5 dm. tall, from creeping rhizomes; leaves 0.5-4 cm. long, ovate to spatulate, 6 in a whorl; flowers 2 or 3 in a cyme in the leaf-axils; fruits covered with white or tan hooked hairs.

Open woods and shrubby areas. Plants fragrant.

Galium verum L. Yellow bedstraw

Perennials; stems slender 1-2.5 dm tall; leaves 1-2 cm. long, in whorls of 6 or more, narrowly linear; flowers yellow in short dense lateral and terminal cymes or panicles; fruit glabrous.

Waste ground and lawns; native of Eurasia.

Sherardia

A monotypic genus.

Sherardia arvensis L.
Field madder

Sherardia arvensis

Annuals; stems weak, erect or somewhat trailing, 0.5-3 dm. long, 4-angled, short hairy; leaves 4-18 mm. long, ovate to elliptic, mucronate, 4-6 in a whorl; flowers pink or lavender, small, several in a head subtended by leaf-like involucre bracts; calyx teeth 4-6, small at first, elongating after flowering; corolla usually 4-parted; fruits separating into 2 parts.

Common on open hillsides and more or less weedy in disturbed sites.

CAPRIFOLIACEAE
Honeysuckle Family

Woody vines, shrubs, trees or rarely herbs; leaves usually opposite, simple or compound; calyx 3-5-lobed; corolla regular or irregular, usually 5-lobed, rarely 4-lobed; stamens generally equal in number to corolla lobes and alternate with them; ovary inferior; fruit a berry, drupe or a capsule.

1a Leaves compound .... **Sambucus**
1b Leaves simple
  2a Fruit a dry capsule; flowers nodding, borne in pairs; plant a slender trailing vine .... **Linnaea**
  2b Fruit fleshy
    3a Corollas rotate; erect shrub .... **Viburnum**
    3b Corollas tubular or funnel-form, regular or irregular
      4a Fruit white; corollas nearly regular .... **Symphoricarpos**
      4b Fruit red or black; corollas usually irregular .... **Lonicera**

Linnaea

Trailing evergreen vines, somewhat woody at base; leaves opposite; flowers borne in pairs, nodding, at the tips of erect peduncles; corolla 5-lobed, bell-shaped to funnel-form; stamens 4, 1 pair longer; capsule 3-loculed, only 1 locule producing a fertile seed.

Linnaea borealis L. var. *longifolia* Torr. Twinflower

Stems slender trailing over the ground or decaying logs, 1.5-3 dm. or more long; leaves small, opposite, somewhat leathery, conspicuously netted-veined, usually toothed above the middle; peduncles 4-9 cm. long, erect, each bearing a pair of bell-shaped flowers; ovary inferior, pubescent, subtended by pubescent bracts; corolla pink and white, pubescent within.

Moist shady woods.

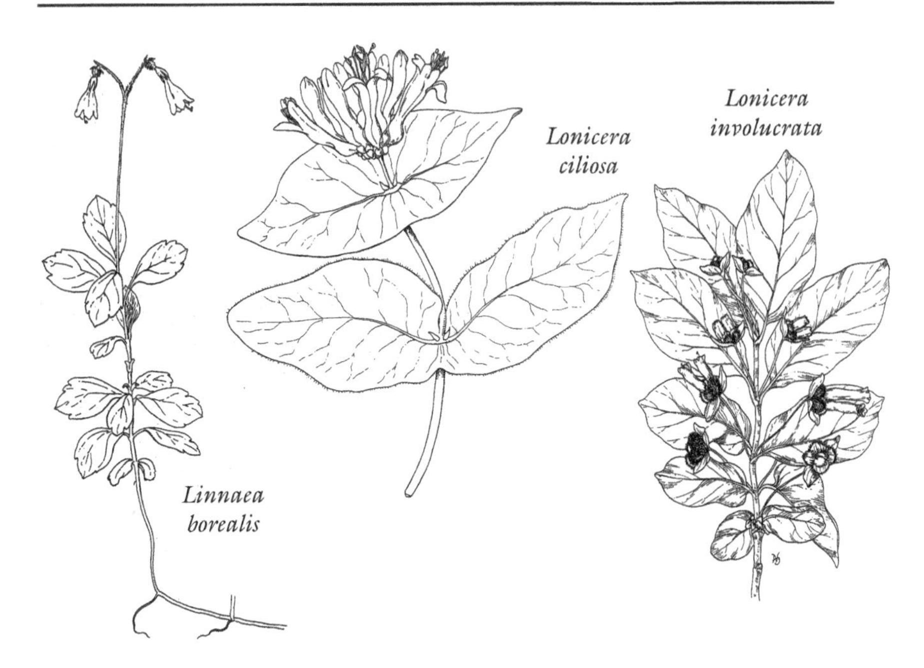

Lonicera

Shrubby plants, erect or twining, with opposite simple entire leaves, the two leaves of a pair sometimes united around the stem below the flowers; flowers in pairs or small clusters in the leaf axils or in larger clusters at the apex of the stem; calyx teeth small or sometimes absent; corolla 5-lobed, nearly regular, or very irregular and 2-lipped, the tube often inflated at one side of the base or spurred; ovary 2-3-loculed; fruit a berry.

1a Erect shrub; leaves never united about stem; berries black... .*L. involucrata*
1b Twining or trailing woody vines; upper leaves united about the stem; berries red or orange
  2a Flowers orange-red, 1.5-4 cm. long; stems glabrous.................*L. ciliosa*
  2b Flowers pink, 1-2 cm. long; stems pubescent......................*L. hispidula*

Lonicera ciliosa Poir. Climbing honeysuckle

Twining about other shrubs to a height of 3-10 m.; leaves elliptic or sometimes ovate, generally obtuse at apex, mucronate, tapering rather abruptly at base to the petioles, green above, bluish beneath, 5-10 cm. long, margins ciliate, the pair of leaves beneath the inflorescence united, the next pair below united at least at base; flowers in a terminal cluster, often accompanied by two or more lateral clusters; corolla 1.5-4 cm. long, orange-red, somewhat 2-lipped; stamens as long as or slightly longer than the corolla; berries red.

Margins of woods.

Lonicera hispidula Dougl. Pink honeysuckle

Stems twining or trailing, 2-6 m. long, pubescent; leaves rigid, 1-5 cm. long, entire, ovate or elliptic, mostly obtuse, glabrous above, pale and pubescent beneath, upper most leaves united around stem, lower leaves with stipule like appendages at base; flowers strongly 2-lipped, 1-2 cm. long, pink; peduncles, bracts, corolla-bases and ovary glandular; fruits red.

In dry rocky places and in woods and thickets. Sometimes plants grow year after year without producing flowers.

Lonicera involucrata (Rich.) Banks ex Spreng. Bush honeysuckle, Twinberry or Inkberry

Erect deciduous bushy shrubs 0.5-4 m. tall; leaves oblong, ovate, or rarely obovate, obtuse and often asymmetrical at base, or sometimes tapering to the petioles, abruptly acute at apex or sometimes rather long-acuminate, 3-14 cm. long, generally pubescent beneath and cilate on the margins; flowers borne in pairs on long peduncles in the leaf axils, the flowers sessile between broad bracts which become reddish or dark red as the fruit matures; calyx obsolete; corolla yellow or reddish, 1-2 cm. long, nearly regular, the tube somewhat swollen to short-spurred at the base, glandular-pubescent; stamens shorter than the corolla; berries black.

Two varieties occur within our area, with var. *involucrata* occurring in the mountains and var. *ledebourii* (Esch.) Jeps. being coastal.

Usually in wet or moist areas, but occasionally in drier sites; common near the coast, along streams or on sand dunes and in the mountains.

Sambucus

Shrubs or small trees; leaves compound, with a terminal leaflet, deciduous; flowers small, white or cream, in a showy terminal cyme, jointed to the pedicels; calyx 5-toothed; corolla rotate, 5-lobed; ovary 3-5-loculed; ovules solitary in each locule, attached at the top; fruit fleshy, a drupe or berry-like, containing several bony nutlets.

Inflorescence flat-topped; fruit appearing blue............................ *S. mexicana*
Inflorescence pyramidal to dome-shaped; fruit red....................... *S. racemosa*

Sambucus mexicana Presl Blue Elderberry

Large shrubs 2-8 m. tall; leaflets 3-11, lanceolate to elliptic, 3-20 cm. long, acute to acuminate, sharply toothed, usually asymmetrical at the base; inflorescence flat-topped; fruits appearing pale-bluish, although actually bluish-black and densely glaucous.

This is an extremely variable species.

(=*Sambucus caerulea* Raf.)

Stream banks and other moist habitats.

Sambucus racemosa L. Red Elderberry

Large shrubs, 2-6 m. tall; leaflets 5-7, lanceolate to lance-ovate, 5-17 cm. long, 2-6 cm. broad, acuminate, sharply toothed, usually asymmetrical at the base; inflorescence pyramidal to dome-shaped; fruits bright red, not glaucous.

(=*Sambucus callicarpa* Greene)

In moist places.

Symphoricarpos

Low shrubs with thin deciduous leaves; flowers white or pink, generally borne in compact terminal or axillary racemes, each flower subtended by 2 minute bracts; calyx tube globose, teeth 4-5; corolla more or less bell-shaped, 4- 5-lobed; stamens attached within the corolla tube; ovary 4-loculed, only 2 locules developing; fruit berry-like, often white (ours).

Erect shrub, corolla densely pubescent within .... *S. albus*
Trailing shrubs; corolla only slightly pubescent within .... *S. mollis*

Symphoricarpos albus (L.) Blake Snowberry

Erect freely branching shrubs 0.5-2.5 m. tall; leaves short-petioled, varying in shape on a single plant from nearly orbicular to ovate or oblong, and entire, toothed, or lobed, obtuse at apex, acute at base, 1.5-3 cm. long (sometimes much larger and deeply lobed on young vegetative shoots), lower surface paler than upper; corolla white to pink, inflated on lower side at base, 5-lobed, densely covered within with white hairs; fruits pulpy, white, globose or nearly so, often deformed by crowding.

Common. The foliage of this shrub is very fragrant in the rain, and the thick clusters of white fruits borne in close association with the scarlet fruits of wild rose, add much to the attractiveness of our road sides in winter.

Symphoricarpos albus

Symphoricarpos mollis Nutt. Mountain snowberry

Stems trailing and creeping, often rooting at the nodes and branches more slender than in *S. albus*; leaves oval to oblong, occasionally lobed or toothed, but mostly entire; flowers pale to deep pink; corolla scarcely pubescent within.

Generally found at higher altitudes than S. *albus.* Common in the Coast Range.

Viburnum

Shrubs or small trees; leaves opposite, simple, deciduous or evergreen; inflorescence axillary or terminal compound cymes; flowers parts in 5's; corolla bell-shaped to rotate; ovary 3-loculed with 2 locules sterile and sometimes nearly obsolete, thus appearing 1-loculed; fruit a drupe, the stone flattened.

Mountain shrubs; most of the leaves shallowly 3-lobed and sharply toothed .... *V. edule*
Lowland shrubs; leaves merely coarsely and usually bluntly toothed and more or less rounded at the apex .... *V. ellipticum*

Viburnum edule (Michx.) Raf. High-bush cranberry

Shrubs 0.5-2.5 m. or more tall; leaf blades 3-10 cm. long, broadly obovate to orbicular with 3 terminal lobes, the lobes sharply toothed, rounded to

Viburnum ellipticum

subcordate and sometimes uneven at the base, with 3 main veins, distinctly petiolate and usually with a pair of glands near the junction of the blade and the petiole, lower leaf surface pubescent, upper nearly glabrous; flowers whitish; fruits red.

Swamps and moist wooded areas in the Cascades.

Viburnum ellipticum Hook.

Western wayfaring tree

Deciduous shrubs 1-3.5 m. tall; young stems and under surface of leaves thinly glandular-pubescent; leaves broadly obovate or elliptic to nearly orbicular, 2.5-8 cm. long, dark green above, lighter beneath, somewhat thick, with 3-5 conspicuous veins from the base, upper half of margins coarsely toothed; cymes 3.5-6 cm. wide; corolla whitish, 5-9 mm. broad; fruit ellipsoid, 1-1.5 cm. long, stone flattened, 2-grooved on one face and 3-grooved on the other.

Along streams and in edges of deciduous woods.

VALERIANACEAE
Valerian family

Annuals or perennials; leaves opposite, simple to compound; flowers in simple or compound cymes or panicles, these sometimes head-like; calyx lobes either inrolled in flower and becoming expanded and plumose in fruit or obsolete; corolla 5-lobed, often 2-lipped, the tube spurred or saccate at the base; stamens 1-3, borne on the corolla tube; ovary inferior; fruit achene-like.

1a Calyx teeth of plume-like bristles incurved in the flower............. **Valeriana**
1b Calyx teeth minute or none, never bristle-like
  2a Stems not forking above; inflorescence head-like..................... **Plectritis**
  2b Stems regularly forking above; inflorescence flat topped..... **Valerianella**

Plectritis

Annuals, usually glabrous; stems angled; leaves opposite, simple, usually entire; inflorescence terminal, head-like or interrupted and spicate; calyx obsolete; corolla 2-lipped or with 5 nearly equal lobes, the tube spurred or saccate at the base; stamens 3; ovary 1-loculed; fruit sometimes winged.

1a Spur, if present, usually less than half the length of the corolla tube; fruit keeled, not grooved
  2a Corolla 3.5-9.5 mm. long, bright pink................................ *P. congesta*
  2b Corolla 1.5-3.5 mm. long, white to pale pink............ *P. brachystemon*

1b Spur over one-half the length of the corolla tube; fruit usually grooved but not keeled
    3a Corolla with 1 or 2 red spots at the base of the middle lobe of the lower lip .... *P. ciliosa*
    3b Corolla more or less regularly 5-lobed, white to pale pink, not spotted .... *P. macrocera*

Plectritis brachystemon Fisch. & Mey.

Stems 2-6 dm. tall; corolla 1.5-3.5 mm. long, white or pale pink, with a short slender spur with an enlarged tip or spurless; fruit keeled, winged or wingless.

[=*Plectritis anomala* (Gray) Suksd.]

Moist, often shady areas, also on coastal bluffs.

Plectritis ciliosa (Greene) Jeps.

Stems slender, unbranched, 1-5 dm. tall; leaves 1-3 cm. long, 3-10 mm. broad, lower leaves obovate, short-petioled, middle and upper cauline leaves oblong and sessile; corolla 2-lipped, 2-8.5 mm. long, dark pink and with 1-2 red spots at the base of the middle lobe of the lower lip, spur long and slender; fruit winged, usually grooved.

Vernally wet areas; wet meadows and shaded to open slopes.

Plectritis congesta

Plectritis congesta (Lindl.) DC.

Stems unbranched, 0.5-6 dm. tall; leaves 1-6 cm. long, 3-22 mm. broad, lower leaves obovate or spatulate and short-petioled, middle and upper cauline leaves oblong to elliptic and sessile; inflorescence dense, head-like or interrupted spike-like cymes; corolla 2-lipped, 3.5-9.5 mm. long, bright pink, spur one-third to nearly one-half as long as the slender tube, tip of the spur enlarged; fruit keeled, winged or not.

Common in vernally wet areas and in drier sites on open hillsides and road sides.

Plectritis macrocera Torr. & Gray

Stems unbranched, 1-6 dm. tall; inflorescence finely glandular; leaves 1-4.5 cm. long, 3-18 mm. broad, lower leaves obovate and short-petioled, middle and upper cauline leaves oblong to elliptic and sessile; corolla more or less regularly 5-lobed, 2-6 mm. long, white to pale pink, the spur long, stout; fruit winged or not, grooved.

Vernally wet areas, stream banks and shaded slopes.

Valeriana

Mostly ill-scented annuals or perennials; leaves mostly basal, the cauline, when present, opposite, simple to pinnately compound; flowers in panicled cymes; calyx teeth 5-20, bristle-like, inrolled in flower but conspicuous in fruit; corolla tubular or funnel-form, 5-lobed, the tube often saccate at the base; stamens 3; ovary with 1 fertile locule, and minute vestiges of 2 lateral sterile locules; fruit an achene.

Valeriana sitchensis Bong. Sitka valerian

Rhizomatous perennials, 1.5-12 dm. tall; leaves 3-20 cm. long, basal leaves simple or deeply lobed to compound, petioled, cauline leaves deeply lobed to pinnately compound, the terminal lobe or leaflet larger than the lateral lobes or leaflets, the upper leaves becoming sessile or nearly so; corolla white or pinkish, 4.5-9 mm. long; achene 3-6 mm. long with 12-20 plumose calyx segments. Two varieties occur in our area: var. *scouleri* (Rydb.) Mey. which is smaller (1.5-7 dm. tall) and with well developed and persistent basal leaves, and var. *sitchensis*, a more robust variety (3-12 dm. tall) with mostly cauline leaves.

In moist places in the mountains.

Valerianella

Erect annuals or biennials; stems regularly forking above; leaves cauline; inflorescence flat-topped, subtended by a ring of bracts; calyx obsolete; corollas minute, funnel-form, nearly regular to 2-lipped, the base of the tube slightly saccate to spurred; stamens 3; ovary 2-3-loculed, only 1 locule fertile, the other(s) empty.

Fruit with a corky enlargement, narrowly and inconspicuously grooved between the sterile locules; common .... *V. locusta*

Fruit without a corky enlargement, with a wide and conspicuous groove between the sterile locules .... *V. carinata*

Valerianella carinata Lois. European corn-salad

Similar to *V. locusta* except for the fruit which does not have a corky mass, but does have a wide conspicuous groove between the sterile locules.

Moist disturbed areas; introduced from Europe; only occasional in our area.

Valerianella locusta (L.) Betcke Blue-flowered corn-salad

Stems slender 1-3 dm. tall, several times forked above, the leafy cymes flat-topped; lower leaves petioled, the upper sessile; corolla 1.5-2 mm. long, with a white tube and pale bluish lobes; fruit with a corky enlargement, but wingless, narrowly and inconspicuously grooved between the sterile locules.

Moist areas, somewhat weedy; introduced from Europe.

Valerianella locusta

DIPSACACEAE
Teasel Family

Annual, biennial, or perennial herbs, with opposite or whorled leaves and perfect flowers in involucrate heads; calyx and corolla 4-5-lobed; stamens 2-4, attached to the corolla and alternate with its lobes; ovary inferior, 1-loculed, 1-ovuled; fruit an achene, crowned with the persistent calyx-limb.

Dipsacus fullonum

Dipsacus

Tall erect biennial or perennial herbs with rough prickly herbage and opposite leaves; leaf bases sometimes united; flowers blue or lilac in dense usually elongated heads surrounded by an involucre of rigid or spine-pointed bracts, the receptacle clothed with spiny bracts, each subtending a flower, each flower also enclosed at base by a somewhat scalloped involucel; calyx 4-lobed; corolla 4-lobed, the lobes unequal; stamens 4; style slender, stigmatic surface oblique or lateral; ovary 1-loculed with one suspended ovule; achene 4-angled.

Dipsacus fullonum L. — Wild teasel

Stout taprooted biennials, 0.5-2 m. tall, the stems striate, sparsely clothed with short, stiff, downward curved prickles; a basal rosette of oblanceolate, crenate leaves produced the first year, cauline leaves up to 3 dm. long, lanceolate, the midribs prickly, the lateral veins many, nearly horizontal and parallel, sometimes prickly, bases of the cauline leaves more or less fused; flower heads terminal and axillary in the upper nodes, showy; the involucral bracts stiff, prickly, and curved upward, sometimes reaching the length of the head; inflorescence 3-10 cm. long on long peduncles; the lilac-colored flowers beginning to open in a circle about the center of the head and continuing both upward and downward; bract subtending the flower, narrowed into a stiff spine generally extending beyond the flower; achene 5-8 mm. long.

(=*Dipsacus sylvestris* Huds.)

Common in waste places; often in moist habitats, but also in drier sites; introduced from Europe.

CUCURBITACEAE
Gourd Family

Mostly climbing or trailing tendril bearing fleshy herbs usually with simple palmately-lobed leaves; stamens and pistils in different flowers, usually on the same plant; calyx-tube united with ovary in pistillate flower, its teeth narrow; corolla 4-7-lobed; stamens usually 3-5, often fused and appearing as 1-3; ovary inferior, 1-6-loculed; fruit a modified berry or an irregularly dehiscent capsule.

Marah

Trailing or climbing perennial herbs from an enlarged woody tuberous root; stems with alternate leaves and branched tendrils; staminate flowers in axillary racemes or panicles, the pistillate solitary in the leaf axils, all small and white to yellow-green; fruit gourd-like at first, becoming dry and bursting irregularly, sometimes tapered to a beak.

Marah oreganus (Torr. & Gray) How.
Old man-in-the-ground or Wild cucumber

Stems trailing or climbing, 1-10 m. long, smooth or slightly roughened, with branching tendrils borne at the leaf-axils; leaves petioled, minutely hispid, thin, broadly heart-shaped, irregularly palmately 5-7-lobed, deep green; corolla white, cup-shaped; gourd like fruits, 4-9 cm. long, broadly ovoid, generally beaked, conspicuously veined, usually dark green striped, prickly near the base, becoming dry and irregularly dehiscent.

Common in open areas, at edges of wooded areas and especially along fence rows.

Marah oreganus

CAMPANULACEAE
Bellflower Family

Herbs (ours) to shrubs and trees; usually with alternate leaves; calyx 5-lobed, generally persistent; corolla usually 5-lobed, regular or irregular; stamens 5; ovary inferior, 1-5-loculed; fruit (in ours) a capsule.

Some genera included here are sometimes separated out into the **Lobeliaceae**.

1a Corolla irregular; filaments and sometimes also the anthers fused into a tube
  2a Flowers sessile in leaf axils (may appear stalked due to the long floral tube)........ **Downingia**
  2b Flowers borne on pedicels; plants aquatic
    3a Annuals; stems naked below; cauline leaves not reduced to filiform bracts........ **Howellia**
    3b Perennials; cauline leaves reduced to filiform bracts........ **Lobelia**
1b Corolla regular; filaments free
  4a Perennials; corolla more or less bell-shaped........ **Campanula**
  4b Annuals
    5a Corolla deeply lobed to below the middle, the lobes spreading........ **Triodanis**
    5b Corolla not lobed to the middle, bell-shaped or cylindrical

6a Capsule opening at the top; corolla bell-shaped .... **Githopsis**
6b Capsule opening near the base; corolla cylindrical .. **Heterocodon**

Campanula

Annuals, biennials or perennials (ours); leaves alternate; flowers regular, blue to white, more or less bell-shaped; stamens free; ovary 3-5-loculed; capsule opening by 2-3 lateral pores.

1a Leaves all entire
  2a Corolla 1 cm. or less in length; tufted alpine species less than 1 dm.tall .. .. *C. scabrella*
  2b Corolla 1-2 cm. long; widespread species, 1.5 to 8 dm. tall.................. .... *C. rotundifolia*
1b Leaves toothed
  3a Corolla lobes spreading or reflexed, stamens and style extending beyond them
    4a Flowers dark blue; cauline leaves nearly sessile ...... *C. prenanthoides*
    4b Flowers pale blue; cauline leaves petioled .... *C. scouleri*
  3b Corolla lobes not spreading; stamens and style included in the corolla
    5a Upper leaves linear, lower leaves orbicular to oblong *C. rotundifolia*
    5b Leaves similar in shape, spatulate to elliptic-oblong............ *C. piperi*

Campanula prenanthoides

Campanula piperi Howell — Piper's bluebell

Perennials up to 1 dm. in height from a woody base; leaves 1-3 cm. long, spatulate or elliptic-oblong, sharply toothed, glabrous, in a basal rosette as well as cauline; flowers 1-3, corolla 12-18 mm. long; capsule nearly globose.

In rocky areas at high elevations in the Olympic Mountains of WA.

Campanula prenanthoides Dur. — Slender bluebell

Taprooted perennials; stems several, erect, branching, 2-15 dm. tall; leaves ovate to lance-shaped, coarsely and sharply toothed, petioles none or very short; flowers in a long loose raceme or panicle; calyx lobes slender, much shorter than corolla; corolla 7-14 mm. long, blue, lobes long, narrow, at first forming a tube, later spreading and curving backward; capsules nearly globose, opening at or below the middle.

Open woods.

Campanula rotundifolia L. — Scotch bluebell

Taprooted perennials; stems slender, branching at the base, weakly ascending, 1.5-8 dm. tall; lower leaves orbicular to oblong, coarsely and irregularly toothed to nearly entire, petioled, cauline leaves mostly linear; flowers 1-2 cm. long, deep blue, nodding on slender pedicels, borne singly or in racemes.

Somewhat locally distributed from sea level to 1,800 m. in the mountains.

Campanula scabrella Engelm. Rough harebell

Tufted perennials up to 8 cm. in height from a woody base; herbage pubescent; leaves firm, 0.5-4 cm. long, entire, linear to oblanceolate, the cauline leaves more narrow and often reduced; flowers 1-5; corolla 6-10 mm. long, light blue; capsule opening between the middle and the apex.

Rocky areas at high elevations in the Cascades.

Campanula scouleri Hook. Pale bluebell

Rhizomatous perennials; stem prostrate at base, erect above, generally not branching, 1.5-3 dm. tall; leaves 1-6 cm. long, ovate to lanceolate, sharply toothed, petioled; flowers in loose racemes or panicles; calyx lobes slender, shorter than corolla; corolla 8-14 mm. long, pale blue or whitish, lobes ovate, curving backward at maturity; capsule opening near the middle.

Moist woods and stream banks.

Downingia

Glabrous, more or less succulent annuals; leaves alternate, simple, sessile; flowers sessile, but appearing stalked due to the long floral tube, axillary or in a terminal spike; corolla 2-lipped, generally twisted; stamens fused, usually with 2 short and 3 long anthers; ovary 1-2-loculed; fruit a capsule.

Stamen tube curved, anthers exserted... *D. elegans*
Stamen tube straight, anthers not exserted .... *D. yina*

Downingia elegans
a. flower (front)
b. flower (side)

Downingia elegans (Dougl. ex Lindl.) Torr.

Stems 1-5 dm. tall; leaves 0.5-2.5 cm. long, to 9 mm. broad, lanceolate to lance-ovate; corolla 5-18 mm. long, white to pink to blue with a large yellow-edged white spot; anthers exserted, incurved; capsule 2-5.5 cm. long.

Vernal pools, wet meadows, lake margins and ditches.

Downingia yina Applegate

Stems erect or decumbent, 0.5-3.5 dm. tall; leaves and bracts 0.7-2.5 cm. long, lanceolate; corolla 7.5-15 mm. long, blue or occasionally pink, with 2 yellow spots inside a larger white spot and with 3 small purple spots or a purple band at the throat; anthers not exserted; ovary 1-loculed; capsule 2-5 cm. long.

Vernal pools, wet meadows, marshes, bogs, lake margins and ditches.

Githopsis

Low annuals with small alternate leaves; flowers solitary; calyx tube slender, 10-ribbed; corolla bell-shaped; ovary inferior, 3-loculed; stigmas 3; fruit a ribbed capsule, bearing at its summit the 5 stiff persistent calyx-lobes.

Githopsis specularioides Nutt. Common blue-cup

Stems 1 to several, simple or sending up branches longer than main stem, minutely pubescent; leaves minute at base, becoming somewhat larger upward, sessile, roundish to narrowly oblong, mostly toothed; calyx tube becoming thick-ribbed, tapering to the base, calyx lobes shorter or longer than corolla; corolla blue to magenta.

Open hillsides; uncommon.

Heterocodon

A monotypic genus.

Heterocodon rariflorum Nutt.

Annuals; stems 4-angled, 5-30 cm. tall; leaves and bracts alternate, 2-10 mm. long, suborbicular-cordate, sharply serrate, sessile and more or less clasping; calyx lobes 2-4 mm. long; corollas of the upper flowers 3-6 mm. long, blue, those of the lower cleistogamous flowers reduced or obsolete; ovary 3-loculed; fruit a capsule, 1.8-3 mm. long, opening by irregular pores.

Vernal pools, stream banks and other moist areas.

Howellia

A monotypic genus.

Howellia aquatilis Gray

Aquatic or subaquatic glabrous annuals, 1-6 dm. tall; stems naked below, branched above, the branches submerged or floating; leaves alternate or sometimes appearing opposite or even whorled, linear, 1-4.5 cm. long, to 1.5 mm. broad, entire or nearly so; flowers axillary, those submerged with undeveloped corollas, those above the surface of the water with conspicuous white 2-lipped corollas 2-2.7 mm. long; fruit 5-13 mm. long.

In lakes and ponds.

Lobelia

Annuals, perennials or shrubs; leaves alternate or all basal; calyx and corolla 5-parted; flowers solitary in axils of upper leaves or in terminal racemes, spikes or panicles; corolla 2-lipped, the tube often split to near the base; stamens 5, fused; ovary inferior, 2-loculed; fruit a capsule.

Lobelia dortmanna L. Water Lobelia

Glabrous perennials, stems normally submerged with only the inflorescence emergent; leaves in a basal rosette, linear, hollow, 2-8 cm. long, cauline leaves reduced to filiform bracts; inflorescence a few-flowered raceme; calyx lobes 1.5-2.5 mm. long; corolla 1-2 cm. long, white to pale blue, lower lip pubescent at the base; capsule 5-10 mm. long.

In shallow water of lakes and ponds; uncommon.

Triodanis

Annuals; stems 4-angled; leaves alternate; flowers of 2 kinds, the earlier with a 3-4-lobed calyx and an undeveloped closed corolla, the later with a 5-lobed calyx and a conspicuous 5-lobed corolla; stamens free; fruit a capsule opening by lateral pores.

Triodanis perfoliata (L.) Nieuwl. Venus'-looking-glass

Stems 1-6 dm. tall, stout, simple or branched, erect at least above; leaves nearly orbicular, toothed, sessile and clasping the stem; corollas 8-12 mm. long, blue.

(=*Specularia perfoliata* DC.)

Road sides, fields and disturbed areas.

ASTERACEAE
Sunflower Family

Herbs, shrubs or trees with small flowers in close heads, subtended at base by a whorl of bracts forming an involucre; ovary wholly inferior, becoming a seed-like achene, crowned by the modified calyx (the pappus) in the form of teeth, scales, bristles, etc., or the pappus absent; stamens 4-5, attached to the corolla, generally united by the anthers; corollas tubular and generally 5-toothed (in disk flowers) or strap-like (in ray or ligulate flowers). The alternate name for the family is **Compositae**.

1a Flowers all strap-shaped; plants with milky juice
  2a Leaves all basal; peduncles naked or with scaly bracts
    3a Pappus of scales, widened at the base, awned above.. **Microseris** (in part)
    3b Pappus, at least in part, of numerous bristles.
      4a Pappus bristles brownish; young heads nodding
        5a Achenes spindle-shaped; inner involucral bracts glabrous or white-hirsute .... **Leontodon**
        5b Achenes not tapering; inner involucral bracts with appressed black hairs.. **Microseris** (in part)
      4b Pappus bristles white; heads typically erect
        6a Outer series of involucral bracts soon reflexed........**Taraxacum**
        6b Outer series of involucral bracts not soon reflexed
          7a Receptacle chaffy with thin scales; peduncles usually branched... **Hypochaeris**
          7b Receptacle naked; peduncles unbranched
            8a Achenes tapering upward into a slender beak .. **Agoseris**
            8b Achenes beakless .... **Nothocalais**
  2b Leaves not all basal
    9a Pappus none; achenes 20-30 ribbed .... **Lapsana**
    9b Pappus present
      10a Pappus of scales, these sometimes awned
        11a Scales broad at base, tapering into an awn.. **Microseris** (in part)
        11b Pappus scales short, blunt .... **Cichorium**
      10b Pappus of hairs or bristles
        12a Achenes strongly flattened
          13a Achenes beaked; pappus bristles separating from the fruit individually .... **Lactuca**
          13b Achenes not beaked; pappus bristles, or at least the inner, separating from the fruit as a group.......................... **Sonchus**
        12b Achenes not strongly flattened
          14a Pappus bristles plumose .... **Tragopogon**
          14b Pappus bristles not plumose
            15a Achenes tapered at both ends; pappus bristles white, soft .... **Crepis**
            15b Achenes cylindrical or only slightly tapered at one end; pappus bristles usually brownish to yellowish-brown

16a Lower leaves oblanceolate to spatulate .... **Hieracium**
16b Lower leaves triangular to deltoid, often hastate at the base .... **Prenanthes**

1b Flowers all tubular or only the marginal (ray) flowers strap-shaped; plants without milky juice

17a Strap-shaped ray flowers absent, though the marginal circle of tubular flowers sometimes enlarged and showy

18a Pappus consisting of bristles, awns or scales also sometimes present

19a Flower heads opening before the foliage leaves; cauline leaves simple and bract-like .... **Petasites** (in part)
19b Flower heads opening with or after the leaves; cauline leaves rarely bract-like

20a Involucral bracts nearly equal, mostly in 1 or 2 series, or with a few basal bractlets

21a Outer flowers pistillate; inner flowers perfect or staminate. .... **Erechtites**
21b Outer flowers, as well as the inner, all perfect and fertile

22a Annuals, or non-rhizomatous perennials; flowers usually many; leaves not palmately cleft .... **Senecio** (in part)
22b Perennials; few-flowered or leaves palmately cleft

23a Leaves deeply palmately cleft; heads 20-50-flowered .... **Cacaliopsis**
23b Leaves entire or nearly so; heads 4-17-flowered

24a Leaves white-woolly, at least below........... **Luina**
24b Leaves glabrous .... **Rainiera**

20b Involucral bracts generally many, in several series, not equal

25a Plants with spiny leaves

26a Pappus bristles roughened; leaves white-blotched......... .... **Silybum**
26b Pappus bristles plumose; leaves not white-blotched...... .... **Cirsium**

25b Plants not spiny-leaved

27a Shrubs; dioecious... **Baccharis**
27b Herbs, sometimes woody at the base

28a Plants woolly; involucral bracts papery; heads small and clustered

29a Stamens and pistils in the same head.................. .... **Gnaphalium**
29b Stamens and pistils typically in separate heads or on separate plants

30a Pappus bristles united at the base, those of the staminate flowers thickened at the apex .... **Antennaria**
30b Pappus bristles free, those of the staminate flowers not thickened at the apex .... **Anaphalis**

28b Plants usually not conspicuously woolly nor bracts papery (except sometimes on the margins); heads larger, seldom clustered

31a Involucre globose, its bracts rigid, with hooked tips .... **Arctium**
31b Involucre and bracts not as above
32a Involucral bracts either spiny or fringed.......**Centaurea** (in part)
32b Involucral bracts neither spiny or fringed
33a Inner pappus bristles plumose..............................**Saussurea**
33b Inner pappus bristles not plumose
34a Leaf blades triangular, coarsely toothed.............**Brickellia**
34b Leaves lanceolate to ovate, entire................ **Heterotheca**
18b Pappus consisting of awns or a thin crown of scales or absent
35a Involucre of pistillate heads, bur-like, armed with spines or tubercles
36a Involucre armed with hooked spines .... **Xanthium**
36b Involucre armed with straight spines or tubercles ....**Ambrosia**
35b Involucre not bur-like
37a Heads sessile in the leaf axils; low plants
38a Leaves compound; achenes awned and usually winged...........
.... **Soliva**
38b Leaves simple; achenes smooth, each enclosed in a woolly bract...**Psilocarphus**
37b Heads not sessile in the leaf axils
39a Receptacle long-bristly ....**Centaurea** (in part)
39b Receptacle not long-bristly, although sometimes chaffy
40a Involucral bracts herbaceous
41a Pappus absent; achenes with stalked glands; heads small
....**Adenocaulon**
41b Pappus of 1-6 awns
42a Receptacle chaffy ....**Bidens** (in part)
42b Receptacle not chaffy, often pitted ....
....**Grindelia** (in part)
40b Involucral bracts largely scarious
43a Achenes, especially the marginal ones, stipitate..**Cotula**
43b Achenes sessile
44a Heads borne in racemes or panicles.........**Artemisia**
44b Heads solitary or borne in cymes or corymbs
45a Receptacle flat or convex ....**Tanacetum** (in part)
45b Receptacle high, conical....................**Matricaria**
17b Strap-shaped ray flowers present
46a Flower stalk appearing before the leaves..................**Petasites** (in part)
46b Flowers appearing with or after the leaves
47a Pappus absent
48a Leaves mostly opposite
49a Fleshy salt marsh and beach plants with overlapping involucral bracts...**Jaumea**
49b Plants not fleshy; occurring in woods, fields and along rivers.
50a Involucral bracts in 2 dissimilar series, not enclosing an achene ....**Coreopsis** (in part)
50b Involucral bracts in 1 series, each bract enclosing an achene
.... **Madia** (in part)
48b Leaves mostly alternate or basal or the lower rarely opposite
51a Both ray and disk flowers yellow

52a Leaves mostly basal, large; heads more than 5 cm. wide .... **Balsamorhiza**
52b Leaves not mainly basal; heads smaller
53a Plants mostly sticky-glandular; achenes usually laterally compressed .... **Madia** (in part)
53b Plants not sticky-glandular; achenes not laterally compressed
54a Heads short-pedunculate; ray flowers 5, 3-5.5 mm. long .... **Lagophylla**
54b Heads long-pedunculate; ray flowers more than 6, up to 20 mm. long .... **Eriophyllum** (in part)
51b Ray flowers white, pink or red
55a Heads borne in terminal corymbs .... **Achillea**
55b Heads borne singly on slender peduncles
56a Receptacle chaffy, at least towards the middle; leaves finely dissected .... **Anthemis** (in part)
56b Receptacle naked; leaves not finely dissected
57a Leaves both basal and cauline .... **Leucanthemum**
57b Leaves mainly basal .... **Bellis**
47b Pappus present
58a Pappus of numerous bristles, sometimes also with scales
59a Involucral bracts generally narrow, equal, or nearly so, in 1 or 2 series
60a Leaves (at least the lower) opposite .... **Arnica**
60b Leaves alternate
61a Receptacle conical .... **Crocidium**
61b Receptacle flattened or convex
62a Ray flowers yellow; heads not solitary; leaves not entire .... **Senecio** (in part)
62b Ray flowers white, purple, pink, or if yellow, leaves entire and heads borne singly
63a Rays white, inconspicuous, heads numerous, 1 cm. or less wide; annuals .... **Conyza**
63b Not as above in all respects .... **Erigeron**
59b Involucral bracts in several series, generally overlapping
64a Ray flowers white, bluish or purple .... **Aster**
64b Ray flowers yellow
65a Heads 5-10 cm. wide; leaves large and velvety beneath . **Inula**
65b Heads less than 3 cm. wide; leaves not velvety
66a Leaves not punctate-glandular, or if so, stems inequably leafy .... **Solidago**
66b Leaves punctate-glandular; stems equably leafy . **Euthamia**
58b Pappus (in ours) of scales and/or awns or short teeth
67a Leaves opposite well up the stem, or throughout
68a Receptacle not chaffy; pappus of scales, or scales and awns .... **Lasthenia**
68b Receptacle chaffy; pappus of 2-4 awns or mintue teeth
69a Awns retrorsely barbed .... **Bidens** (in part)
69b Pappus minute, not barbed
70a Leaves, at least the lower, pinnately dissected into linear segments .... **Coreopsis** (in part)

70b Leaves entire or merely toothed .... **Madia** (in part)
67b Leaves alternate or all basal (or opposite only below in *Eriophyllum*)
71a Involucre strongly sticky-resinous, tips of the bracts firm, recurved to spreading .... **Grindelia** (in part)
71b Involucre not sticky-resinous, although sometimes glandular-pubescent
72a Pappus of 2 or 4 awns .... **Boltonia**
72b Pappus of scales, these sometimes awn-tipped or merely a minute crown
73a Receptacle bristly or chaffy
74a Chaffy scales of the receptacle folded around the disk ovaries and achenes .... **Wyethia**
74b Scales or bristles of the receptacle not folded around the ovaries or achenes
75a Ray flowers 3-cleft; involucral bracts not scarious .... **Gaillardia**
75b Ray flowers entire or only shallowly notched; involucral bracts scarious .... **Anthemis** (in part)
73b Receptacle naked or minutely horny-toothed
76a Receptacle minutely horny-toothed; plants of talus slopes .... **Hulsea**
76b Receptacle naked
77a Involucral bracts few (5-15), keeled, subequal in 1 series; plants woolly .... **Eriophyllum** (in part)
77b Involucral bracts many (mostly 20 or more), in 2-3 series
78a Leaves shallowly toothed to nearly entire ..... **Helenium**
78b Leaves pinnately dissected .... **Tanacetum** (in part)

Achillea

Strongly-scented perennial herbs with alternate serrate or pinnately dissected leaves; heads small, white or rarely pink, yellow, or red, borne in terminal flat-topped corymbs and bearing both ray and disk flowers; involucre cylindrical, ovoid, or subglobose, the bracts narrow, in a few series; ray-flowers pistillate, few; achenes flattened, with a thickened margin; pappus none.

Achillea millefolium L. Yarrow or Milfoil

Stems erect, often tufted, generally simple, 1-15 dm. tall, somewhat pubescent, the hairs white; leaves pinnately divided into numerous narrow segments, the basal leaves larger and at first tufted; heads many, small, white or rarely pink, in a terminal corymb; ray flowers usually 3-8.

A highly variable species.

Common in a variety of habitats; sometimes weedy.

Achillea millefolium

Adenocaulon

Annual or perennial herbs with slender stems; leaves mostly at or near the base, alternate, large, thin, green above, densely white-woolly beneath; heads small, whitish, in loose terminal glandular panicle-like cymes; involucral bracts few, in 1 series; ray flowers none; outer circle of flowers pistillate, the inner staminate; achenes club-shaped, with stalked glands; pappus absent.

Adenocaulon bicolor Hook. Pathfinder

Perennials 1.5-10 dm. tall, more or less woolly below, sticky-glandular above; leaves mostly basal, blades triangular, the lower blades often 1 dm. long and equally as broad, green above, densely white-woolly beneath, heart-shaped to hastate at the base, the margins coarsely toothed or wavy, the petioles narrowly winged; heads small, borne in terminal loose widely branched glandular panicle-like cymes; pistillate flowers 1-7, but mostly 3-4; staminate flowers 3-10; mature achenes 5-8 mm. long, thickened at the apex and loosely covered above with stalked glands.

Common in open to shady woods.

Adenocaulon bicolor
a. head showing
(i) developing fruits,
(ii) old staminate flowers

Agoseris

Annual or more often perennial (ours) herbs with a long, blackish taproot; leaves usually all basal, entire to pinnately lobed or divided, pubescent with white hairs; heads solitary, narrowly or broadly bell-shaped, more or less pubescent, especially about the base of the involucre; involucral bracts overlapping in 2-4 series, the outer ones usually broader and shorter than the narrow, tapering inner ones; flowers all strap-shaped, yellow or orange, often drying purplish; achenes spindle-shaped, tapering above into a short to very long and slender beak; pappus of numerous fine, white, non-plumose bristles; achenes at maturity forming a globose head, the involucre shriveling and reflexed.

1a Flowers orange, drying purplish; heads narrow; beak of achene about as long as the body .... *A. aurantiaca*
1b Flowers yellow, the outer ones with a pinkish tint when dry; heads and beaks of achenes various
  2a Achenes with a short beak; alpine plants........ *A. glauca*
  2b Achenes with a long slender beak; plants of lower elevations
    3a Beak of achene 2-4 times as long as body........ *A. grandiflora*
    3b Beak of achene less than twice as long as body
      4a Plants 3-7 dm. tall; heads over 2 cm. tall in fruit; not coastal plants .... *A. elata*
      4b Plants 0.5-4 dm. tall; heads less than 2 cm. tall in fruit; coastal plants .... *A. apargioides*

Agoseris apargioides (Less.) Greene — Woolly Agoseris

Plants less than 4 dm. tall, often with several rosettes from a single deep taproot; leaves usually spatulate, the tip blunt to acute, usually coarsely toothed, glabrous or pubescent; heads bell-shaped, pubescent about the base; outer involucral bracts shorter and more blunt than the inner ones; beak of achene about as long as body or a little longer; heads many-flowered.

Seashore dunes and bluffs.

Agoseris aurantiaca (Hook.) Greene — Orange Agoseris

Plants 1.5-5 dm. tall, glabrous or pubescent; leaves linear to oblanceolate, entire or somewhat toothed or laciniate, often purplish on the petiole; heads narrow, pubescent at the base, all the involucral bracts tapering and acute; flowers orange, drying purplish; beak about equaling the achene body or somewhat longer.

Meadows and moist areas in the foothills to moderately high elevations.

Agoseris elata (Nutt.) Greene — Tall Agoseris

Plants 3-7 dm. tall; scapes pubescent; leaves oblanceolate, entire or with a few lobes; involucral bracts acute, ciliate, often glandular; head many-flowered; body of the achenes 7-10 mm. long, the beaks about the same length.

Meadows and wooded slopes; low to moderate elevations; infrequent.

Agoseris glauca (Pursh) Raf. — Pale Agoseris

Plants mostly 1-5 dm. tall; leaves usually entire or wavy margined, or rarely lobed, usually evenly pubescent with short hairs; involucral bracts in 3-4 graduated series, glabrous or with long, often glandular, hairs; flowers yellow; body of achenes 4-10 mm. long, beaks much shorter.

Alpine meadows and slopes; in the Olympic and Cascade Mountains.

Agoseris grandiflora (Nutt.) Greene — Large-flowered Agoseris

Plants 2-8.5 dm. tall, often with many peduncles from a single rosette; leaves long, tapering to an acute (rarely blunt) tip, frequently with narrow, sharp lobes, usually pubescent at least on the petiole; heads broadly bell-shaped in flower, 2-4 cm. long when fruits are nearly mature; outer involucral bracts acute or abruptly tapering, broader and shorter than the linear inner bracts, more or less densely pubescent with white hairs; flowers numerous in the head, yellow; body of achene 3.5-7 mm. long, sharply ribbed, beak more than twice as long.

Dry open woods, meadows and road sides, somewhat shaded by shrubbery; widespread at low to moderate elevations.

Ambrosia

Annual or perennial herbs to shrubs, generally monoecious; leaves opposite, usually alternate above, rarely entire, usually lobed or dissected; flower heads small, greenish, the staminate nodding, in narrow racemes or spikes; pistillate generally in clusters at base of the staminate inflorescence or in the upper leaf axils; staminate involucre cup-shaped, the bracts fused nearly their entire length, enclosing many flowers; pistillate involucre 1-flowered, top-shaped, with several spines or tubercles.

Annuals; stems erect.. *A. artemisiifolia*
Perennials; stems more or less prostrate.................................. *A. chamissonis*

Ambrosia artemisiifolia L. Common Ragweed
Annuals, 1-12 dm. tall, hispid, simple or branched; leaves opposite below, ovate in outline, deeply lobed to several times pinnately divided; staminate heads nodding, in long racemes; pistillate heads in the axils of leafy bracts at the bases of the staminate racemes; fruits with a circle of several short spines or tubercles.

One of the most serious causative agents of hay-fever and asthma in the eastern and midwestern states.

Now established in several counties of western Oregon.

Ambrosia chamissonis (Less.) Greene Cutleaf beach-bur
Stems more or less prostrate, whole plant silky or the stems woolly; leaves thick, oval or spatulate, petioled, coarsely toothed or the lower dissected, or the leaves once or twice dissected in var. ***bipinnatisecta*** (Less.) J. T. Howell, silvery silky; bur somewhat hairy, covered with short flattened spines.

(=*Franseria chamissonis* Less.)

Coastal beaches.

Anaphalis

More or less white-woolly perennial herbs with simple erect leafy stems and entire leaves; flower heads small, white, in terminal panicles, typically dioecious, but with a few staminate or perfect flowers generally, in the center of the pistillate heads; involucral bracts white, papery, numerous, closely imbricated; staminate and perfect flowers with a tubular corolla, enlarged above and 5-lobed, the style, if present, included or slightly exserted; corolla of pistillate flowers tubular-filiform, minutely 5-toothed, not enlarged above, the 2-cleft style conspicuously exserted; pappus white, with minutely barbed bristles, occurring on all three types of flowers.

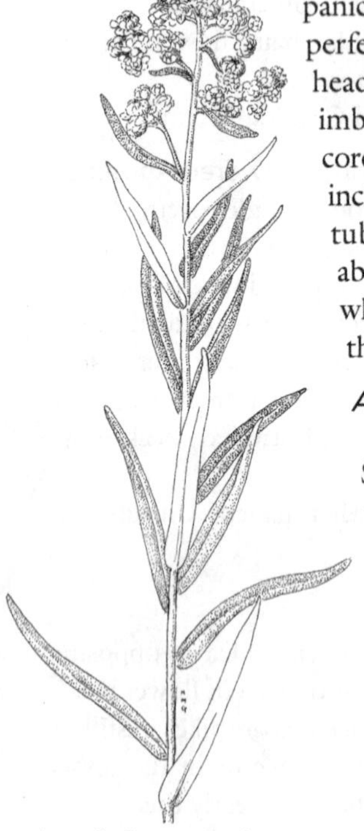

Anaphalis margaritacea

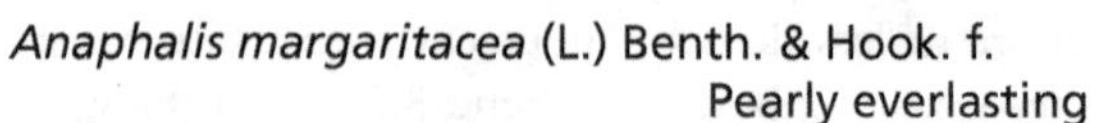

Anaphalis margaritacea (L.) Benth. & Hook. f.
Pearly everlasting

Stems erect, generally clustered, 2-12 dm. tall, white-woolly or silky, at least at first; leaves 3-12.5 cm. long, green or gray above, white-woolly beneath, linear to lanceolate, entire, somewhat clasping at base, sometimes involute; panicles 2.5-15 cm. in diameter.

There is considerable variation in the leaf shape and pubescence of the upper surface of the leaves.

Common, along road sides and in woods, especially in burned-over forest land.

Antennaria

Low tufted woolly dioecious perennials with alternate leaves and small heads solitary or in cyme-like clusters; involucral bracts papery, overlapping in several series, the lower generally woolly; pappus of staminate flowers thickened at the apex, pappus of the pistillate flowers slender, united at the base in a ring.

1a Plants without stolons or offsets .... *A. lanata*
1b Plants with stolons
  2a Leaves green above, mostly 3.5-5 cm. long .... *A. howellii*
  2b Leaves mostly 2.5 cm. long or less, white- or gray-woolly on both surfaces
    3a Involucral bracts deep green or brown .... *A. media*
    3b Involucral bracts more or less rose-colored .... *A. rosea*

Antennaria howellii Greene — Howell's pussytoes

Stoloniferous perennials; stems slender, 1.5-4.5 dm. tall from a rosette of basal leaves on a crown; plant short-woolly except for the upper surface of the leaves; basal leaves obovate, densely gray-woolly beneath, obtuse and mucronate at apex, midrib and 2 lateral parallel veins visible for a distance, cauline leaves much reduced, bract-like, narrowly lanceolate to linear; heads in a compact cluster, often 3 cm. or more wide; involucral bracts narrowly lanceolate, acute or acuminate, the outer woolly at base, inner white and papery at apex.

[=*Antennaria neglecta* Greene var. *howellii* (Greene) Cronq.]

Open woods and rocky slopes.

Antennaria lanata (Hook.) Greene — Woolly pussytoes

Tufted perennials; stems slender, 7.5-15 cm. tall; whole plant densely whitish-woolly; lower leaves oblanceolate, upper lanceolate to linear, becoming reduced above; inflorescence appearing like one compact head, very woolly at base; outer involucral bracts brownish, inner shining white, papery.

In rocky areas; alpine or subalpine.

Antennaria media Greene — Alpine pussytoes

Stoloniferous perennials; stems 3.5-10 cm. tall, mostly densely tufted, whole plant gray-woolly; basal leaves oblanceolate to spatulate, cauline leaves linear; heads few in a compact cluster; involucral bracts deep green or brown above, woolly below; pappus bristles of the staminate flowers somewhat dilated at apex.

[=*Antennaria alpina* (L.) Gaertn. var. *media* (Greene) Jeps.]

Alpine or subalpine meadows and slopes.

Antennaria rosea Greene — Rosy pussytoes

Stems 0.5-4.5 dm. tall, slender, erect or ascending from a leafy stoloniferous base; herbage with woolly or silky pubescence; basal leaves spatulate, silky on both sides or the upper surface becoming nearly glabrous, cauline leaves mostly lanceolate or linear, sessile; heads in a compact inflorescence, woolly at base, outer bracts obtuse, inner acute, pale to deep rose color.

A highly variable species.

Dry open areas from the foothills to the mountains.

Anthemis

Branching annual or perennial herbs with finely dissected alternate leaves and solitary heads terminating the branches; ray flowers generally showy, white or yellow, at first spreading, later turned downward about the peduncle; disk flowers yellow; receptacle with narrow chaffy bracts; involucral bracts scarious; achenes ribbed or ridged; pappus none, or minute.

Anthemis cotula L. — Dogfennel or Mayweed

Ill-smelling generally much-branched annuals 1.5-6 dm. tall; leaves finely dissected; ray flowers sterile, white, becoming reflexed; receptacle becoming cone-shaped; disk flowers yellow; involucral bracts green with brown scarious margins; achenes rough-surfaced.

A common weed in various habitats.

In addition to the above, two other introduced species may sometimes be found in our limits: *A. arvensis* L. (Corn chamomile) resembling *Anthemis cotula*, but with smooth achenes, heads generally slightly larger and more showy, and plant not ill-smelling and *A. tinctoria* L. (Yellow chamomile) with above general characters but with yellow rays.

Arctium

Coarse biennial herbs with large heart-shaped or ovate leaves; heads purple or pink, many-flowered, solitary or clustered; involucre globose, its bracts rigid, narrow, spreading, with hooked tips; ray flowers none; disk flowers perfect; pappus of short roughened bristles.

Inflorescence raceme-like; heads 1-2.5 cm. wide .... *A. minus*
Inflorescence corymb-like; heads 2.5-4 cm. wide .... *A. lappa*

Arctium lappa L. — Great burdock

Plants up to 3 m. tall; leaves to 4 dm. long, broadly ovate, glabrous or nearly so above, tomentose beneath; heads 2.5-4 cm. wide in a corymb-like inflorescence.

Occasional in waste places; introduced from Europe.

Arctium minus (Hill) Bernh. — Common burdock

Stems stout, erect, branched, 6-15 dm. tall; leaves broadly ovate, to 4 dm. or more in length, more or less pubescent on both surfaces, but often becoming glabrous above, petioled, the lower leaves cordate; heads 1-2.5 cm. wide, in racemes, the hooked prickles of the involucre slender; mature flowers longer than the bracts.

Common in disturbed habitats; introduced from Europe.

Arnica

Erect perennial often aromatic herbs; leaves opposite or the reduced upper ones sometimes alternate; heads yellow, generally consisting of both ray and disk flowers; involucre generally bell-shaped, its bracts narrow, nearly equal, in 1 or 2 series; ray flowers pistillate, fertile, the corolla entire or several-toothed; disk flowers perfect, fertile, the corolla 5-lobed; achenes slender; pappus a single series of rough, stiff or subplumose bristles.

1a Basal leaves long-petioled
  2a Lower cauline leaves sessile, or on very short broad petioles. *A. latifolia*
  2b Lower cauline leaves slender-petioled .... *A. cordifolia*
1b Basal leaves short-petioled
  3a Cauline leaves in 5-12 pairs, these scarcely reduced upwards .... *A. amplexicaulis*
  3b Cauline leaves in 3-5 pairs, these distinctly reduced upwards... *A. mollis*

Arnica amplexicaulis Nutt. — Clasping Arnica

Stems tufted at the base, from short rhizomes; stems 3-8 dm. tall, leaves broadly or narrowly ovate, the upper clasping, the lower narrowing to short petioles, all somewhat sticky-glandular above, the margins toothed; heads several, about 4 cm. in diameter, long-peduncled; involucral bracts pubescent, erect, the tips spreading; achenes pubescent; pappus brownish.

In moist places, especially at low altitudes in the mountains.

Arnica cordifolia Hook. — Heart-leaf Arnica

Stems 1-6 dm. tall, solitary or clustered, somewhat glandular; leaves large, nearly entire to coarsely toothed, cordate at the base, mostly petioled; heads 1 or few, long-peduncled.

Meadows and woods in the Cascades.

Arnica latifolia Bong. — Mountain Arnica

Stems erect, 1.5-6 dm. tall, from slender rhizomes, lower part generally glabrous, glandular and pubescent above; leaf pairs distant, few, the blades ovate to oblong, coarsely and often irregularly toothed, basal leaves petioled, lower cauline leaves sessile or with very short broad petioles; heads 1 or few, about 4 cm. wide, including the rays, mostly long-peduncled; involucral bracts and upper end of peduncle sparingly glandular.

Moist places in the high Cascades.

Arnica mollis Hook. — Hairy Arnica

Stems 2-6 dm. tall from a stout rhizome, pubescent or glabrous below; inflorescence glandular; basal leaves short-petioled, the blades oblanceolate or elliptic, cauline leaves in 3-5 pairs, ovate, sessile, entire or toothed; heads 1-several; involucre glandular and pubescent.

Wet ground at middle altitudes in the Cascades.

Artemisia

Mostly aromatic herbs or shrubs with alternate leaves, and small discoid heads in panicles or racemes; flowers yellowish, brownish or greenish, the outer pistillate and the inner perfect or staminate, or all perfect; pistillate flowers, when present, smaller, the corollas 2-3-toothed; corollas of disk flowers 5-toothed; bracts of the involucre dry, imbricated; pappus none or minute.

1a Leaves entire, toothed or merely lobed
  2a Leaves usually less than 1 cm. wide; plants growing below the high water level of the Columbia River; plants without well developed rhizomes.......... .... *A. lindleyana*
  2b Leaves usually over 1 cm. wide; plants rhizomatous
    3a Heads cylindrical, becoming brown; disk flowers 2-8 ... *A. suksdorfii*
    3b Heads subglobose, gray or green; disk flowers 10-25 .... *A. douglasiana*
1b Leaves deeply divided
  4a Leaves green and glabrous; annuals or biennials .... *A. biennis*
  4b Leaves pubescent, at least on the under surface; mostly perennials
    5a Basal and lower cauline leaves long-petioled; disk flowers sterile .... *A. campestris*
    5b Leaves sessile, all cauline; disk flowers fertile
      6a Divisions of the lower leaves 5-15 mm. wide; pistillate flowers 6-9 .... *A. vulgaris*
      6b Divisions of the lower leaves less than 5 mm. wide; pistillate flowers 9-12 .... *A. michauxiana*

Artemisia biennis Willd. Biennial wormwood

Glabrous annuals or biennials 3-30 dm. tall; leaves 4-15 cm. long, finely divided, the lobes toothed; inflorescence compact, leafy; heads numerous; involucre 2-3 mm. high.

Waste places and stream banks.

Artemisia campestris L.

Taprooted biennials or usually perennials; stems 1-10 dm. tall; basal leaves crowded, 2-10 cm. long, including the petioles, 2-3-times pinnatifid into linear segments, cauline leaves smaller and less divided; involucre 2-4.5 mm. high; outer flowers pistillate and fertile, inner flowers sterile; achenes glabrous.

This is a variable species with several named varieties.

Northern mountains and in the Columbia River Gorge.

Artemisia douglasiana Besser Douglas' sagewort or mugwort

Rhizomatous perennials; stems tufted and ridged, 5-15 dm. tall or more; leaves variable, 7-15 cm. long, usually green above, white-tomentose beneath, margins entire and sometimes involute, or toothed or lobed; heads small, subglobose, borne in terminal panicles; disk flowers 10-25, pistillate flowers 6-10.

Common in the Willamette Valley especially along streams and in the mountain passes of Oregon and southern Washington.

Artemisia lindleyana Besser Riverbank wormwood

Stems 2-7 dm. tall, clustered from a woody base; leaves mostly entire or with a few teeth or lobes, 2-5 cm. long, narrow, usually green above, white-tomentose on the under surface; inflorescence narrow; involucre about 3 mm. high; disk flowers perfect.

Just below the high water level; entering our area along the Columbia River.

Artemisia michauxiana Besser — Lemon sagewort

Perennials, usually rhizomatous; stems 2-4 dm. tall, rarely taller; leaves bipinnatifid, the segments again toothed, upper leaves reduced and often entire, upper surfaces generally green and glabrous, white-tomentose beneath at least when young; involucre subglobose, 3.5-4 mm. high; pistillate flowers 9-12, disk flowers 15-35, fertile.

Usually in rocky areas, such as talus slopes; subalpine to alpine.

Artemisia suksdorfii Piper — Coastal mugwort

Rhizomatous perennials; stems usually unbranched, stiff, brittle, erect, 0.5-2 m. tall, very leafy; leaves lanceolate, entire to toothed or lobed, white-tomentose beneath, green above; inflorescence densely flowered; heads cylindrical, 3-4 mm. high, becoming shining brown; disk flowers 2-8, pistillate flowers 3-7.

Common along the Columbia River and the coast on bluffs and beaches, also in the Willamette Valley, extending into the mountain passes of northern Washington.

Artemisia vulgaris L. — Common wormwood

Rhizomatous perennials 0.5-1.5 m. tall; leaves more or less obovate in outline, deeply bipinnatifid, 5-15 cm. long, green and glabrous above, white-tomentose beneath; involucre bell-shaped; 3-4.5 mm. high; pistillate flowers 6-9, disk flowers 13-20.

Disturbed ground, especially near sea ports; introduced from Europe.

Aster

Annual, biennial or perennial herbs with alternate simple, entire or toothed leaves; heads in panicles or cymes or rarely solitary; involucral bracts in several ranks, leaf-like or leathery, green-tipped; ray flowers usually present, purple, violet, blue, or white; disk flowers yellow, sometimes turning brownish or purplish; achenes usually ribbed; pappus consisting of many hair-like bristles.

1a Involucre bracts closely appressed, stiff, purplish .... *A. ledophyllus*
1b Involucre bracts generally spreading, not stiff
  2a Leaves mostly basal; cauline leaves bract-like .... *A. alpigenus*
  2b Leaves not mostly basal
    3a Mountain species
      4a Stems glandular above.. *A. modestus*
      4b Stems not at all glandular.. *A. foliaceus*
    3b Lowland species
      5a Leaves leathery; stems stiffly erect............................ *A. radulinus*
      5b Leaves thin or scarcely leathery; stems weak
        6a Ray flowers white or pale lavender
          7a Ray flowers only 1-3 .... *A. curtus*

7b Ray flowers more numerous .... *A. hallii*
6b Ray flowers violet, purple, or bluish
8a Outer involucral bracts obtuse; plants coastal .... *A. chilensis*
8b Outer involucral bracts acute .... *A. subspicatus*

Aster alpigenus (Torr. & Gray) Gray — Alpine Aster

Stems 1 to several, 5-40 cm. tall, often reddish, usually bearing a few reduced leaves; basal leaves linear to narrowly oblanceolate, sessile, 5-25 cm. long with several parallel-appearing veins; heads solitary, 2.5-4 cm. wide; involucral bracts narrow, pubescent; ray flowers numerous, reddish-purple to light purple.

Alpine or subalpine meadows in the Cascades.

Aster chilensis Nees — Pacific Aster

Rhizomatous perennials; stems often decumbent to ascending up to 1 m. in length, pubescent above; leaves oblanceolate to obovate, entire to finely toothed, nearly glabrous to pubescent; heads borne in cymes; outer involucral bracts obtuse; ray flowers numerous, violet; achenes pubescent.

Common only along the coast, in salt marshes, meadows and bluffs.

Aster curtus Cronq. — Rigid white-topped Aster

Rhizomatous perennials; stems 1-5 dm. tall, usually unbranched; lower leaves early deciduous, leaves just above the base the largest, oblanceolate, only gradually reduced upwards, margins and sometimes the midribs harshly-ciliate, otherwise glabrous; inflorescence a nearly flat-topped compact cluster of heads; ray flowers only 1-3, white; disk flowers few, light yellow; pappus white.

Prairies; uncommon in western Oregon, more abundant in western Washington.

Aster foliaceus Lindl.

Stems single or tufted, stout, 3-8 dm. tall, usually glabrous; leaves 7-18 cm. long, conspicuous, long-lanceolate or oblanceolate, or broadly elliptic, entire, lower narrowed into a petiole, upper clasping; heads 3.2-3.8 cm. wide, borne singly or in loose corymbs, sometimes with axillary peduncles below, the larger heads generally solitary, outer bracts leaf-like; ray-flowers dark reddish-purple.

In the mountains.

Aster hallii Gray — Hall's Aster

Stems 3-8 dm. tall, slender, erect or often weakly ascending, very leafy, pubescent at least above; leaves linear to narrowly lanceolate or elliptic, 2-7 cm. long, entire, scabrous-margined; heads small, numerous, about 12.5 mm. high, white to pale lavender, the buds sometimes pinkish-tipped, in raceme-like or panicle-like arrangements on the ends of main stems and branches; achenes usually pubescent.

[=*Aster chilensis* Nees ssp. *hallii* (Gray) Cronq.]

Rather common in ground very wet in spring and dry in summer.

Aster ledophyllus Gray — Cascade Aster

Stems 3-7.5 dm. tall, very leafy to the top; leaves reduced upward, oblong to elliptic, sessile, entire or slightly toothed, glabrous or nearly so above, tomentose beneath; involucre bracts stiff, appressed, generally purplish; heads spreading, 4 cm. in diameter or less; ray flowers lavender to purple.

Meadows and woods in the mountains.

Aster modestus Lindl. Few-flowered Aster

Rhizomatous perennials; stems 4-10 dm. tall, glandular, at least above and sometimes also villous, nearly glabrous below; leaves 5-13 cm. long, 1-4 cm. broad, lanceolate, acuminate, entire or toothed, sessile and auriculate-clasping; heads 2.5-3.2 cm. broad including rays; involucral bracts narrow, glandular; ray flowers purple to violet; pappus brownish.

Stream banks and other moist to wet areas.

Aster radulinus Gray Rough-leaved Aster

Rhizomatous perennials; stems 1-7 dm. tall, rigid; leaves firm, leathery, obovate to oblanceolate, sharply toothed, harshly pubescent, tapering at base but scarcely petioled, 5-10 cm. long; heads somewhat bell-shaped, 2 cm. or more wide (including rays), borne in terminal flat-topped cymes or panicles, involucral bracts firm, pubescent; ray flowers dirty-white to lavender.

Dry wooded areas.

Aster subspicatus Nees Douglas' Aster

Plants variable, glabrous or pubescent; stems erect, simple or somewhat branched, 3-12 dm. tall, slightly ridged, often with vertical lines of minute hairs; leaves lanceolate or linear or the lower narrowly oblong, somewhat narrowed at both ends, usually toothed, the lower 15 cm. or less long, becoming smaller upward; the stem and branches terminating in leafy panicles of violet, purple or blue rayed heads; bracts of the involucre narrow, green-tipped, or the outer entirely green and leaf-like, acute; ray flowers about 12 mm. long.

Very common in a variety of habitats.

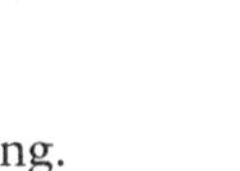

Aster subspicatus
a. basal leaf

Baccharis

Dioecious perennial herbs or shrubs, usually with glandular foliage; leaves alternate; flowers of 2 kinds (pistillate and functionally staminate), borne in small heads, the heads in small tight clusters; corolla of the pistillate flowers filiform, shorter than the style; pappus present, usually hair-like in the pistillate flowers; stiff bristles in the staminate flowers.

Baccharis pilularis DC. Chaparral broom

A much-branched shrub, 3-25 dm. tall, with grooved, resinous stems; leaves numerous, thick, 1-5.5 cm. long, resinous, cuneate, entire or with a few coarse teeth; bracts of the involucre overlapping, resinous; pappus of the pistillate flowers brown-silky; of the staminate, slightly dilated at the tips.

Bluffs and hillsides, mostly near the coast.

Balsamorhiza

Taprooted perennials with mostly basal leaves; stems bearing 1 or several large yellow heads of both ray and disk flowers; involucral bracts in 2-4 series; receptacle chaffy, the bracts surrounding the achenes; pappus none.

Leaves densely covered with a felt-like pubescence........................*B. sagittata*
Leaves green, only sparingly pubescent.......................................*B. deltoidea*

Balsamorhiza deltoidea Nutt. Deltoid balsamroot

Basal leaves and stems clustered, 2-9 dm. tall; basal leaves up to 6 dm. long and 2 dm. wide, broadly more or less arrow-shaped, with rounded lobes, long-petioled, often glandular, rough, the margins entire, wavy, or coarsely toothed below, cauline leaves petioled, narrowly ovate or lanceolate, tapering to the petiole; heads sunflower-like, 6-8 cm. broad, with glandular leaf-like involucre bracts mostly reflexed.

Open ground or partial shade.

Balsamorhiza sagittata (Pursh) Nutt. Arrowleaf balsamroot

Stems 2-6 dm. tall, glandular-pubescent; basal leaves broadly triangular up to 5 dm. long, entire, arrowhead-shaped at the base, densely pubescent, cauline leaves few, narrowly lanceolate; heads large, yellow, 5-10 cm. broad.

Open areas.

Balsamorhiza deltoidea

Bellis

Annual or perennial herbs with mostly, or all, basal simple leaves; heads solitary; ray flowers white, pink or rose-colored; disk flowers yellow; involucral bracts in 2 equal series; pappus absent.

Bellis perennis L. English daisy

Pubescent perennials 5-25 cm. tall; leaf blades 1-5 cm. long, obovate to orbicular, minutely toothed, petioles nearly equal to the blades in length; heads solitary; rays white to rose-colored; disk flowers yellow.

Introduced from Europe; widely distributed; abundant west of the Cascades, especially in lawns.

Bellis perennis

Bidens

Annuals, perennials or shrubs; leaves opposite, simple to compound; heads many-flowered; rays 3-10 or absent; involucral bracts in 2 dissimilar series, the outer series leafy, the inner membranous; receptacle chaffy; achenes flattened or 4-sided; pappus, if present, of 1-6 awns, usually barbed.

1a Perennials with dimorphic leaves, the submerged leaves filiformly dissected. .... *B. beckii*
1b Annuals; leaves similar, none of them filiformly dissected
  2a Leaves simple; achenes generally 4-awned .... *B. cernua*
  2b Leaves pinnately compound; achenes generally 2-awned
    3a Heads orange; outer involucral bracts 5-8 .... *B. frondosa*
    3b Heads pale yellow; outer involucral bracts 10-16 .... *B. vulgata*

Bidens beckii Torr. ex Spreng. Water marigold

Partially submerged perennials; submerged leaves filiformly dissected and appearing whorled, 2-4 cm. long; aerial leaves simple, lanceolate to ovate, serrate, sessile; head solitary; ray flowers 1-2 cm. long, yellow; outer involucral bracts green; pappus of 3-6 awns retrorsely barbed at the summit.

[= *Megalodonta beckii* (Torr. ex Spreng.) Greene]

Principally in ponds, slow moving streams and irrigation ditches east of the Cascades, but occasionally reported within our limits.

Bidens beckii

Bidens cernua L. Nodding beggar-ticks

Annuals; stems erect, 1-10 dm. tall, simple or somewhat branched, slender or stout, usually roughened by short stiff hairs; leaves simple, glabrous or slightly pubescent, 4-20 cm. long, lanceolate, acuminate, toothed to nearly entire, sessile and the bases fused around the stem; heads generally nodding in fruit; outer involucral bracts 5-8, leafy and an inner involucre of short yellowish membranous bracts; ray-flowers few, short, light yellow, 8-15 mm. long, or rarely absent; achenes wedge-shaped, 4-angled, with 4 barbed awns.

Moist areas, along ponds, lakes and stream banks.

Bidens frondosa L. Sticktight or Leafy beggar-ticks

Annuals; stems 2-12 dm. tall, 4-angled, glabrous or with a few short hairs on the stem and the under surface of the leaves; leaves petiolate, pinnately compound with 3-5 leaflets, leaflets 2-10 cm. long, lanceolate, acuminate, serrate; ray flowers, if present, few, orange; outer

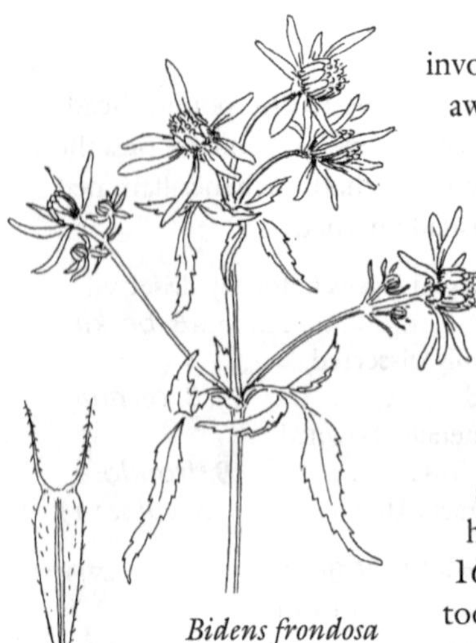
Bidens frondosa

involucral bracts 5-8, leafy; pappus of 2 barbed awns.

Wet habitats, often in disturbed sites, or occasionally in dry areas.

Bidens vulgata Greene Tall beggar-ticks
Annuals, erect, 8-15 dm. tall, glabrous or slightly pubescent; stems often purplish; leaves somewhat bluish-green above, paler beneath, pinnately compound with 3-5 leaflets, leaflets 2-8 cm. long, lanceolate, acuminate, toothed; heads slender, 6-12 mm. high; outer involucral bracts generally 10-16, spreading, slender, oblanceolate, minutely toothed, usually much longer than the erect, scarious-margined inner ones; ray flowers none or inconspicuous; achenes flat, generally sparsely covered with upward pointing stiff hairs, the 2 awns with downward pointing barbs.

Low moist land, particularly along the coast; native of eastern North America.

Boltonia

Erect perennial glabrous herbs, branching or simple below, the stems more or less ridged, at least above; leaves alternate, entire, sessile; heads numerous, aster-like, somewhat showy, with both ray and disk flowers; involucre bell-shaped or hemispheric, the bracts in several series, overlapping or nearly equal, scarious margined; ray-flowers pistillate; disk flowers perfect; style- branches flattened; achenes flattened, margins winged or thickened; pappus of short bristles and 2-4 awns.

Boltonia asteroides (L.) L'Her.
Stems stout, simple below the inflorescence, 0.5-2 m. tall, glabrous, hollow, and swollen below with conspicuous leaf-scars nearly encircling the stem, inconspicuously ridged above; lower leaves deciduous before flowering time, upper leaves glabrous, lanceolate to oblanceolate, more or less clasping at the base, 2.5-14 cm. long; heads 2.5 cm. or more in diameter, aster-like in large loose corymbose cymes; ray flowers pink, pale lavender or white; disk flowers yellow; awns of the pappus well developed.

On tide flats of Clatsop County, Oregon; nearly submerged when the tide is in.

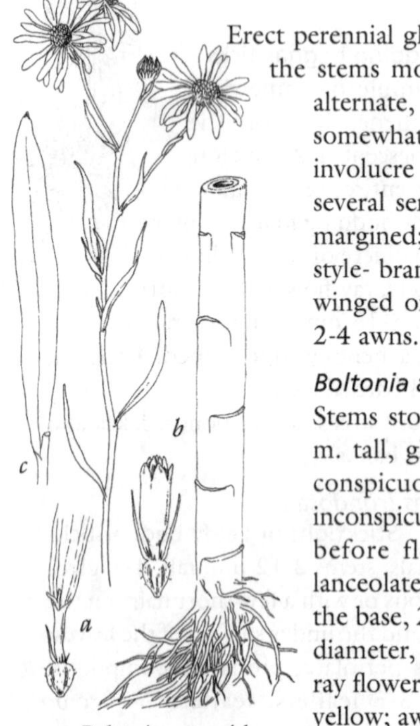

Boltonia asteroides
a. ray-flower; b. disk-flower, c. larger leaf

Brickellia

Herbs or shrubs with alternate or opposite leaves, and white, yellowish, or pink heads of disk flowers in cymes or panicles, rarely solitary; involucre bell-shaped, the bracts conspicuously veined, arranged in several series, overlapping, the inner longer; pappus a circle of bristles.

Brickellia grandiflora (Hook.) Nutt. Large-flowered tasselflower

Stems erect, branched, 3-8 dm. tall, glabrous or somewhat pubescent; leaves petioled, the blades broadly or narrowly triangular, coarsely toothed, acute at apex, 3-12 cm. long, resinous-dotted; heads 12.5-18 mm. long, drooping, in paniculate cymes; involucral bracts closed about flowers at first, spreading or reflexed in fruit, conspicuously veined; flowers whitish; achenes slender, ribbed, slightly pubescent.

Along the Columbia River Highway; uncommon in our limits.

Cacaliopsis

A monotypic genus.

Cacaliopsis nardosmia (Gray) Gray Silvercrown

Rhizomatous perennials; stems erect, 3-12 dm. tall; lower leaves long-petiolate, blades deeply palmately cleft then coarsely toothed or again lobed, leaves greatly reduced upwards, upper surface of the leaves green, lower surface tomentose or becoming glabrate in age; heads several; involucre bell-shaped, bracts in 1-2 series; ray flowers absent; disk flowers yellow; pappus of numerous fine white bristles.

[=*Luina nardosmia* (Gray) Cronq.]

Meadows and open woods.

Centaurea

Annual or perennials herbs with alternate leaves; heads generally showy; involucral bracts in several series, overlapping, generally fringed, toothed, or spiny; only tubular flowers present, but the outer circle of these sometimes larger; pappus of stiff unequal bristles or scales.

1a Heads yellow; involucre spiny .... *C. melitensis*
1b Heads not yellow
  2a Leaves, except sometimes the uppermost, deeply pinnately parted into narrow segments .... *C. maculosa*
  2b Leaves entire to lobed, but not deeply pinnately parted
    3a Annuals; heads purple, blue, pink or white .... *C. cyanus*
    3b Perennials; heads rose-purple
      4a Involucral bracts irregularly cleft .... *C. jacea*
      4b Involucral bracts regularly fringed
        5a Fringes of the bracts black; heads usually without enlarged sterile marginal flowers .... *C. nigra*
        5b Fringes brown; heads with enlarged sterile marginal flowers .... *C. xpratensis*

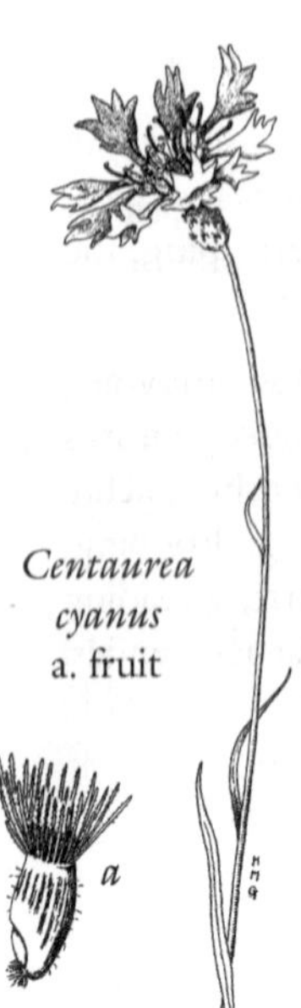
Centaurea cyanus
a. fruit

Centaurea cyanus L. Bachelor's button or Cornflower

Annuals; stems erect, 2-12 dm. tall, somewhat loosely woolly; lowermost leaves often toothed or with a few narrow lobes, all other leaves linear; heads solitary and terminal on main stem and branches; involucral bracts fringed, usually margined with red or black; flowers blue, purple, pink, or white, those on the margin much enlarged, ray-like, and showy.

Escaped from cultivation; common along road sides and other disturbed sites; weedy in grain fields.

Centaurea jacea L. Brown knapweed

Perennials, branched from the base, stems reaching 1 m. in height when shaded and crowded by other vegetation, leafy to the top; herbage somewhat hispid; lower leaf blades 7.5-10 cm. long, 2.5 cm. wide, long-elliptic, tapering at both ends, entire or slightly dentate or rarely lobed, petioles 10-13 cm. long, upper leaves sessile, mostly lanceolate, mostly entire but sometimes lobed at the base; heads purple or rosy-lavender 3-4 cm. in diameter, broadly ovoid before opening, solitary on the tips of branches; involucral bracts with broad, brown, irregularly fringed appendages (the upper silvery and merely toothed); flowers not crowded in the head, the marginal flowers larger, ray-like.

Introduced from Europe; occasional in disturbed habitats.

Centaurea jacea
a. involucral bract
b. upper leaf
c. basal leaf

Centaurea maculosa Lam. Spotted knapweed

Biennials or short-lived perennials; herbage gray-tomentose when young; stems 3-15 dm. tall; leaves 0.5-1.5 dm. long, deeply pinnatifid, or the upper nearly entire, resinous-dotted; involucral bracts with dark fringed tips; flowers pink to purple, or rarely white, the marginal ones only slightly enlarged.

Introduced from Europe; in disturbed habitats and fields, but not common in our area.

Centaurea melitensis L. Maltese starthistle or Tocalote

Annuals 1-9 dm. tall; stems branched; whole plant hispid or slightly woolly; basal leaves pinnately lobed, terminal lobe large, rounded, lateral lobes oblong, the basal leaves often dried and gone by flowering time, cauline leaves linear or lanceolate, entire or sparingly toothed, more or less decurrent on the stem; heads terminal and lateral; involucral bracts ending in slender yellow usually branched spines; flowers yellow, all alike.

Introduced from Europe; weedy in waste places and fields.

Centaurea nigra L. Black knapweed

Somewhat resembling *C. jacea*, but the flowers generally all alike, involucral bracts more slender, nearly black, fringed with slender bristly segments.

Introduced from Europe; occasional in disturbed habitats.

Centaurea xpratensis Thuill. Meadow knapweed

Resembling *C. jacea* and *C. nigra* in general appearance; involucral bracts brown-fringed, the teeth of the fringe about equaling in length the body of the bract in width; marginal sterile flowers showy, purple.

A hybrid between *C. jacea* and *C. nigra*.

Waste places west of the Cascades.

Cichorium

Erect stiffly branching annual or perennial herbs with milky juice; leaves mostly basal, those of the stem alternate, reduced and bract-like above; flowers blue, purple, pink, or white, in showy heads, the corollas all strap-shaped and 5-toothed; involucre of 2 series of bracts, the outer spreading, the inner erect; pappus of short blunt scales.

Cichorium intybus L. Common chicory

Perennials from a large taproot; stems 3-15 dm. tall, stiff, short-branched, somewhat ridged, glabrous to pubescent; basal leaves in a spreading rosette, 7.5-15 cm. or more long, oblanceolate, pinnately lobed to nearly entire, the petioles and midribs often reddish, upper leaves much smaller, mostly entire, clasping, auriculate at the base; heads numerous, showy, 2.5-4 cm. broad, usually deep blue, rarely white.

Introduced from Europe; common along roadsides and in fields.

Cirsium

Annual, biennial or perennial herbs with alternate, toothed or pinnately lobed or cleft spine-toothed leaves; heads discoid, usually showy, red, yellow or white, terminal; involucral bracts overlapping, in many ranks, at least the outer typically spinose-tipped; receptacle with hairs or bristles interspersed with the flowers; flowers all tubular and perfect except in *C. arvense* which is a dioecious species; pappus of a series of plumose or barbed bristles which are united at the base, all falling together; achenes glabrous.

1a Dioecious perennials; involucre less than 2 cm. long .... *C. arvense*
1b Flowers bisexual; biennials or short-lived perennials; involucre often longer than 2 cm.
  2a Leaves hispid; stems conspicuously spiny-winged from the decurrent leaf bases; heads large .... *C. vulgare*
  2b Leaves with soft pubescence or glabrous; stems not conspicuously spiny-winged, although the petioles sometimes spiny-winged
    3a Involucral bracts broad, conspicuously fringed .... *C. callilepis*
    3b Involucral bracts narrow, not conspicuously fringed
      4a Flowers cream-colored .... *C. remotifolium*
      4b Flowers purple or magenta.

5a Style included in the corolla or exserted less than 1 mm.; corolla tube 10-20 mm. long .... *C. brevistylum*

5b Style exserted 2 mm. or more from the corolla; corolla tube 5-11 mm. long .... *C. edule*

Cirsium arvense (L.) Scop. Canada thistle

Perennials; stems 0.5-2 m. tall, slender, from a deep creeping root-system sending up adventitious branches; leaves lanceolate or oblanceolate, sometimes somewhat woolly beneath, generally lobed, the lobes edged with very sharp yellowish prickles of various lengths, or leaves sometimes nearly entire, lower leaves usually narrowed slightly at base; staminate and pistillate flowers on different plants; heads less than 2 cm. long, pistillate heads somewhat bell-shaped, staminate globose at first, flowers purple.

A serious weed of disturbed sites; introduced from Europe.

Cirsium arvense

Cirsium brevistylum Cronq. Short-styled thistle

Taprooted biennials or short-lived perennials; stems to 0.5-2.5 m. tall, woolly, especially below the inflorescence; leaves coarsely toothed to pinnatifid, with weak spines, green above, gray-pubescent below at least when young, lower leaves tapering to a spiny-winged petiole, the upper decurrent or clasping the stem; involucral bracts in several series, overlapping, the outer spine-tipped; flowers purple.

Meadows and other moist to wet areas from the west slope of the Cascades westward.

Cirsium callilepis (Greene) Jeps. Mountain thistle

Taprooted short-lived perennials; stems less than 1.5 m. tall, spiny, pubescent; heads small, involucre 1.5-2.5 cm. tall, the bracts overlapping, scarious and fringed at the tip, each of the outer ones tipped with a short spine; flowers cream-colored to pink or purple.

This species is sometimes included in *C. remotifolium*, with which it hybridizes.

Woods and open areas in the Cascades and westward.

Cirsium edule Nutt. Edible thistle

Stems usually simple, pubescent, 0.4-3 m. tall, leafy to the top; leaves more or less woolly below, becoming glabrate above, deeply pinnately lobed, and densely spiny on the margins, the basal leaves 15-25 cm. long, the upper smaller and auriculate-clasping; heads few, commonly more or less clustered; lower involucral bracts pinnately-lobed and spined, the upper simple without glands, ciliate on the margin, each tipped with a spine, sometimes purplish; corolla purple or magenta, the lobes shorter than the throat, thickened at the tips.

Moist, somewhat shaded places, in the mountains and on the coast.

Cirsium remotifolium (Hook.) DC. Few-leaved thistle

Taprooted short-lived perennials; stems erect, 3-18 dm. tall, sparingly branched, woolly or becoming nearly glabrous; leaves woolly beneath, often becoming glabrous above, deeply lobed or divided, spiny, the lower borne on spiny petioles, the upper becoming sessile; involucral bracts lanceolate, the outer spine-tipped and papery-margined, sometimes also slightly fringed; flowers cream-colored.

Meadows and other open places to moist shaded woods.

Cirsium vulgare (Savi) Ten. Bull or Common thistle

Biennials; stems leafy, stout, 0.3-2 m. tall, generally branched, somewhat woolly; leaves elongated, irregularly lobed, the lobes narrow, terminal lobe much longer than the laterals, all tipped by stout yellowish spine-like prickles, leaf surfaces with appressed stiff hairs, the lower also somewhat woolly, leaf bases decurrent, forming spiny wings on the stem; heads large; involucral bracts many, narrow, sharply spine-tipped; flowers rose-colored.

A common weed; introduced from Europe.

Conyza

Annuals or perennials, resembling *Erigeron*; stems leafy; leaves alternate; heads small, most flowers tubular; the rays minute (ours) or absent in some species; involucral bracts in 2-3 overlapping series; pappus of slender bristles.

Midrib of the involucral bracts brown, filled with resin; common .... *C. canadensis*

Midrib of the involucral bracts green or purple, not filled with resin; uncommon .... *C. bonariensis*

Conyza bonariensis (L.) Cronq.

Similar to *Conyza canadensis,* it differs mainly in having larger heads (about 1 cm. wide), and midribs of the involucral bracts green or purple and not resinous.

Only occasionally found in our area where it has been introduced from South America.

Conyza canadensis (L.) Cronq. Canadian fleabane or Horseweed

Annuals; stems erect, simple below, the larger stems generally much branched above, 2-25 dm. tall, usually hispid or glabrous above; lower leaves oblanceolate, entire or dentate to somewhat lobed, upper leaves linear, ciliate; midrib of the involucral bracts brown and filled with resin; heads small, less than 6 mm. wide, numerous in a generally elongated panicle; rays inconspicuous, white, shorter than the pappus.

(=*Erigeron canadensis* L.)

A common native weed.

Coreopsis

Annuals, perennials or shrubs; leaves opposite or rarely alternate, simple or compound; heads with both ray and disk flowers; involucral bracts in two dissimilar series; pappus, if present, of 2 short awns or teeth.

Coreopsis tinctoria Nutt. var. *atkinsoniana* (Dougl.) H. M. Parker
Glabrous annuals or biennials; stems 3-13 dm. tall; leaves opposite, pinnately dissected into linear segments or the upper leaves entire; ligulate flowers 1-2 cm. long, showy, orange-yellow, generally reddish-brown at the base; fruit narrowly winged, pappus, if present, of 2 small teeth.

(=*Coreopsis atkinsoniana* Dougl.)

Along the Columbia River and otherwise locally distributed.

Cotula

Annuals or perennials; leaves alternate, entire to pinnately dissected; heads solitary, rays none, but outer circle of flowers pistillate, without a corolla; disk flowers with a 4-toothed corolla; involucral bracts in 2-several series, usually scarious-margined; pappus usually absent.

Cotula coronopifolia L. — Brass buttons
Stems weakly ascending or prostrate below, succulent, often reddish, 5 dm. or less long; leaves lanceolate or oblong, entire or more often once or twice pinnately cut into linear lobes, especially near the apex, the base forming a short sheath around the stem; heads somewhat flattened, golden yellow, about 1 cm. wide; involucral bracts in 3 slightly unequal series; pistillate flowers forming the first circle within the involucre, the achenes winged; pappus absent.

Marshes and tidal flats along the coast; native to South Africa.

Crepis

Taprooted annual, biennial, or perennial herbs with milky juice; leaves entire or toothed, mostly basal, but a few on the stem; heads in a branching inflorescence; flowers all strap-shaped, yellow, at least at first; involucre a single or double circle of bracts; receptacle naked; achenes oblong or spindle-shaped, sometimes beaked, 10-20-ribbed; pappus of soft, hair-like bristles.

1a Low glabrous and glaucous alpine plants .... *C. nana*
1b Not glabrous and glaucous nor alpine
  2a Beak of the achenes about as long as the body; plants stiff-bristly; involucre bristly, not glandular .... *C. setosa*
  2b Achenes beakless or nearly so; plants soft pubescent at least at the base; inner involucral bracts glandular-pubescent .... *C. capillaris*

Crepis capillaris (L.) Wallr. — Smooth hawksbeard
Annuals or biennials; stems 0.5-9 dm. tall, often purplish and soft-pubescent at least below; leaves elongated, deeply and irregularly lobed, or merely toothed, basal leaves petioled, cauline leaves clasping, often pubescent with yellow hairs at least on the midrib of the under surface; inner involucral bracts glandular-pubescent; achenes beakless or nearly so.

Common weed; introduced from Europe.

Crepis nana Richards. Dwarf hawksbeard

Dwarf perennials; stems 2-15 cm. tall, tufted at base, whole plant glabrous and glaucous, sometimes purplish below; leaves mostly basal, the few cauline leaves not reduced, obovate, narrowed to long winged petioles, entire or toothed or pinnately lobed; involucre narrowly cylindrical, each bract dark-striped in the center.

Talus slopes or gravelly areas; alpine or subalpine.

Crepis setosa Haller f. Bristly hawksbeard

Annuals; stems 1.5-8 dm. tall, erect, angled, more or less stiffly branched, sometimes purplish at the base, sparsely covered with yellowish spreading bristles; leaves dull to dark green, more or less pubescent, the basal leaves with a large terminal toothed lobe, the teeth pointing downward, the lower half of leaf narrowed, pinnately lobed or divided, cauline leaves sessile, long-acuminate, broadest at base with several sharp upward-pointing lobes; inflorescence loosely branched into ultimate cymose arrangements of frequently 3 heads; lower circle of involucral bracts spreading, upper erect, all with yellow bristles; heads yellow; achenes 10-ribbed, stiff-bristly, long-beaked; pappus white.

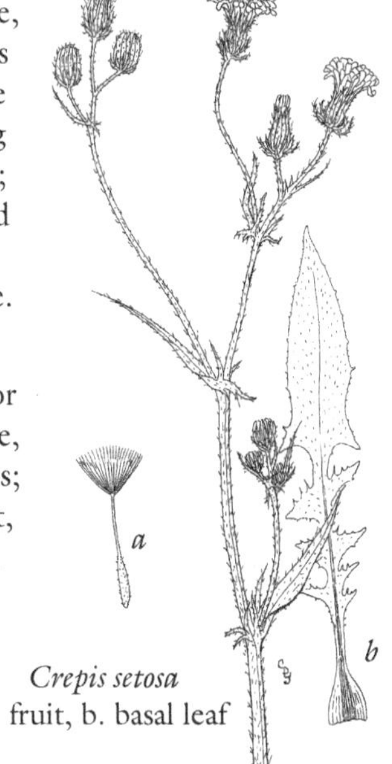

Crepis setosa
a. fruit, b. basal leaf

Introduced from Europe; a common weed in the Willamette Valley.

Crocidium

A monotypic genus.

Crocidium multicaule Hook. Spring gold

Annuals; stems 5-25 cm. tall, arising singly or in a cluster and bearing solitary bright golden-yellow heads; leaves basal and alternate, oblanceolate to linear; involucral bracts thin, green, in a single series; ray flowers present, usually lacking a pappus; pappus of disk flowers of white barbed bristles in a single series.

Moist open places, flowering very early in the spring. Abundant locally, and widely distributed, but with many gaps between stations.

Erechtites

Rank-scented annuals or perennials with alternate leaves; heads usually small, numerous consisting of disk-flowers only, these of two kinds; outer circle of flowers pistillate with very slender tubes scarcely dilated at apex, inner flowers perfect or staminate, slender-tubed, dilated above, 4-5-lobed; pappus of slender bristles.

Leaves merely sharply and irregularly toothed .... *E. minima*
Leaves deeply pinnately lobed to pinnatifid................................ *E. glomerata*

Erechtites glomerata (Desf. ex Poir.) DC. New Zealand burnweed

Erect annuals 0.5-2 m. tall, pubescent or becoming glabrous; leaves deeply pinnately lobed to pinnatifid, the margins often also toothed, lower leaves petioled, the upper becoming sessile and usually auriculate; heads small and numerous; flowers yellow.

(=*Erechtites arguta* DC.)

Introduced from Australia; weedy in disturbed sites along the coast; mostly south of our range.

Erechtites minima (Poir.) DC. Australian burnweed

Stems coarse, erect, somewhat simple or generally much-branched, 0.5-2 m. tall; leaves 7-20 cm. long, narrowly lanceolate, acute at apex and narrowed downward, but widening at the auriculate base, margins sharply and irregularly toothed, green above, somewhat white-woolly beneath, at least at first; panicle much-branched, broad; heads slender, about 6-8 mm. long; achenes about 2.5 mm. long, purplish-brown, nearly cylindrical, with vertical rows of short white hairs.

Erechtites minima

[=*Erechtites prenanthoides* (A. Rich.) DC.]

Native of Australia; abundant along the coast, less common inland.

Erigeron

Aster-like herbs but many species flowering in the spring; leaves alternate, usually entire; involucral bracts in only 1 or 2 series, nearly equal in length, scarcely overlapping; both ray and disk flowers commonly present, the ray-flowers generally numerous; achenes usually flattened and 2-ridged; pappus usually present, of slender bristles or narrow scales.

1a Ray flowers without a pappus, pappus of the disk flowers of both bristles and scales; ray white; plants annual or biennial .... *E. strigosus*
1b Ray flowers with a pappus
  2a Leaves once or twice 3-cleft into narrow segments .... *E. compositus*
  2b Leaves entire, toothed or only shallowly lobed
    3a Ray flowers yellow .... *E. aureus*
    3b Ray flowers never yellow
      4a Plants thick and fleshy; found on coastal bluffs and dunes; heads usually solitary; ray flowers pinkish-lavender to purple ...... *E. glaucus*
      4b Leaves not thick and fleshy; if coastal, heads not solitary
        5a Rays 4 mm. or less in length; numerous, but inconspicuous .... *E. acris*
        5b Rays 5-25 mm. long, conspicuous
          6a Rays averaging over 150, often as many as 400, pink or white .... *E. philadelphicus*
          6b Rays less than 150

7a Basal leaves coarsely toothed or lobed, upper cauline leaves entire; plants of the Columbia River Gorge .... *E. oreganus*
7b Leaves entire or nearly so or plants montane
  8a Rays less than 2 mm. wide
    9a Cauline leaves lanceolate to elliptic; heads 30 mm. broad or more .... *E. speciosus*
    9b Cauline leaves linear
      10a Pubescence of the stem spreading....................... *E. corymbosus*
      10b Pubescence of the stem appressed .... *E. decumbens*
  8b Rays 2-4 mm. wide
    11a Rays white; plants of the Columbia River Gorge............ *E. howellii*
    11b Rays pink, violet or purple; plants montane
      12a Involucre densely pubescent below, glandular above... *E. aliceae*
      12b Involucre glandular-pubescent throughout .... *E. peregrinus*

Erigeron acris L. Northern daisy
Biennials or perennials; stems 0.5-8 dm. tall, generally several from a tufted base; lower leaves spatulate, slender petioled, upper oblong to lanceolate, sessile, more or less pubescent; heads 1.3-2 cm. wide in loose corymbs or some long-peduncled in the leaf axils below, rarely solitary; the narrow involucral bracts pubescent, at least at first; ray flowers inconspicuous, very narrow, pinkish or purplish, equalling or somewhat exceeding the involucre; pappus slightly tawny.

Rocky areas in the mountains.

Erigeron aliceae How. Eastwood's daisy
Perennials, often from a short rhizome; stems few-branched, nearly glabrous to pubescent, 3-10 dm. tall; lower leaves somewhat toothed, narrowly oblanceolate, petioled, upper leaves lanceolate, sessile; heads one or few, 3-3.5 cm. or more broad, rays white, pink or lavender.

Meadows and other open areas in the Cascade and Olympic Mountains.

Erigeron aureus Greene Alpine yellow daisy
Perennials; stems 2-15 cm. tall, several from a much-branched woody base; herbage grayish pubescent; leaves mostly basal, long-spatulate, upper leaves reduced, sessile or nearly so, lanceolate to linear, all entire; heads yellow, borne singly on the stems; involucral bracts gray woolly.

Alpine areas in the Cascades of Washington.

Erigeron compositus Pursh Cutleaf daisy
Perennials; stems 2.5-25 cm. tall, from a generally much-branched woody base, glandular-pubescent to nearly glabrous; leaves mostly basal, once or twice 3-cleft into narrow segments, upper leaves linear, mostly entire; heads about 2 cm. broad, borne singly on the stems, the narrow involucral bracts gray-pubescent; rays white to reddish-purple.

Rocky places in the mountains.

Erigeron corymbosus Nutt. Foothill daisy

Perennials; stems many from base, slender, more or less erect, 1.5-5 dm. tall, grayish-pubescent, more or less branched above; leaves many, linear to narrowly lanceolate or the lower slightly oblanceolate, principally 3-veined from near the base; heads about 1.5 cm. wide, in small corymbs, the rays lavender or pinkish.

Dry, open places; not common in our area.

Erigeron decumbens Nutt. Pacific fleabane

Decumbent perennials; stems 1.5-7 dm. tall, often purplish at the base, pubescent; basal leaves up to 20 cm. long, including the long petiole, 3-veined, narrowly oblanceolate, the cauline leaves linear, only gradually reduced upwards; heads 1-many; involucre 3-6 mm. long, rays 20-50, filiform, 6-12 mm. long, blue.

Only occasional in prairies in the Willamette Valley.

Erigeron glaucus Ker-Gawl. Seaside daisy

Rhizomatous perennials; stems from 5 cm. to rarely 30 cm. tall from a tufted base, plants usually pubescent, somewhat fleshy; leaves broadly spatulate or obovate, lower short-petioled, upper sessile, entire or nearly so; heads about 3.5 cm. wide, generally borne singly on the stem; ray flowers many, pinkish-lavender to purple.

Bluffs and dunes on the seacoast, common.

Erigeron howellii (Gray) Gray Howell's daisy

Rhizomatous perennials; stems 1.5-5 dm. tall, 1 or several from base, plants glabrous except under the heads; leaves thin, spatulate to ovate, basal long-petioled, uppermost sessile and clasping at the base, mostly entire; heads borne singly on the stem; involucre glandular; ray flowers showy, white.

Columbia River Gorge.

Erigeron oreganus Gray Gorge daisy

Stems 1-3 dm. tall, several to many from a tufted base, plants somewhat pubescent; leaves spatulate, the basal coarsely toothed or lobed, 2.5-9 cm. long, the upper mostly entire, shorter; heads about 2 cm. wide, in loose corymbs or borne singly; ray flowers 30-60, pink to white or rarely blue.

Rocky areas in the Columbia River Gorge.

Erigeron peregrinus (Pursh) Greene

Perennials often from a short rhizome; stems 0.5-7 dm. tall; lower leaves long-spatulate, narrowed to a petiole, upper leaves lanceolate, oval, or ovate, sessile, all leaves glabrous or slightly pubescent, generally with marginal cilia; heads 3-4.5 cm. wide, usually borne singly on the stems, involucral bracts sticky-glandular and pubescent; ray flowers white, pink or purple.

Meadows and stream banks in the mountains; abundant.

Erigeron philadelphicus L. Philadelphia daisy

Biennials or short-lived perennials; stems erect, 2-9 dm. tall, more or less pubescent; lower leaves petioled, obovate, toothed or entire, 7.5-15 cm. long, upper leaves sessile, clasping the stem, sharply toothed; heads on rather long peduncles in corymbose cymes, showy; rays many, to 1 cm. long, pink to white, numerous, linear.

In a wide range of habitats, but most often in moist places.

Erigeron speciosus (Lindl.) DC. Showy daisy

Perennials; stems tufted, 2-8 dm. tall, leafy, sometimes forming large clumps, plants nearly glabrous below the inflorescence; leaves oblanceolate or long-elliptic, entire, lower petioled, upper sessile; heads 3-5 cm. wide, few in a corymb; ray flowers narrow, numerous, blue-violet.

Dry soil, usually in open woods; not common. One of our most showy species.

Erigeron strigosus Muhl. ex Willd. Daisy fleabane

Annuals or biennials, erect, branched above, reaching 3-7 dm. tall, somewhat pubescent; lower leaves petioled, narrowly to broadly lanceolate or elliptic, upper leaves becoming gradually smaller and sessile; heads in loose corymbose cymes; involucre broad, the bracts very slender with a few long hairs and obscurely glandular; ray flowers numerous, filiform, white; pappus of ray flowers generally absent; pappus of disk-flowers double, consisting of a few long bristles and a crown of short scales.

Weedy in disturbed sites, but also in moist prairies and on hillsides.

Eriophyllum

Woolly herbs or sometimes woody at the base; leaves mostly alternate; both ray and disk flowers generally present, usually yellow; involucral bracts free or fused in a single series; pappus of chaffy scales, or none.

Eriophyllum lanatum (Pursh) Forbes
Common woolly sunflower

Stems more or less decumbent at the base, 1-8 dm. tall, woolly at least at first; leaves 1-8 cm. long, variable but mostly pinnately lobed, except the lowermost and uppermost, which are narrowly lanceolate or oblanceolate, densely white-woolly beneath, less so above; heads on long terminal or axillary peduncles, orange-yellow, 2-4.5 cm. in diameter, the involucre woolly.

This is a variable species with numerous named varieties. Common on hillsides and in somewhat dry fields.

Eriophyllum lanatum

Euthamia

Rhizomatous perennials; leaves alternate, sessile, entire, 3-5-veined from the base, gland-dotted; involucral bracts in several series, overlapping; ray and disk flowers yellow; pappus of bristles in a single series.

Euthamia occidentalis Nutt. Western goldenrod

Stems leafy, 1 m. or more tall, simple below; leaves 6-10 cm. long, linear or narrowly lanceolate, entire, gland-dotted, 3-veined; heads 4-5 mm. long; rays 1.5-2.5 mm. long, narrow, many.

[=*Solidago occidentalis* (Nutt.) Torr. & Gray]

Moist places, especially along streams and in ditches.

Gaillardia

Annual or perennial herbs with alternate or basal leaves and large peduncled showy heads; receptacle bristly; involucral bracts in 2-3 series, spreading or turned downward; ray flowers yellow, red, purple, or variegated (sometimes absent), sterile or pistillate flowers usually deeply 3-lobed; disk flowers fertile, the corollas 5-toothed, the teeth pubescent; achenes top-shaped, with long hairs at least at the base; pappus of chaffy scales.

Gaillardia aristata Pursh Blanket flower

Erect perennials, usually branching from base, 2-7 dm. tall, pubescent; lower leaves 5-15 cm. long, oblanceolate, long-petioled, upper leaves reduced, sessile, lanceolate, all entire or somewhat lobed, hispid; heads 1-2, showy, about 6 cm. in diameter, long-peduncled; involucral bracts slender, acuminate, spreading, pubescent, the outer about 1 cm. long, the inner shorter; ray flowers deep yellow, often red or purplish at base, 3-lobed; disk flowers purplish-brown, pubescent; achenes densely pubescent; pappus scales awn-tipped.

Prairies. Not common in our limits.

Gnaphalium

Woolly annual to perennial herbs with alternate leaves, and with small discoid heads; outer flowers of the head pistillate, inner perfect, all fertile; flowers white or yellow; involucral bracts scarious; pappus bristles slender, in a single series.

1a Heads in a terminal bracted spike; pappus bristles falling in a ring .... *G. purpureum*

1b Heads not in terminal bracted spikes; pappus bristles falling separately

  2a Heads small, borne in clusters nearly hidden by the leafy bracts; involucres 4 mm. or less high; plants usually less than 2 dm. tall

    3a Pubescence appressed; leaves linear to linear-oblanceolate .... *G. uliginosum*

    3b Pubescence loosely woolly; leaves oblong to spatulate ..... *G. palustre*

  2b Heads larger, not hidden by the leafy bracts; involucres mostly over 4 mm. long; plants usually over 2 dm. tall

    4a Leaves glandular-pubescent, usually on both surfaces, or the upper surface becoming nearly glabrous .... *G. californicum*

    4b Leaves gray, tomentose on both surfaces, but not glandular

5a Upper leaves decurrent; short-lived perennials .... *G. canescens*
5b Upper leaves auriculate; annuals or biennials .... *G. stramineum*

Gnaphalium californicum DC. California cudweed

Annuals or biennials; stems 2-9 dm. tall, branched above, glandular, sometimes sparingly woolly; leaves linear to oblanceolate, sessile and decurrent on the stem, green, at least above, glandular on both surfaces; inflorescence branched, heads many; involucral bracts white to yellowish or pink-tinged; pappus bristles free.

Dry hillsides; mainly coastal, but occasionally inland.

Gnaphalium canescens DC. subsp. *thermale* (E. Nels.) Stebb. & Keil
Slender or White cudweed

Short-lived perennials 2-10 dm. tall; stems usually freely branched above; basal leaves tufted, oblanceolate, 3-10 cm. long, the cauline leaves mostly linear and shortly decurrent on the stem; heads many in small clusters forming a broad panicle-like inflorescence; involucre 4-7 mm. high, the bracts white to tawny, sometimes woolly at the base; pappus bristles falling separately.

[=*Gnaphalium microcephalum* Nutt. var. *thermale* (E. Nels.) Cronq.]

In rocky or gravelly soil, usually in dry open areas, often along road sides.

Gnaphalium palustre Nutt. Lowland cudweed

Annuals, woolly throughout; stems much-branched, 2-20 cm. tall, rarely taller; leaves oblong to spatulate, 0.5-3 cm. long, clasping to short decurrent; heads in leafy clusters at the ends of the branches, very woolly; bracts of the involucre greenish below, shining scarious at margins and tips; achenes somewhat spindle-shaped, greenish, minute; pappus bristles falling separately.

Common in bottoms of dried pools, and in moist places.

Gnaphalium purpureum L. Purple cudweed

Annuals or biennials; stems erect or the base horizontal, slender, usually simple, 5-60 cm. tall, silvery pubescent; basal leaves spatulate with long petiole-like bases, 1.5-12 cm. long, 2-16 mm. wide, densely white-tomentose on under surface, becoming green above, cauline leaves narrower; heads in a dense or sometimes more open bracted spike; involucral bracts brownish or purplish, shining; achenes slightly scabrous; pappus bristles united into a ring and falling together.

Common in open, often disturbed, places.

Gnaphalium stramineum Kunth Cotton batting plant

Annuals or biennials; stems stout, 1.5-7 dm. tall, erect, simple or somewhat branched, woolly; leaves linear above, lanceolate to oblanceolate below, woolly; heads somewhat broad, 6 mm. or less long, generally in a few dense cymes; involucral bracts shining white to yellow; pappus bristles falling separately.

(=*Gnaphalium chilense* Spreng.)

Dry places.

Gnaphalium uliginosum L. Marsh cudweed

Somewhat resembling G. *palustre* in size and habit, but appressed-woolly; leaves linear to linear-oblanceolate; involucral bracts brown to nearly white.

Moist places and dried pools.

Grindelia

Annuals, biennials, perennials or subshrubs, glabrous to pubescent or resinous; leaves alternate, glandular-punctate; flowers yellow; involucre bell-shaped or hemispheric, the bracts in many series, firm, often with slender spreading green tips, resinous; both ray and disk flowers usually present, the rays yellow, fertile, narrow, generally many; pappus of 1-6 firm deciduous awns.

Species of *Grindelia* tend to readily hybridize.

1a Tips of the involucral bracts strongly recurved .... *G. nana*
1b Tips of bracts spreading or scarcely recurved
  2a Leaves not fleshy, oblanceolate....................................... *G. integrifolia*
  2b Leaves thick and fleshy, oblong to lanceolate; coastal .... *G. stricta*

Grindelia integrifolia DC. Willamette Valley gumweed

Taprooted perennials, 1.5-8 dm. tall, glabrous or usually villous; basal leaves oblanceolate, toothed or entire, up to 4 dm. long and 4 cm. wide, cauline leaves sessile and often clasping; involucral bracts more or less resinous, the green tips long and slender, scarcely recurved; ray flowers 10-40, 8-20 mm. long.

Grindelia integrifolia and *Grindelia nana* var. *nana* readily hybridize.

Common in vernally wet areas and dry pastures.

Grindelia nana Nutt. Columbia gumweed

Glabrous biennials or perennials; stems 1.5-8 dm. tall; leaves oblanceolate to linear-oblanceolate, up to 1 dm. long and 1.5 cm. broad, nearly entire to serrate, lower petioled, cauline sessile; involucres resinous, appearing varnished, tips of bracts stiff, strongly recurved.

Variety *discoidea* (Nutt.) Gray differs from var. *nana* (which is usually east of our area) in that it lacks ray flowers. It is found along the Columbia River in our area; elsewhere along stream banks in sandy or gravelly soil.

[=*Grindelia columbiana* (Piper) Rydb.]

Grindelia nana

Grindelia stricta DC. Coastal gumweed

Perennials, often somewhat woody at the base; 2-15 dm. tall, more or less succulent, glabrous to tomentose; leaves fleshy, oblong to lanceolate, 1-20 cm. long and to 4 cm. broad, entire or sharply toothed, lower leaves petioled, cauline leaves sessile and sometimes more or less clasping; involucral bracts in 4-6 series, spreading to slightly recurved; ray flowers 12-25 mm. long.

Salt marshes, tidal flats, sloughs, dunes and coastal bluffs.

Helenium

Erect simple or branching annual to perennial herbs with gland-dotted foliage; leaves alternate, often forming wings down the stem, at least the upper leaves sessile; heads on naked terminal or axillary peduncles; both ray and disk flowers yellow, the latter sometimes turning brownish or purplish, ray flowers often drooping, few to several, sometimes sterile, disk flowers small, numerous, perfect; involucral bracts linear, generally becoming reflexed; receptacle globose or nearly so; achenes top-shaped, ribbed; pappus, if present, of several chaffy scales.

Helenium autumnale L. var. *grandiflorum* (Nutt.) Torr. & Gray
Large-flowered sneezeweed

Erect perennials, 2-12 dm. tall, minutely pubescent or becoming glabrous; leaves oblong or broadly lanceolate, somewhat toothed, acute, 2.5-15 cm. long, the bases extending down the stem to form wings; centers of heads globose, about 2 cm. broad, the rays 10-20, showy, somewhat drooping, 1-2.5 cm. long, both ray and disk corollas pubescent and glandular; achenes pubescent on the ribs; pappus scales awn-tipped.

Common in moist areas along the Columbia River.

Heterotheca

Annual to perennial herbs; stems bristly and with many alternate glandular leaves; heads in a flat-topped inflorescence; ray flowers, when present, fertile; involucre bell-shaped to hemispherical, the bracts narrow, overlapping in 3-7 series; pappus, when present, of brownish or reddish long slender bristles generally surrounded by a basal circle of small scales.

Heterotheca oregona (Nutt.) Shinners — Oregon golden-aster

Perennials; stems tufted, much-branched, 2-8 dm. tall; herbage pubescent; leaves lanceolate to oblanceolate to ovate, entire, the upper sessile; heads up to 14 mm. long; involucral bracts narrow, in several series; ray flowers absent; pappus present.

[=*Chrysopsis oregona* (Nutt.) Gray]

Gravelly stream margins.

Hieracium

Perennials with milky juice and entire or toothed leaves; heads generally in panicles or cymes; involucre cylindric or bell-shaped, the bracts in 1-4 series with a few shorter ones at the base; flowers yellow or white, all strap-shaped; achenes slender, ribbed, not beaked; pappus of 1 or 2 circles of roughened dull white, yellowish or brownish bristles.

1a Flowers orange-red.. *H. aurantiacum*
1b Flowers white or yellow
  2a Alpine plants; leaves scarcely pubescent or the hairs short....... *H. gracile*
  2b Not alpine; leaves conspicuously long-hairy
    3a Flowers white .... *H. albiflorum*
    3b Flowers yellow... *H. scouleri*

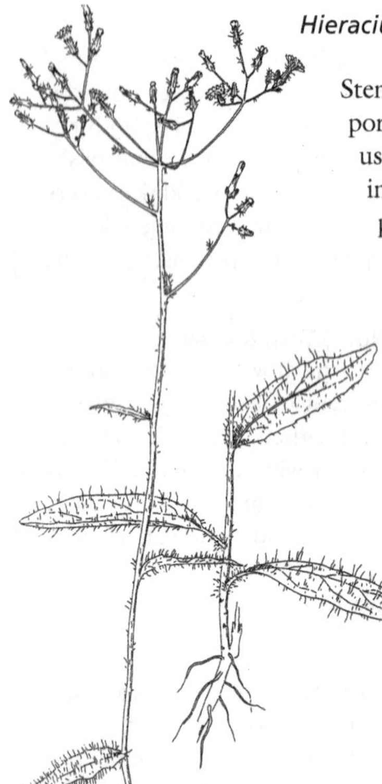
Hieracium albiflorum

Hieracium albiflorum Hook.
White-flowered hawkweed
Stems simple, 4-10 dm. tall, villous on the lower portion; leaves mostly basal, entire or few-toothed, usually villous at least on the under surface; heads in panicles or corymbs of cymes; flowers white; pappus dull white to yellowish-brown.

Open woods.

Hieracium aurantiacum L.
Orange hawkweed
Stoloniferous and rhizomatous; stems 2-7 dm. tall, densely pubescent; leaves mostly basal, oblanceolate to elliptic, entire to toothed, soft-pubescent; heads in cymes; involucre glandular, pubescent with black hairs; flowers orange-red; pappus white.

Disturbed habitats; introduced from Europe.

Hieracium gracile Hook.
Alpine hawkweed
Stems 1-3 dm. tall, with minute grayish hairs, hairs denser and darker in the inflorescence; leaves mostly basal, oblong or oblanceolate, narrowed at the base to slender petioles, scarcely pubescent; heads few, small, the peduncles and involucral bracts densely rough pubescent, glandular, some of the hairs generally black; flowers yellow; pappus yellowish-brown.

Alpine meadows and other open areas high in the mountains.

Hieracium scouleri Hook. Scouler's hawkweed
Stems stout, 3-7 dm. tall, generally leafy, whole plant usually covered with long whitish or brownish bristly hairs, or these sometimes reduced above; leaves lanceolate to oblanceolate, the upper sessile; heads many; involucral bracts generally with long simple and minute branched hairs as well as glands; flowers yellow; pappus yellowish-brown.

(=*Hieracium albertinum* Farr)

Woods and prairies.

Hulsea

Glandular-pubescent, balsam-scented annuals or perennials chiefly with basal or sometimes alternate leaves; heads yellow or purple to red; involucre hemispherical, the bracts glandular, green, narrow, loose, in 2-3 series; ray flowers many; disk flowers orange or yellow, the corolla enlarged above the tube; achenes pubescent; pappus of 4 usually ragged-edged scales.

Hulsea nana Gray

Perennials; stems 4-20 cm. tall; leaves mostly near the base, lower leaves oblong, coarsely toothed to lobed, the uppermost entire, bract-like; heads 1.5-2.5 cm. broad, yellow, borne singly on the stems.

Talus slopes in the Cascades.

Hypochaeris

Annual or perennial herbs with milky sap; leaves in a basal rosette, pinnately parted or toothed; peduncles several-headed; involucral bracts in several series, all erect; flowers all strap-shaped; achenes slender, beaked, or the outer beakless; pappus of plumose bristles.

Perennials; leaves hispid .... *H. radicata*
Annuals; leaves glabrous.. *H. glabra*

Hypochaeris glabra L. Smooth cat's ear

Glabrous annuals 1-4 dm. tall; leaves basal, oblanceolate, nearly entire or toothed or shallowly pinnately lobed; rays short, about equaling the involucre; inner achenes long-beaked, outer ones beakless.

Introduced from Europe; weedy in waste places.

Hypochaeris radicata L. Rough cat's ear, False dandelion or Gosmore

Perennials; scapes 1.5-8 dm. tall from a cluster of basal leaves; leaves hispid, pinnately lobed or pinnatifid, the lateral lobes mostly oblong, entire to toothed or lobed; scape branched, each branch bearing 2 or more heads, the head which terminates the primary stem opening first; rays conspicuous; achenes slender, roughened toward the apex, long beaked, bearing a plumose pappus.

A very common lawn and garden weed. Often mistaken for Dandelions.

Inula

Coarse perennials or shrubs with alternate and basal leaves and large heads usually consisting of both ray and disk flowers; involucre hemispheric or bell-shaped, its bracts in several series, the outer often leaf-like; receptacle naked; anthers tailed at the base; style-branches linear, obtuse; achenes 4-5-ribbed; pappus of rough slender, often unequal, bristles.

Inula helenium L. Elecampane

Stems more or less tufted, usually simple or rarely branched above, 6-18 dm. tall; basal leaves long-petioled, elliptic, sometimes reaching 7.5 dm. in length, rough above, velvety beneath, the margins dentate, cauline leaves smaller, ovate, sessile, clasping the base; heads large, spreading 5-10 cm. in diameter, solitary or few; outer involucral bracts ovate, leaf-like, velvety beneath, inner bracts smaller, spatulate; ray flowers numerous, golden-yellow, slender; achenes glabrous, 4-sided.

Introduced from Europe; usually found along road sides, but not common west of the Cascades.

Jaumea

Glabrous succulent perennials; leaves opposite, simple, entire, sessile; heads usually with ray and disk flowers; receptacle cone-shaped, naked; involucral bracts in 3-5 unequal series; flowers yellow; pappus, when present, of short bristles or scales.

Jaumea carnosa (Less.) Gray

Rhizomatous perennials; stems simple or somewhat branched, prostrate to ascending, up to 3 dm. long, rooting at the lower nodes; leaves succulent, about 2 cm. long; heads terminal; involucre cylindrical, of several fleshy overlapping bracts; rays 6-10, inconspicuous; disk-flowers more numerous; achenes glabrous; pappus absent.

Salt marshes and tidal flats.

Lactuca

Tall leafy annuals to perennials with milky juice, alternate leaves, and yellow, white or blue flowers; bracts of the involucre of several lengths, the outer shorter; corollas all strap-shaped; achenes somewhat flattened, ribbed, narrowed at the summit into a beak; pappus of simple slender white or brownish bristles.

1a Flowers blue or purplish or rarely cream-colored
  2a Rhizomatous perennials; pappus white.................................. *L. tatarica*
  2b Taprooted biennials; pappus brownish.................................. *L. biennis*
1b Flowers yellow
  3a Beak of the achene much shorter than the body..................... *L. muralis*
  3b Beak of the achene equal to or longer than the body
    4a Leaves broad in outline; heads 13-27-flowered.................. *L. serriola*
    4b Leaves mostly linear in outline; heads 8-12-flowered......... *L. saligna*

Lactuca biennis (Moench) Fern. — Tall blue lettuce

Taprooted biennials; stems erect, simple, 1-3.5 m. tall, glabrous, pale green or often purplish below, hollow; leaves mostly deeply pinnately lobed, the lobes toothed, the lower leaves often 1-3 dm. or more long, narrowed into a long petiole which is widened at the base, the upper leaves reduced, sessile and clasping; panicle long and narrow; involucre pale green, tips of the bracts purplish; rays blue or cream-colored; achenes ribbed, roughened, with a short stout beak; pappus brownish.

In moist ground, especially toward the coast.

Lactuca muralis (L.) Fresen. — Wall lettuce

Glabrous annuals or biennials; stems slender, 3-9 dm. tall; leaves pinnatifid, the basal and lower cauline with a large terminal triangular lobe, the bases clasping the stem, middle and upper cauline leaves reduced, the leaf blades all glaucous on the under surface; heads several to many; involucres narrow; heads 5-flowered, rays yellow; achenes short-beaked.

Introduced from Europe; usually found in moist habitats.

Lactuca saligna L. Willow lettuce

Annuals; stems 3-10 dm. tall; leaves linear-sagittate or with a few spreading or retrorse linear lobes, the clasping bases often toothed; inflorescence narrow; heads small, 8-12-flowered, rays yellow; achenes long-beaked; pappus white.

Waste places, mainly to the south of our area.

Lactuca serriola L. Prickly lettuce

Stems 0.5-2 m. tall, simple below, branched above, glabrous and glaucous above, often somewhat prickly below; leaves oblong to lanceolate, entire to deeply pinnately lobed, prickly on margins and midrib beneath, sessile and usually clasping; flowers pale yellow; fruit long-beaked; pappus brownish.

A common weed; introduced from Europe.

This plant is also sometimes called "compass-plant" from the habit of the leaves which generally point approximately north and south. It is said this position is so constant that the plant can be used to determine directions.

Lactuca serriola
a. achene

Lactuca tatarica (L.) C. A. Mey. subsp. *pulchella* (Pursh) Stebb. Blue lettuce

Rhizomatous perennials, 4-10 dm. tall; lower leaves sometimes pinnatifid, upper leaves generally entire or merely dentate, more or less clasping the stem, often glaucous beneath; flower heads 10-15-flowered, blue or purplish, large and showy; achenes short-beaked; pappus white.

[=*Lactuca pulchella* (Pursh) DC.]

Occasional in fields of western Oregon; more abundant east of the Cascades.

Lagophylla

Pubescent annuals with slender brittle stems and mostly alternate leaves, or the lower opposite, generally entire; heads small, yellow, the ray flowers about 5, the ovaries enclosed by the involucral bracts which fall with them; disk flowers 6; achenes dorsally compressed; pappus none.

Lagophylla ramosissima Nutt. Slender hareleaf

Stems slender, erect, simple or sparingly branched, 1-10 dm. tall; leaves linear or oblanceolate, grayish-silky, the lower toothed, the middle and upper leaves entire; heads terminal and sometimes axillary; yellow ray flowers scarcely longer than the gray-silky involucre.

Dry prairies.

Lapsana

Leafy annual or perennial herbs with milky juice; leaves alternate; heads yellow, few-flowered, borne in panicles; flowers all strap-shaped; involucre cylindrical, its main bracts in 1 series, with a few smaller ones at the base; achenes somewhat flattened, 20-30-ribbed; pappus none or rarely of a few short awns.

Lapsana communis L. — Nipplewort

Annuals; stems slender, erect, simple or branched above, 2-8 dm. tall; leaves thin, broadly oval or suborbicular, usually with a few small lower lobes and coarsely toothed or the uppermost leaves linear, the lower long-petioled; heads few; rays yellow; pappus absent.

Introduced from Europe; common in disturbed habitats.

Lasthenia

Annual or perennial herbs with opposite leaves; heads terminal, solitary or in cymes; involucral bracts separate to partly or wholly united; ray flowers pistillate; disk flowers perfect; pappus, if any, of scales or awns, or both.

Involucral bracts united .... *L. glaberrima*
Involucral bracts separate .... *L. maritima*

Lasthenia glaberrima DC. — Smooth goldfields

Glabrous annuals; stems 0.5-3.5 dm. long, weakly ascending, usually rooting at the lower nodes; leaves linear, 3-10 cm. long, entire, the pairs usually fused at the base; involucral bracts green, fused, cup-shaped with short teeth; ray flowers 6-13, inconspicuous, pale yellow; corollas of the disk flowers usually 4-lobed, yellow or greenish; pappus of 5-10 scales.

Vernal pools and other wet, often muddy areas.

Lasthenia maritima (Gray) Vasey — Seaside goldfields

Succulent annuals; stems 1-2.5 dm. tall, usually freely branched, weak and usually spreading; leaves 1-9 cm. long, linear, deep green, entire or irregularly toothed; heads numerous; involucral bracts separate; ray flowers 7-12; pappus of 4-9 brown awns and smaller scales.

[=*Lasthenia minor* (D.C.) Ornduff subsp. *maritima* (Gray) Ornduff]

Isolated stations on the Pacific coast, usually on off shore islands.

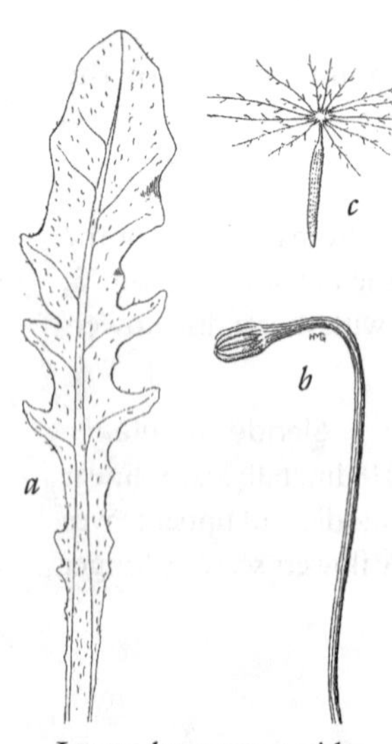

Leontodon taraxacoides
a. leaf; b. bud; c. achene

Leontodon

Annual or perennial herbs with milky juice and all basal leaves; flowers yellow, all strap-shaped; involucral bracts principally in one series, but several smaller bracts at the base; achenes slender, spindle-shaped; pappus of at least the inner flowers composed of plumose bristles somewhat widened at the base, pappus of outer flowers sometimes reduced to a chaffy crown.

Leontodon taraxacoides (Vill.) Merat — Hairy hawkbit

Perennials 1-3.5 dm. tall; leaves usually forming a dense rosette flattened against the ground, 2.5-15 cm. long, 0.5-2.5 cm. broad, widest above the middle, usually deeply and coarsely toothed or lobed, pubescent; peduncles unbranched, erect or often horizontal, somewhat pubescent; young buds nodding; involucre urn-shaped; rays yellow, outer flowers longer than inner, often purplish or brownish on back; pappus of outer flowers a crown of short scales, pappus of inner flowers consisting of barbed or plumose hairs.

[=*Leontodon nudicaulis* (L.) Merat]

In disturbed ground or a lawn weed often associated with *Hypochaeris*; introduced from Europe.

Leucanthemum

Perennials with large showy heads solitary on long peduncles; leaves alternate, entire to pinnately lobed; ray flowers usually white; disk flowers yellow; achenes 10-ribbed; pappus a short crown of scales or lacking.

Leucanthemum vulgare Lam. — Oxeye daisy

Rhizomatous perennials; stems generally many, 2-7.5 dm. tall, somewhat branched, the branches bearing solitary terminal heads; leaves obovate to spatulate, the lower with short-lobed blades tapering to long petioles, the upper with long blades, mostly sessile, all more or less pinnately lobed to parted; heads 3-7 cm. broad, ray flowers white; pappus absent.

(=*Chrysanthemum leucanthemum* L.)

Introduced from Europe; very common along road sides, in fields and other disturbed habitats.

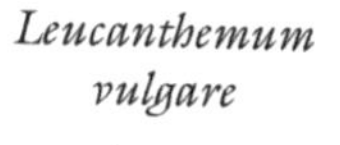

Leucanthemum vulgare

Luina

Pubescent perennials with alternate leaves; heads in terminal flat-topped clusters; involucral bracts stiff, narrow, nearly equal in 1-2 series; ray flowers none; disk-flowers yellow or cream-colored, the corolla 5-cleft; pappus of numerous soft white or tawny slender bristles.

Luina hypoleuca Benth. — Silverback Luina

Stems stiff, erect, leafy, simple below, 1.5-5 dm. tall, whole plant white-woolly, or upper surface of leaves becoming nearly glabrous and shining; leaves 2.5-6 cm. long, ovate to long-elliptic, entire or nearly so, sessile, veins conspicuous above; heads long-peduncled in simple or compound corymbs; flowers creamy-white to yellowish; pappus tawny or white.

Rocky places in the mountains.

Madia

Sticky-glandular, strong-scented annual or perennial herbs; leaves often opposite below, alternate above; heads usually with both ray and disk flowers, the ray achenes compressed, each enclosed by an involucral bract, these in 1 series, and falling with the achenes; a circle of bracts between ray and disk flowers remains to form a cup; pappus generally lacking.

1a Heads 5 mm. or less high; disk flowers usually 1 or 2
  2a Leaves mostly opposite .... *M. minima*
  2b Leaves, at least the middle and upper, alternate...................... *M. exigua*
1b Heads over 5 mm. high
  3a Biennials or short-lived perennials; leaves, all but the upper most, opposite.. *M. madioides*
  3b Annuals; leaves mostly alternate or basal
    4a Ray flowers showy, 1 cm. or more long, 3-lobed, generally red at the base.. *M. elegans*
    4b Ray flowers 7 mm. or less long
      5a Heads slender, longer than wide............................ *M. glomerata*
      5b Heads more or less globose, at least as wide as long
        6a Heads long-peduncled in loose racemes or cymes ... *M. gracilis*
        6b Heads sessile or short-peduncled, crowded............... *M. sativa*

***Madia elegans* D. Don ex Lindl.** — Showy tarweed

Annuals; stems erect, simple below the inflorescence, 2-15 dm. tall, pubescent and sticky-glandular above; lower leaves often densely crowded, linear to lanceolate, 3-20 cm. long, entire or obscurely toothed; upper leaves reduced, linear; inflorescence a loose corymbose or paniculate cyme; involucre bell-shaped, the bracts with long tips, usually glandular-pubescent; ray flowers spreading, showy, deep yellow, generally red at the base, deeply 3-lobed; ray achenes somewhat flattened, the surface minutely roughened; disk achenes abortive.

Extremely variable.

Common in prairies and open woodlands, but also often along road sides.

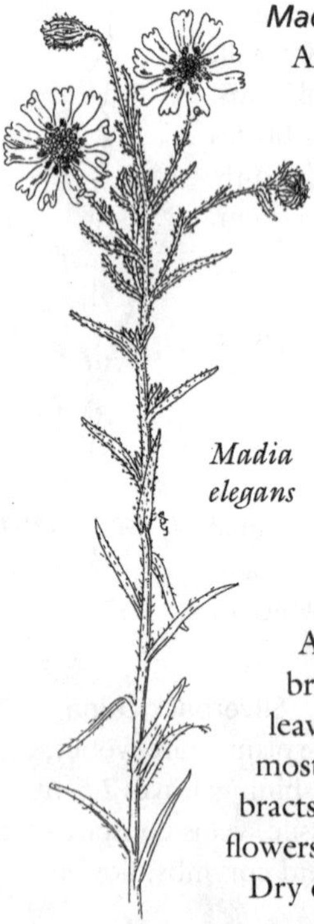

Madia elegans

***Madia exigua* (Sm.) Gray** — Little tarweed

Annuals; stems slender, 0.5-5 dm. tall, simple or generally branched above, whole plant pubescent, glandular above; leaves linear; heads 5 mm. or less high, somewhat broader, mostly borne singly or on long slender axillary peduncles; bracts 4 to 8, each enclosing an inconspicuous ray flower; disk flowers generally 1, rarely 2.

Dry open woods and prairies.

***Madia glomerata* Hook.** — Cluster tarweed

Annuals; stems erect, slender or stout, 1-10 dm. tall, simple or with stiffly ascending branches, whole plant pubescent, glandular above; leaves linear;

heads slender, longer than wide, glandular-pubescent; ray flowers usually 3 or less in some heads; disk flowers few.

Common in dry open areas, often in disturbed ground.

Madia gracilis (Sm.) Keck — Slender tarweed

Annuals; stems 1-10 dm. tall, branching above, simple below, pubescent below, glandular above; leaves linear to narrowly lanceolate, 2-11 cm. long, glandular-pubescent; heads 6-12 mm. high, subglobose or ovoid, rather long-peduncled in loose racemose or corymbose cymes; outer involucral bracts glandular, long-fringed on margins, inner bracts fringed, glandular where exposed; ray and disk flowers 2-13 each, corolla tubes of disk flowers narrow, pubescent; both ray and disk achenes minutely roughened, somewhat flattened and curved, the former slightly broader.

Common in dry ground, often along road sides.

Madia madioides (Nutt.) Greene — Woodland tarweed

More or less villous biennials or short-lived perennials; stems erect, slender, simple or somewhat branched, 2-7 dm. tall, glandular-pubescent above; leaves opposite, except the uppermost, narrowly lanceolate or oblanceolate, obscurely toothed, 3-12 cm. long, the 2 or 3 upper alternate leaves much shorter; involucre bristly-glandular; heads about 12 mm. in diameter, arranged in a loose branching cymose inflorescence, each head long-peduncled in the axils of bracts, smaller bracts borne on the peduncles; ray flowers 5-15, corollas mostly deeply 3-toothed; disk flowers sterile; achenes of the ray flowers somewhat curved, flattened.

In open woods.

Madia minima (Gray) Keck — Least tarweed

Small annuals, stems 2.5-15 cm. high; leaves 1-2 cm. long, mainly opposite, linear, sometimes toothed, pubescent; heads small, yellow, solitary or in small terminal cymes; achenes of the ray flowers dorsally compressed and curved, those of the disk flowers usually club-shaped.

Dry ground, often in woods or on open prairies; uncommon.

Madia sativa Mol. — Coast tarweed

Strongly-scented annuals; stems generally stout, 2-15 dm. tall, stiffly erect, simple or sometimes branched, pubescent and glandular throughout or only above; leaves 2-15 cm. long, linear to narrowly lanceolate, pubescent and usually glandular; heads generally 6-15 mm. high and about as broad, sessile or on very short axillary and terminal peduncles, sometimes very crowded near the top of the stem; bracts deciduous with the ray achenes; achenes flattened, only slightly curved.

Common in disturbed habitats.

Matricaria

Glabrous branching annual to perennial herbs with alternate pinnately dissected leaves; heads solitary or in cymes or corymbs; rays, if present, white; disk flowers with a 4-6-toothed corolla; involucral bracts in several series with scarious margins, overlapping; receptacle naked; achenes glabrous; pappus, if present, a short crown.

Matricaria discoidea DC. Rayless dogfennel or Pineapple weed
Annuals 0.5-4 dm. tall; leaves finely pinnately dissected; receptacle conical; heads greenish-yellow; rays lacking, involucre spreading; pappus present.
[=*Matricaria matricarioides* (Less.) Porter and *Chamomilla suaveolens* (Pursh) Rydb.]
A common weed, usually in hard-packed ground.

Microseris

Perennial herbs with pale fleshy roots or fibrous-rooted annuals; milky juice present; leaves principally basal (some species with leaves or bracts on the stem), entire to pinnately lobed, glabrous or more or less covered with minute mealy scales, never white-hairy in the species of our area; heads nodding in the bud, borne on usually leafless peduncles, these arising from the basal rosette of leaves or from a branching leafy stem; involucre bell-shaped, the bracts black-hairy on their overlapping margins, forming one main series with much shorter outer bracts at the base, or several series gradually reduced in size; flowers all strap-shaped, yellow to orange; achenes columnar, tapering only toward the base, distinctly ribbed, filled by the seed; pappus in our species consisting of 5-10 white or brownish scales that taper into long bristles, or of numerous brownish bristles only; fruits expanding into a globose head when mature.

1a Plants annual; pappus of 5 bristle-tipped scales........................ *M. bigelovii*
1b Plants perennial; pappus scales or bristles more numerous
  2a Heads on leafless peduncles arising from the basal rosette; pappus of numerous bristles.. *M. borealis*
  2b Heads on peduncles arising from a branching, leafy stem; pappus of 6-10 bristle-tipped scales... *M. laciniata*

Microseris bigelovii (Gray) Sch.-Bip. Coast Microseris
Annuals, usually 4 dm. tall or less, often prostrate or decumbent, glabrous to more or less mealy; leaves all basal, blunt, often with broad or narrow pinnate lobes; heads on naked peduncles, nodding until the fruit is nearly mature; involucral bracts with a red stripe on the midvein, forming one main series with some much smaller outer bracts at the base; flowers small and inconspicuous, yellow or orange; achenes brownish, 2-5 mm. long; pappus scales 1-3 mm. long, white or brownish, somewhat shorter than the attached bristle.
Occasional, on windswept coastal bluffs.

Microseris borealis (Bong.) Sch.-Bip. Bog Microseris
Perennials 1.5-5 dm. tall, with short rhizomes and fleshy roots, glabrous throughout or finely mealy when young; leaves in a basal rosette, oblanceolate, acute, entire or with a few small teeth; heads on long, naked peduncles, many-flowered, the involucral bracts mostly in one series with some shorter

and narrower outer ones, all glabrous except for blackish hairs on the inner bracts; flowers yellow; achenes reddish-brown, 4-8 mm. long; pappus of 30-60 rather coarse, brownish bristles.

[*Apargidium boreale* (Bong.) Torr. & Gray]

Boggy meadows, in the Cascade Mountains.

Microseris laciniata (Hook.) Sch.-Bip. subsp. *laciniata* Cutleaf Microseris
Perennials 2-10 dm. tall, with pale, parsnip-like roots, the stem branching toward the base or higher up, distinctly leafy; leaves highly variable, broad to narrow, blunt or acute, often pinnatifid with narrow, sharp lobes, glabrous or mealy; heads on long peduncles, many-flowered, bell-shaped; involucral bracts in several series, abruptly narrowed above, the outer ones shorter than the inner and 2.5 mm. or more wide, the inner bracts tapering, black-hairy; flowers yellow, showy; achenes pale brown, 3.5-6.5 mm. long; pappus scales ovate, much shorter than the attached bristle.

Meadows and open places, at low elevations in the interior valleys.

Subspecies *leptosepala* (Nutt.) Chambers (=*Scorzonella leptosepala* Nutt.) Like the above, but less robust; involucre frequently mealy, the outer bracts much smaller than the inner and less than 2.5 mm. wide.

Plains of the lower Willamette Valley and northern Oregon coast.

Nothocalais

Perennial herbs with thick, corky caudex and deep taproot; leaves all basal, linear to oblanceolate or spatulate, entire to pinnately lobed or divided, glabrous or pubescent on the margins and midrib; heads single on leafless peduncles, glabrous (white hairy in one species outside our area); involucral bracts in several series, all about the same length, the outer ones like the inner or somewhat broader; flowers all strap-shaped, yellow; achenes spindle-shaped, distinctly ribbed, nearly filled by the seed and lacking a beak; pappus of numerous white bristles (or of narrow white scales tapering into bristles, in species not of our area); fruits expanding into a globose head when ripe.

Nothocalais alpestris (Gray) Chambers Alpine lake Agoseris
Plants 3-25 cm. tall, glabrous throughout or rarely finely hairy on the petioles; leaves highly variable, linear to spatulate, entire or with coarse to narrow teeth or lobes; heads bell-shaped, many-flowered (few flowered in dwarfed specimens), the outer involucral bracts as long as the inner but broader, evenly speckled with fine purple dots; flowers yellow, drying pinkish; achenes 5-10 mm. long, filled by the seed or the upper 1-3 mm. empty.

Usually in pumice soil, alpine areas of the Cascade Mountains. Often confused with alpine varieties of *Agoseris glauca*.

Petasites

Rhizomatous perennials; basal leaves large, palmately-veined, cauline leaves bract-like, appearing parallel-veined, alternate; heads in more or less dense, generally corymbose, clusters; plants functionally dioecious, the staminate heads sometimes bearing a few perfect but sterile flowers; minute rays sometimes present; involucral bracts in one series and nearly equal; some heads almost entirely pistillate, other heads perfect; achenes ribbed; pappus of capillary bristles.

Petasites frigidus (L.) Fries
Colt's-foot
Flowering stalks arising very early in spring, 2-6 dm. tall; basal leaves orbicular to reniform in outline, 1-4 dm. broad, coarsely-toothed to palmately-lobed and then shallowly pinnately-lobed or toothed, long petiolate, glabrous or short pubescent above, loosely white-tomentose on the under surface; heads several to many; flowers white to pink, pistillate flowers with short inconspicuous rays. Two varieties occur in our area: var. ***nivalis*** (Greene) Cronq. with leaves longer than broad and the more common var. ***palmatus*** (Ait) Cronq. with leaves usually broader than long.

Swamps, wet meadows and moist wooded areas.

Petasites frigidus

Prenanthes

Perennial herbs with milky juice; leaves alternate; heads in cymes, corymbs or panicles, ours with dull-colored flowers; flowers all strap-shaped; achenes ridged; pappus of slender bristles.

Prenanthes alata (Hook.) Dietr. Western rattlesnake root
Stems 1-8 dm. tall; leaves triangular or deltoid, often hastate at the base, mostly acute at the apex, sharply and irregularly toothed, narrowed abruptly or gradually into petioles, or the upper nearly sessile; heads in corymbs or panicles of cymes; flowers purplish or more often white.

Margins of streams in the mountains and along the Columbia River.

Psilocarphus

Low white-woolly annuals; leaves simple, mostly opposite, or the upper alternate, entire; heads solitary in leaf axils or forks of branches or at tips of branches, or sometimes clustered; only disk flowers present; pistillate flowers each enclosed by a chaffy bract; staminate flowers few, borne in the center of the head, not subtended by chaffy bracts; pappus lacking.

Heads loosely-woolly; stems spreading to erect .... *P. elatior*
Heads with appressed pubescence; stems prostrate or decumbent ... *P. tenellus*

Psilocarphus elatior (Gray) Gray Tall Woollyheads
Thinly and loosely white-tomentose annuals; stems usually erect, 1-15 cm. tall; leaves oblanceolate to linear-oblong, 1-3.5 cm. long, 2-9 mm. broad;

heads globose, loose-woolly, usually surpassed by the subtending leaves; achenes cylindrical.

Vernal pools and other moist areas, often in fields.

Psilocarpus tenellus Nutt. Slender Woollyheads

Tomentose, decumbent or prostrate annuals, forming mats 0.5-3 dm. in diameter; leaves oblong to spatulate, 4-15 mm. long, 1-5 mm. broad, heads 4 mm. or less in diameter, surpassed by the subtending leaves; achenes less than 1.5 mm. long, oblong.

Vernal pools and other, often drier, sites.

Rainiera

A monotypic genus.

Rainiera stricta (Greene) Greene

Perennials, sometimes rhizomatous; stems mostly 4.5-12 dm. tall; herbage glabrous except in the inflorescence; lower leaves broadly oblanceolate, entire or nearly so, upper leaves reduced and sessile; inflorescence elongate; heads few-flowered; ray flowers lacking; disk flowers yellow; pappus of capillary bristles.

[=*Luina stricta* (Greene) Rob.]

Meadows and open slopes in the high mountains.

Saussurea

Perennial herbs with alternate leaves; heads many, in corymbs, the involucral bracts overlapping; ray flowers lacking; disk flowers blue to purple, all perfect; pappus usually double, outer bristles shorter than inner and not plumose.

Saussurea americana Eat. American sawwort

Stems stout, erect, 4-12 dm. tall, very leafy; leaves ovate to lanceolate, generally sharply toothed, acuminate, the lower petioled, the petioles becoming shorter above to generally none; heads narrowly bell-shaped, 10-18 mm. long, numerous; involucral bracts thick, dark-tipped and dark-margined; flowers purple; outer pappus falling separately from the inner.

Alpine in the Cascade and Olympic Mountains of WA.

Senecio

Annuals, biennials, perennials, shrubs or sometimes trees; leaves alternate or all basal; head usually with ray as well as disk flowers, involucral bracts in a single equal series, but often subtended by much-reduced bracts at the base; flowers usually yellow or orange; pappus of entire to finely barbed bristles.

1a Annuals
  2a Outer bracts of the involucre black-tipped .... *S. vulgaris*
  2b Outer bracts of the involucre not black-tipped.................. *S. sylvaticus*
1b Perennials or biennials
  3a Leaves not much reduced upwards
    4a Leaf-blades merely toothed, triangular-hastate............ *S. triangularis*
    4b Leaf-blades pinnately divided.. *S. jacobaea*

3b Leaves reduced upwards, basal and often the lower cauline leaves well developed.
5a Herbage pubescent at flowering time, often densely so
6a Roots fibrous; involucral bracts with long black tips .*S. integerrimus*
6b Plants from a stout caudex or rhizomatous
7a Plants appearing gray from the dense pubescence .... *S. canus*
7b Plants not gray, the pubescence thinner.....................*S. macounii*
5b Herbage glabrous or nearly so, at least at flowering time
8a None of the leaves pinnatifid, either entire or merely toothed
9a Herbage glaucous; leaves fleshy; plants to 2 m. tall; plants of alkaline soils .... *S. hydrophilus*
9b Herbage not glaucous; leaves if thick, not fleshy; plants 1 m. tall or less
10a Heads many; plants 3-10 dm. tall; stems hollow ....
.... *S. hydrophiloides*
10b Heads few; plants 1-3 dm. tall.................. *S. cymbalarioides*
8b At least some of the leaves pinnatifid, sometimes just the upper
11a Basal leaves, or some of them usually compound with a large terminal leaflet and smaller lateral ones
12a Cauline leaves present, but reduced; plants 1.5-7 dm. tall .....
....*S. bolanderi*
12b Cauline leaves few; plants 0.5-4 dm. tall; plants found in the mountains ....*S. flettii*
11b Basal leaves not as above
13a Upper cauline leaves lacerate-pinnatifid towards the clasping base, basal leaves usually cordate at the base; heads usually many....
....*S. pseudoaureus*
13b Upper cauline leaves not lacerate-pinnatifid, basal leaves not cordate; heads few .... *S. cymbalarioides*

Senecio bolanderi Gray — Bolander's groundsel

Perennials from slender rhizomes; stems erect or ascending, 1.5-7 dm. tall, glabrous or nearly so; lower leaves in a basal tuft, often simple, orbicular to oblong in outline, lobed or crenate, long-petioled, upper leaves shorter petioled to nearly sessile, mostly pinnately lobed or divided, the terminal segment much larger than the lateral; heads few to many in a terminal corymbose cyme, peduncled; involucre bracts about 14, slender, glabrous or pubescent; ray flowers 6-18 mm. long, deep yellow; disk flowers numerous; achenes slender, glabrous.

(Includes *Senecio harfordii* Greenm.)

On moist cliffs along the ocean and inland in open woods.

Senecio canus Hook. — Woolly groundsel

Perennials 1-4 dm. tall, herbage gray to white tomentose or the upper surface of the leaves nearly glabrous; leaves lanceolate to ovate, entire to dentate or pinnatifid, petiolate, the upper much reduced and bract-like; ray flowers 5-13 mm. long; disk flowers numerous; achenes glabrous.

Dry open areas, often where rocky; from the valleys to alpine areas.

Senecio cymbalarioides Buek — Alpine meadow groundsel

Rhizomatous perennials; stems 0.5-3 dm. tall; basal leaves obovate to broadly ovate, crenately toothed, 1-3 cm. long, to 2 cm. broad, long-petiolate; cauline leaves much reduced, sometimes pinnatifid; ray flowers 7-15 mm. long.

Wet meadows in the mountains.

Senecio flettii Wiegand — Flett's groundsel

Rhizomatous perennials 0.5-4 dm. tall; usually glabrous or pubescent at the base of the stem and in the leaf axils; basal leaves long-petioled, usually pinnatifid, often with the terminal lobe the largest, the segments again toothed or lobed, upper leaves much reduced and with narrower segments; heads several in a dense cluster; ray flowers 5-10 mm. long.

Talus slopes and other rocky areas, from the foothills to the high mountains.

Senecio hydrophiloides Rydb. — Stout meadow groundsel

Glabrous biennials or perennials; stems 3-10 dm. tall, stout, hollow; leaves thick but not fleshy, elliptic to oblanceolate, usually toothed, 5-25 cm. long, 2-7 cm. broad, basal and lower cauline leaves long-petioled, middle and upper leaves reduced and becoming sessile; ray flowers, if present, few, 5-10 mm. long.

Wet meadows.

Senecio hydrophilus Nutt. — Water groundsel

Glabrous and glaucous biennials or perennials; stems 4-20 dm. tall, hollow; leaves thick and more or less fleshy, narrowly elliptic to oblanceolate, entire or less often shallowly toothed, basal and lower cauline leaves long-petioled, the blades 1-3 dm. long, 2-8 cm. broad, middle and upper cauline leaves much reduced and becoming sessile; heads numerous, ray flowers, if present, few, 3-8 mm. long.

Swampy areas, tolerant of salt and alkali.

Senecio integerrimus Nutt.

Stems erect, simple, 1.5-12 dm. tall, more or less woolly; lower leaf blades oblong or ovate, narrowed to a long petiole, leaves becoming narrower upward, the uppermost lanceolate to linear, sessile, blades of both lower and upper leaves entire or somewhat toothed; heads 6-12.5 mm. high, often many in a crowded corymb-like cyme; involucral bracts generally black-tipped; rays usually 6-15 mm. long.

Variable and with several named varieties.

Frequently found in our region in a variety of habitats.

Senecio jacobaea L. — Tansy ragwort

Biennials or short-lived perennials from short thickened taproots; stems erect, 3-12 dm. tall, simple at least below, somewhat woolly at first, often reddish or purplish, more or less ridged; leaves 5-25 cm. long, pinnately divided, the lobes often again pinnate, the basal leaves with a large terminal lobe, lower leaves petioled, the upper sessile, somewhat clasping; heads many in large corymbs; involucre narrowly bell-shaped, the bracts slender, green, usually tipped with black, a few shorter bracts generally found at the base; heads 12-25 mm. broad, ray flowers about 12, 6-12 mm. long, narrow, spreading, deep yellow; disk flowers brownish-yellow; achenes of ray flowers glabrous, those of the disk flowers minutely pubescent.

A native of Europe, in waste places, pastures and road sides, a troublesome weed. Poisonous to livestock.

Senecio jacobaea

Senecio macounii Greene
Perennials; stems erect 3-4.5 dm. tall; herbage woolly, at least at first; lower leaves, including long petioles, 7-18 cm. long, spatulate or oblanceolate, blades entire or shallowly toothed, upper leaves much reduced, linear or narrowly lanceolate, sessile; heads 6-14, in a corymb- like cyme; ray flowers 8-10 mm. long.

Open areas in the woods, disturbed sites and sometimes along streams.

Senecio pseudoaureus Rydb. Streambank groundsel
Perennials, stems 3-7 dm. tall, usually pubescent when young; leaves thin, the blades broadly lanceolate to ovate, 2-10 cm. long, basal leaves long-petioled, toothed, cauline leaves variously pinnately incised at least towards the base, becoming sessile; ray flowers usually present, 6-12 mm. long.

Wet meadows, stream banks and moist woods.

Senecio sylvaticus L. Woodland groundsel
Annuals 1.5-8 dm. tall, pubescent to nearly glabrous; leaves pinnatifid and irregularly toothed; involucre 5-7 mm. high; ray flowers inconspicuous, 2-6 mm. long.

Introduced from Europe; locally abundant in disturbed sites and waste ground; often in logged and burned areas in the forests.

Senecio triangularis Hook. Arrowleaf Groundsel
Perennials; stems 3-15 dm. tall, glabrous or puberulent; leaves not much reduced upwards, blades triangular, the upper often elongated, 4-20 cm. long, 2-10 cm. broad, toothed, the lower long-petioled, the upper short-petioled or becoming sessile; ray flowers 5-8, 7-15 mm. long.

Stream banks and wet meadows in the mountains, but also occasionally near the coast.

Senecio vulgaris L. Old-man-in-the-spring or Common groundsel
Annuals; stems erect or slightly decumbent, ribbed, branched, 1-5 dm. tall, smooth or more or less long white-hairy; leaves pinnately lobed or parted,

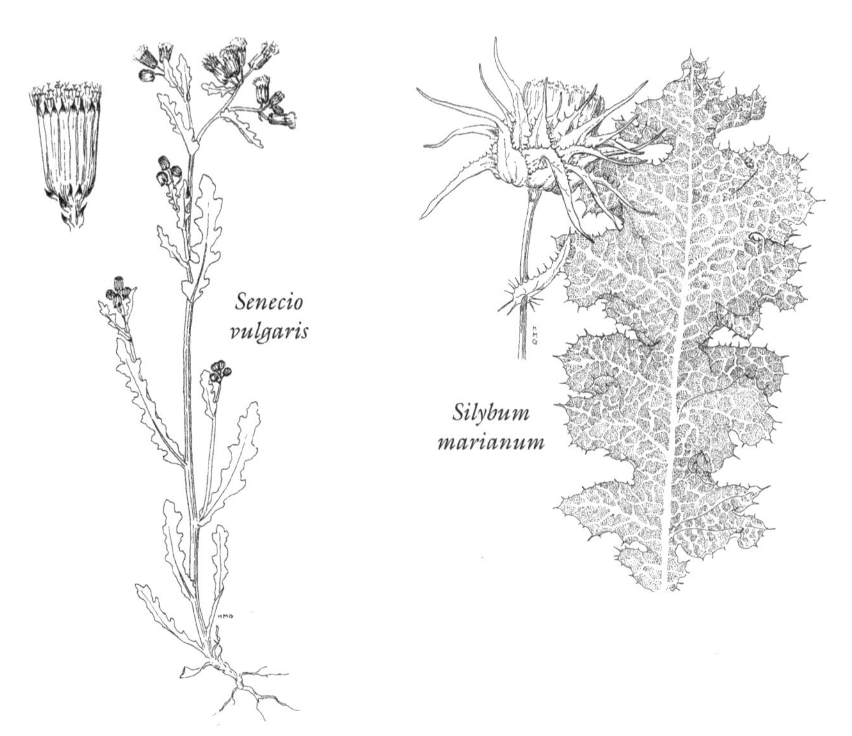

more or less toothed; heads cylindrical, about 1 cm. long, without ray flowers, borne in small crowded cymes; involucre bracts narrow, black-tipped, forming a cup nearly as long as the flowers.

A common garden weed, but also found in other disturbed sites; introduced from Europe. Toxic to livestock.

Silybum

Annual or biennial thistle-like plants with alternate spiny-toothed leaves, these dark green and blotched with white; involucral bracts overlapping, rigid, spreading at the spiny-fringed tips; heads large, pink or purple; ray flowers lacking; pappus of many slender bristles.

Silybum marianum (L.) Gaertn. Milk thistle

Annuals or biennials 0.5-2.5 m. tall; leaves large, pinnately lobed, prickly-margined, dark green with conspicuous irregular white blotches; heads large, purple, solitary; only disk-flowers present, these perfect, fertile; achenes glabrous; pappus bristles many, flattened, roughened, falling together.

Native to the Mediterranean region; weedy in disturbed habitats.

Solidago

Perennial herbs sometimes woody at the base; leaves alternate, heads small, variously arranged in panicles or corymbs; involucral bracts narrow, imbricate to nearly equal in several series; both disk and ray flowers yellow (in ours), the latter inconspicuous; pappus consisting of one series of white scabrous bristles; achenes cylindrical or angled.

1a Middle cauline leaves the largest
  2a Stems minutely pubescent throughout............................. *S. canadensis*
  2b Stems glabrous and glaucous except in the inflorescence..... *S. gigantea*
1b Lower cauline or basal leaves the largest
  3a At least the lower leaves and petioles with long cilia on the margins; alpine plants.. *S. multiradiata*
  3b None of the leaves with on cilia on the margins
    4a Plants not appearing varnished; not occurring on dunes .... *S. missouriensis*
    4b Plants appearing varnished; plants of coastal dunes........... *S. simplex*

Solidago canadensis L. var. *salebrosa* (Piper) M.E. Jones
Canada goldenrod

Stems simple, erect, mostly tufted, 3-15 dm. tall, minutely pubescent; leaves variable, long-elliptic or oblanceolate, acute, narrowed at the base, margins of the upper half usually sharply serrate and scabrous, glabrous or somewhat pubescent, paler beneath, typically 3-veined from the base; heads small (less than 5 mm. high), generally in a dense showy panicle; involucral bracts narrow.

Fields and road sides; very common in our area.

Solidago gigantea Ait. Late goldenrod

Rhizomatous perennials; stems 6-20 dm. tall, glaucous; middle cauline leaves the largest, lanceolate to narrowly elliptic, 5-15 cm. long, toothed at least above the middle, prominently 3-veined; heads numerous in a pyramidal panicle; ray flowers 8-16, 2-2.5 mm. long.

Wet meadows, stream banks and lake shores; only occasional west of the Cascades.

Solidago missouriensis Nutt. Missouri goldenrod

Rhizomatous perennials; stems 1.5-8 dm. tall, nearly glabrous except in the inflorescence; leaves oblanceolate, usually reduced upwards and becoming sessile, 3-veined, often weakly so; inflorescence often with recurved branches; involucre 3-5 mm. high; ray flowers 5-13, 2-3 mm. long.

Dry, open areas.

Solidago multiradiata Ait. Northern goldenrod

Stems usually clustered, 0.5-4.5 dm. tall, leafy; leaves firm, deep green, oblanceolate to spatulate, acute, ciliate on the margins otherwise nearly glabrous, basal leaves petioled, upper leaves reduced, narrowed at the base; heads few to many, 4-7 mm. long; ray flowers 2-5.5 mm. long.

Alpine meadows to open rocky areas.

Solidago simplex Kunth var. *spathulata* (DC.) Cronq. Dune goldenrod

Stems 1-6 dm. tall, the whole plant, or parts of it, appearing varnished; leaves firm, spatulate or oblanceolate to narrowly oblong or the upper lanceolate, serrate above the middle, or the uppermost leaves entire, basal leaves petioled, 5-15 cm. long; heads about 7 mm. or less high, in compact terminal panicles; involucral bracts 4-7 mm. long, obtuse, sticky-glandular.

(=*Solidago spathulata* DC.)

On dunes along the coast.

Soliva

Soliva sessilis

Low short-branched annuals with entire to pinnately dissected leaves; heads small, sessile in axillary cymes; outer circle of pistillate flowers without corollas, the disk flowers sterile, with 4-toothed corollas; bracts of involucre in 1-2 series, green or scarious; achenes compressed dorsally, tipped by a persistent sharp spine; pappus none.

Soliva sessilis Ruiz & Pav.

Plants prostrate to ascending, forming mats up to 2 dm. in diameter; minutely pubescent; leaves pinnately dissected; heads small, greenish, sessile; achenes winged or not.

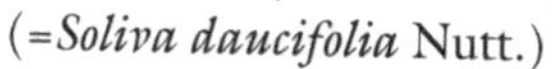
(=*Soliva daucifolia* Nutt.)

Introduced from South America; in disturbed ground or a weed in lawns; not common.

Sonchus

Annual or perennial herbs with milky juice; herbage mostly glabrous and glaucous, the upper parts sometimes glandular; stems leafy, the leaves often prickly-margined; heads yellow, enlarged at base, in a corymbose or umbel-like inflorescence; flowers all strap-shaped; achenes usually flattened dorsally, ribbed, not beaked; pappus of soft white hairs generally falling together.

1a Rhizomatous perennials; flower heads 3-5 cm. broad .... *S. arvensis*
1b Annuals or biennials; flower heads usually less than 3 cm. broad
  2a Clasping leaf bases acute; achenes transversely winkled .... *S. oleraceus*
  2b Clasping leaf bases rounded; achenes not transversely winkled .. *S. asper*

Sonchus arvensis L. — Perennial sow thistle

Rhizomatous perennials, 0.5-2 m. tall; leaves 3-15 cm. long, nearly entire to pinnatifid, the middle and upper leaves clasping the stem, margins prickly-toothed; peduncles and involucre with glandular bristles; heads large, bright orange-yellow; achenes flattened, ribbed longitudinally, transversely wrinkled.

Introduced from Europe. A weedy pest east of the Cascades, and sporadically occurring west of the mountains.

Sonchus asper (L.) Hill — Prickly sow thistle

Somewhat resembling *S. oleraceus,* but leaves often not lobed, margins sharply toothed, the teeth needle-like, clasping leaf bases rounded; achenes without transverse wrinkles.

A common weed; introduced from Europe.

Sonchus oleraceus L. — Common sow thistle

Annuals; stems 1-10 dm. tall, erect, generally simple; herbage bluish-green; leaves pinnately cleft, the lateral lobes small, the terminal large, triangular, all toothed, the teeth weak, clasping base of the upper leaves acute; heads

1.5-2.5 cm. broad, receptacle swollen at base; achenes longitudinally ribbed and transversely wrinkled.

Common weed in waste areas and gardens; introduced from Europe. Sometimes used in salads.

Tanacetum

Perennial herbs with strong-scented foliage, leaves alternate; heads few to many; ray flowers often lacking; achenes ribbed or angled, with broad apex; pappus a short crown of scales, or lacking.

Plants pubescent, occurring in sand dunes............................*T. camphoratum*
Plants glabrous or nearly so; an escape from cultivation .... *T. vulgare*

Tanacetum camphoratum Less. Seaside or Dune tansy

Rhizomatous perennials; stems stout, erect or more often decumbent, 2-5 dm. tall, somewhat woolly; leaves thick, pinnately 2 or 3 times divided, 7-25 cm. in length, and 3-5 cm. or more in width, pubescent and often glandular; heads hemispheric, 8-15 mm. broad; flowers a tarnished yellow; pappus a short irregular toothed crown of scales.

(=*Tanacetum douglasii* DC.)

Coastal dunes.

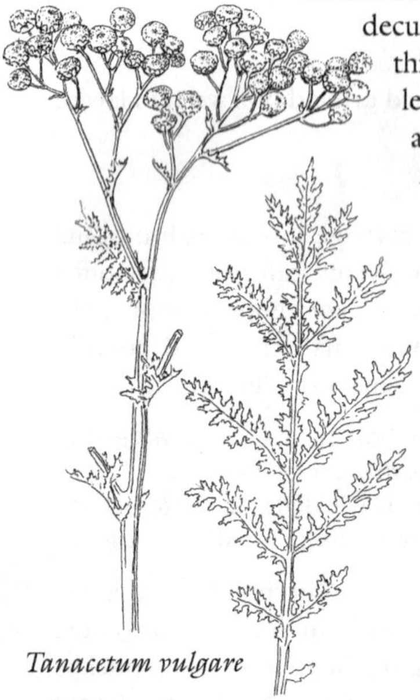

Tanacetum vulgare

Tanacetum vulgare L. Garden or Common tansy

Rhizomatous perennials; stems stout, erect, usually simple except at the summit, glabrous or nearly so, 4-15 dm. tall; leaves pinnately divided into narrow toothed segments, the basal leaves sometimes reaching 3 dm. in length; heads about 6-10 mm. in diameter, golden or tawny, many in a corymbose inflorescence, the peduncles 2 to 3 times the length of the head; involucre hemispheric, the bracts narrow, overlapping, sometimes slightly pubescent; pappus a short crown of scales.

An escape from gardens, locally abundant especially along road sides and stream banks.

Taraxacum

Taprooted perennials with milky juice and all basal leaves; heads yellow, terminal, single, on hollow scapes; leaves toothed to pinnately parted; outer bracts of involucre usually reflexed, inner erect; flowers all strap-shaped; achenes ribbed, the ribs with minute spines toward the apex, slender-beaked, bearing a pappus of soft white slender bristles.

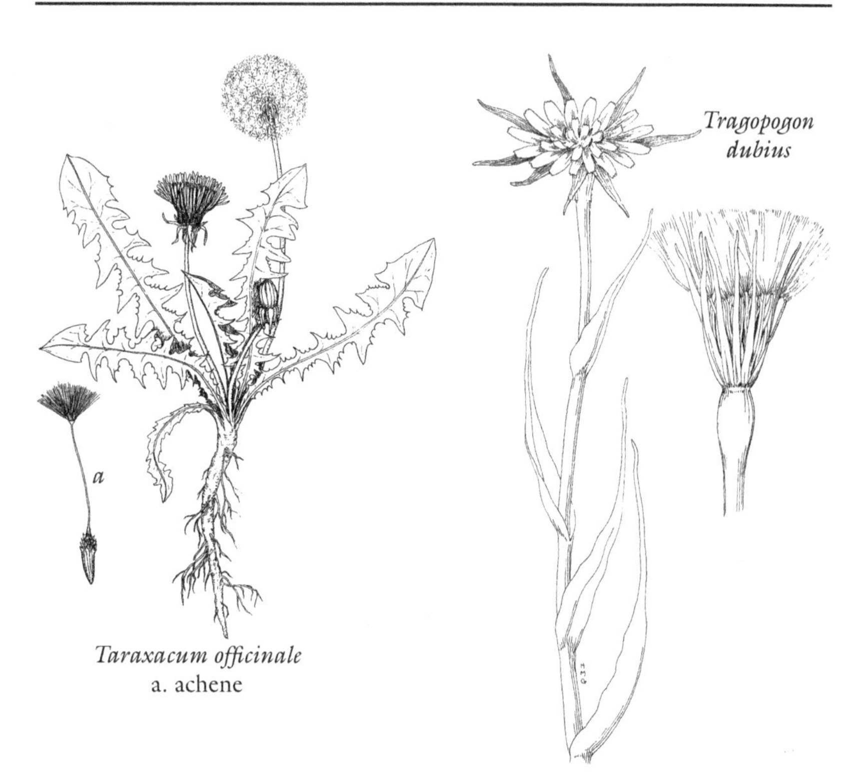

Taraxacum officinale
a. achene

Taraxacum officinale Weber ex Wigg. Common dandelion

Leaves clustered at base of plant; scapes 40 cm. or more tall; leaves thin, deep green, usually pinnately lobed, the larger segments somewhat triangular, entire or with slender soft teeth, glabrous or rarely sparsely pubescent; outer bracts of the involucre reflexed; heads deep yellow, large; achenes short spiny near upper end, beaked; pappus white or brownish.

A common weed flowering very early in the spring, and continuing through the summer; introduced from Europe.

Tragopogon

Taprooted annual, biennial or perennial herbs with milky juice; leaves alternate, entire, appearing parallel veined; heads large, showy, solitary on long peduncles; flowers all strap-shaped, yellow or purple; involucral bracts slender, equal, in 1 series; achenes with a long stout beak; pappus of stout, plumose bristles.

1a Flowers purple .... *T. porrifolius*
1b Flowers yellow
  2a Peduncles much enlarged just below the heads; flowers pale yellow .... *T. dubius*
  2b Peduncles not greatly enlarged below the heads; flowers bright yellow .... *T. pratensis*

Tragopogon dubius Scop. Yellow salsify

Annuals or biennials 3-10 dm. tall; leaves 2-5 dm. long, long-tapering to the apex, often floccose pubescent when young; peduncles greatly enlarged just below the heads; flowers pale yellow; pappus whitish.

Native of Europe; more or less weedy.

Tragopogon porrifolius L. Purple salsify or Vegetable oyster

Biennials; stems stiffly erect from a fleshy tap root, 4-12 dm. tall; leaves lanceolate, long-tapering to apex, the lower 2-4 dm. long, becoming shorter upward; heads 2.5-7 cm. high, the peduncles somewhat enlarged below the heads; flowers purple; involucral bracts longer than the flowers; achenes slender, roughened, long-beaked; pappus tawny.

Common along road sides and in other disturbed sites. An escape from gardens. The flower heads generally close during the day.

Tragopogon pratensis L. Meadow salsify

Biennials; stems erect, 1.5-8 dm. tall; leaves up to 4 dm. long, narrow; involucral bracts equal to or shorter than the bright yellow rays; achenes 15-25 mm. long, short-beaked; pappus white.

Fields and waste places.

Wyethia angustifolia

Wyethia

Taprooted perennials; leaves alternate; heads large, yellow, solitary or several on a stem; receptacle with chaffy scales; involucral bracts in 2-3 series; usually both ray and disk flowers present, yellow or white; pappus a crown of scales or absent.

Wyethia angustifolia (DC.) Nutt. Narrow-leaved sunflower or Mule's ears

Stems clustered, 2-9 dm. tall; basal leaves to 5 dm. long, oblanceolate to long-elliptic, tapering at both ends, petioled, entire or slightly toothed, rough, upper leaves similar but much smaller and often sessile; heads sunflower-like, 6-7.5 cm. broad, golden-yellow to nearly orange, the involucral bracts leaf-like, pubescent and ciliate on the margins, sometimes purplish at the base; pappus of 1-4 awn-tipped scales.

Hillsides and moist meadows.

Xanthium

Annuals; stems stout, widely branching, with alternate and petioled leaves; flowers of 2 kinds, the staminate heads mostly in terminal clusters, the receptacle with chaffy bracts; pistillate heads axillary, the bracts forming a 2-loculed enclosure, each locule containing one flower, the bracts later forming a bur-like fruit covered with hooked prickles; achenes thick; pappus none.

Leaf axils with a 3-forked spine .... *X. spinosum*
Leaf axils without a spine .... *X. strumarium*

Xanthium spinosum L. Spiny cocklebur

Stems 3-12 dm. tall; leaves lanceolate, usually narrowly lobed, white-veined, white-silky pubescent beneath, with axillary 3-parted yellow spines, 1-2.5 cm. long; burs about 1.5 cm. long, covered sparsely with hooked bristles.

Disturbed areas; rarely entering our region.

Xanthium spinosum

Xanthium strumarium L. Common cocklebur

Stems 2-20 dm. tall, purplish-spotted, rough; leaves broadly ovate to triangular, irregularly toothed and lobed, rough on both sides with short hairs, glandular above; burs generally clustered in the leaf axils, about 1-3.5 cm. long, closely covered with slender but rigid prickles with hooked tips, the prickles densely brown-hairy from base to above middle, burs with 2 stout incurved beaks at the apex.

Disturbed areas, especially along streams.

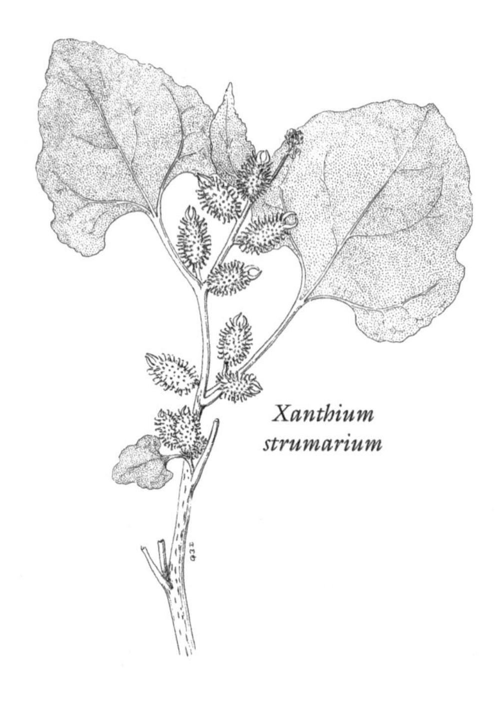

Xanthium strumarium

Glossary

Aborted, said of parts which are nonfunctional, as stamens consisting of filaments only.
Achene, a small dry, 1-seeded fruit which does not split open at maturity.
Acrid, sharp, burning, irritating.
Acuminate, tapering to a long narrow tip or sharp point.
Acute, sharply pointed (as to shape, not texture), but not long-tapering.
Adventitious, occurring in unusual places; as a root which originates from a stem.
Aggregate fruit, derived from separate but closely associated ripened ovaries from a single flower; the individual fruits may be achenes, drupes, etc. Examples: strawberry, blackberry.
Alternate, said of leaves which arise singly from each node; situated between organs, as stamens alternate with the petals.
Amphibious, living both on land and in the water, especially where plants start out in the water and continue to live on land after the water recedes.
Annual, a plant which completes its life-history and dies within a single growing season.
Anther, the pollen-bearing part of a stamen.
Anthesis, the period during which the flower is fully developed, open and fertilization takes place.
Apex, the free end or tip.
Apiculate, having a short sharp point; ending abruptly in a point.
Appressed, flattened or pressed against another body.
Aril, an outgrowth from the placenta which covers or surrounds a seed.
Auriculate, with small projecting lobes or appendages; ear-like.
Awn, a slender, usually terminal bristle.
Awned, having an awn.
Axil, the angle between a leaf and a stem.
Axile, the type of placentation where the ovary is divided by partitions and the ovules are borne on the central axis.
Axillary, borne in an axil.
Axis, a central support.

Banner, upper petal of a pea flower or pea-like flower.
Basal, situated at the base.
Beaked, tipped with a point; said of fruits which end in a slender tip on the thicker body.
Berry, a fruit in which the entire ovary wall becomes fleshy. Examples: grape, gooseberry.
Bi, Latin prefix, meaning two.
Biennial, a plant which completes its life history and dies in 2 years; flowers and fruits are usually only produced the second year.

Bifid, two forked.

Bilabiate, two lipped.

Bipinnate, twice pinnate, both first and second divisions of the leaf pinnate.

Blade, the expanded part of an organ as a leaf or a petal.

Bloom, a waxy coating covering a surface and making it appear lighter than it's true color; also referring to the flower or blossom or the act of flowering.

Bract, a modified or much reduced leaf in a flower cluster or on the flower stalk; in gymnosperms it is an appendage of the cone axis.

Bracteate, bearing bracts.

Bractlet, diminutive of bract; a very small modified leaf at the base of a flower or on its pedicel.

Bulb, a compact, mostly underground cluster of fleshy leaves on a short stem, for food storage and propagation.

Bulbil or Bulblet, a small bulb, sometimes produced in unusual places on the plant, as in the leaf axils or in place of flowers.

Caespitose, growing in dense, low tufts.

Callosity, a thickened or hardened structure.

Calyx, the outer set of parts in a complete flower, consisting of separate or more or less united sepals.

Campanulate, bell-shaped.

Canescent, pale or gray from a close white pubescence.

Capillary, more or less hair-like; fine and slender.

Capitate, head-like or in a head.

Capsule, a dry fruit from a compound ovary, splitting at maturity to release the seeds.

Carpel, a single simple pistil, or a single unit of a compound pistil. The number of carpels can generally be determined by the number of placentae, or the number of compartments in a pistil, or the number of styles and stigmas. The blending of parts in a compound pistil may obscure this point.

Caryopsis, the fruit of members of the grass family, differs from an achene in that the seed coat is attached to the ripened ovary wall.

Catkin, a scaly spike of unisexual apetalous flowers, often drooping, which in the case of staminate catkins generally falls as a unit after pollination.

Caudex, a short, often woody, persistent stem at the surface of the ground or just beneath it, from which new stems arise each year.

Cauline, pertaining to the stem; cauline leaves are attached along the length of the stem as opposed to basal leaves.

Chaffy, having thin, dry scales.

Chlorophyll, the green pigment of plants necessary for photosynthesis.

Ciliate, fringed with hairs.

Circinate, rolled or coiled with the apex as the center. Common in young fern fronds.

Circumscissile, splitting horizontally in a circle, so that the top comes off as a cap.

Clavate, club-shaped; thicker towards the apex than at the base.

Claw, the narrowed base of some petals and sepals.
Clawed, having a narrow base.
Cleft, cut about to the middle.
Cleistogamous flowers, small, inconspicuous, closed, self-pollinating flowers, usually borne in addition to ordinary flowers.
Column, in orchids, the organ formed of the united stamens and styles in the center of the flower.
Complete, said of a flower with sepals, petals, stamens and one or more pistils.
Compound, said of an inflorescence when it is branched; of a pistil when it consists of more than one carpel; of a leaf when it is divided into leaflets.
Compressed, flattened, either dorsally (front to back) or laterally (side to side).
Coniferous, referring to fir, pine, or other gymnosperms.
Connate, united, grown together or attached; the term is usually applied only to like organs.
Constricted, narrowed.
Convolute, arranged in such a way that each petal or each sepal has one edge covered and the other edge exposed.
Copious, abundant; profuse.
Cordate, heart-shaped, with a notch at the base.
Coriaceous, leathery; tough in texture.
Corm, a bulb-like fleshy stem base, generally under ground, but unlike a bulb it is solid; serves as food storage and propagation.
Corolla, the set of parts second from the outside in a complete flower, consisting of separate or more or less united petals; serves as protection to the stamens and pistils, and to attract pollinators.
Corrugated, wrinkled.
Corymb, a flat-topped cluster of flowers due to pedicels of unequal length, in which the outer flowers blossom first.
Corymbose, arranged in corymbs.
Crenate, with rounded teeth; appearing scalloped.
Crenulate, with minute rounded teeth.
Crest, a ridge or projection.
Crisped, irregularly curled; with undulations.
Cruciform, in the shape of a cross.
Culm, the stem of a grass or a sedge.
Cuneate, wedge-shaped, with the narrow end at the point of attachment.
Cuspidate, tipped with an abrupt, sharp, rigid point.
Cyme, a flower cluster in which the central flower blossoms first.
Cymose, consisting of cymes.

Deciduous, falling after maturity; or said of trees or shrubs whose leaves fall at the approach of winter.
Decompound, divided several times.
Decumbent, lying on the ground at the base, but rising at the tip.
Decurrent, forming wings down the stem.
Deflexed, turned or bent outward.
Dehiscent, splitting open at maturity to release the contents; said of fruits and anthers.
Deltoid, more or less the shape of an equilateral triangle.

Dentate, having spreading, pointed teeth.
Denticulate, with minute teeth pointing outward.
Depressed, somewhat flattened above.
Determinate inflorescence, where the upper or inner flowers open before the lower or outer flowers.
Dichotomous, branching into two parts of about equal size.
Dilated, broadened.
Dimorphic, of two distinct forms.
Dioecious, with unisexual flowers and with the staminate and pistillate flowers on separate plants.
Discoid, of disk-flowers only, or disk-like.
Disk, an enlargement of the stem around the base of the pistil.
Disk flower, the tubular flower in members of the sunflower family.
Dissected, deeply cut into many small or narrow segments.
Distinct, referring to parts which are free or not fused to each other.
Divided, cut nearly to the base or the midrib.
Dorsal, referring to the back.
Drupe, a fruit with a fleshy outer part and a stony inner part enclosing the seed. Example: plum, cherry.
Drupelet, a minute drupe forming a unit of an aggregate fruit, as in a blackberry.

e-, ex-, Latin prefix, meaning without, or away from.
Ebracteate, without bracts.
Elliptic, the shape of an ellipse; more or less oblong with rounded ends.
Emarginate, with a small notch at the apex.
Entire, having margins that are smooth, not cut or toothed.
Epigynous, said of flowers with an inferior ovary.
Epiphyte, a plant growing upon another plant but not parasitic upon it.
Equitant, said of leaves borne in two ranks whose bases over lap the ones above them. Example: Iris.
Erose, unevenly toothed, as if gnawed.
Exfoliating, peeling off in layers.
Exserted, projecting out from a surrounding structure.
Exstipulate or estipulate, without stipules.

False indusium, in ferns where the reflexed margins of the leaves cover the sori.
Fascicle, a bundle or a cluster.
Fertile, capable of sexual reproduction.
Fibrous, consisting of fibers or small slender structures; said of root systems where the roots are similar in size.
Filament, the anther bearing stalk of a stamen.
Filiform, thread-like, very thin and slender.
Fimbriate, fringed; with ragged or finely cut margins.
Foliaceous, having the nature of a foliage leaf.
Foliage, the leaves of a plant.
Foliate, having leaves; for example: **trifoliate,** having three leaves, as Trillium.
Foliolate, having leaflets; **trifoliolate,** having 3 leaflets, as most clovers.

Follicle, a dry fruit from a simple pistil which splits open at maturity only on one side.
Fornix, one of a set of appendages in the throat of a corolla as in the Borage family; plural, **fornices.**
Free-central, that form of placentation which consists of an ovule bearing column arising from the base of a compound, 1-loculed ovary.
Fruit, a ripened ovary, together with any other parts developing with it.
Funnel-form, widening from the base more or less gradually.
Fusiform, spindle-shaped; thickened in the middle and tapering at both ends.

Gametophyte, in plants which have alternation of generations, the generation which bears the sex organs; in ferns, the gameophyte is reduced to a minute thallus-like body.
Gamopetalous, having the petals united.
Gamosepalous, having the sepals united.
Gelatinous, jelly-like.
Geniculate, abruptly bent.
Glabrate, nearly without hairs or becoming glabrous.
Glabrous, without hairs.
Gland, a small structure or depression that secretes a sticky substance, embedded in or on the surface of organs or at the tip of a hair.
Glandular, bearing glands, generally sticky.
Glaucous, covered with a waxy or powdery substance, giving a whitish or bluish cast to the surface.
Globose, nearly spherical.
Glochidia, hair-like structures with downward pointing barbs at the tip.
Glomerule, a dense, compact cyme where the inner flowers open first.
Glume, one of a pair of empty bracts that form the lowest part of a spikelet in the grass family.
Granulose, covered with minute bumps.

Hastate, arrowhead-shaped but with the basal lobes somewhat divergent.
Head, a close, compact inflorescence in which the outer flowers open first.
Helicoid cyme, an inflorescence in which the main axis is curved because the lateral branches all arise from the same side.
Herb, a plant whose stem above ground dies back at end of the growing season.
Herbaceous, like an herb; not woody.
Hirsute, with coarse stiff hairs, often bent, but not as short and firm and sharp as hispid.
Hispid, with short coarse, stiff hairs that are usually sharp.
Hoary, gray from a minute hairy covering.
Hyaline, thin and transparent or translucent.
Hygroscopic, sensitive to change in moisture content, and reacting by movements, or by swelling or shrinking.
Hypanthium, floral tube or cup formed from the fusion of the bases of the calyx, corolla and stamens. The part of the apparent calyx tube below the attachment of the petals.
Hypogaeous, underground.

Hypogynous, said of flowers where the sepals, petals and stamens are attached directly to the receptacle.

Imbricate, overlapping with a shingle-like arrangement.
Imperfect, said of unisexual flowers.
Incised, deeply and sharply cut.
Incomplete, said of a flower lacking one or more of the floral organs.
Indehiscent, remaining closed, as fruits which do not open up to release the seeds.
Indeterminate inflorescence, where the lower, or outer flowers, open before the upper or inner flowers.
Indusium, outgrowth of a leaf, or its margin, which protects a cluster of sporangia; plural, **indusia.**
Inferior, said of an ovary which is below the apparent attachment of the calyx and is fused with the bases of sepals, petals and stamens.
Inflorescence, the arrangement of flowers on the floral axis.
Internode, part of stem between nodes or "joints".
Involucel, bracts of an umbellet in a compound umbel.
Involucral, pertaining to involucre.
Involucrate, having an involucre.
Involucre, a circle of bracts at base of a flower cluster.
Involute, rolled inward.
Iso-diametric, all diameters equal.

Keel, a ridge; the 2 lower petals of a pea-like flower.
Keeled, having a ridge.

Laciniate, cut into narrow, irregular segments.
Lactiferous, containing a milky latex.
Lanceolate, broadest below the middle and gradually tapering to apex.
Lateral, at the side or referring to the side.
Leaflet, a leaf-like division of a compound leaf.
Lenticel, a corky pore, usually on stems.
Ligulate flower, the strap shaped, ray flower of the Asteraceae.
Ligule, the flattened part of the ray flower in the Asteraceae, or the appendage at the junction of the sheath and the blade in some Monocots.
Limb, the expanded part of a petal or corolla or calyx.
Linear, long and narrow, with parallel sides.
Lip, one of the 2 parts, or sets of lobes, of an unequally divided corolla or calyx.
Lobe, a shallow division of a leaf or other organ.
Lobed, cut into shallow segments.
Locule, the seed-cavity or chamber of an ovary or carpel.
Loculicidal, opening directly into the locules through the ovary wall near the center of each locule.

Megasporangium, a sporangium bearing only megaspores.
Megaspore, a large asexual spore giving rise to a female gametophyte.
Membranous, thin and more or less transparent.

Mericarp, one unit of a schizocarp Example: the fruit of the Apiaceae which splits into two parts, each part is termed a mericarp.
-merous, Greek suffix, referring to the number of parts, usually proceeded with a numerical prefix.
Mesic, moist, but neither very wet nor very dry.
Microsporangium, a sporangium bearing only microspores.
Microspore, typically smaller than a megaspore, and giving rise to a male gametophyte.
Midrib, main vein of a leaf.
Monoecious, with unisexual flowers and with staminate and pistillate flowers borne on the same plant.
Monotypic, containing a single type. Example: a genus with a single species.
Mucronate, with a short, straight, slender, sharp point at the tip.

Node, the place on a stem where a leaf is attached.
Nomenclature, a system of naming plants or animals.
Nutlet, a small hard thick-walled indehiscent 1-seeded fruit, typically formed from one half a carpel, as in the mint family.

ob-, Greek prefix, meaning in reverse direction, or inversion of shape.
Obcordate, shaped like a heart with the notch at the apex.
Oblanceolate, widest above the middle, tapering gradually to the base, several times longer than wide.
Oblique, with unequal sides or an asymmetrical base.
Oblong, longer than wide with parallel sides and rounded base and apex.
Obovate, widest near the apex, narrowing rather abruptly to the base.
Obtuse, blunt or rounded.
Opposite, paired structures; located directly across from each other, as at a node, like leaves, or directly in front of another organ, as stamens opposite the petals.
Orbicular, nearly circular in outline.
Ovary, the enlarged basal, seed-bearing part of the pistil.
Ovate, broadest near the base, narrowing rather abruptly to the apex.
Ovoid, egg-shaped.
Ovule, an immature seed.

Palmate, spreading from a common point.
Panicle, a branched inflorescence in which the lower outer flowers open first.
Paniculate, consisting of panicles.
Papilionaceous, butterfly shaped; said of flowers of the pea family that have a banner, 2 wings and 2 keel petals.
Papillae, small, blunt protuberances.
Papillose, covered with small protuberances.
Pappus, the bristles, hairs, scales, etc., arising at the top of the inferior ovary in the sunflower family; the modified calyx.
Parasite, in botany it is a plant which secures its food from a living plant or animal, and is typically without chlorophyll, so is not cabable of making its own food.

Parietal, placentation type where the ovules are borne on the walls or sides of a single loculed ovary.
Parted, deeply cut to below the middle but not to the base or midrib.
Pedicel, the stalk of a single flower.
Pediceled, said of a flower having a pedicel.
Pedicellate, having a pedicel
Peduncle, the main stalk of a flower cluster, or of an individual flower when solitary.
Peltate, having the stalk attached on the underside near the middle, umbrella-like.
Pendent, hanging or drooping.
Pendulous, hanging.
Perennial, a plant which lives and fruits for more than two years, usually year after year.
Perfect, having both stamens and pistils in the same flower.
Perfoliate, surrounding a stem as if pierced by it.
Perianth, the sepals and petals collectively, particularly applied when these are similar.
Perigynous, said of a flower with a floral tube (hypanthium) and a superior ovary.
Persistent, remaining after flowering.
Petal, one of the second set of whorls in a complete flower; a unit of the corolla; generally colored.
Petaloid, petal-like in color or texture.
Petiolate, having a petiole.
Petiole, the stalk of a leaf.
Petioled, said of a leaf having a stalk.
Pilose, with long straight, usually soft, spreading hairs.
Pinna, a leaflet of a pinnate leaf; in ferns, one of the segments of a leaf; plural, **pinnae**.
Pinnate, arranged along the sides of an axis like the parts of a feather.
Pinnatifid, pinnately cleft.
Pinnule, an ultimate segment of a pinna.
Pistil, the central organ of a perfect flower, consisting of ovary, style, and stigma.
Pistillate, flowers bearing pistils but without functional stamens.
Placenta, that part of the ovary to which the seeds are attached; plural, **placentae**.
Plumose, feathery.
Pollen, the powdery substance borne in the anther of the stamen.
Polygamous, bearing both perfect and unisexual flowers.
Pome, a fleshy fruit in which the outer fleshy layer is derived from the floral cup and with an inner "core" of papery or cartilaginous material surrounding the seeds. Examples: apples, pears.
Procumbent, lying or trailing on the ground.
Prostrate, lying flat against the ground.
Puberulent, with fine, short hairs.
Pubescent, with hairs of any type.
Pungent, sharp-pointed; or as applied to an odor, acrid.

Raceme, a flower cluster in which the flowers are borne along a stem on individual pedicels of about equal length, the lower flowers blooming before the terminal ones.
Racemose, in racemes.
Rachilla, the axis of the spikelet in grasses and sedges.
Rachis, the main axis of a flower-cluster or of a compound leaf.
Ray flowers, the flowers in the sunflower family having strap-like (ligulate) corollas.
Receptacle, that end of the stem (or pedicel or peduncle) to which a flower is attached.
Recurved, curved backward.
Reflexed, abruptly bent backward.
Regular, said of flower when all the petals and/or all the sepals are of the same size and shape.
Reniform, kidney-shaped.
Resinous, exuding a semi-solid substance; gummy, sticky.
Reticulate, net-like or netted.
Retrorse, turned backward or downward.
Retuse, with a small apical notch on an otherwise blunt tip.
Revolute, rolled backward.
Rhizomatous, having rhizomes.
Rhizome, an underground (or under water) perennial stem which is generally prostrate, and sends new shoots above ground each year, and roots below.
Rhombic, 4-sided, the sides oblique.
Rootstock, a rhizome.
Rosette, a cluster of leaves or other organs arranged in a basal circle.
Rotate, saucer-shaped, flat and circular in outline.
Rugose, wrinkled.

Saccate, sac-shaped or with a pouch.
Sagittate, arrowhead-shaped with the basal lobes more or less in line with the main lobe.
Salverform, tubular below, abruptly spreading saucer-shaped above.
Samara, a dry indehiscent winged fruit, typically 1-seeded.
Saprophyte, a plant which secures its food from dead plant or animal substances and is typically without chlorophyll.
Scabrous, rough to the touch.
Scale, a thin, membranaceous, woody, or leathery structure, generally foliar in origin.
Scape, a leafless peduncle arising from the ground.
Scapose, with the flowers borne on a scape.
Scar, a mark left after the falling of a leaf or other organ.
Scarious, thin and dry, not green, often like tissue paper.
Schizocarp, a fruit splitting into separate units, the units not splitting open to release the seeds.
Scorpioid, coiled or curved inward.
Scorpioid cyme, an inflorescence with a zigzag rachis caused by the successive lateral branches arising on different sides.
Semi-sagittate, shaped like half an arrowhead.

Sensitive, closing at a touch.
Sepal, a unit of the calyx, generally green; one member of the outermost whorl in a complete flower.
Septate, with partitions.
Septicidal, opening through the partitions between the locules.
Serrate, saw-toothed; the margins with sharp forward-pointing teeth.
Serrulate, diminutive of serrate; with minute sharp teeth.
Sessile, attached directly at the base without a stalk.
Setose, bearing bristles.
Sheath, part of a leaf which enfolds the stem.
Sheathing, enfolding the stem.
Shrub, a woody plant smaller than a tree, and generally without a pronounced trunk.
Silicle, modified capsule in which 2 valves separate, leaving a false partition, not much longer than wide.
Silique, modified capsule in which 2 valves separate, leaving a false partition, definitely longer than wide.
Simple, unbranched or undivided, or (in case of a pistil) consisting of a single carpel.
Sinus, the cleft formed between two connective lobes.
Solitary, only one in a place.
Sorus, a cluster of sporangia in ferns; plural, **sori.**
Spadix, a spike with a fleshy axis.
Spathe, a large bract enclosing a flower cluster, generally a spadix.
Spatulate, narrow at base and rounded and widest at the apex; spoon-shaped.
Spike, a flower cluster in which sessile flowers are borne along a stem, the lower flowers blooming before the terminal ones.
Spikelet, the smallest aggregation of florets, in grasses a spikelet includes the empty glumes.
Spinulose, having very small spines.
Sporangium, a case or sac in which asexual spores are borne; plural, **sporangia.**
Spore, a generally single-celled reproductive body, capable of producing a new plant; the reproductive unit of ferns and fern allies.
Sporocarp, a spore-bearing body; in Azolla, one of the capsule-like masses of sporangia.
Spur, a hollow projection, usually of a sepal or a petal and usually functioning as a nectary.
Spurred, extended into a spur.
Stamen, the pollen-bearing organ of a flower, typically consisting of anther and filament.
Staminate, having stamens but without pistils.
Staminode or staminodium, a modified or aborted stamen which does not produce pollen; plural, **staminodia**.
Stellate, star-shaped.
Sterile, not capable of performing its normal function, as a stamen or pistil; or not producing flowers, as a shoot.
Stigma, the pollen-receiving part of the pistil.
Stipe, a stalk.

Stiped, stalked.
Stipitate, having a stalk.
Stipule, an appendage of the leaf usually borne in pairs at the base of the petiole.
Stolon, an elongate creeping stem which runs along the surface of the ground and which roots at the nodes and sends up new shoots and ultimately produces new plants.
Stoloniferous, having stolons or "runners".
Striate, marked with lines or ridges.
Strigose, bearing stiff appressed hairs or bristles.
Strobilus, a cone or cone-like body; plural, **strobili**.
Style, the stalk-like part of the pistil between the stigma and the ovary.
Stylopodium, enlarged base of the styles in the Apiaceae.
Sub-, Latin prefix meaning under, or almost, or not quite.
Subcordate, slightly cordate.
Submerged, under water.
Subtended by, bearing at its base.
Subulate, awl-shaped; slender and tapering to a point.
Succulent, fleshy and juicy.
Superior, said of an ovary which is not attached to the sepals, petals or stamens, but borne free on the receptacle in the center of the flower.
Symbiont, one of two dissimilar organisms living in a close association which is generally of mutual advantage.
Symmetrical flower, one having all the parts of a whorl similar in size and shape.

Taproot, the main descending root, forming a direct continuation of the stem.
Tawny, yellowish-brown.
Tendril, a slender coiling stem or modified leaf or portion of a leaf by which a climbing plant supports itself.
Terete, round in cross section, cylindrical.
Terminal, at the end.
Ternate, in threes.
Terrestrial, of the earth; growing on the ground.
Thallus, a plant body not differentiated into roots, stems and leaves.
Throat, the upper expanded portion of a corolla tube.
Tomentose, covered by dense, matted, woolly hairs.
Tomentum, a covering of dense, woolly hairs.
Trailing, growing over the ground or over other plants, but not rooting at the nodes.
Translucent, nearly transparent.
Transverse, at right angles to the longitude.
Truncate, square-cut.
Tuber, a short, thickened part of an underground stem used for food storage.
Tuberculate, covered by wart-like protuberances.
Tuberous, like a tuber or producing tubers.
Tufted, in compact clusters.
Turion, a small, often bulb-like, structure produced at or near ground level that acts as a means of vegetative propagation.

Two-lipped, bilabiate; said of an irregular calyx or corolla which has distinct upper and lower divisions.

Umbel, a flat-topped inflorescence whose pedicels all appear to arise from the same level and with the outer flowers opening before the inner ones.
Umbellate, consisting of umbels.
Umbellet, a small umbel terminating each ray of a compound umbel.
Umbo, a rounded projection at the end of a solid organ, as on the scales of pine cones.
Utricle, a small inflated dry indehiscent 1-seeded fruit.

Valves, the parts into which a dehiscent fruit splits at maturity.
Vascular bundles, the fiber-like strands of cells, in plants,which carry raw materials and food; the veins in a leaf, for example.
Veins, the ribs in a leaf; the vascular bundles of a leaf.
Venation, the vein arrangement.
Vernation, the arrangement of foliage leaves within the bud.
Vestiges, remains.
Villous, clothed with long, slender soft, often bent hairs.
Vine, a plant with trailing, climbing, or running stems.

Whorl, a circle, or apparent circle, of plant organs or inflorescences, as a whorl of leaves, or a whorl of cymes.
Whorled, borne in whorls.
Wing, one of the lateral petals of a pea flower or pea-like flower; the flattened appendage of some fruits or the flat extension of a surface or margin, as in winged stems.

Zygomorphic, irregular; capable of being divided into two equal parts only by one longitudinal plane passing through its center.

Index

Synonyms are in italics.